An Invitation to Mathematical Physics and Its History

Jont Allen

An Invitation to Mathematical Physics and Its History

Springer

Jont Allen
Electrical & Computer Engineering
University of Illinois at Urbana Champaign
Urbana, IL, USA

ISBN 978-3-030-53761-6 ISBN 978-3-030-53759-3 (eBook)
https://doi.org/10.1007/978-3-030-53759-3

This Springer imprint is published by the registered company Springer Nature Switzerland AG
The registered company address is: Gewerbestrasse 11, 6330 Cham, Switzerland

Preface

Science has evolved over thousands of years. It began out of curiosity about how the world around us works as well as a need to know how to make things work better. From water management to space travel, science is essential for success.

The evolution of science is layered: Early science depended mostly on critical observation, thus the early scientist was considered a philosopher, who created a theory. Soon experiments were designed to test these theories. This process is successful when a new principle is established, leading to a deeper and reproducible experimental observation. This typically takes years. A key component of science is making and testing models that are designed to evaluate the results of experiments quantitatively in a mathematical framework. Good science is observation and experimentation. Great science is the art of making models that explain experimental results. This always results in a deeper question, suggesting new experiments. Each generation has its geniuses. One of these was Galileo, who was a philosopher, experimentalist, and mathematician.

An understanding of physics requires a knowledge of mathematics. The converse is not true. By definition, pure mathematics contains no physics. Yet historically, mathematics has a rich history filled with physical applications. Mathematics was developed by individuals who intend to make things work. As an engineer, I see these creators of early mathematics as budding engineers. This book is an attempt to tell the story of the development of mathematical physics as viewed by an engineer.

There are two distinct ways to learn mathematics: by learning definitions and relationships or by associating each mathematical concept with its physical counterpart. Students of physics and engineering best learn mathematics based on the underlying physical concepts. Students of pure mathematics are taught via definitions of abstract structures, not from the history of mathematical physics. These two teaching methods result in very different understandings of the material.

There is a deep common thread between physics and mathematics: the chronological development or the history of mathematics. This is because much of mathematics was developed to solve physical problems. Most early mathematics evolved from attempts to understand the world, with the goal of navigating it. Pure mathematics followed as generalizations of these physical concepts.

Around 1638 Galileo stated that, based on his experiments with balls rolling down inclined planes and pendulums, the height of a falling object is given by

$$h(t) = \frac{1}{2}Gt^2, \tag{1}$$

where t is time and G is a constant. This formula leads to a constant acceleration $a(t)$ of the object since

$$a(t) = \frac{d^2}{dt^2}h(t) = G$$

is independent of time. It follows that the force on a body is proportional to its acceleration a defined as G, namely, $F = a \equiv G$. Thus G must be the object's mass, which must be a constant. If the object has a constant forward velocity, then the object will have a parabolic trajectory. The relative mass may be measured using a balance scale. I believe Galileo understood all this.

Years later, following up on the observations from Galileo's study of pendulums and falling objects, Newton showed that differential equations were necessary to explain gravity, and that the force of gravity is proportional to the masses of the two objects, divided by the square of the reciprocal of the distance between them:

$$\frac{d^2}{dt^2}r(t) = G\frac{mM}{r^2(t)}.$$

To find $r(t)$ we must integrate this equation. For an object at height $h(t)$ above the surface of the earth, $r(t) = R_e + h(t) \approx R_e$, where R_e is the radius of earth. In this case, the force is effectively constant, since $h \ll R_e$. Newton's equation says the acceleration is constant,

$$\frac{d^2h(t)}{dt^2} = G\frac{mM}{R_e^2},$$

but different from Galileo's G (a simple mass). Yet it seems clear that the physics behind Newton's formula for the acceleration $a(t)$ of two large masses (Sun and Earth or Earth and Moon) and Galileo's physics for balls rolling down inclined planes are the same.[1] The difference is that Newton's proportionality constant is a significant generalization of Galileo's. But, other than the constant, which defines the acceleration, the two formulas are the same.

This is not a typical mathematics book; rather, it is about the relationship between math and physics, presented roughly in chronological order via their history. To teach mathematical physics in an orderly way, our treatment requires a step backward in terms of the mathematics, but a step forward in terms of the physics.

[1] https://physicstoday.scitation.org/do/10.1063/PT.6.3.20191002a/full/.

Historically speaking, mathematics was created by individuals such as Galileo who, by modern standards, may be viewed as engineers. This book contains the basic information that well-informed engineers need to know, as best I can provide.

Let the reader beware that engineering and physics texts do not intend to be rigorous, in the mathematical sense. In some ways, mathematics cleans up the mess by proving theorems, which frequently start with speculations in physics and even engineering. The cleanup is a slow, tedious process. Just because something seems obvious based on the known physical facts does not make it a fundamental theorem of mathematics.

Although there are similarities between this book and that of Graham et al. (1994), the differences are notable. First, Graham's *Concrete Mathematics* presents an impossible standard to be measured against. Second, it is clearly a math book, brilliantly written and targeted at computer science students. This book is not just a math book—it is a mathematical physics text, which depends much on underlying math. I would like to believe there are similarities in (1) the broad range of topics, (2) the in-depth discussion, and (3) the use of historical context.

Organization: As discussed in Sect. 1.2.2 and Fig. 1.6 (p. xx), the book is divided into three mathematical themes, called streams, presented as five chapters: Introduction, Number Systems, Algebraic Equations, Scalar Calculus, and Vector Calculus. Appendices are used to isolate complex self-contained topics and large tables, such a those for Laplace transforms.

Chapter 2, Number Systems, introduces two key concepts, the greatest common divisor (GCD) and the continued fraction algorithm (CFA). When we deal with simple electrical networks composed of inductors, resistors, and capacitors (Fig. 3. 8), or mechanical networks consisting of masses, dashpots, and springs, or their equivalent, pendulums, as used by Galileo in his studies of gravity (Figs. 1.3, and 3. 11), the system may be modeled as a Brune impedance, defined as the ratio of polynomials of the Laplace frequency $s = \sigma + \omega j$ (see Sect. 3.2.3, and Sect. 4.4.2). Of special importance is the development of ordinary differential equations (Sect. 3. 4.2, and Eq. 3.4.4) which under generalized symmetry conditions, called postulates (Sect. 3.10.2), characterize the Brune impedances (Brune, 1931a).

Using the CFA (Sect. 2.4.4), we can generalize the Brune impedance. This generalization results in a transmission line, that describes wave propagation in horns, dealt with in Chaps. 4 and 5 (Cauer et al. 1958a, Cauer 1958). This topic is both physically and mathematically important (Cauer, 1932).

The material is delivered in numbered sections (e.g., Sect. 1.1) spread out over a semester of 15 weeks, three lectures per week, with a three-lecture time-out for administrative duties. Eleven problem sets are provided for weekly assignments.

Many students have rated these assignments as the most important part of the course. There is a built-in interplay between these assignments and the lectures. When a student returns an assignment or exam, the full solution is provided while it is still fresh in their mind (a teaching moment).

Author's Personal Statement

An expert is someone who has made all possible mistakes in a small field. I don't know if I would be called an expert, but I certainly have made my share of mistakes. I openly state that I love making mistakes because I learn so much from them. One might call that the "expert's corollary."

This book has been written out of my love for the topic of mathematical physics, a topic that provides many insights that lead to a deep understanding of important physical concepts. Over the years I have developed a *physical sense* of math along with a related *mathematical sense* of physics. While doing my research,[2] I believe that math can be physics and physics math. I have come across what I feel are certain conceptual holes that need filling, and I sense many deep relationships between math and physics that remain unidentified. What we presently teach is not wrong, but it is missing these relationships. What is lacking is an intuition for how math "works." Good scientists "listen" to their data. In the same way, we need to start listening to the language of mathematics. We need to let mathematics guide us toward our engineering goals.

As summarized in Fig. 1, this union of math, engineering, and physics (MEP)[3] helps us make progress in understanding the physical world. We must turn to mathematics and physics when trying to understand the universe. My views follow from a lifelong attempt to understand human communication—that is, the perception and decoding of human speech sounds. This research arose from my 32 years at Bell Labs in the Acoustics Research Department. There such lifelong pursuits not only were possible but also were openly encouraged. The idea was that if you are successful at something, take it as far as you can, but, on the other hand, you should not do something well that's not worth doing. People got fired for the latter. I should have left for a university after a mere 20 years at Bell Labs,[4] but the job was just too cushy.

In this text, it is my goal to clarify conceptual errors while telling the story of physics and mathematics. My views have been inspired by classic works, as documented in the Bibliography. This book was inspired by my reading of Stillwell (2002) through his Chap. 21. Somewhere in Chap. 22. I switched to the third edition (Stillwell 2010), at which point I realized I had much more to master. It became clear that by teaching this material to first-year engineers, I could absorb the advanced material at a reasonable pace. This book soon followed.

[2] `https://auditorymodels.org/index.php/Main/Publications`.

[3] MEP is a focused alternative to STEM.

[4] I started around December 1970, fresh out of graduate school, and retired on December 5, 2002.

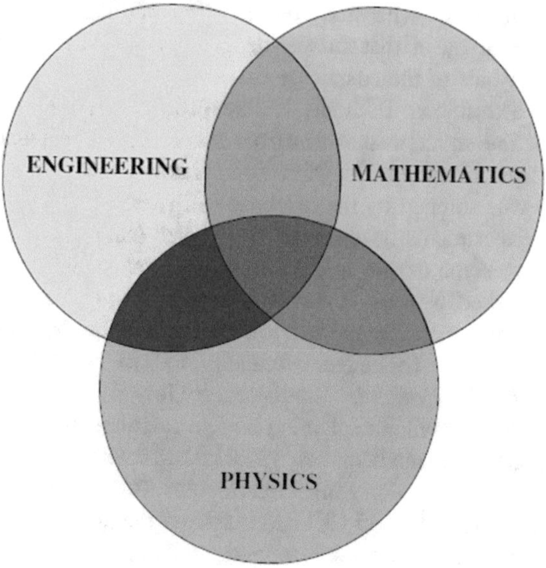

Fig. 1 There is a natural symbiotic relationship among mathematics, engineering, and physics (MEP), depicted in the Venn diagram. Mathematics provides the method and rigor. Engineering transforms the method into technology. Physics explores the boundaries. While these three disciplines work well together, there is poor communication, in part due to the vastly different vocabularies. But style may be more at issue. For example, mathematics rarely uses a system of units, whereas physics and engineering depend critically on them. Mathematics strives to abstract the ideas into proofs. Physics rarely uses a proof. When they attempt rigor, physicists and engineers typically get into difficulty. An important observation by Felix Klein about pure mathematicians, regarding the unavoidable inaccuracies in physics: "It may be said that the idea [of inaccuracy] is usually so repulsive to [mathematicians] that its recognition sooner or later spoils their interest in natural science" [Condon and Morse, 1929, p. 19]

Summary

This is foremost a math book, but not the typical math book. *First*, this book is for the engineering minded, for those who need to understand math to do engineering, to learn how things work. In that sense, the book is more about physics and engineering than mathematics. Math skills are essential for making progress in building things, be it pyramids or computers, as clearly shown by the great civilizations of the Chinese, Egyptians, Mesopotamians, Greeks, and Romans.

Second, this is a book about the math that developed to explain physics, to enable people to engineer complex things. To sail around the world, one needs to know how to navigate. This requires a model of the planets and stars. You can know where you are on the Earth once you understand where the Earth is relative to the Sun, planets, Milky Way, and distant stars. The answer to such a cosmic question depends on whom you ask. Who is qualified to answer such a question? It is best answered by those who study mathematics applied to the physical world. The utility and accuracy of that answer depend critically on the depth of understanding of the physics of the cosmic clock.

The English astronomer Edmond Halley (1656–1742) asked Isaac Newton (1643–1727) for the equation that describes the orbit of the planets. Halley was obviously interested in comets. Newton immediately answered, "an ellipse." It is said that Halley was stunned by the response (Stillwell, 2010, p. xxx), as this was what had been experimentally observed by Kepler (ca. 1619), and he knew that Newton must have some deeper insight. Both were eventually knighted.

When Halley asked Newton to explain how he knew, Newton responded, "I calculated it." But when challenged to show the calculation, Newton was unable to reproduce it. This open challenge eventually led to Newton's grand treatise, *Philosophiae Naturalis Principia Mathematica* (July 5, 1687). It had a humble beginning, as a letter to Halley explaining how to calculate the orbits of the planets. To do this Newton needed mathematics, a tool he had mastered. It is widely accepted that Newton and Gottfried Leibniz invented calculus. But the early record shows that perhaps Bhaskara II (1114–1185 CE) had mastered the art well before Newton.[5]

Third, the main goal of this book is to teach mathematics to motivated engineers, in a way that it can be understood, mastered, and remembered. How can this impossible goal be achieved? The answer is to fill in the gaps with *Who did what, and when?* Compared with the math, the historical record is easily mastered.

To be an expert in a field, one must know its history. This includes who the people were, what they did, and the credibility of their story. Do you believe the Pope or Galileo on the roles of the Sun and the Earth? The observables provided by science are clearly on Galileo's side. Who were those first engineers? They are names we all know: Archimedes, Pythagoras, Leonardo da Vinci, Galileo, Newton, and so on. All of these individuals had mastered mathematics. This book presents the tools taught to every engineer. Rather than memorizing complex formulas, make the relationships "obvious" by mastering each simple underlying concept.

Fourth, when most educators look at this book, their immediate reactions are: *Each lecture is a topic we spend a week on (in our math/physics/engineering class)* and *You have too much material crammed into one semester*. The first sentence is correct, the second is not. Tracking the students who have taken the course, looking at their grades, and interviewing them personally demonstrate that the material presented here is appropriate for one semester.

To write this book I had to master the language of mathematics. I had already mastered the language of engineering and a good part of physics. One of my secondary goals was to build this scientific Tower of Babel by unifying the terminology and removing the jargon.

[5] `https://www-history.mcs.st-and.ac.uk/Projects/Pearce/Chapters/Ch8_5.html`.

Acknowledgments

I would like to acknowledge John Stillwell for his brilliant and constructive historical summary of mathematics as well as my close friend and long-time (40 years) colleague Steve Levinson, who somehow drew me into this project without my even knowing it. Next, my brilliant graduate student Sarah Robinson was constantly at my side, first repairing blunders in my first-draft homeworks and then grading these and the exams and tutoring the students. Without her, I would never have survived the first semester. Her proofreading skills are amazing. Thank you, Sarah, for your infinite help. Without Kevin Pitts, this work never could have been started, as he provided early funding when the project was a germ of an idea. Matt Ando's (math) and Michael Stone's (physics) encouragement was psychologically important in helping me think I might actually write a book. Finally, I would like to thank John D'Angelo for his highly critical comments, in response to my thousands of silly math questions. When it comes to the heavy hitting, John was always there to provide a brilliant explanation that I could easily translate into engineerese (matheering?) (i.e., engineer language).

My delightful friend Robert Fossum Emeritus Professor of Mathematics from the University of Illinois kindly pointed out flawed mathematical terminology. Jerome Colburn and James (Jamie) Hutchinson's precise use of the English language dramatically raised the bar on my more than occasionally casual writing style. To each of you, thank you! Swati Meherishi of the Springer staff was perpetually positive about this project and always steered me in the right directions regarding the details of making a manuscript into a book (they are not the same thing).

Finally, I would like to thank my wife, Sheau Feng Jeng (Patricia Allen), for her unbelievable support and love. She delivered constant peace of mind, without which this project could never have been started, much less finished. Many others, including the many students who took courses based on this book, played important roles, but given their large numbers, sadly they must remain anonymous.[6]

C'est par la logique qu'on démontre, c'est par l'intuition qu'on invente.
It is by logic that we prove, but by intuition that we discover
—Henri Poincaré

Mahomet, IL, USA
March 2020

Jont Allen

[6] https://www.istem.illinois.edu/news/jont.allen.html.

Contents

About the Author

Jont Allen is a Professor in the Department of Electrical and Computer Engineering, University of Illinois. After completing his Ph.D. from the University of Pennsylvania, Philadelphia in 1970, he went to Bell Labs, where he enjoyed a 32 year AT&T Bell Labs career. At AT&T Allen specialized in nonlinear cochlear modeling, auditory and cochlear speech processing, and speech perception. Since joining University of Illinois in 2003, he has taught and worked with his students on the theory and practice of human speech recognition, for both normal and hearing impaired hearing as well as reading disabilities in young children. Prof Allen has more than 20 US patents on hearing aids, signal processing and middle ear measurement diagnostics.

Chapter 1
Introduction

Much of early mathematics, say before 1600 BCE, involved the love of art and music, the sensations of light and sound. Our psychological senses of color and pitch are determined by the frequencies (i.e., wavelengths) of light and sound. The Chinese and later the Pythagoreans are well known for their early contributions to music theory.

We are largely ignorant of exactly what Chinese scholars knew. The best record of early mathematics comes from Euclid, who lived in the fourth century BCE, after Pythagoras. Thus we can trace early mathematics back to the Pythagoreans in the sixth century (580–500 BCE), who focused on the Pythagorean theorem and early music theory.

Pythagoras strongly believed that "all is number," meaning that every number, and every mathematical and physical concept, could be explained by integral (integer) relationships, mostly based on either ratios or the Pythagorean theorem. It is likely that his belief was based on Chinese mathematics from thousands of years earlier. It is also believed that his ideas about the importance of integers followed from music theory. The musical notes (pitches) obey natural integral ratio relationships based on the octave (a factor of two in frequency). The western 12-tone scale breaks the octave into 12 ratios called *semitones*. Today this has been rationalized to be the 12th root of 2, which is approximately equal to $18/17 \approx 1.06$ or 0.0833 [octave]. Our innate sense of frequency ratios comes from the physiology of the auditory organ (the cochlea), with a fixed distance along the organ of Corti, the sensory organ of the inner ear.

As acknowledged by Stillwell (2010, p. 16), the Pythagorean view is still relevant today:

> With the digital computer, digital audio, and digital video coding everything, at least approximately, into sequences of whole numbers, we are closer than ever to a world in which "all is number."

J. Allen, *An Invitation to Mathematical Physics and Its History*,
https://doi.org/10.1007/978-3-030-53759-3_1

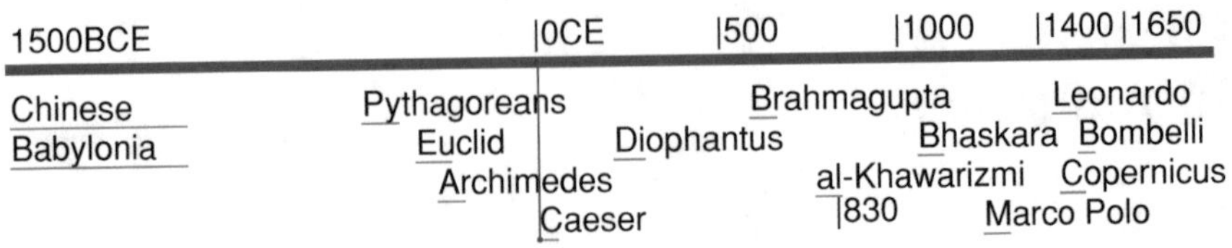

Fig. 1.1 Timeline between 1500 BCE and 1650 CE. The European renaissance is considered to have occurred between the fifteenth and seventeenth centuries CE

1.1 Early Science and Mathematics

Although early Asian mathematics has been lost, it clearly defined its course for at least several millennia. The first recorded mathematics was from the Chinese (5000–1200 BCE) and the Egyptians (3300 BCE). Some of the best early records were left by the people of Mesopotamia (Iraq, 1800 BCE).[1] While the first 5000 years of math are not well documented, the basic timeline is clear, as shown in Fig. 1.1.

Thanks to Euclid, and later Diophantus (ca. 250 CE), we have some basic (but vague) understanding of Chinese mathematics. For example, Euclid's formula (Eq. 2.6.6) provides a method for computing Pythagorean triplets, a formula believed to be due to the Chinese.[2]

Chinese bells and stringed musical instruments were exquisitely developed with tonal quality, as documented by ancient physical artifacts (Fletcher and Rossing 2008). In fact this development was so rich that one must ask why the Chinese failed to initiate the Industrial Revolution. Specifically, why did European innovation eventually dominate when it was the Chinese who were responsible for such extensive early invention?

Our best insight into the scientific history of China came from Joseph Needham, an American chemist and scholar from Cambridge, England. Needham learned Chinese from a colleague[3] and ended up researching early Chinese science and technology for the U.S. government (Winchester 2009).

According to Lin (1995), the Needham question is:

> Why did modern science, the mathematization of hypotheses about Nature, with all its implications for advanced technology, take its meteoric rise only in the West at the time of Galileo[, but] had not developed in Chinese civilization or Indian civilization?

As discussed by Lin (1995) and Apte (2009), Needham cites the many developments in China:

> Gunpowder, the magnetic compass, and paper and printing, which Francis Bacon considered as the three most important inventions facilitating the West's transformation from the Dark Ages to the modern world, were invented in China.

---

[1] See Fig. 2.8.

[2] One might reasonably view Euclid's role as that of a mathematical messenger.

[3] Whom he later fell in love with Winchester (2009).

Needham's works attribute significant weight to the impact of Confucianism and Taoism on the pace of Chinese scientific discovery, and emphasize what it describes as the "diffusionist" approach of Chinese science as opposed to a perceived independent inventiveness in the western world. Needham held that the notion that the Chinese script had inhibited scientific thought was "grossly overrated." (Grosswiler 2004)

Lin (1995) focused on military applications, missing the importance of nonmilitary contributions. A large fraction of mathematics was developed to better understand the solar system, acoustics, musical instruments, and the theory of sound and light. Eventually the universe became a popular topic, as it still is today.

Regarding the Needham question, I suspect the answer is now clear. In the end, China withdrew from its several earlier expansions because of internal politics (Menzies 2004, 2008).

History of Mathematics to the Sixteenth Century CE

20th	Chinese (primes; quadratic equation; Euclidean algorithm (GCD))	
18th	Babylonians (Mesopotamia/Iraq) (quadratic solution)	
6th	Thales of Miletus (first Greek geometry) (624)	
5th	Pythagoras and the Pythagorean "tribe"(570)	
4th	Euclid; Archimedes	
3rd	Eratosthenes (276–194)	BCE
3rd	Diophantus (ca. 250)	CE
4th	Library of Alexandria destroyed by fire (391)	
7th	Brahmagupta (negative numbers; quadratic equation) (598–670)	
10th	al–Khwarizmi (algebra) (830); Hasan Ibn al–Haytham (Alhazen) (965–1040)	
14th	Bhaskara (calculus) (1114–1183)	
15th	Leonardo da Vinci (452–1519); Michelangelo (1475–1564); Copernicus (1473–1543)	
16th	Tartaglia (cubic solution); Bombelli (1526–1572); Galileo Galilei (1564–1642)	

1.1.1 *The Pythagorean Theorem*

Thanks to Euclid's *Elements* (written ca. 323 BCE) we have a historical record tracing the progress in geometry as established by the Pythagorean theorem, which states that *for any right triangle* having sides of lengths $(a, b, c) \in \mathbb{R}$ that are either positive-real numbers or, more interesting, integers $c > [a, b] \in \mathbb{N}$ such that $a + b > c$,

$$c^2 = a^2 + b^2. \tag{1.1}$$

Early integer solutions were likely found by trial and error rather than by an algorithm.

If a, b, c are lengths, then a^2, b^2, c^2 are each the area of a square. Equation 1.1 says that the area a^2 plus the area b^2 equals the area c^2. Today a simple way to prove this is to compute the magnitude of the complex number $c = a + b\jmath$, which forces the right angle

$$|c|^2 = (a + b\jmath)(a - b\jmath) = a^2 + b^2. \tag{1.2}$$

However, complex arithmetic was not an option for the Greek mathematicians, since complex numbers and algebra had yet to be discovered.

Almost 700 years after Euclid's *Elements*, the Library of Alexandria was destroyed by fire (391 CE), taking with it much of the accumulated Greek knowledge. As a result, one of the best technical records remaining is Euclid's *Elements*, along with some sparse mathematics due to Archimedes (ca. 300 BCE) on geometrical series, the volume of a sphere, the area of a parabola, and elementary hydrostatics. In about 1572 a copy Diophantus's *Arithmetic* was discovered by Bombelli in the Vatican library (Burton 1985; Stillwell 2010, p. 51). This book became an inspiration for Galileo, Descartes, Fermat, and Newton.

Early number theory: Well before Pythagoras, the Babylonians (ca. 1,800 BCE) had tables of triplets of integers $[a, b, c]$ that obey Eq. 1.1, such as [3, 4, 5]. However, the triplets from the Babylonians were larger numbers, the largest being $a = 12{,}709$ and $c = 18{,}541$. A clay tablet (Plimpton-322) dating back to 1800 BCE was found with integers for $[a, c]$. Given such sets of two numbers, which determined a third positive integer $b = 13{,}500$ such that $b = \sqrt{c^2 - a^2}$, this table is more than convincing that the Babylonians were well aware of Pythagorean triplets (PTs), but less convincing that they had access to Euclid's formula, a formula for PTs (Eq. 2.6.6).

It seems likely that Euclid's *Elements* was largely the source of the fruitful era of the Greek mathematician Diophantus (215–285) (see Fig. 1.1), who developed the field of discrete mathematics, now known as *Diophantine analysis*. The term means that the solution, not the equation, is integer. The work of Diophantus was followed by fundamental changes in mathematics, possibly leading to the development of algebra but at least including these discoveries:

1. Negative numbers
2. Quadratic equations (Brahmagupta, seventh century)
3. Algebra (al-Khwarizmi, nineth century)
4. Complex arithmetic (Bombelli, fifteenth century)

These discoveries overlapped with the European Middle Ages (also known as the Dark Ages). Although Europe went "dark," presumably European intellectuals did not stop working during these many centuries.[4]

[4]It would be interesting to search the archives of the monasteries, where the records were kept, to determine exactly what happened during this religious blackout.

1.1.2 *What Is Science?*

Science is a process that quantifies hypotheses to build truths.[5] It has evolved from early Greek philosophers, Plato and Aristotle, into a method that uses statistical tests to either validate or prove wrong the null hypothesis. Scientists use the term *null hypothesis* to describe the supposition that there is no difference between two intervention groups or no effect of a treatment on some measured outcome. The measure of the likelihood that an outcome occurred by chance is called the p-value. From the p-value we can have some confidence that the null hypothesis is either true (the treatment causes no difference between two groups) or false (the probability p of a difference is greater than chance). The p-value is the present standard of scientific truth, but it is not ironclad and must be used with care. For example, not all experimental questions may be reduced to a single binary test. Does the Sun revolve around the Moon or around the Earth? There is no test of this question, as it is nonsense. To even say that the Earth revolves around the Sun is, in some sense, nonsense because all the planets are involved in the orbital motion.

Yet science works quite well. Thanks to mathematics, we have learned many deep secrets regarding the universe over the last 5000 years.

1.1.3 *What Is Mathematics?*

It seems strange when people complain that they "can't learn math"[6] but then claim to be good at languages. Before high school, students tend to confuse arithmetic with math. One does not need to be good at arithmetic to be good at math (but it doesn't hurt).

Math is a language with symbols taken from various languages–not so different from other languages. Today's mathematics is a written language with an emphasis on symbols and glyphs, primarily Greek letters, obviously due to the popularity of Euclid's *Elements*. The specific evolution of these symbols is interesting (Mazur 2014). Each symbol is assigned a meaning appropriate for the problem being described. These symbols are then assembled to make sentences. In the Chinese language, the spoken and written versions are different across dialects. Similarly, mathematical sentences may be read out loud in any language (dialect), but the symbols (like Chinese characters) are universal.

Learning languages is an advanced social skill. However, the social outcomes of learning a language and learning math are very different. Learning a new language is fun because it opens doors to other cultures. Math is different due to the rigor of the grammar (rules of the language) as well as the way it is taught (i.e., not as a language). A third difference between math and language is that math evolved from physics and has important technical applications.

---

[5] https://physicstoday.scitation.org/do/10.1063/PT.6.3.20191018a/full/.

[6] "It looks like Greek to me."

As with any language, the more mathematics you learn, the easier it is to understand, because mathematics is built from the bottom up. It's a continuous set of concepts, much like the construction of a house. If you try to learn calculus and differential equations but skip simple number theory, the lessons will be more difficult to understand. You will end up memorizing instead of understanding, and as a result you will likely soon forget it. When you truly understand something, it can never be forgotten. A nice example is the solution to a quadratic equation: If you learn how to complete the square (see Sect. 3.1.1), you don't need to know the quadratic formula.

Mathematical topics need to be learned in order, just as in the case of building the house. You can't build a house if you don't know about screws or cement (plaster). Likewise in mathematics, you can't learn to integrate if you have failed to understand the difference between integers, complex numbers, polynomials, and their roots.

A short list of topics in mathematics includes numbers ($\mathbb{N}, \mathbb{Z}, \mathbb{Q}, \mathbb{I}, \mathbb{C}$), algebra, derivatives, antiderivatives (i.e., integration), differential equations, vectors and the spaces they define, matrices, matrix algebra, eigenvalues and vectors, solutions of systems of equations, and matrix differential equations and their eigensolutions. Learning is about understanding, not memorizing.

The rules of mathematics are formally defined by algebra. For example, the sentence $a = b$ means that the number a has the same value as the number b. The sentence is read as "a equals b." The numbers are nouns and the equal sign says they are equivalent; it plays the role of a verb or action symbol. Following the rules of algebra, this sentence may be rewritten as $a - b = 0$. Here the symbols for minus and equal indicate two types of actions (verbs).

Sentences can become arbitrarily complex, such as the definition of the integral of a function or a differential equation. But in each case, the mathematical sentence is written down, may be read out loud, has a well-defined meaning, and may be manipulated into equivalent forms following the rules of algebra and calculus. This language of mathematics is powerful, with deep consequences, first known as algorithms but eventually as theorems.

The writer of an equation should always translate (explicitly summarize the meaning of the expression), so the reader will not miss the main point. This is a simple matter of clear writing.

Just as math is a language, so language may be thought of as mathematics. To properly write correct English it is necessary to understand the construction of the sentence. It is important to identify the subject, verb, object, and various types of modifying phrases. For example, look up the interesting distinction between *that* and *which*.[7] Thus, like math, language has rules. Most individuals use language that "sounds right," but if you're learning English as a second language, you must understand the rules, which are arguably easier to master than its foreign speech sounds.

Context can be critical, and the most important context for mathematics is physics. Without a physical problem to solve, there can be no engineering mathematics. People needed to navigate the Earth and weigh things, which required an understanding of

---

[7] https://en.oxforddictionaries.com/usage/that-or-which.

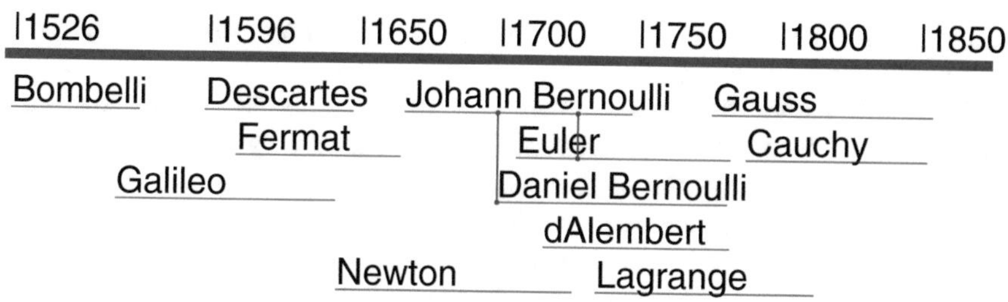

Fig. 1.2 Timeline covering the two centuries from 1596CE to 1855CE, including the development of modern theories of analytic geometry, calculus, differential equations, and linear algebra. Newton was born about one year after Galileo died and thus was heavily influenced by his many discoveries. The vertical red lines indicate mentor–student relationships. Note the significant overlap among Newton, the Bernoullis, and Euler, a nucleation point for modern mathematics. Lagrange had a key role in the development of linear algebra. Gauss had the advantage of input from Newton, Euler, d'Alembert, and Lagrange. Likely Cauchy had a significant contemporary influence on Gauss as well. Finally, note that Fig. 1.1 ends with Bombelli while this figure begins with him. He famously discovered a copy of Diophantus's book in the Vatican library. This was the same book where Fermat noted that the margin was too small to hold the "proof" of his "last theorem"

gravity. Many questions about gravity were deep, such as Where is the center of the universe?[8] But church dogma goes only so far. Mathematics along with a heavy dose of physics finally answered this huge question. Someone needed to perfect the telescope, put satellites into space, and view the cosmos. Without mathematics none of this would have happened.

1.1.4 Early Physics as Mathematics: Back to Pythagoras

We have established that math is the language of science. There is an additional answer to the question What is mathematics? The answer, the creation of algorithms and theorems, comes from studying its history, beginning with the earliest records. This chronological view starts, of course, with the study of numbers. First there is the taxonomy of numbers. It took thousands of years to realize that numbers are more than the counting numbers $\mathbb{N}$, to create a symbol for nothing (i.e., zero), and to invent negative numbers. With the invention of the abacus, a memory aid for manipulating complex sets of real integers, one could do very detailed calculations. But this required the discovery of algorithms (procedures) to add, subtract, multiply (many additions of the same number), and divide (many subtractions of the same number), such as the Euclidean algorithm for the greatest common divisor (GCD). Eventually it became clear to the experts (early mathematicians) that there were

[8] Actually this answer is simple: Ask the Pope and he will tell you. (I apologize for this inappropriate joke.).

natural rules to be discovered; thus books (e.g., Euclid's *Elements*) were written to summarize this knowledge.

The role of mathematics is to summarize algorithms (i.e., sets of rules) and formalize an idea as a theorem. Pythagoras and his followers, the Pythagoreans, believed that there was a fundamental relationship between mathematics and the physical world. The Asian civilizations were the first to capitalize on the relationship between science and mathematics, to use mathematics to design things for profit. This may have been the beginning of capitalizing technology (i.e., engineering), based on the relationship between physics and math. This influenced commerce in many ways–map making, tools, implements of war (the wheel, gunpowder), art (music), water transport, sanitation, secure communication, food, namely, all aspects of human existence. Of course it was the Asian cultures to first to master many of these early technologies.

The Pythagorean theorem (Eq. 1.1) did not begin with Euclid or Pythagoras; rather they appreciated its importance and documented its proof. Why is Eq. 1.1 called a *theorem*? Theorems require a proof. What exactly needs to be proved? We do not need to prove that (a, b, c) obeys this relationship, since this condition is observed. We do not need to prove that a^2 is the area of a square, as this is the definition of an area. What needs to be proved is that the relationship $c^2 = a^2 + b^2$ holds *if, and only if,* the angle between the two shorter sides is 90°.

In the end, because they instilled fear in the neighbor, the Pythagoreans were burned out and murdered, likely the result of mixing technology with politics:

> [It is] said that when the Pythagoreans tried to extend their influence into politics they met with popular resistance. Pythagoras fled, but he was murdered in nearby Mesopotamia in 497 BCE. (Stillwell 2010, p. 16)

1.2 Modern Mathematics

Modern mathematics (what we practice today) was born in the fifteenth and sixteenth centuries in the minds of Leonardo da Vinci, Bombelli, Galileo, Descartes, Fermat, and many others (Burton 1985). Many of these early masters were, like the Pythagoreans, extremely secretive about how they solved problems. This soon changed with Galileo, Mersenne, Descartes, and Newton, which caused mathematics to blossom. Developments during this time may seem hectic and disconnected, but this is a wrong impression. The developments were dependent on new technologies, such as the telescope (optics) and more accurate time and frequency measurements, due to Galileo's studies of the pendulum, and a better understanding of the relationship $f\lambda = c_o$ among frequency f, wavelength λ, and wave speed c_o.

1.2.1 *Science Meets Mathematics*

Early studies of vision and hearing: Since light and sound (music) played such a key role in the development of the early science, it was important to understand fully the mechanism of our perception of light and sound. There are many outstanding examples where physiology impacted mathematics. Leonardo da Vinci (1452–1519) is well known for his early studies of human anatomy, the knowledge of which was key when it came to drawing and painting the human form.

Galileo: In 1589 Galileo Galilei (1564–1642) famously conceptualized an experiment in which he suggested dropping two different masses from the Leaning Tower of Pisa. He suggested that both must take the same time to hit the ground.

Conceptually this is a mathematically sophisticated experiment, driven by a mathematical argument in which Galileo considered the two masses to be connected by an elastic cord (a spring) or rolling down a frictionless inclined plane (see Fig. 1.3). His studies resulted in the concept of conservation of energy, one of the cornerstones of physical theory since that time.

Being joined with an elastic cord, the masses become one. If the velocity were proportional to the mass, as Archimedes believed, the sum of the two masses would necessarily fall even faster. This results in a logical fallacy: How can two masses fall faster than either mass alone? This also violates the concept of conservation of energy, as the total energy of the two masses would be greater than that of the parts. In fact, Galileo's argument may have been the first time that the principle of conservation of energy was clearly stated.

It seems likely that Galileo was attracted to this model of two masses connected by a spring because he was also interested in planetary motion, which consists of masses (Sun, Earth, Moon) also mutually attracted by gravity (i.e., the spring).

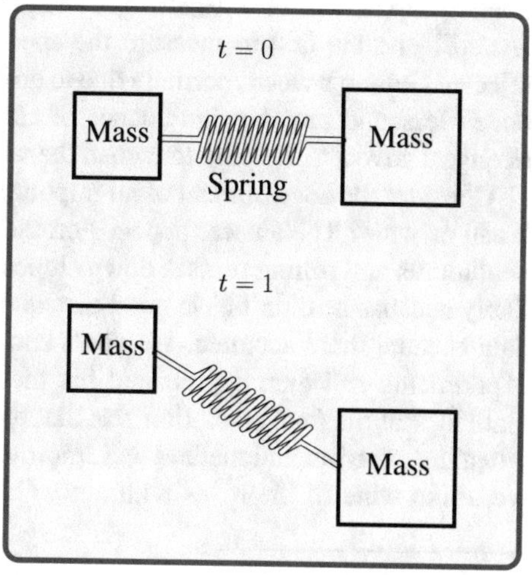

Fig. 1.3 Depiction of Galileo's argument (from his unpublished book of 1638) as to why objects of different masses (i.e., weights) must fall with the same velocity, contrary to what Archimedes had proposed in about 250 BCE

Galileo performed related experiments on pendulums, where he varied the length l, mass m, and angle θ of the swing. By measuring the period[9] he was able to formulate precise relationships between the variables. This experiment also measured the force exerted by gravity, so the experiments were related, but in very different ways. The pendulum served as the ideal clock, as it needed very little energy to keep it going, due to its very low friction (energy loss).

In a related experiment, Galileo measured the length of a day by counting the number of swings of the pendulum in 24 h, measured precisely by the daily period of a star as it crossed the tip of a church steeple. The number of seconds in a day is $24{\cdot}60{\cdot}60 = 86,400 = 2^7 3^3 5^2$ [s/day]. Since 86,400 is the product of the first three primes, it is highly composite and thus may be expressed in many equivalent ways. For example, the day can be divided evenly into 2, 3, 4, or 5 parts and remain the same in terms of the number of seconds that transpire. It would be interesting to know who was responsible for this highly composite number of seconds per day.

Factoring the number of days in a year ($365 = 5 \cdot 73$) is a poor choice, since it may not be decomposed into many small primes. For example, if the year were taken as $364 = 2^2 \cdot 7 \cdot 13$ days, it would make for shorter years (by 1 day), 13 months per year, perfect quarters, $28 = 4 \cdot 13$ day months, and $52 = 4 \cdot 13$ weeks. Every holiday would always fall on the same day, every year. It would be a calendar that humans could better understand. As with second per day, it would be fascinating to know why the number of days per year was so poorly chosen.

Galileo also studied on the relationship between the wavelength and frequency of a sound wave in musical instruments. Galileo also greatly improved the telescope, which he needed for his observations of the planets and their Moons.

Many of Galileo's contributions resulted in new mathematics, leading to Newton's discovery of the wave equation (1687), followed 60 years later by its one-dimensional general solution by d'Alembert (1747).

Mersenne: Marin Mersenne (1588–1648) contributed to our understanding of the relationship between the wavelength and the dimensions of musical instruments and is said to be the first to measure the speed of sound. At first Mersenne strongly rejected Galileo's views, partially due to errors in Galileo's reports of his results. But once Mersenne saw the significance of Galileo's conclusion, he became Galileo's strongest advocate, helping to spread the word (Palmerino 1999).

Consider the development of an important theorem of nature. Are data more like bread or wine? The answer depends on the data. Galileo's original experiments on pendulums and rolling masses down slopes were flawed by inaccurate data. This is likely because he didn't have good clocks. But he soon solved that problem and the data became more accurate. We don't know whether Mersenne repeated Galileo's experiments and then appreciated his theory, or whether he communicated with Galileo. But the final resolution was that the early data were like bread (it rots), but when the experimental method was improved with a better clock, the corrected data were like wine (it improves with age). Galileo claimed that the distance to reach

[9] The term *period* refers to the duration in units of time of a periodic function. For example, the periods of the Moon and the Sun are 28 days and 1 year, respectively.

the ground is proportional to the square of the time. This expression is equivalent to $F = m_o a$ assuming constant mass m_o.

Mersenne was also a decent mathematician, inventing in 1644 the Mersenne prime (MP) π_m of the form

$$\pi_m = 2^{\pi_k} - 1,$$

where π_k $(k < m)$ denotes the kth prime (see Sect. 2.1). As of December 2018, 51 MPs are known.[10] The first MP is $3 = \pi_2 = 2^{\pi_1} - 1$, and the largest known prime is an MP, which is $\pi_{12251} = 2^{\pi_7} - 1$.

Newton: With the closure of Cambridge University due to the plague of 1665, Isaac Newton (1642–1726) returned home to Woolsthorpe-by-Colsterworth (95 miles north of London) to work by himself for over a year.[11] It was during this solitary time that he did his most creative work.

Exploring our physiological senses requires a scientific understanding of the physical processes of vision and hearing, first considered by Newton, but researched later in much greater detail by Helmholtz (Stillwell 2010, p. 261). While Newton may be best known for his studies on light and gravity, he was the first to predict the speed of sound. However, his theory was in error by $\sqrt{c_p/c_v} = \sqrt{1.4} = 1.183$.[12] This famous issue would not be corrected for 129 years, awaiting the formulation of thermodynamics and the equipartition theorem by Laplace in 1816 (Britannica 2004).

Just 11 years prior to Newton's 1687 *Principia*, there was a basic understanding that sound and light traveled at very different speeds, due to the experiments of Ole Rømer (Feynman 1968, 2019, *Google for Feynman videos*).

> Ole Rømer first demonstrated in 1676 that light travels at a finite speed (as opposed to instantaneously) by studying the apparent motion of Jupiter's moon Io. In 1865, James Clerk Maxwell proposed that light was an electromagnetic wave, and therefore traveled at the speed c_o appearing in his theory of electromagnetism. (Wikipedia: Speed of Light 2019)

The idea behind Rømer's discovery was that due to the large distance between Earth and Io, there was a difference between the period (phase) of the Moon when Jupiter was closest to Earth and when it was farthest from Earth. This difference in distance caused a delay or advance in the observed eclipse of Io as it went behind Jupiter, delayed by the difference in time due to the difference in distance. This is like watching a video of a clock's motion. When the video is delayed or slowed down, the time will be inaccurate (it will indicate an earlier time).

The amazing Bernoulli family: The first individual who seems to have openly recognized the importance of mathematics, enough to actually teach it, was Jacob Bernoulli (Fig. 1.4). Jacob worked on what is now viewed as the standard package

---

[10] https://mathworld.wolfram.com/MersennePrime.html.

[11] Because the calendar was modified during Newton's lifetime, his birth date is no longer given as Christmas 1642 (Stillwell 2010, p. 175).

[12] The square root of the ratio of the specific heat capacity at constant pressure c_p to that at constant volume c_v.

Fig. 1.4 Above left: Jacob (1655–1705) and right: Johann (1667–1748) Bernoulli, both painted by their portrait painter brother, Nicolaus. Below left: Leonhard Euler (1707–1783) and right: Jean le Rond d'Alembert (1717–1783). Euler was blind in his right eye, hence the left portrait view

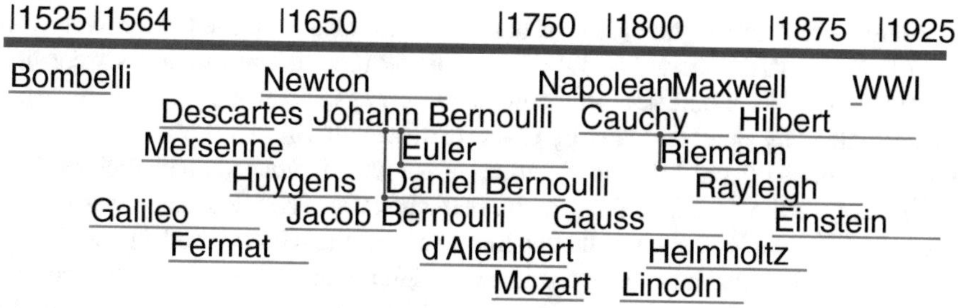

Fig. 1.5 Timeline for the sixteenth through twentieth centuries covering Bombelli to Einstein. As noted in the caption for Fig. 1.2, it seems likely that Bombelli's discovery of Diophantus's book *Arithmetic* in the Vatican library triggered many of the ideas presented by Galileo, Descartes, and Fermat, followed by others (i.e., Newton). Thus Bombelli's discovery might be considered a magic moment in mathematics. The vertical red lines indicate mentor–student relationships. For orientation Mozart is indicated along the bottom and Napoleon at the top. Napoleon hired Fourier, Lagrange, and Laplace to help with his many bloody military campaigns. See Figs. 1.1, 1.2, and 3.1 for additional timelines

of analytic "circular" (i.e., periodic) functions: $\sin(x)$, $\cos(x)$, $\exp(x\jmath)$, $\log(x)$.[13] Eventually the full details were developed (for real variables) by Euler (Sect. 3.2.5).

From Fig. 1.5 we may conclude that Jacob (1654–1705), the oldest brother, would have been strongly influenced by Newton.[14] Newton would have been influenced by Fermat, Descartes, and Galileo, who died 1 year before Newton was born[15] (White 1999).

Jacob Bernoulli, like all successful mathematicians of the day, was largely self-taught. Yet Jacob was in a new category of mathematicians because he was an effective teacher. Jacob taught his sibling Johann, who then taught his son Daniel. But most important, Johann taught Euler, the most prolific (thus influential) of all mathematicians, including Gauss. It is most significant that all four mathematicians published their methods and findings. Much later, Jacob studied with students of Descartes[16] (Stillwell 2010, pp. 268–69).

Euler: Leonhard Euler's mathematical talent went far beyond that of the Bernoulli family (Burton 1985). Another special strength of Euler was his large number of publications. First he would master a topic, and then he would publish. Once the tools of mathematics were openly published, largely by Euler, mathematics grew exponentially.[17] His papers continued to appear long after his death (Calinger 2015). It is also interesting that Euler was a contemporary of Mozart (see Fig. 1.5).

[13]The log and tan functions are related by Eq. 4.1.2.

[14]For a similar timeline see https://www.famousscientists.org/joseph-louis-lagrange/.

[15]https://www-history.mcs.st-andrews.ac.uk/Biographies/Newton.html.

[16]It seems clear that Descartes was also a teacher.

[17]There are at least three useful exponential scales: factors of 2, factors of $e \approx 2.7$, and factors of 10. The octave and decibel use factors of 2 (6 [dB]) and 10 (20 [dB]). Information theory uses factors of 2 (1 [bit]), 4 (2 [bits]). Circuit theory uses all three scales.

D'Alembert: Another individual of that time who also published extensively was Jean la Rond d'Alembert (Fig. 1.4). Some of the most innovative ideas were first proposed by d'Alembert. Unfortunately, and perhaps somewhat unfairly, his rigor was criticized by Euler and later by Gauss (Stillwell 2010).
Gauss: Figures 1.2 and 1.4 show timelines of the most famous mathematicians. This was one of the most creative times in mathematics. Carl Friedrich Gauss was born at the end of Euler's long and productive life. I suspect that Gauss owed a great debt to Euler; surely he must have been a scholar of Euler. One of Gauss's most important achievements may have been his contribution to solving the open question about the density of prime numbers, and his use of least-squares.[18]
Cauchy: Augustin-Louis Cauchy (1789–1857) was the son of a well-to-do family but had the misfortune of being born during the time of of the French Revolution, which likely had its origins in the Seven Years' War, around 1756. Today the French still celebrate Bastille Day (July 14, 1789), which is viewed as a celebration of the revolution. The French Revolution left Cauchy with a lifelong scorn for French politics that deeply influenced his life. But Cauchy had an unmatched intellect for mathematics. His most obvious achievement was complex analysis, for which he proved many key theorems.
Helmholtz: Hermann von Helmholtz (1821–1894) was educated and experienced as a military surgeon. He also mastered classical music, acoustics, physiology, vision, hearing (Helmholtz 1863b), and, most important of all, mathematics. He was the first person to measure the speed of a neural spike in a patch of neuron, which he correctly found to be 27 [m/s]. Gustav Kirchhoff frequently expanded on Helmholtz's acoustic contributions. It is reported that Lord Rayleigh learned German so he could read Helmholtz's great works, which he amplified in his famous treatise on acoustics (Rayleigh 1896).

Helmholtz's studies and theories of music and perception of sound are fundamental scientific contributions (Helmholtz 1863a). His best known mathematical contribution is known as the fundamental theorem of vector calculus or simply "Helmholtz theorem" (Sect. 5.6.3).
Lord Kelvin: Lord Kelvin (William Thompson, 1824–1907) was one of the first true engineer-scientists, equally acknowledged as a mathematical physicist, well known for his interdisciplinary research, and knighted by Queen Victoria in 1866.[19] Lord Kelvin coined the term *thermodynamics*, a science more fully developed by Maxwell (the same Maxwell of electrodynamics).[20]

The history during this time is complex. For example, in 1850 Lord Kelvin wrote a letter to George Stokes, suggesting that Stokes try to prove what is today known as "Stokes's theorem." As a result, Stokes posted a reward (Smith's Prize), searching for a proof of "Lord Kelvin's theorem," which was finally achieved by Hermann

[18]https://www-history.mcs.st-andrews.ac.uk/Biographies/Gauss.html.

[19]Lord Kelvin was one of a half dozen interdisciplinary mathematical physicists, all working about the same time, who made a fundamental change in our scientific understanding. Others include Helmholtz, Stokes, Green, Heaviside, Rayleigh, and Maxwell.

[20]Thermodynamics is another topic that warrants an analysis along historical lines Kuhn (1978).

Hankel (1839–1873).[21] Many new concepts were being proved and appreciated over this productive period. Maxwell had published his famous equations, which were later reformatted in modern vector notation by Oliver Heaviside, J. Willard Gibbs, and Heinrich Hertz. Figure 1.5 should put to rest the myth that one's best work is done in the early years. Many of these scientists were fully productive in old age. Those who were not died early due to poor health or accidents.

James Clerk Maxwell (1831–1879) In 1869 a Cambridge senate committee was formed to create the Cambridge Physics Laboratory and "the founding of a special Professorship." The Chancellor of Cambridge was the seventh Duke of Devonshire and a distant relative of Henry Cavendish, his family name. Thus the new laboratory became known as the Cavendish.

> There was, naturally, much speculation about the choice of the new Professor of Experimental Physics. [Lord] Kelvin was the most likely candidate, but on being approached in private, refused in order to stay in Glasgow. Another likely candidate was Lord Rayleigh, a brilliant mathematician and physicist who had left Cambridge to work in his private laboratory at his country seat in Essex. When the appointment was eventually announced, the reaction was, if anything, one of disappointment. The new Professor, James Clerk Maxwell, was relatively unknown.
>
> He was a much respected mathematician, but he had not since made any great name for himself — his major and astounding books on Electricity and Kinetic Theory had yet to be published. Moreover, the six years before his appointment had been spent in isolation at his Scottish home.
>
> His appointment was announced on March 8th 1871, and in spite of the initial disappointment, his inaugural lecture was looked forward to by his likely students as much as by the rest of the Cambridge scientists.
>
> When, a few days later, Maxwell began his first course with a lecture on Heat, his students had the delight of seeing the lecture room packed with their tutors, lecturers, professors and all the important personages of the University. Thinking that this was his first public appearance they sat, in their formal academic dress, while Maxwell, "with a perceptible twinkle in his eye," gravely expounded the difference between Fahrenheit and Centigrade, and the principle of the air thermometer.
>
> It was felt afterward that Maxwell had done it on purpose, perhaps out of modesty, perhaps out of his later well-known sense of humor, or perhaps because he knew of the still considerable opposition his new laboratory had to face. As he had written to his friend Lord Rayleigh, "if we succeed too well, and corrupt the minds of youth till they observe vibrations and deflections and become Senior Ops. instead of Wranglers, we may bring the whole University and all the parents about our ears.
>
> However, Maxwell made only a casual announcement of his inaugural lecture which was not to be in the Senate House, as expected, but in an out-of-the-way lecture room. Consequently only his students got to hear of it and it was to them, rather than a general gathering, that he delivered an exciting and interesting lecture, mapping out his plans for the future of Cambridge physics. . . .."[22]

Lord Rayleigh (William Strutt): Lord Rayleigh (1842–1919) wrote a classic text (1896) that is widely read even today by those who study acoustics. In 1904 he received the Nobel Prize in Physics for his investigations of the densities of the most important gases and for his discovery of argon in connection with these studies.

[21] https://en.wikipedia.org/wiki/Hermann_Hankel.

[22] https://www.phy.cam.ac.uk/history/years/firstten, Moralee (1995).

1.2.2 Three Streams from the Pythagorean Theorem

From the outset of his presentation, Stillwell (2010, p. 1) defines "three great *streams* of mathematical thought: *Numbers, Geometry and Infinity*" that flow from the Pythagorean theorem, as summarized in Fig. 1.6. This is a useful concept, based on reasoning not as obvious as one might think. Many factors are in play here. One of these is the strongly held opinion of Pythagoras that all mathematics should be based on integers. The rest is tied up in the long, necessarily complex history of mathematics, as best summarized by the fundamental theorems (see Sect. 2.3.1), each of which is discussed in detail in a later chapter.

As shown in Fig. 1.6, Stillwell's concept of three streams, following from the Pythagorean theorem, is the organizing principle behind this book.

This chapter: The *Introduction* is a historical survey of pre-college mathematical physics, presented in terms of the three main Pythagorean streams (stream 1–stream 3), leading to the book's five chapters. Stream 3 is split into 3 A and 3B.

Chapter 2: *Number systems* presents some important ideas from number theory, starting with prime numbers, complex numbers, vectors, and matrices. Five classic number theory problems are discussed: the Euclidean algorithm (GCD), continued fractions (CFA), Euclid's formula, Pell's equation, and the Fibonacci difference equation. The general solution of these problems leads to the concept of the eigenfunction analysis, which is introduced in Sect. 2.6.2.

Chapter 3: *Algebraic equations* discusses Algebra and its development, as we know it today. The chapter presents the theory of real and complex equations and functions of real and complex variables. Newton's method for finding complex roots of polynomials, poles vs. zeros, and the Gauss–Lucas theorem (bounds on the root locations of the derivative of a polynomial). Complex impedance $Z(s)$ of complex frequency $s = \sigma + \omega \jmath$ is covered with some care, developing the topic that is needed for engineering mathematics.

While the algebra of real and complex functions is identical, the calculus is fundamentally different. This leads to the concepts of complex analytic functions, complex

1. **Numbers**
    - $\mathbb{N}$ *counting numbers*, $\mathbb{Q}$ *rationals*, $\mathbb{P}$ *primes (6th century BCE)*
    - $\mathbb{Z}$ *common integers*, $\mathbb{I}$ *irrationals (5th century BCE)*
    - *zero* $\in \mathbb{Z}$ *(7th century CE)*
2. **Geometry** *(e.g., lines, circles, spheres, toroids, other shapes)*
    - *Composition of polynomials (Descartes, Fermat),*
    - *Euclid's geometry and algebra* $\Rightarrow$ *analytic geometry (17th century CE)*
    - *Fundamental theorem of algebra (18th century CE)*
3. **Infinity** *($\infty \rightarrow$ sets)*
    - *Taylor series, functions, calculus (Newton, Leibniz) (17th and 18th century CE)*
    - $\mathbb{R}$ *real*, $\mathbb{C}$ *complex (19th century CE)*
    - *Set theory (20th century CE)*

Fig. 1.6 Three streams that follow from the Pythagorean theorem: numbers, geometry, and infinity

Taylor series, the Cauchy–Riemann conditions, branch points, branch cuts, and Riemann sheets. All of these ideas are fundamental to impedance functions that describe the linear relationships between force and flow in the complex frequency domain (i.e., impedance $\in \mathbb{C}$).

Chapter 4: *Scalar calculus* (Stream 3A) covers ordinary differential equations and integral theorems of simple physical systems (mass-springs, inductors-capacitors, heat dynamics), solutions to scalar differential equations that have constant coefficients, colorized mappings of complex analytic functions, multivalued functions, Cauchy's theorems, and inverse Laplace transforms.

Chapter 5: *Vector calculus* (Stream 3B) introduces vector partial differential equations, as well as gradient, divergence, and curl differential operators, Stokes's and Green's theorems, and Maxwell's equations.

Chapter 2
Stream 1: Number Systems

Number theory (the study of numbers) was a starting point for many key ideas. For example, in Euclid's geometrical constructions the Pythagorean theorem for real $[a, b, c]$ was accepted as true, but the emphasis in the early analysis was on integer constructions, such as Euclid's formula for Pythagorean triplets (Eq. 2.6.6).

As we shall see, the derivation of the formula for Pythagorean triplets is the first of a rich body of mathematical constructions—such as solutions of Pell's equation (Sect. 2.6.2),[1] and recursive difference equations, such as solutions of the Fibonacci recursion formula $f_{n+1} = f_n + f_{n-1}$ (see Sect. 2.6.3)—that goes back at least to the Chinese (2000 BCE). These are early pre-limit forms of calculus, best analyzed using an eigenfunction (e.g., eigenmatrix) expansion, a geometrical concept from linear algebra, as an orthogonal set of normalized unit-length vectors (see Appendix B.3).

It is hard to imagine that anyone who uses an abacus would not appreciate the concept of zero and negative numbers. It does not take much imagination to go from counting numbers $\mathbb{N}$ to the set of all integers $\mathbb{Z}$ including zero. On an abacus, subtraction is obviously the inverse of addition. Subtraction to obtain zero abacus beads is no different from subtraction from zero, which gives *negative* beads. To assume that the Romans, who first developed counting sticks, or the Chinese, who then deployed the concept using beads, did not understand negative numbers is impossible. However, understanding the concept of zero (and negative numbers) is not the same as having a symbolic notation. The Roman number system has no such symbols. The first recorded use of a symbol for zero is said to be by Brahmagupta in 628 CE.[2]

[1]Heisenberg, an inventor of the matrix algebra form of quantum mechanics, learned mathematics by studying Pell's equation (Sect. 2.6.2) by eigenvector and recursive analysis methods. https://www.aip.org/history-programs/niels-bohr-library/oral-histories/4661-1.

[2]https://news.nationalgeographic.com/2017/09/origin-zero-bakhshali-manuscript-video-spd/; https://www.nytimes.com/2017/10/07/opinion/sunday/who-invented-zero.html.

J. Allen, *An Invitation to Mathematical Physics and Its History*,
https://doi.org/10.1007/978-3-030-53759-3_2

However, this is likely wrong, given the notation developed by the Mayan civilization, which existed from 2000 BCE to 900 CE.[3] There is speculation that the Mayans cut down so much of the Amazon jungle that it eventually resulted in global warming, possibly leading to their demise.

The definition of zero depends on the concept of subtraction, which formally requires the creation of algebra (ca. 830 CE; see Fig. 1.1). But apparently it took more than 600 years from the time Roman numerals were put into use, without any symbol for zero, to the time the symbol for zero is first documented. Likely this delay was more about the political situation, such as government rulings, than mathematics.

The concept that caused much more difficulty was ∞ or infinity, first resolved by Bernhard Riemann in 1851 with the development of the extended plane, which mapped the plane to a sphere (see Fig. 3.15). His construction made it clear that the point at ∞ is simply another point on the open complex plane, since rotating the sphere (extended plane) moves the point at ∞ to a finite point on the plane, thereby closing the complex plane.

2.1 The Taxonomy of Numbers: $\mathbb{P}, \mathbb{N}, \mathbb{Z}, \mathbb{Q}, \mathbb{F}, \mathbb{I}, \mathbb{R}, \mathbb{C}$

Once symbols for zero and negative numbers were accepted, progress could be made. To fully understand numbers, a transparent notation is required. First one must identify the different classes (genus) of numbers, providing a notation that defines each of these classes along with their relationships. It is logical to start with the most basic counting numbers, which we indicate with the double-bold symbol $\mathbb{N}$ (Appendix A.1.3). For easy access, notation is summarized in Appendix A.

Counting numbers $\mathbb{N}$: These are known as the *natural numbers* $\mathbb{N} = \{1, 2, 3, \ldots\}$ and denoted by the double-bold symbol $\mathbb{N}$. For clarity we shall refer to the natural numbers as *counting numbers*, since *natural*, which means *integer*, is vague. The mathematical sentence "$2 \in \mathbb{N}$" is read as "2 is a member of the set of counting numbers." The word *set* is defined as the collection of any objects that share a specific property. Typically a set may be defined either by a sentence or by example.

Primes $\mathbb{P}$: A number is prime ($\pi_n \in \mathbb{P}$) if its only factors are 1 and itself. The set of primes $\mathbb{P}$ is a subset of the counting numbers ($\mathbb{P} \subset \mathbb{N}$). A somewhat amazing fact, well known to the earliest mathematicians, is that every integer may be written as a unique product of primes. A second key idea is that the density of primes $\rho_\pi(N) \sim N/\log(N)$; that is, $\rho_\pi(N)$ is inversely proportional to the log of N, an observation first quantified by Gauss (Goldstein 1973). A third is that there is a prime between every integer $N \geq 2$ and $2N$.

We shall use the convenient notation π_n for the prime numbers, indexed by $n \in \mathbb{N}$. The first 12 primes are $\{n|1 \leq n \leq 12\} = \{\pi_n|2, 3, 5, 7, 11, 13, 17, 19, 23, 29, 31, 37\}$. Since $4 = 2^2$ and $6 = 2 \cdot 3$ may be factored, $4, 6 \notin \mathbb{P}$ (read as "4 and 6 are not in the set of primes"). Given this definition, multiples of a prime—that is,

[3] https://www.storyofmathematics.com/mayan.html.

$[2, 3, 4, 5, \ldots] \times \pi_k$ of any prime π_k—cannot be prime. It follows that all primes except 2 must be odd, and every integer N is unique in its prime factorization.

Exercise #1 Write the first 20 integers in prime-factored form.
Sol: $1, 2, 3, 2^2, 5, 2 \cdot 3, 7, 2^3, 3^2, 2 \cdot 5, 11, 3 \cdot 2^2, 13, 2 \cdot 7, 3 \cdot 5, 2^4, 17, 2 \cdot 3^2, 19, 2^2 \cdot 5$ ■

Exercise #2 Write the integers 2 to 20 in terms of π_n. Here is a table to assist you:

n	1	2	3	4	5	6	7	8	9	10	11	...
π_n	2	3	5	7	11	13	17	19	23	29	31	...

Sol:

n	2	3	4	5	6	7	8	9	10	11	12	13	14	...
$\Pi\pi_n$	π_1	π_2	π_1^2	π_3	$\pi_1\pi_2$	π_4	π_1^3	π_2^2	$\pi_1\pi_3$	π_5	$\pi_1^2\pi_2$	π_6	$\pi_1\pi_4$	...

■

Coprimes are two relatively prime numbers that have no common (i.e., prime) factors. For example, $21 = 3 \cdot 7$ and $10 = 2 \cdot 5$ are coprime, whereas $4 = 2 \cdot 2$ and $6 = 2 \cdot 3$, which have 2 as a common factor, are not. By definition all unique pairs of primes are coprime. We shall use the notation $m \perp n$ to indicate that m and n are coprime. The ratio of two coprimes is reduced, as it has no factors to cancel. The ratio of two numbers that are not coprime may always be reduced by canceling the common factors. This is called the *reduced form* or an *irreducible fraction*. When we do numerical work, for computational accuracy it is always beneficial to work with coprimes. Generalizing this idea, we could define *triprimes* as three numbers with no common factor, such as $\{\pi_3, \pi_9, \pi_2\}$.

The fundamental theorem of arithmetic states that each integer may be uniquely expressed as a unique product of primes. The prime number theorem estimates the mean density of primes over $\mathbb{N}$.

Integers $\mathbb{Z}$: The integers include positive and negative counting numbers and zero. Notionally we might indicate this using set notation as $\mathbb{Z} = -\mathbb{N} \cup \{0\} \cup \mathbb{N}$. We read this as: "The integers are in the set composed of the negative natural numbers $-\mathbb{N}$, zero, and $\mathbb{N}$."

Rational numbers $\mathbb{Q}$: These are defined as numbers formed from the ratio of two integers. Given two numbers $n, d \in \mathbb{N}$, we have $n/d \in \mathbb{Q}$. Since d may be 1, it follows that the rationals include the counting numbers as a subset. For example, the rational number $3/1 \in \mathbb{N}$.

The main utility of rational numbers is that that they can efficiently approximate any number on the real line, to any precision. For example, the rational approximation $\pi \approx 22/7$ has a relative error of $\approx 0.04\%$ (see Sect. 2.4.4).

Fractional number $\mathbb{F}$: A fractional number $\mathbb{F}$ is defined as the ratio of signed coprimes. If $n, d \in \pm\mathbb{P}$, then $n/d \in \mathbb{F}$. Given this definition, $\mathbb{F} \subset \mathbb{Q} = \mathbb{Z} \cup \mathbb{F}$.

Because of the powerful approximating power of rational numbers, the fractional set $\mathbb{F}$ has special utility. For example, $\pi \approx 22/7$, $1/\pi \approx 7/22$ (to 0.13%), $e \approx 19/7$ to 0.4%, and $\sqrt{2} \approx 7/5$ to 1.4%.

Irrational numbers $\mathbb{I}$: Every real number that is not rational is irrational ($\mathbb{Q} \perp \mathbb{I}$). Irrational numbers include π, e, and the square roots of primes. These are decimal numbers that never repeat, thus requiring infinite precision in their representation. Such numbers cannot be represented on a computer, as they would require an infinite number of bits (precision).

The rationals $\mathbb{Q}$ and irrationals $\mathbb{I}$ split the reals ($\mathbb{R} = \mathbb{Q} \cup \mathbb{I}$, $\mathbb{Q} \perp \mathbb{I}$); thus each is a subset of the reals ($\mathbb{Q} \subset \mathbb{R}$, $\mathbb{I} \subset \mathbb{R}$). This relationship is analogous to that of the integers $\mathbb{Z}$ and fractionals $\mathbb{F}$, which split the rationals ($\mathbb{Q} = \mathbb{Z} \cup \mathbb{F}$, $\mathbb{Z} \perp \mathbb{F}$) (thus each is a subset of the rationals ($\mathbb{Z} \subset \mathbb{Q}$, $\mathbb{F} \subset \mathbb{Q}$)).

Irrational numbers $\mathbb{I}$ were famously problematic for the Pythagoreans, who incorrectly theorized that all numbers were rational. Like ∞, irrational numbers required mastering a new and difficult concept before they could even be defined: It was essential to understand the factorization of counting numbers into primes (i.e., the fundamental theorem of arithmetic) before the concept of irrationals could be sorted out. Irrational numbers could be understood only once limits were mastered.

As shown in Fig. 2.6, fractionals can approximate any irrational number with arbitrary accuracy. Integers are also important, but for a very different reason. All numerical computing today is done with $\mathbb{Q} = \mathbb{F} \cup \mathbb{Z}$. Indexing uses integers $\mathbb{Z}$, while the rest of computing (flow dynamics, differential equations, etc.) is done with fractionals $\mathbb{F}$ (i.e., IEEE-754). Computer scientists are trained on these topics, and computer engineers need to be at least conversant with them.

Real numbers $\mathbb{R}$: Reals are the union of rational and irrational numbers—namely, $\mathbb{R} = \mathbb{Q} \cup \mathbb{I} = \mathbb{Z} \cup \mathbb{F} \cup \mathbb{I}$. Lengths in Euclidean geometry are reals. Many people assume that IEEE-754 floating-point numbers (ca. 1985) are real (i.e., $\in \mathbb{R}$). In fact, they are rational ($\mathbb{Q} = \{\mathbb{F} \cup \mathbb{Z}\}$) approximations to real numbers, designed to have a very large dynamic range. The hallmark of fractional numbers ($\mathbb{F}$) is their power in making highly accurate approximations of any real number.

Using Euclid's compass and ruler methods, one can make the length of a line proportionally shorter or longer or (approximately) the same. A line may be made be twice as long, or an angle can be bisected. However, the concept of an integer length in Euclid's geometry was not defined.[4] Nor can one construct an imaginary or complex line, as all lines are assumed to have real lengths. The development of analytic geometry was an analytic extension of Euclid's simple (but important) geometrical methods.

Real numbers were first fully accepted only after set theory was developed by Georg Cantor in 1874 (Stillwell 2010, p. 461). At first blush, this seems amazing, given how widely accepted real numbers are today. In some sense they were accepted by the Greeks as lengths of real lines.

[4] As best I know.

Complex numbers $\mathbb{C}$: Complex numbers are best defined as ordered pairs of real numbers.[5] For example, if $a, b \in \mathbb{R}$ and $\jmath = -\imath = \pm\sqrt{-1}$, then $c = a + b\jmath \in \mathbb{C}$. The word *complex*, as used here, does *not* mean that the numbers are complicated or difficult. The $b\jmath$ is known as the *imaginary* part of c. This does not mean the number disappears.

Complex numbers are quite special in engineering mathematics, for example, as roots of polynomials. The most obvious example is the quadratic formula for the roots of polynomials of degree 2 that have coefficients $\in \mathbb{C}$. All real numbers have a natural order on the real line. Complex numbers do not have a natural order. For example, $\jmath > 1$ makes no sense.

Today the common way to write a complex number is using the notation $z = a + b\jmath \in \mathbb{C}$, where $a, b \in \mathbb{R}$. Here $1\jmath = \sqrt{-1}$. We also define $1\imath = -1\jmath$ to account for the two possible signs of the square root. Accordingly $1\jmath^2 = 1\imath^2 = -1$.

Cartesian multiplication of complex numbers follows the basic rules of real algebra—for example, the rules for multiplying two monomials and polynomials. Multiplication of two first-degree polynomials (i.e., monomials) gives

$$(a + bx)(c + dx) = ac + (ad + bc)x + bdx^2.$$

If we substitute $1\jmath$ for x and use the definition $1\jmath^2 = -1$, we obtain the Cartesian product of two complex numbers:

$$(a + b\jmath)(c + d\jmath) = ac - bd + (ad + bc)\jmath.$$

Thus multiplication and division of complex numbers obey the usual rules of algebra.

However, there is a critical extension: Cartesian multiplication holds only when the angles sum to less than $\pm\pi$—namely, the range of the complex plane. When the angles add to more than $\pm\pi$, one must use polar coordinates, where the angles add for angles beyond $\pm\pi$ (Boas 1987, p. 8). This is particularly striking for the Laplace transform of a delay (see Appendix 3.9).

Complex numbers can be challenging and may provide unexpected results. For example, it is not obvious that $\sqrt{3 + 4\jmath} = \pm(2 + \jmath)$.

Exercise #3 Verify that $\sqrt{3 + 4\jmath} = \pm(2 + \jmath)$.
Sol: Squaring the left side gives $\sqrt{3 + 4\jmath}^2 = 3 + 4\jmath$. Squaring the right side gives $(2 + \jmath)^2 = 4 - \jmath^2 + 4\jmath = 3 + 4\jmath$. Thus the two are equal. ■

Exercise #4 What is special about the above example (Exercise #3)?
Sol: Note this is a $\{3, 4, 5\}$ triangle. Can you find another example like this one? Namely, how does one find integers that obey Eq. 1.1? ■

An alternative to Cartesian multiplication of complex numbers is to work in polar coordinates. The polar form of the complex number $z = a + b\jmath$ is written in terms of its magnitude $\rho = \sqrt{a^2 + b^2}$ and angle $\theta = \angle z = \tan^{-1} z = \arctan z$ as

---

[5] A polynomial $a + bx$ and a 2-vector $[a, b]^T = \begin{bmatrix} a \\ b \end{bmatrix}$ are also examples of ordered pairs.

$$z = \rho e^{\theta \jmath} = \rho(\cos\theta + \jmath \sin\theta).$$

From the definition of the complex natural log function, we have

$$\ln z = \ln \rho e^{\theta \jmath} = \ln \rho + \theta \jmath,$$

which is important, even critical, in engineering calculations. When the angles of two complex numbers are greater that $\pm\pi$, one must use polar coordinates. It follows that for computing the phase, the log function is different from the single- and double-argument $\angle\theta = \arctan(z)$ function.

The polar representation makes clear the utility of a complex number: Its magnitude scales while its angle θ rotates. The property of scaling and rotating is what makes complex numbers useful in engineering calculations. This is especially obvious when dealing with impedances, which have complex roots with very special properties, as discussed on Sect. 4.4.

Matrix representation: An alternative way to represent complex numbers is in terms of 2×2 matrices. This relationship is defined by the mapping from a complex number to a 2×2 matrix:

$$a + b\jmath \leftrightarrow \begin{bmatrix} a & -b \\ b & a \end{bmatrix}, \quad 1 \leftrightarrow \begin{bmatrix} 1 & 0 \\ 0 & 1 \end{bmatrix}, \quad 1\jmath \leftrightarrow \begin{bmatrix} 0 & -1 \\ 1 & 0 \end{bmatrix}, \quad e^{\theta \jmath} \leftrightarrow \begin{bmatrix} \cos(\theta) & -\sin(\theta) \\ \sin(\theta) & \cos(\theta) \end{bmatrix}. \tag{2.1.1}$$

The *conjugate* of $a + b\jmath$ is then defined as $a - b\jmath \leftrightarrow \begin{bmatrix} a & b \\ -b & a \end{bmatrix}$. By taking the inverse of the 2×2 matrix (assuming $|a + b\jmath| \neq 0$), we can define the ratio of one complex number by another. Until you try out this representation, it may not seem obvious or even possible.

This representation proves that $1\jmath$ is not necessary when defining a complex number. What $1\jmath$ can do is to conceptually simplify the algebra. It is worth your time to become familiar with the matrix representation, to clarify any possible confusions you might have about multiplication and division of complex numbers. This matrix representation can save you time, heartache, and messy algebra. Once you have learned how to multiply two matrices, it's a lot simpler than doing the complex algebra. In many cases we will leave the results of our analysis in matrix form, to avoid the algebra altogether.[6] Thus both representations are important.

Exercise #5 Using MATLAB/Octave, verify that

$$\frac{a + b\jmath}{c + d\jmath} = \frac{ac + bd + (bc - ad)\jmath}{c^2 + d^2} \longleftrightarrow \begin{bmatrix} a & -b \\ b & a \end{bmatrix} \begin{bmatrix} c & -d \\ d & c \end{bmatrix}^{-1} = \begin{bmatrix} a & -b \\ b & a \end{bmatrix} \begin{bmatrix} c & d \\ -d & c \end{bmatrix} \frac{1}{c^2 + d^2}. \tag{2.1.2}$$

Sol: The typical solution may use numerical examples. A better solution is to use the MATLAB/Octave symbolic code:

[6] Sometimes we let the computer do the final algebra, numerically, as 2×2 matrix multiplications.

```
syms a b c d A B
A=[a -b;b a];
B=[c -d;d c];
C=A*inv(B)
```

■

History of complex numbers: It is notable that complex numbers were not accepted until 1851 even though Bombelli introduced them in the sixteenth century. One might have thought that the solution of the quadratic, known to the Chinese, would have settled this question. It seems that complex integers (Gaussian integers) were accepted before nonintegral complex numbers. Perhaps this was because real numbers ($\mathbb{R}$) were not accepted (i.e., proved to exist, thus mathematically defined) until the development of real analysis in the late nineteenth century, thus providing a proper definition of the real numbers.

2.1.1 Numerical Taxonomy

A simplified taxonomy of numbers is given by the mathematical sentence

$$\pi_k \in \mathbb{P} \subset \mathbb{N} \subset \mathbb{Z} \subset \mathbb{Q} \subset \mathbb{R} \subset \mathbb{C}.$$

This sentence says:

1. Every prime number π_k is in the set of primes $\mathbb{P}$,
2. which is a subset of the set of counting numbers $\mathbb{N}$,
3. which is a subset of the set of integers $\mathbb{Z} = -\mathbb{N}, \{0\}, \mathbb{N}$,
4. which is a subset of the set of rationals $\mathbb{Q}$,
5. which is a subset of the set of reals $\mathbb{R}$,
6. which is a subset of the set of complex numbers $\mathbb{C}$.

The rationals $\mathbb{Q}$ may be further decomposed into the fractionals $\mathbb{F}$ and the integers $\mathbb{Z}$ ($\mathbb{Q} = \mathbb{F} \cup \mathbb{Z}$), and the reals $\mathbb{R}$ into the rationals $\mathbb{Q}$ and the irrationals $\mathbb{I}$ ($\mathbb{R} = \mathbb{I} \cup \mathbb{Q}$). This classification nicely defines all the numbers (scalars) used in engineering and physics.

> The taxonomy structure may be summarized with the single compact sentence, starting with the prime numbers π_k and ending with complex numbers $\mathbb{C}$:
>
> $$\pi_k \in \mathbb{P} \subset \mathbb{N} \subset \mathbb{Z} \subset (\mathbb{Z} \cup \mathbb{F} = \mathbb{Q}) \subset (\mathbb{Q} \cup \mathbb{I} = \mathbb{R}) \subset \mathbb{C}.$$

As discussed in Appendix A, all numbers may be viewed as complex; that is, every real number is complex if we take the imaginary part to be zero (Boas 1987).

For example, $2 \in \mathbb{P} \subset \mathbb{C}$. Likewise, every purely imaginary number (e.g., $0 + 1\jmath$) is complex with zero real part.

Finally, note that complex numbers $\mathbb{C}$, much like vectors, do not have rank order, which means that one complex number cannot be larger or smaller than another. It makes no sense to say that $\jmath > 1$ or $\jmath = 1$ (Boas 1987). The real and imaginary parts, and the magnitude and phase, have order. Order seems restricted to $\mathbb{R}$. If time t were complex, there could be no yesterday and tomorrow.[7]

2.1.2 Applications of Integers

The most relevant question at this point is Why are integers important? First, we count with them so that we can keep track of "how many." But there is much more to numbers than counting: We use integers for any application where absolute accuracy is essential, such as in banking transactions (making change), the precise computing of dates (Stillwell 2010, p. 70) and locations ("I'll meet you at 34th and Vine at noon on Jan. 1, 2034"), and the construction of roads and pyramids out of bricks (objects built from a unit size).

To navigate we need to know how to predict the tides and the location of the Moon and Sun. Integers are important precisely because they are precise: Once a month there is a full Moon, easily recognizable. The next day it's slightly less than full. If we could represent our position as integers in time and space, we would know exactly where we were. But such an integral representation of our position or time is not possible since time $t \in \mathbb{I}$.

The Pythagoreans claimed that everything was integer. From a practical point of view, when precision is critical, they were right. Today all computers compute floating-point numbers as fractionals. However, in theory the Pythagoreans were wrong. The error (difference) is a matter of precision.

Numerical Representations of $\mathbb{I}, \mathbb{R}, \mathbb{C}$: When doing numerical work, one must consider how to compute within the set of reals (i.e., which contain irrationals). There can be no irrational number representation on a computer. The international standard of computation, IEEE floating-point numbers,[8] is based on rational approximation. The mantissa and the exponent are both integers, having sign and magnitude. The size of each integer depends on the precision of the number being represented. An IEEE floating-point number is rational because it has a binary (integer) mantissa, multiplied by 2 raised to the power of a binary (integer) exponent. For example, $\pi \approx \pm a 2^{\pm b}$ with $a, b \in \mathbb{Z}$. In summary, IEEE floating-point rational numbers cannot be irrational because irrational representations would require an infinite number of bits.

True floating-point numbers contain irrational numbers, which must be approximated by fractional numbers. This leads to the concept of fractional representation,

---

[7]One may meaningfully define $\xi = x + \jmath\, c_o t$ to be complex ($x, t \in \mathbb{R}, \xi \in \mathbb{C}$), with x in meters [m], t is in seconds [s], and the speed of light c_o [m/s].

[8]IEEE-754: https://www.h-schmidt.net/FloatConverter/IEEE754.html.

which requires the definition of the *mantissa*, *base*, and *exponent*, where both the mantissa and the exponent are signed. Numerical results must not depend on the base. One could dramatically improve the resolution of the numerical representation by the use of the fundamental theorem of arithmetic (see Sect. 2.3.2). For example, one could factor the exponent into its primes and then represent the number as $2^a 3^b 7^c$ $(a, b, c \in \mathbb{Z})$. Such a representation would improve the resolution of the representation. But even so, the irrational numbers would be approximate. For example, base ten is natural using this representation, since $10^n = 2^n 5^n$.[9] Thus

$$\pi \cdot 10^5 \approx 314,159.27\ldots = 3 \cdot 2^5 5^5 + 1 \cdot 2^4 5^4 + 4 \cdot 2^3 5^3 + \cdots + 9 \cdot 2^0 5^0 + \not{2} \cdot \not{2}^{-1} 5^{-1} \ldots .$$

Exercise #6 If we work in base 2 and use the approximation $\pi \approx 22/7$, then according to the MATLAB/Octave DEC2BIN() routine, show that the binary representation of $\hat{\pi}_2 \cdot 2^{17}$ is

$$\pi \cdot 2^{17} \approx 411,940_{10} = 64,924_{16} = 1,100,100,100,100,100,100_2.$$

Sol: First we note that this must be an approximation, since $\pi \in \mathbb{I}$, which cannot have an exact representation $\in \mathbb{F}$. The fractional $(\in \mathbb{F})$ approximation to π is

$$\hat{\pi}_2 = 22/7 = 3 + 1/7 = [3; 7], \tag{2.1.3}$$

where `int64(fix(2^17*22/7))` = 411,940 and `dec2hex(int64(fix(2^17*22/7)))` = 64,924, and where 1 and 0 are multipliers of powers of 2, which are then added together:

$$411,940_{10} = 2^{18} + 2^{17} + 2^{14} + 2^{11} + 2^8 + 2^5 + 2^2.$$

■

Computers keep track of the decimal point using the exponent, which in Exercise #6 is the scale factor $2^{17} = 131,072_{10}$. The concept of the decimal point is replaced by an integer that has the desired precision and a scale factor of any base (radix). This scale factor may be thought of as moving the decimal point to the right (larger number) or left (smaller number). The mantissa fine-tunes the value about a scale factor (the exponent). In all cases the number actually is an integer. Negative numbers are represented by an extra sign bit.

Exercise #7 Using MATLAB/Octave and base 16 (i.e., hexadecimal) numbers, with $\hat{\pi}_2 = 22/7$, find (a) $\hat{\pi}_2 \cdot 10^5$ and (b) $\hat{\pi}_2 \cdot 2^{17}$.

1. $\hat{\pi}_2 \cdot 10^5$
   Sol: (a) Using the command `dec2hex(fix(22/7*1e5))` we get $4cbad_{16}$, since $22/7 \times 10^5 = 314,285.7\ldots$ and `hex2dec('4cbad')`= 314,285. (b) $2^{18} \cdot 11_{16}/7_{16}$. ■

---

[9] Base 10 is the accepted standard, simply because we have 10 fingers that we count with.

2. $\hat{\pi}_2 \cdot 2^{17}$ **Sol:** $2^{18} \cdot 11_{16}/7_{16}$. ■

Exercise #8 Write the first 11 primes, base 16.
Sol: The Octave/MATLAB command `dec2hex()` provides the answer:

n	dec	1	2	3	4	5	6	7	8	9	10	11	...
π_n	dec	2	3	5	7	11	13	17	19	23	29	31	...
π_n	hex	2	3	5	7	0B	0D	11	13	17	1D	1F	...

■

Exercise #9 $x = 2^{17} \times 22/7$, using IEEE-754 double precision:[10]

$$\begin{aligned} x &= 411,940.5625_{10} \\ &= 2^{54} \times 1,198,372 \\ &= 0,10010,00,110010,010010,010010,010010_2 \\ &= 0x48c92,492_{16}. \end{aligned}$$

The exponent is 2^{18} and the mantissa is $4,793,490_{10}$. Here the commas in the binary (0,1) string are to help visualize the quasiperiodic nature of the bitstream. The numbers are stored in a 32-bit format, with 1-bit for the sign, 8 bits for the exponent, and 23 bits for the mantissa.

Perhaps a more instructive number is

$$\begin{aligned} x &= 4,793,490.0 \\ &= 0,100,1010,100,100,100,100,100,100,100,100_2 \\ &= 0x4a924,924_{16}, \end{aligned}$$

which has a repeating binary bit pattern of $((100))_3$, broken by the scale factor $0x4a$. Even more symmetrical is

$$\begin{aligned} x &= 0x24924924_{16} \\ &= 00,100,100,100,100,100,100,100,100,100,100_2 \\ &= 6.344,131,191,146,900 \times 10^{-17}. \end{aligned}$$

Here the scale factor is 2×24. In this example the repeating pattern is clear in the hex representation as a repeating $((942))_3$, as represented by the double brackets, with the subscript indicating the period—in this case, three digits. As before, the commas are to help with readability and have no other meaning.

The representation of numbers is not unique. For example, irrational complex numbers have approximate rational representations (i.e., $\pi \approx 22/7$). A better example is complex numbers $z \in \mathbb{C}$, which have many representations, as a pair of reals (i.e., $z = (x, y)$), or by Euler's formula, and matrices ($\theta \in \mathbb{R}$):

[10] https://www.h-schmidt.net/FloatConverter/IEEE754.html.

$$e^{j\theta} = \cos\theta + j\sin\theta \leftrightarrow \begin{bmatrix} \cos\theta & -\sin\theta \\ \sin\theta & \cos\theta \end{bmatrix}.$$

At a higher level, differentiable functions (analytic functions) may be represented by a single-valued Taylor series expansion (see Sect. 3.2.3), limited by its region of convergence (RoC).

Pythagoreans and Integers: The integer is the cornerstone of the Pythagorean doctrine—so much so that it caused a fracture within the Pythagoreans when it was discovered that not all numbers are rational. One famous proof of such irrational numbers comes from the spiral of Theodorus, as shown in Fig. 2.1, where the radius of each triangle has length $b_n = \sqrt{n}$ with $n \in \mathbb{N}$, and the long radius (the hypotenuse) is $c_n = \sqrt{1 + b_n^2} = \sqrt{1+n}$. This figure may be constructed using a compass and ruler by maintaining right triangles.

Public-key security: Most people assume encryption is done by a personal login and passwords. But passwords are fundamentally insecure because it is relatively easy for a computer to search all possible 2 to 10 combinations of all possible keyboard characters. The metric of security is entropy, a measure of randomness. One-bit of entropy is two possible outcomes. Your first guess would be correct half the time, and your second (the other guess) would be correct. With 10 bits of entropy there 1024 possible passwords, so an exhaustive search is quite feasible. 20 bits is a million possibilities. In the realm of 100 is 10^{30}. 100 bits is 10^{301} possibilities, much harder to guess.

An important application of prime numbers is public-key encryption, which is essential for Internet security applications (e.g., online banking). Decryption depends on factoring large integers formed from products of primes having thousands of bits.[11] The security is based on the relative ease of multiplying large primes along with the virtual impossibility of factoring them.

A *trapdoor function* is one where a computation is easy in one direction but its inverse is very difficult. As an example think of a very long list of numbered (indexed) items. If I give you the index, you can look up the item quickly. But if I give you the item as a randomized set of characters, it would be hard (but not impossible) to find the index of the item.

Public-key encryption is based on a trapdoor function. If everyone were to switch from passwords to public-key encryption, the Internet would be much more secure.[12]

---

[11] As a simple but concrete example, a public-key encryption could work by starting with two numbers with a common prime, such as 5*3 and 5*11, and then using the Euclidean algorithm, the greatest common divisor (GCD) could be worked out (5 in my example). One of the integers could be the public-key and the second could be the private key.

[12] https://fas.org/irp/agency/dod/jason/cyber.pdf.

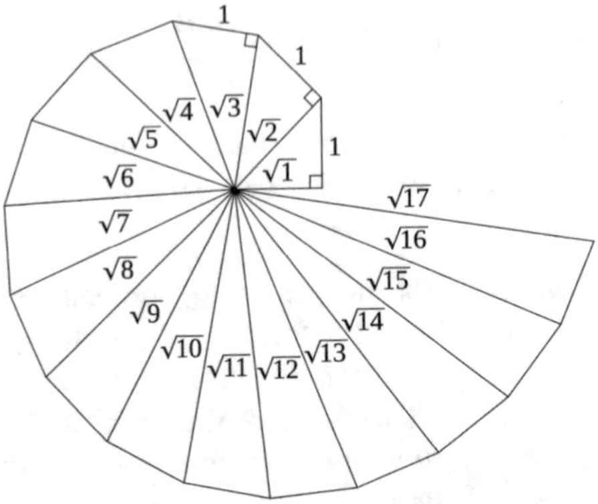

Fig. 2.1 The spiral of Theodorus, made from contiguous right triangles having lengths $a = 1, b_n = \sqrt{n}$, with $n \in \mathbb{N}$, and $c_n = \sqrt{n+1}$. In this way, each value of $c_n^2 = b_n^2 + a^2 \in \mathbb{N}$. This sequence of triangles generate the set $\{\sqrt{n}\} \in \mathbb{I}$, with $n \in \mathbb{N}$, and is easily generated using a compass and a ruler (Adapted from https://en.wikipedia.org/wiki/spiral_of_Theodorus)

Exercise #10 Consider the following method of generating a trap door: The private key is an irrational number `PRIV` (for example: π, π^π, $\log_\pi(\sqrt{\pi_{101}})$. The public-key `PUBL` is then a sequence of say 10 integers generated from `PRIV` using the CFA (see problem #8 of NS-2, Sect. 2.5.5).

How would you propose to recover `PRIV` given `PUBL`?

Sol: In my view this is impossible because the space of irrational numbers is more than huge. Knowing the first 10 CFA integers tells you nothing about `PRIV`.

It is likely that NSA would not accepted such a scheme because it is "too good." It is known that the NSA requires an encryption method that only the NSA can crack in a fixed amount of time, using the world's most powerful computer. The problem with this view is that the *bad guys* can use their own scheme rather than the RSA method. ■

Puzzles: Another application of integers is imaginative problems that use integers. An example is the classic Chinese *four stone problem:* Find the weights of four stones that can be used with a scale to weigh anything (e.g., salt, gold) between 0 and 40 grams (Assignment AE-2, Problem #5) The answer is not as interesting as the method, since the problem may be easily recast into a related one. This type of problem can be found in airline magazines as amusement on a long flight. This puzzle is best cast as a linear algebra problem with integer solutions. Again, once you know the trick, it is "easy."[13]

[13] Whenever someone tells you something is "easy," you should immediately appreciate that it is very hard, but once you learn a concept, the difficulty evaporates.

2.2 Problems NS-1

2.2.1 *Topic of This Homework*

Introduction to MATLAB/Octave (see the MATLAB or Octave tutorial for help)
Deliverables: Report with charts and answers to questions. Hint: Use LaTeX.[14]

2.2.2 *Plotting Complex Quantities in Octave/MATLAB*

Problem #1 *Consider the functions $f(s) = s^2 + 6s + 25$ and $g(s) = s^2 + 6s + 5$.*
–1.1: Find the zeros of functions $f(s)$ and $g(s)$ using the command `roots()`.
–1.2: Show the roots of $f(s)$ as red circles and of $g(s)$ as blue plus signs. The x-axis should display the real part of each root, and the y-axis should display the imaginary part. Use `hold on` *and* `grid on` *when plotting the roots.*
–1.3 Give your figure the title "Complex Roots of f(s) and g(s)." Label the x- and y-axes "Real Part" and "Imaginary Part." Hint: Use `xlabel`, `ylabel`, `ylim([-10 10])`, and `xlim([-10 10])` to expand the axes.

Problem #2 *Consider the function $h(t) = e^{\jmath 2\pi f t}$ for $f = 5$ and* `t=[0:0.01:2]`.
–2.1: Use `subplot` *to show the real and imaginary parts of $h(t)$. Make two graphs in one figure. Label the x-axes "Time (s)" and the y-axes "Real Part" and "Imaginary Part."*
–2.2: Use `subplot` *to plot the magnitude and phase parts of $h(t)$. Use the command* `angle` *or* `unwrap(angle())` *to plot the phase. Label the x-axes "Time (s)" and the y-axes' 'Magnitude" and "Phase (radians)."*

2.2.3 *Prime Numbers, Infinity, and Special Functions in Octave/MATLAB*

Problem #3 *Prime numbers, infinity, and special functions.*
–3.1: Use the MATLAB/Octave function `factor` *to find the prime factors of 123, 248, 1767, and 999,999.*
–3.2: Use the MATLAB/Octave function `isprime` *to determine whether 2, 3, and 4 are prime numbers. What does the function* `isprime` *return when a number is prime or not prime? Why?*

[14] https://www.overleaf.com.

–3.3: Use the MATLAB/Octave function `primes` *to generate prime numbers between 1 and* 10^6. *Save them in a vector* `x`*. Plot this result using the command* `hist(x)`.

–3.4: Now try `[n,bincenters] = hist(x)`*. Use* `length(n)` *to find the number of bins.*

–3.5: Set the number of bins to 100 by using an extra input argument to the function `hist`*. Show the resulting figure, give it a title, and label the axes.* Hint: `help hist` and `doc hist`.

Problem #4 `Inf`, `NaN`, *and logarithms in Octave/MATLAB.*

–4.1: Try `1/0` *and* `0/0` *in the Octave/MATLAB command window. What are the results? What do these "numbers" mean in Octave/MATLAB?*

–4.2: Try `log(0)`, `log10(0)`, *and* `log2(0)` *in the command window.] In MATLAB/Octave, the natural logarithm* $\ln(\cdot)$ *is computed using the function* `log`*. Functions* $\log_{10}$ *and* $\log_2$ *are computed using* `log10` *and* `log2`.

–4.3: Try `log(1)` *in the command window. What do you expect for* `log10(1)` *and* `log2(1)`*?*

–4.4: Try `log(-1)` *in the command window. What do you expect for* `log10(-1)` *and* `log2(-1)`*?*

–4.5: Explain how MATLAB/Octave arrives at the answer in problem 4.4. Hint: $-1 = e^{i\pi}$.

–4.6: Try `log(exp(j*sqrt(pi)))` *(i.e.,* $\log e^{j\sqrt{\pi}}$*) in the command window. What do you expect?*

–4.7: What does inverse *mean in this context? What is the inverse of* $\ln f(x)$*?*

–4.8: What is a decibel? (Look up decibels *on the Internet.)*

Problem #5 *Very large primes on Intel computers. Find the largest prime number that can be stored on an Intel 64-bit computer, which we call* $\pi_{\max}$. Hint: As explained in the MATLAB/Octave command `help flintmax`, the largest positive integer is 2^{53}; however, the largest integer that can be factored is $2^{32} = 2^{54} - 6$. Explain the logic of your answer. Hint: `help isprime()`.

Problem #6 *We are interested in primes that are greater than* $\pi_{\max}$*. How can you find them on an Intel computer (i.e., one using IEEE floating point)?* Hint: Consider a sieve that contains only odd numbers, starting from 3 (not 2). Since every prime number greater than 2 is odd, there is no reason to check the even numbers. $n_{odd} \in \mathbb{N}/2$ contain all the primes other than 2.

Problem #7 *The following identity is interesting. Can you find a proof?*

$$
\begin{aligned}
1 &= 1^2 \\
1+3 &= 2^2 \\
1+3+5 &= 3^2 \\
1+3+5+7 &= 4^2 \\
1+3+5+7+9 &= 5^2 \\
&\vdots \\
\sum_{n=0}^{N-1} 2n+1 &= N^2.
\end{aligned}
$$

2.3 The Role of Physics in Mathematics

Integers arose naturally in art, music, and science. Examples include the relationships between musical notes, the natural eigenmodes (tones) of strings, and other musical instruments. These relationships were so common that Pythagoras believed that to explain the physical world (the universe), one needed to understand integers. As discussed on Sect. 1.1, "all is number" was a seductive song.

As we will discuss on Sect. 3.1, it is best to view the relationships among acoustics, music, and mathematics as historical, since these topics inspired the development of mathematics. Today integers play a key role in quantum mechanics, again based on eigenmodes, but in this case, eigenmodes follow from solutions of the Schrödinger equation, with the roots of the characteristic equation being purely imaginary numbers $\in \mathbb{N}$. If there were a real part (i.e., damping), the modes would not be integers.

As discussed by Vincent Salmon (1946, p. 201), Schrödinger's equation follows directly from the Webster horn equation. While Philip Morse (1948, p. 281) (a student of Arnold Sommerfeld) fails to make the direct link, he comes close to the same view when he shows that the real part of the horn resistance goes exactly to zero below a cutoff frequency. He also discusses the trapped modes inside musical instruments created by the horn flare. One may assume Morse read Salmon's paper on horns, since he cites Salmon (Morse 1948, footnote 1, p. 271).

Engineers are so accustomed to working with real (vs. complex) numbers that they rarely acknowledge the distinction between real (i.e., irrational) and fractional numbers. Integers arise in many contexts. One cannot master computer programming without understanding integer, hexadecimal, octal, and binary representations, since all numbers in a computer are represented in numerical computations in terms of rationals ($\mathbb{Q} = \mathbb{Z} \cup \mathbb{F}$).

The primary reason integers are so important is their absolute precision. Every integer $n \in \mathbb{Z}$ is unique[15] and has the indexing property, which is essential for making lists that are ordered, so that one can quickly look things up. The alphabet also has this property (e.g., a book's index).

Because of the integer's absolute precision, the digital computer quickly overtook the analog computer once it was practical to make logic circuits that were fast. From 1946 the first digital computer was thought to be the University of Pennsylvania's ENIAC. We now know that the code-breaking effort in Bletchley Park, England, under the guidance of Alan Turing, created the first digital computer (the Colossus), which was used to break the World War II German Enigma code. Due to the high secrecy of this war effort, the credit was acknowledged only in the 1970s when the project was finally declassified.

There is zero possibility of analog computing displacing digital computing because of the importance of precision (and speed). But even with binary representation, there is a nonzero probability of error—for example, on a hard-drive—due to physical noise. To deal with this, error-correcting codes have been developed, reducing the error by many orders of magnitude. Today error correction is a science, and billions of dollars are invested to increase the density of bits per area to increasingly larger factors. A few years ago the terabyte drive was unheard of; today it is standard. In a few years petabyte drives will certainly become available. It is hard to comprehend how these will be used by individuals, yet they are essential for online (cloud) computing.

2.3.1 The Three Streams of Mathematics

Modern mathematics is built on a hierarchical construct of fundamental theorems, as summarized in the following boxed material on this section. The importance of such theorems cannot be overemphasized.

Gauss's and Stokes's laws play a major role in understanding and manipulating Maxwell's equations. Every engineering student needs to fully appreciate the significance of these key theorems. If necessary, memorize them. But memorization will not do over the long run, as each and every theorem must be fully understood. Fortunately most students already know several of these theorems, but perhaps not by name. In such cases, it is a matter of mastering the vocabulary.

---

[15] Check out the history of $1729 = 1^3 + 12^3 = 9^3 + 10^3$.

The three streams of mathematics

1. **Number systems: Stream 1**
   - Arithmetic
   - Prime numbers
2. **Geometry: Stream 2**
   - Algebra
3. **Calculus: Stream 3** spscitepFlanders80
   - Leibniz $\mathbb{R}^1$
   - Complex $\mathbb{C} \subset \mathbb{R}^2$
   - Vectors $\mathbb{R}^3, \mathbb{R}^n, \mathbb{R}^\infty$
     - Gauss's law (divergence theorem)
     - Stokes's law (curl theorem, or Green's theorem)
     - Vector calculus (Helmholtz's decomposition theorem)

The fundamental theorems are naturally organized and may be thought of in terms of the three streams of Stillwell (2010). For Stream 1 we have the fundamental theorem of arithmetic and the prime number theorem. For Stream 2 there is the fundamental theorem of algebra, and for Stream 3 there are a host of theorems on calculus, ordered by their dimensionality. Some of these theorems seem trivial (e.g., the fundamental theorem of arithmetic). Others are more challenging, such as the fundamental theorem of vector calculus and Green's theorem.

Complexity should not be confused with importance. Each of these theorems, as stated, is fundamental. Taken as a whole, they are a powerful way of summarizing mathematics.

2.3.2 *Stream 1: Prime Number Theorems*

There are two easily described fundamental theorems about primes:

1. *The fundamental theorem of arithmetic*: This states that every integer $n \in \mathbb{Z}$ may be uniquely factored into prime numbers. This raises the question of the meaning of *factor* (split into a product). The product of two integers $m, n \in \mathbb{Z}$ is $mn = \sum_m n = \sum_n m$. For example, $2 \times 3 = 2 + 2 + 2 = 3 + 3$.
2. *The prime number theorem:* One would like to know how many primes there are. That is easy: $|\mathbb{P}| = \infty$ (the size of the set of primes is infinite). A better way of asking this question is What is the average density of primes in the limit as $n \to \infty$? This question was answered, for all practical purposes, by Gauss, who in his free time computed the first three million primes by hand. He discovered

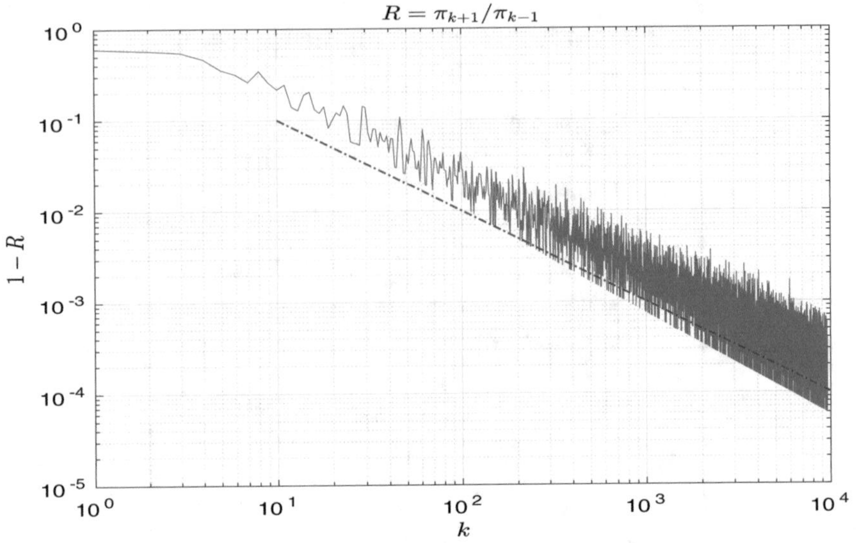

Fig. 2.2 The prime ratios $R_k = \pi_{k+1}/\pi_{k-1} \to 1$ as $k \to \infty$; thus we must plot the difference $1 - R_k$ on a log–log scale. While the minimum values of $1 - R$ are linear on a log–log scale, they decrease slightly faster than $1/k$, as indicated by the the red dashed line. Note that R is a random variable. Thus one must estimate both the mean and the standard deviation, which increases with k

that, to a good approximation, the primes are equally likely on a log scale. This is nicely summarized by the couplet:

> Chebyshev said, and I say it again: There is always a prime between *n* and *2n*.

This alludes to the mathematician Pafnuty Chebyshev, who proved the prime number theorem in a novel way (Stillwell 2010, p. 585).
When the ratio of two frequencies (pitches) is 2, the relationship is called an *octave*. With a slight stretch of terminology, we could say there is at least one prime per octave.

Exercise #11 An interesting extension of Chebyshev's observation is: As $n \to \infty$, what is the limiting density in primes per fraction of an octave? More specifically let $\pi < \pi_n < \pi_{k+1}$; then what is the limit of $\log_2 \pi_{n+1}/\pi_{n-1}$?
Sol: The answer is shown in Fig. 2.2. Note that the graph has a well-defined dense lower bound with a slope slightly steeper than $1/k$, and a random upper bound. ■

In modern western music the octave is further divided into 12 ratios called *semitones*, equal to $\sqrt[12]{2}$. Twelve semitones is an octave. In the end, it is a question of the density of primes on a log–log (i.e., ratio) scale. One might wonder about the maximum number of primes per octave in the neighborhood of N, or ask for the fractions of an octave (factors of 2) for π_k as k becomes large. The maximum value of $R_k < 0.5$; thus Chebyshev's bound of 2 is conservative. As $k \to \infty$ the bound is exponentially tightened. The results of this calculation are shown in Fig. 2.2. For reference, when $k = 9,592$, $\pi_k = 99,991$, $\pi_{k-1} = 99,989$, $\pi_{k+1} = 100,003$, thus $P(k+1)/P(k-1) - 1 = 1.4 \times 10^{-4}$.

2.3.3 Stream 2: Fundamental Theorem of Algebra

This theorem states that every polynomial in x of degree N,

$$P_N(x) = \sum_{k=0}^{N} a_k x^k, \tag{2.3.1}$$

has at least one root (see Sect. 3.4). When a common root is factored out, the degree of the polynomial is reduced by 1. Applied recursively, a polynomial of degree N has N roots. Note there are $N+1$ coefficients (i.e., $[a_N, a_{N-1}, \ldots, a_0]$). If we are interested only in the roots of $P_N(x)$, it is best to define $a_N = 1$, which defines the monic polynomial. If the roots are fractional numbers ($\in \mathbb{F}$), this might be possible. However, when the roots are irrational numbers (likely), a perfect factorization is at least unlikely, if not impossible.

2.3.4 Stream 3: Fundamental Theorems of Calculus

In Sects. 5.6 and 5.6.6 we will deal with each of the theorems for Stream 3. We consider the several fundamental theorems of integration, starting with Leibniz's formula for integration on the real line ($\mathbb{R}$), then progressing to complex integration in the complex plane ($\mathbb{C}$) (Cauchy's theorem), which is required for computing the inverse Laplace transform. Gauss's and Stokes's laws for $\mathbb{R}^2$ require closed and open surfaces, respectively. One cannot manipulate Maxwell's equations, fluid flow, or acoustics without understanding these theorems. Any problem that deals with the wave equation in more than one dimension requires an understanding of these theorems. They are the basis of the derivation of the Kirchhoff voltage and current laws, as first proposed by Newton for mechanics and acoustics.

Finally we define three vector operations based on the gradient operator,

$$\nabla \equiv \hat{\mathbf{x}}\frac{\partial}{\partial x} + \hat{\mathbf{y}}\frac{\partial}{\partial y} + \hat{\mathbf{z}}\frac{\partial}{\partial z}, \tag{2.3.2}$$

pronounced "del" (preferred) or "nabla," which are the gradient $\nabla()$, divergence $\nabla\cdot()$, and curl $\nabla\times()$.

Second-order operators such as the scalar Laplacian $\nabla\cdot\nabla() = \nabla^2()$, the divergence of the gradient (**DoG**), may be constructed from first-order operators. The most important of these is the vector Laplacian $\nabla^2()$, the gradient of the divergence (**gOd**) which is required when working with Maxwell's wave equations.[16]

The first three operations are defined in terms of integral operations on a surface in one, two, or three dimensions by taking the limit as that surface, or the volume contained within, goes to zero. These three differential operators are essential to fully understand Maxwell's equations, the crown jewel of mathematical physics. Hence mathematics plays a key role in physics, as does physics in math.

2.3.5 Other Key Mathematical Theorems

In addition to the widely recognized fundamental theorems, there are a number of equally important theorems that have not yet been labeled "fundamental."[17]

The widely recognized Cauchy integral theorem is an excellent example, since it is a steppingstone to Green's theorem and the fundamental theorem of complex calculus. In Sect. 4.5 we clarify the contributions of each of these special theorems.

Once these fundamental theorems of integration (Stream 3) have been mastered, the student is ready for the complex frequency domain, which takes us back to Stream 2 and the complex frequency plane ($s = \sigma + \omega \jmath \in \mathbb{C}$). While the Fourier and Laplace transforms are taught in mathematics courses, the concept of complex frequency is rarely mentioned. The complex frequency domain (see Sect. 3.10) and causality are fundamentally related (see Sects. 4.5.2–4.7.3), and are critical for the analysis of signals and systems, especially when dealing with the concept of impedance (see Sect. 4.4).

Without the concept of time and frequency, we cannot develop an intuition for the Fourier and Laplace transforms, especially within the context of engineering and mathematical physics. The Fourier transform covers signals, while the Laplace transform $\mathcal{LT}$ describes systems. Separating these two concepts, based on their representations as Fourier and Laplace transforms, is an important starting place for understanding physics and the role of mathematics. However, these methods, by themselves, do not provide the insight into physical systems that we need to be productive or, better, creative with these tools. We need to master the tools of differential equations and then partial differential equations to fully appreciate the world that they describe. Electrical networks composed of inductors, capacitors, and resistors are isomorphic to mechanical systems composed of masses, springs, and dashpots. Newton's laws are analogous to those of Kirchhoff, which are the rules needed to analyze simple physical systems composed of linear (and nonlinear) subcomponents. When lumped-element systems are taken to the limit in several dimensions, we obtain Maxwell's partial differential equations, the laws of continuum mechanics, and beyond.

---

[16] See Sect. 5.6.6 for the definitions of **DoG** and **gOd**.

[17] It is not clear what it takes to reach this more official sounding category.

The ultimate goal of this text is to make you aware of and productive in using these tools. This material can be best absorbed by treating it chronologically through history, so you can see how this body of knowledge came into existence, through the minds and hands of Galileo, Newton, Maxwell, and Einstein. Perhaps one day you too can stand on the shoulders of the giants who went before you.

2.4 Applications of Prime Numbers

If someone asked you for a theory of counting numbers, I suspect you would laugh and start counting. It sounds like either a stupid question or a bad joke. Yet integers are a rich topic, so the question is not even slightly dumb. It is somewhat amazing that even birds and bees can count. While I doubt birds and bees can recognize primes, cicadas, and other insects crawl out of the ground only in prime number cycles (e.g., 13- or 17-year cycles). If you have ever witnessed such an event (I have), you will never forget it. Somehow they know. Finally, there is the Euler zeta function, first introduced by Euler based on his analysis of sieve of Eratosthenes, now known as the Riemann zeta function $\zeta(s)$, that is *complex analytic*, with its poles at the logs of the prime numbers. The properties of this function are truly amazing, even fun. Many of the questions and answers about primes go back to at least the early Asian mathematicians (ca. 1500 BCE).

2.4.1 *The Importance of Prime Numbers*

While each prime perfectly predicts multiples of that prime, but there seems to be no regularity in predicting primes. It follows that prime numbers are the key to the theory of numbers because of the fundamental theorem of arithmetic (FTA).

It is likely that the first insight into the counting numbers started with the sieve shown in Fig. 2.3. A sieve answers the question How can one identify the prime numbers? The answer comes from looking for irregular patterns in the counting numbers.

Starting from $\pi_1 = 2$, we strike out all even numbers $2(2, 3, 4, 5, 6, \ldots)$ but not 2. By definition, the multiples are products of the target prime (2 in our example) and every other integer ($n \geq 2$). In this way all the even numbers are removed in this first iteration. The next remaining integer (3 in our example) is identified as the second prime π_2. Then all the multiples of $\pi_2 = 3$ are removed. The next remaining number is $\pi_3 = 5$, so all multiples of $\pi_3 = 5$ are removed (i.e., $\cancel{10}, \cancel{15}, \cancel{25}, \ldots$). This process is repeated until all the numbers of the list have been either canceled or identified as prime.

The recursive "sieve" method for finding primes was first devised by the Greek Eratosthenes (O'Neill, 2009).

1. Write $N-1$ integers n, starting from 2: $n \in \{2, 3, \ldots, N\}$ (e.g., $N = 4, n \in \{2, 3, 4\}$). Note that the first element $\pi_1 = 2$ is the first prime. Cross out all multiples of π_1; that is, cross out $n \cdot \pi_1 = 4, 6, 8, 10, \ldots, 50$, or all n such that mod(n, π_1) = 0.

	2	3	~~4~~	5	~~6~~	7	~~8~~	9	~~10~~
11	~~12~~	13	~~14~~	15	~~16~~	17	~~18~~	19	~~20~~
21	~~22~~	23	~~24~~	25	~~26~~	27	~~28~~	29	~~30~~
31	~~32~~	33	~~34~~	35	~~36~~	37	~~38~~	39	~~40~~
41	~~42~~	43	~~44~~	45	~~46~~	47	~~48~~	49	~~50~~

2. Let $k = 2$ and note that $\pi_2 = 3$. Cross out $n\pi_2$ $3 \cdot (2, 3, 4, 5, 6, 7, \ldots, 45)$, that is, all n such that mod(n, π_2) = 0.

	2	3	~~4~~	5	~~6~~	7	~~8~~	~~9~~	~~10~~
11	~~12~~	13	~~14~~	~~15~~	~~16~~	17	~~18~~	19	~~20~~
~~21~~	~~22~~	23	~~24~~	25	~~26~~	~~27~~	~~28~~	29	~~30~~
31	~~32~~	~~33~~	~~34~~	35	~~36~~	37	~~38~~	~~39~~	~~40~~
41	~~42~~	43	~~44~~	~~45~~	~~46~~	47	~~48~~	49	~~50~~

3. Let $k = 3, \pi_3 = 5$. Cross out $n\pi_3 \cdot (25, 35)$ (mod(n, 5)=0).

	2	3	~~4~~	5	~~6~~	7	~~8~~	~~9~~	~~10~~
11	~~12~~	13	~~14~~	~~15~~	~~16~~	17	~~18~~	19	~~20~~
~~21~~	~~22~~	23	~~24~~	~~25~~	~~26~~	~~27~~	~~28~~	29	~~30~~
31	~~32~~	~~33~~	~~34~~	~~35~~	~~36~~	37	~~38~~	~~39~~	~~40~~
41	~~42~~	43	~~44~~	~~45~~	~~46~~	47	~~48~~	~~49~~	~~50~~

4. Finally let $k = 4, \pi_4 = 7$ (mod(n, 7)=0). Cross out $n\pi_4$: (49). Thus there are 15 primes less than $N = 50$: $\pi_k = 2, 3, 5, 7, 11, 13, 17, 19, 23, 29, 31, 37, 41, 43, 47$ (highlighted in red). Above 2, all end in odd numbers, and above 5, all end with 1, 3, 7, or 9.

Fig. 2.3 Sieve of Eratosthenes for $N = 50$

As the word *sieve* implies, this process takes a heavy toll on the integers, rapidly pruning the non-primes. In four iterations of the sieve algorithm, all the primes less than $N = 50$ are identified in red. The final set of primes is displayed in Step 4 of Fig. 2.3.

Once a prime greater than $N/2$ has been identified (25 in the example), the recursion stops, since twice that prime is greater than N, the maximum number under consideration. Thus once $\sqrt{49}$ has been reached, all the primes have been identified (this follows from the fact that the next prime π_n is multiplied by an integer $n = 1, \ldots, N$).

When we use a computer, memory efficiency, and speed are the main considerations. There are various schemes for making the sieve more efficient. For example, the recursion $n\pi_k = (n-1)\pi_k + \pi_k$ will speed up the process by replacing the multiplication with an addition of π_k.

2.4.2 *Two Fundamental Theorems of Primes*

Early theories of numbers revealed two fundamental theorems (see Sects. 2.3.3 and 2.3.4). The first of these is the fundamental theorem of arithmetic, which says that every integer $n \in \mathbb{N}$ greater than 1 may be uniquely factored into a product of primes:

$$n = \prod_{k=1}^{K} \pi_k^{\beta_k}, \tag{2.4.1}$$

where $k = 1, \ldots, K$ indexes the integer's K prime factors $\pi_k \in \mathbb{P}$. Typically prime factors appear more than once—for example, $25 = 5^2$. To make the notation compact we define the *multiplicity* β_k of each prime factor π_k. For example, $2312 = 2^3 \cdot 17^2 = \pi_1^3\, \pi_7^2$ (i.e., $\pi_1 = 2, \beta_1 = 3; \pi_7 = 17, \beta_7 = 2$) and $2313 = 3^2 \cdot 257 = \pi_3^2\, \pi_{55}$ (i.e., $\pi_2 = 3, \beta_3 = 2; \pi_{55} = 257, \beta_{55} = 1$). Our demonstration of this is empirical, using the MATLAB/Octave `factor(N)` routine, which factors N.[18]

What seems amazing is the unique nature of this theorem. Each counting number is uniquely represented as a product of primes. No two integers can share the same factorization. Once you multiply the factors out, the result is unique (N). Note that it's easy to multiply integers (e.g., primes) but expensive to factor them. And factoring the product of three primes is significantly more difficult than factoring two.

Factoring is much more expensive than division. This is not due to the higher cost of division over multiplication, which is less than a factor of 2.[19] Dividing the product of two primes, given one, is trivial, slightly more expensive than multiplying. Factoring the product of two primes is nearly impossible, as one needs to know what to divide by. Factoring means dividing by some integer and obtaining another integer with remainder zero.

This brings us to the prime number theorem (PNT). The security problem is the reason these two theorems are so important: (1) Every integer has a unique representation as a product of primes, and (2) the density of primes is large (see the discussions on Sect. 2.3.2). Thus security reduces to the "needle in the haystack problem" due to the cost of a deep search. One could factor a product of primes $N = \pi_k \pi_l$ by doing M divisions, where M is the number of primes less than $\sqrt{N}$. This assumes the primes less than $\sqrt{N}$ are known. However, most integers are not a simple product of two primes.

But the utility of using prime factorization has to do with their density. If we were simply looking up a few numbers from a short list of primes, it would be easy to

---

[18] If you wish to be a mathematician, you need to learn how to prove theorems. If you're a physicist, you are happy that someone else has already proved them, so that you can use the result.

[19] https://streamcomputing.eu/blog/2012-07-16/how-expensive-is-an-operation-on-a-cpu/.

factor them. But given that their density is logarithmic ($\gg$1 per octave, as shown in Fig. 2.2), factoring comes at a very high computational cost compared to a table lookup.

2.4.3 *Greatest Common Divisor (Euclidean Algorithm)*

The Euclidean algorithm is a systematic method known to all educated people, identical to how we perform long division. The largest common integer factor k between two integers n and m (divisor and dividend) is denoted $k = \text{gcd}(n, m)$, where $n, m, k \in \mathbb{N}$ (Graham et al. 1994). For example, $15 = \text{gcd}(30, 105)$ since, when factored, $(30, 105) = (2 \cdot 3 \cdot 5,\ 7 \cdot 3 \cdot 5) = 3 \cdot 5 \cdot (2, 7) = 15 \cdot (2, 7)$. Thus the GCD is 15. Two integers are said to be *coprime* if their GCD is 1 (i.e., they have no common prime factor). The Euclidean algorithm was known to the Chinese (i.e., not discovered by Euclid) (Stillwell 2010, p. 41).

Examples of the GCD: $l = \text{gcd}(m, n)$

- *Examples ($m, n, l \in \mathbb{Z}$):*
  - $5 = \text{gcd}(13 \cdot 5, 11 \cdot 5)$. *The GCD is the common factor 5.*
  - $(13 \cdot 10, 11 \cdot 10) = 10\,\text{gcd}(130, 110) = 10 = 2 \cdot 5$ *is not prime*
  - $\text{gcd}(1234, 1024) = 2$, *since* $1234 = 2 \cdot 617$, $1024 = 2^{10}$
  - $\text{gcd}(\pi_k \pi_m, \pi_k \pi_n) = \pi_k$
  - $l = \text{gcd}(m, n)$ *is the part that cancels in the fraction* $m/n \in F$
  - $m/\text{gcd}(m, n) \in \mathbb{Z}$
- Coprimes ($m \perp n$) *are numbers that have no distinct common factors; that is,* $\text{gcd}(m, n) = 1$
  - *The GCD of two primes is always 1:* $\text{gcd}(13, 11) = 1$, $\text{gcd}(\pi_m, \pi_n) = 1$ $(m \neq n)$
  - $m = 7 \cdot 13, n = 5 \cdot 19 \Rightarrow (7 \cdot 13) \perp (5 \cdot 19)$
  - *If* $m \perp n$, *then* $\text{gcd}(m, n) = 1$.
  - *If* $\text{gcd}(m, n) = 1$, *then* $m \perp n$.

The Euclidean algorithm is best explained by a trivial example: Consider the two numbers 6 and 9. At each step the smaller number (6) is subtracted from the larger (9) and the smaller number and the difference (the remainder) are saved. This process continues until the two resulting numbers are equal, which is the GCD. For our example, $9 - 6 = 3$, leaving the smaller number 6 and the difference 3. Repeating this, we get $6 - 3 = 3$, leaving the smaller number 3 and the difference 3. Since these two numbers are the same, we are done; thus $3 = \text{gcd}(9, 6)$. We can verify this result by factoring [e.g., $(9, 6) = 3(3, 2)$]. The value may also be numerically verified using the MATLAB/Octave GCD command `gcd(6,9)`, which returns 3. Thus the GCD reduces to the definition of long division.

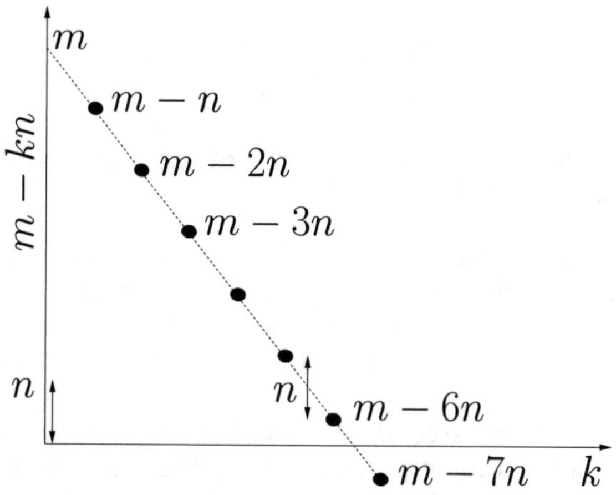

Fig. 2.4 The Euclidean algorithm recursively subtracts n from m until the remainder $m - kn$ is less than either n or zero. Note that this recursion is the same as `mod(m,n)`. Thus the GCD recursively computes `mod(m,n)` until the remainder `rem(m,n)` is less than n, which is called the GCD's *turning-point*. It then swaps m and n, so that $n < m$. This repeats until it terminates on the GCD. Due to its simplicity this is called the *direct method* for finding the GCD. For the case depicted here, the value of k renders the *remainder* $m - 6n < n$. If one more step were taken beyond the turning-point ($k = 7$), the remainder would become negative. Thus the turning-point satisfies the linear relationship $m - \alpha n = 0$ with $\alpha \in \mathbb{R}$

Direct matrix method: The GCD may be written as a matrix recursion given the starting vector $(m_0, n_0)^T$. The recursion is then

$$\begin{bmatrix} m_{k+1} \\ n_{k+1} \end{bmatrix} = \begin{bmatrix} 1 & -1 \\ 0 & 1 \end{bmatrix} \begin{bmatrix} m_k \\ n_k \end{bmatrix}. \tag{2.4.2}$$

This recursion continues until $m_{k+1} < n_{k+1}$, at which point m and n must be swapped. The process is repeated until $m_k = n_k$, which equals the GCD. We call this the *direct method* (see Fig. 2.4). The direct method is inefficient because it recursively subtracts n_k many times until the resulting m_k is less than n_k. It also must test for $m \leq n$ after each subtraction and then swap them if $m_k < n_k$. If they are equal, we are done.

The GCD's *turning-point* may be defined using the linear interpolation $m - \alpha n = 0, \alpha \in \mathbb{R}$, where the solid line cross the abscissa in Fig. 2.4. If, for example, $l = 6 + 43/97 \approx 6.443298\ldots$, then $6 = \lfloor m/n \rfloor < n$ and $\alpha \in \mathbb{F} \in \mathbb{R}$. This is nonlinear (truncation) arithmetic, which is a natural property of the GCD. The `floor()` functions finds the turning-point, where we swap the two numbers, since by definition, $m > n$. In this example, $6 = \lfloor l \rfloor$.

Exercise #12 Show that

$$\begin{bmatrix} 1 & -1 \\ 0 & 1 \end{bmatrix}^n = \begin{bmatrix} 1 & -n \\ 0 & 1 \end{bmatrix}.$$

Sol: To prove this let $n = 2$ and then 3. Each recursive multiplication adds 1 to the upper right corner. ■

Why is the GCD important? The utility of the GCD algorithm arises directly from the fundamental difficulty in factoring large integers. Computing the GCD using the Euclidean algorithm costs less than factoring when finding the coprime factors, which is extremely expensive. The utility surfaces when the two numbers are composed of very large primes.

When two integers have no common factors, they are said to be *coprime* and their GCD is 1. The ratio of two integers that are coprime is automatically in reduced form (they have no common factors). For example, $4/2 \in \mathbb{Q}$ is not reduced, since $2 = \text{gcd}(4, 2)$ (with a zero remainder). Canceling out the common factor 2 gives the reduced form $2/1 \in \mathbb{F}$. Thus if we wish to form the ratio of two integers, we first compute the GCD, then remove it from the two numbers to form the ratio. This ensures that the rational number is in its reduced form ($\in \mathbb{F}$ rather than $\in \mathbb{Q}$). If the GCD were 10^3 digits, it is obvious that any common factor would need to be removed, thus greatly simplifying further computation. This can make a huge difference when using IEEE-754.

The floor function and the GCD are related in an important way, as discussed next.

Indirect matrix method: A much more efficient method uses the `floor()` function, which is called *division with rounding* or simply the *indirect method*. Specifically the GCD may be written in one step as

$$\begin{bmatrix} m \\ n \end{bmatrix}_{k+1} = \begin{bmatrix} 0 & 1 \\ 1 & -\left\lfloor \frac{m}{n} \right\rfloor \end{bmatrix} \begin{bmatrix} m \\ n \end{bmatrix}_k . \tag{2.4.3}$$

This matrix is Eq. 2.4.2 to the power $\lfloor m/n \rfloor$, followed by swapping the inputs (the smaller number must always be on the bottom).

The GCD and multiplication: Multiplication is simply recursive addition, and finding the GCD takes advantage of this fact. For example, $3 * 2 = 3 + 3 = 2 + 2 + 2$. Since division is the inverse of multiplication, it must be recursive subtraction.

The GCD and long division: When we learn how to divide a smaller number into a larger one, we must learn how to use the GCD. For example, suppose we wish to compute $110 \div 6$ (110/6). We start by finding out how many times 6 goes into 11. Since $6 \times 2 = 12$, which is larger than 11, the answer is 1. This is of course the floor function (e.g., $\lfloor 11/6 \rfloor = 1$). We then subtract 6 from 11 to find the remainder 5.

Example Start with the two integers [873, 582]. In factored form these are $[\pi_{25} \cdot 3^2, \pi_{25} \cdot 3 \cdot 2]$. Given the factors, we see that the largest common factor is $\pi_{25} \cdot 3 = 291$ ($\pi_{25} = 97$). When we take the ratio of the two numbers, this common factor cancels:

$$\frac{873}{582} = \frac{\cancel{\pi_{25}} \cdot \cancel{3} \cdot 3}{\cancel{\pi_{25}} \cdot \cancel{3} \cdot 2} = \frac{3}{2} = 1.5.$$

Of course if we divide 582 into 873, we numerically obtain the answer $1.5 \in \mathbb{F}$.

Exercise #13 What does it mean to reach the turning-point when using the Euclidean algorithm?
Sol: When $m/n - \lfloor m/n \rfloor < n$, we have reached a turning-point. When the remainder is zero (i.e., $m/n - \lfloor m/n \rfloor = 0$), we have reached the GCD. ■

Exercise #14 Show that in MATLAB/Octave `rat(873/582)` $= 2 + 1/(-2)$ gives `rats(873/582)=3/2`, which is the wrong answer. Hint: Factor the two numbers and cancel out the GCD.
Sol: Since

$$\texttt{factor(873)} = 3 \cdot 3 \cdot 97 \text{ and } \texttt{factor(582)} = 2 \cdot 3 \cdot 97,$$

the GCD is $3 \cdot 97$. Thus `3/2 = 1 + 1/2` is the correct answer. But due to rounding methods, it is not 3/2. As an example, in MATLAB/Octave `rat(3/2)= 2+1/(-2)`. MATLAB'S `rat()` command uses rounding rather than the floor function, which explains the difference. When the `rat(·)` function produces negative numbers, rounding must have been employed. ■

Exercise #15 Divide 10 into 99. The floor function (`floor(99/10)`) must be used, followed by the remainder function (`rem(99,10)`).
Sol: When we divide a smaller number into a larger one, we must first find the floor followed by the remainder. For example, 99/10 = 9 + 9/10 has a floor of 9 and a remainder of 9/10. ■

Graphical description of the GCD: The Euclidean algorithm is best viewed graphically. In Fig. 2.4 we show what is happening as one approaches the turning-point, at which point the two numbers must be swapped to keep the difference positive, which is addressed by the upper row of Eq. 2.4.3.

Here is a simple MATLAB/Octave code to find `l=gcd(m,n)` based on the Stillwell (2010) definition:

```
 %~/M/gcd0.m
function k = \gcd(m,n)
 while m ~=0
  A=m; B=n;
  m=max(A,B); n=min(A,B); %m>n
  m=m-n;
 endwhile %m=n
  k=A;
```

This program loops until $m = 0$.

Coprimes: When the GCD of two integers is 1, the only common factor is 1. This is of key importance when trying to find common factors between the two integers. When `1=gcd(m,n)`, the two integers are said to be *coprime* or *relatively prime*, which is may be written as $m \perp n$. By definition, the largest common factor of coprimes is 1. But since 1 is not a prime ($\pi_1 = 2$), the two integers have no common primes. It can be shown (Stillwell 2010, pp. 41–4) that when $a \perp b$, there exist $m, n \in \mathbb{Z}$ such that

$$am + bn = \gcd(a, b) = 1.$$

This linear equation may be related to the addition of two fractions that have coprime numerators ($a \perp b$). For example,

$$\frac{a}{m} + \frac{b}{n} = \frac{an + bm}{mn}.$$

It is not obvious that this is simply $1/mn = 1$.

Exercise #16 Show that

$$\begin{bmatrix} 0 & 1 \\ 1 & -\lfloor \frac{m}{n} \rfloor \end{bmatrix} = \begin{bmatrix} 0 & 1 \\ 1 & 0 \end{bmatrix} \begin{bmatrix} 1 & -1 \\ 0 & 1 \end{bmatrix}^{\lfloor \frac{m}{n} \rfloor}$$

.

Sol: This exercise uses the results of the earlier Exercise # 10, times the row-swap matrix. ■

2.4.4 Continued Fraction Algorithm

As shown in Fig. 2.5, the continued fraction algorithm (CFA) starts from a single real decimal number $x_o \in \mathbb{R}_o$ and recursively expands it as a fraction $x \in \mathbb{F}$ (Graham et al. 1994). Thus the CFA may be used for forming rational approximations to any real number. For example, $\pi \approx 22/7$, an excellent approximation well known to Chinese mathematicians.

The Euclidean algorithm (i.e., GCD), on the other hand, operates on a pair of integers $m, n \in \mathbb{N}$ and returns their greatest common divisor $k \in \mathbb{N}$, such that $m/k, n/k \in \mathbb{F}$ are coprime, thus reducing the ratio to its irreducible form (i.e., $m/k \perp n/k$). Note this is done without factoring m and n.

Despite this seemingly irreconcilable difference between the GCD and CFA, the two are closely related—so close that Gauss called the Euclidean algorithm for finding the GCD the continued fraction algorithm (CFA) (Stillwell 2010, p. 48).

At first glance it is not clear why Gauss would be so "confused." One is forced to assume that Gauss had some deeper insight into this relationship. If so, it would be valuable to understand that insight.

Definition of the CFA

1. Start with $n = 0$ and the positive input target $x_0 \in \mathbb{R}^+$. — $n = 0, m_0 = 0, x_0 = \pi$
2. Rounding: Let $m_n = \lfloor x_n \rceil \in \mathbb{N}$. — $m_0 = \lfloor \pi \rceil = 3$
3. The input vector is then $[m_n, x_n]^T$. — $[3, \pi]^T$
4. Remainder: $r_n = x_n - m_n \qquad (-0.5 \le r_n \le 0.5)$ — $r_0 = \pi - 3 \approx 0.1416$
5. Reciprocate:

$$x_{n+1} = \begin{cases} 1/r_n, \; n \leftarrow n+1; \textit{ go to step 2} & r_n \neq 0 \\ 0, \quad \textit{terminate} & r_n = 0 \end{cases} \qquad \begin{array}{l} x_2 = 1/0.14159 = 7.06\ldots \\ \textit{Output: } [m_n, x_{n+1}]^T = [3, 7.06]^T \end{array}$$

Fig. 2.5 Definition of the CFA of any positive number, $x_0 \in \mathbb{R}^+$,. Numerical values for $n = 0, x_0 = \pi, m_0 = 0$ are on the right. For $n = 1$ the input vector is $[m_1, x_2]^T = [3, 7.0626]^T$. If at any step the remainder is zero, the algorithm terminates (step 5). Convergence is guaranteed. The recursion may continue to any desired accuracy, and terminates if $r_n = 0$. Alternative rounding schemes are given on Appendix A.1.5

Since Eq. 2.4.3 may be inverted, the process may be reversed, which is closely related to the CFA as discussed in Fig. 2.5. This might be the basis behind Gauss's insight.

Notation: Writing out all the fractions can become tedious. For example, expanding $e = 2.7183\ldots$ using the MATLAB/Octave command `rat(exp(1))` gives the approximation

$$\begin{aligned}\exp(1) &= 3 + 1/(-4 + 1/(2 + 1/(5 + 1/(-2 + 1/(-7))))) - o\left(1.75 \times 10^{-6}\right) \\ &= [3; -4, 2, 5, -2, -7] - o(1.75 \times 10^{-6}).\end{aligned}$$

Here we use a compact bracket notation, $\hat{e}_6 \approx [3; -4, 2, 5, -2, -7]$, where $o()$ indicates the error of the CFA expansion.

Since entries are negative, we deduce that MATLAB/Octave is using rounding arithmetic (but this is not documented). Note that the leading integer part m_0 may be indicated by an optional semicolon.[20] If the steps are carried further, the values of $m_n \in \mathbb{Z}$ give increasingly more accurate rational approximations. The five rounding schemes are discussed in Appendix A.1.5.

Exercise #17 Let $x_0 \equiv \pi \approx 3.14159\ldots$. As shown in Fig. 2.6, $a_0 = 3$, $r_0 = 0.14159$, $x_1 = 7.065 \approx 1/r_0$, and $a_1 = 7$. If we were to stop here, we would have

$$\widehat{\pi}_2 = 3 + \frac{1}{7 + \cancel{0.0625}\ldots} = 3 + \frac{1}{7} = \frac{22}{7}. \tag{2.4.4}$$

[20]Unfortunately MATLAB/Octave does not support the bracket notation.

Rational approximation examples

$$\hat{\pi}_2 = \frac{22}{7} = [3; 7] \qquad \approx \hat{\pi}_2 + o(1.3 \times 10^{-3})$$

$$\hat{\pi}_3 = \frac{355}{113} = [3; 7, 16] \qquad \approx \hat{\pi}_3 - o(2.7 \times 10^{-7})$$

$$\hat{\pi}_4 = \frac{104,348}{33,215} = [3; 7, 16, -249] \qquad \approx \hat{\pi}_4 + o(3.3 \times 10^{-10})$$

Fig. 2.6 The expansion of π to various orders, using the CFA, along with the order of the error of each rational approximation, with rounding. For example, $\hat{\pi}_2$ =22/7 has an absolute error ($|22/7 - \pi|$) of about 0.13%

This approximation of $\widehat{\pi}_2 = 22/7$ has a relative error of 0.04%:

$$\frac{22/7 - \pi}{\pi} \approx 4 \times 10^{-4}.$$

Exercise #18 For a second level of approximation we continue by reciprocating the remainder $1/0.0625 \approx 15.9966$, which rounds to 16, giving a negative remainder of $\approx -1/300$:

$$\widehat{\pi}_3 \approx 3 + 1/(7 + 1/16) = 3 + 16/(7 \cdot 16 + 1) = 3 + 16/113 = 355/113.$$

With rounding, the remainder is -0.0034, resulting in a much more rapid convergence. If floor rounding is used ($15.9966 = 15 - 0.9966$) the remainder is positive and close to 1, resulting in a much less accurate rational approximation for the same number of terms. It follows that there can be a dramatic difference depending on the rounding scheme, which, for clarity, should be specified, not inferred.

Exercise #19 Find the CFA using the floor function, to the 12th order.
Sol: $\hat{\pi}_{12} = [3; 7, 15, 1, 292, 1, 1, 1, 2, 1, 3, 1]$. Octave/MATLAB will give a different answer due to the use of rounding rather than floor. ■

Exercise #20 MATLAB/Octave's `rat(pi,1e-16)` gives:
`3 + 1/(7 + 1/(16 + 1/(-294 + 1/(3 + 1/(-4 + 1/(5 + 1/(-15 + 1/(-3)))))))).`
In bracket notation,

$$\hat{\pi}_9 = [3; 7, 16, -294, 3, -4, 5, -15, -3].$$

Because the sign changes, it is clear that MATLAB/Octave uses rounding rather than the floor function. It follows that the error will be smaller with rounding than with truncation, which is 266.764×10^{-9}.

Exercise #21 Based on the several examples given above, which rounding scheme is the most accurate? Explain why.

Sol: Rounding results in a smaller remainder at each iteration and thus results in a smaller net error and faster convergence. Using the floor truncation the CFA always gives positive coefficients, which could have useful applications. ■

When the CFA is applied and the expansion terminates ($r_n = 0$), the target is rational. When the expansion does not terminate (which is not always easy to determine, as the remainder may be ill-conditioned due to small numerical rounding errors), the number is irrational. Thus the CFA has important theoretical applications with irrational numbers. You may explore this using MATLAB'S `rats(pi)` command.

In addition to these five basic rounding schemes, there are two other important $\mathbb{R} \to \mathbb{N}$ functions (i.e., mappings) that will be needed later: `mod(x,y)` and rem(x,y) with $x, y \in \mathbb{R}$. The base-10 numbers may be generated from the counting numbers using `y=mod(x,10)`.

Exercise #22 1. Show how to generate a base-10 real number $y \in \mathbb{R}$ from the counting numbers $\mathbb{N}$ using the `m=mod(n,10)`$+ k10$ with $n, k \in \mathbb{N}$.
Sol: Every time n reaches a multiple of 10, m is reset to 0 and the next digit to the left is increased by 1 by adding 1 to k, generating the digit pair km. Thus the `mod()` function forms the underlying theory behind decimal notation. ■

2. How would you generate binary numbers (base 2) using the `mod(x,b)` function?
Sol: Use the same method as in part 1, but with $b = 2$. ■
3. How would you generate hexadecimal numbers (base 16) using the `mod(x,b)` function? **Sol:** Use the same method as in part 1, but with $b = 16$. ■
4. Write the first 19 numbers in hex notation, starting from zero. **Sol:** 0, 1, 2, 3, 4, 5, 6, 7, 8, 9, A, B, C, D, E, F, 10, 11, 12. Recall that $10_{16} = 16_{10}$, thus $12_{16} = 18_{10}$, resulting in a total of 19 numbers if we include 0. ■
5. What is FF_{16} in decimal notation? **Sol:** `hex2dec('ff')` $= 255_{10}$ ■

Symmetry: A continued fraction expansion can have a high degree of recursive symmetry. For example, consider the CFA

$$R_1 \equiv \frac{1+\sqrt{5}}{2} = 1 + \frac{1}{1 + \frac{1}{1+\ldots}} = 1.618033988749895\ldots. \tag{2.4.5}$$

Here a_n in the CFA is always 1 ($R_1 \equiv [1; 1, 1, \ldots]$), thus the sequence cannot terminate, proving that $\sqrt{5} \in \mathbb{I}$. A related example is $R_2 \equiv$ `rat(1+sqrt(2))`, which gives $R_2 = [2; 2, 2, 2, \ldots]$.

When we expand a target irrational number ($x_0 \in \mathbb{I}$) and the CFA is truncated, the resulting rational fraction approximates the irrational target. For the example above, if we truncate at three coefficients ([1; 1, 1]), we obtain

$$1 + \frac{1}{1 + \frac{1}{1+0}} = 1 + 1/2 = 3/2 = 1.5 = \frac{1+\sqrt{5}}{2} + 0.118\ldots.$$

Truncation after six steps gives

$$[1; 1, 1, 1, 1, 1, 1] = 13/8 \approx 1.6250 = \frac{1+\sqrt{5}}{2} + 0.0070\ldots.$$

Because all the coefficients are 1, this example converges very slowly. When the coefficients are large (i.e., the remainder is small), the convergence will be faster. The expansion of π is an example of faster convergence.

In summary: Every rational number $m/n \in \mathbb{F}$, with $m > n > 1$, may be uniquely expanded as a continued fraction, with coefficients a_k determined using the CFA. When the target number is irrational ($x_0 \in \mathbb{Q}$), the CFA does not terminate; thus each step produces a more accurate rational approximation, converging in the limit as $n \to \infty$.

Thus the CFA expansion is an algorithm that can, in theory, determine when the target is rational, but with an important caveat: One must determine whether the expansion terminates. This may not be obvious. The fraction $1/3 = 0.33333\ldots$ is an example of such a target, where the CFA terminates yet the fraction repeats. It must be that

$$1/3 = 3 \times 10^{-1} + 3 \times 10^{-2} + 3 \times 10^{-3} + \ldots.$$

Here $3 \cdot 3 = 9$. As a second example,[21]

$$1/7 = 0.142857, 142857, 142857, 142857 \cdots = 14,2857 \times 10^{-6} + 142,857 \times 10^{-12} + \ldots.$$

There are several notations for repeating decimals, such as $1/7 = 0.1\overline{142857}$ and $1/7 = 0.1((142857))$. Note that $142,857 = 999,999/7$. Related identities include $1/11 = 0.090909\ldots$ and $11 \times 0.090909 = 999,999$. When the sequence of digits repeats, the sequence is predictable and it must be rational. But it is impossible to be sure that it repeats because the length of the repeat can be arbitrarily long.

Exercise #23 Discuss the relationship between the CFA and the transmission line modeling method on Sect. 3.8.
Sol: The solution is detailed in Appendix F. ■

---

[21] Taking the Fourier transform of the target number, represented as a sequence, could help to identify an underlying periodic component. The number $1/7 \leftrightarrow [[1, 4, 2, 8, 5, 7]]_6$ has a 50 [dB] notch at 0.8π [rad] due to its six-digit periodicity, carried to 15 digits (MATLAB/Octave maximum precision), Hamming-windowed, and zero padded to 1024 samples.

2.5 Problems NS-2

2.5.1 *Topic of This Homework*

Prime numbers, greatest common divisors, the continued fraction algorithm

2.5.2 *Prime Numbers*

Problem #1 *Every integer may be written as a product of primes.*

–1.1: Write the numbers 1, 000, 000, 1, 000, 004, *and* 999, 999 *in the form* $N = \prod_k \pi_k^{\beta_k}$. Hint: Use MATLAB/Octave to find the prime factors.

–1.2: Give a generalized formula for the natural logarithm of a number $\ln(N)$ *in terms of its primes* π_k *and their multiplicities* β_k*. Express your answer as a* sum *of terms.*

Problem #2 *Using the computer*

–2.1: Explain why the following brief MATLAB/Octave program returns the prime numbers π_k *between 1 and 100.*

```
n=2:100;
k = isprime(n);
n(k)
```

–2.2: How many primes are there between 2 *and* $N = 100$*?*

Problem #3 *Prime numbers may be identified using a sieve (see Fig. 2.3).*

–3.1: By hand, complete the sieve of Eratosthenes for $n = 1, \ldots, 49$*. Circle each prime p, then cross out each number that is a multiple of p.*

–3.2: What is the largest number you need to consider before only primes remain?

–3.3: Generalize: For $n = 1, \ldots, N$*, what is the largest number you need to consider before only the primes remain?*

–3.4: Write each of these numbers as a product of primes: 22, 30, 34, 43, 44, 48, 49.

–3.5: Find the largest prime $\pi_k \le 100$*. Do not use MATLAB/Octave other than to check your answer.* Hint: Write the numbers starting with 100 and count backward: 100, 99, 98, 97, ... Cross off the even numbers, leaving 99, 97, 95, ... Pull out a factor (only one is necessary to show that it is not prime).

–3.6: Find the largest prime $\pi_k \le 1000$*. Do not use MATLAB/Octave other than to check your answer.*

–3.7: Explain why $\pi_k^{-s} = e^{-s \ln \pi_k}$.

2.5.3 Greatest Common Divisors

Consider using the Euclidean algorithm to find the greatest common divisor (GCD; the largest common prime factor) of two numbers. Note that this algorithm may be performed using one of two methods:

Method	Division	Subtraction
On each iteration...	$a_{i+1} = b_i$ $b_{i+1} = a_i - b_i \cdot \text{floor}(a_i/b_i)$	$a_{i+1} = \max(a_i, b_i) - \min(a_i, b_i)$ $b_{i+1} = \min(a_i, b_i)$
Terminates when	$b = 0$ (GCD $= a$)	$b = 0$ (GCD $= a$)

The division method (Eq. 2.1, Sect. 2.1.2, Chap. 2) is preferred because the subtraction method is much slower.

Problem #4 *Understanding the Euclidean algorithm (GCD)*

–*4.1: Use the Octave/MATLAB command* `factor` *to find the prime factors of* $a = 85$ *and* $b = 15$.

–*4.2: What is the greatest common prime factor of* $a = 85$ *and* $b = 15$?

–*4.3: By hand, perform the Euclidean algorithm for* $a = 85$ *and* $b = 15$.

–*4.4: By hand, perform the Euclidean algorithm for* $a = 75$ *and* $b = 25$. *Is the result a prime number?*

–*4.5: Consider the first step of the GCD division algorithm when* $a < b$ *(e.g.,* $a = 25$ *and* $b = 75$*). What happens to a and b in the first step? Does it matter if you begin the algorithm with* $a < b$ *rather than* $b < a$?

–*4.6: Describe in your own words how the GCD algorithm works. Try the algorithm using numbers that have already been divided into factors (e.g.,* $a = 5 \cdot 3$ *and* $b = 7 \cdot 3$*).*

–*4.7: Find the GCD of* $2 \cdot \pi_{25}$ *and* $3 \cdot \pi_{25}$.

Problem #5 *Coprimes*

–*5.1: Define the term* coprime.

–*5.2: How can the Euclidean algorithm be used to identify coprimes?*

–*5.3: Give at least one application of the Euclidean algorithm.*

–*5.4: Write a MATLAB function,* `function x = my_gcd(a,b)`, *that uses the Euclidean algorithm to find the GCD of any two inputs* `a` *and* `b`. *Test your function on the (a, b) combinations from the previous problem. Include a printout (or hand-write) your algorithm to turn in.*

Hints and advice:

- Don't give your variables the same names as MATLAB functions! Since `gcd` is an existing MATLAB/Octave function, if you use it as a variable or function name, you won't be able to use `gcd` to check your `gcd()` function. Try `clear all` to recover from this problem.
- Try using a "while" loop for this exercise (see MATLAB documentation for help).
- You may need to use some temporary variables for `a` and `b` in order to perform the algorithm.

2.5.4 Algebraic Generalization of the GCD (Euclidean) Algorithm

Problem #6 *In this problem we are looking for integer solutions $(m, n) \in \mathbb{Z}$ to the equations $ma + nb = gcd(a, b)$ and $ma + nb = 0$ given positive integers $(a, b) \in \mathbb{Z}^+$.*

Note that this requires that either m or n be negative. These solutions may be found using the Euclidean algorithm only if (a, b) are coprime $(a \perp b)$. Note that integer (whole number) polynomial relations such as these are known as Diophantine equations. *Such equations (e.g., $ma + nb = 0$) are linear Diophantine equations, possibly the simplest form of such relations.*

Example: `gcd(2,3)= 1`**:** For $(a, b) = (2, 3)$, the result is

$$\begin{bmatrix}1\\0\end{bmatrix} = \begin{bmatrix}0 & 1\\1 & -2\end{bmatrix}\begin{bmatrix}0 & 1\\1 & -1\end{bmatrix}\begin{bmatrix}0 & 1\\1 & 0\end{bmatrix}\begin{bmatrix}2\\3\end{bmatrix} = \underbrace{\begin{bmatrix}-1 & 1\\3 & -2\end{bmatrix}}_{m \quad n}\begin{bmatrix}2\\3\end{bmatrix}.$$

Thus from the above equation we find the solution (m, n) to the integer equation

$$2m + 3n = \texttt{gcd}(2, 3) = 1,$$

namely, $(m, n) = (-1, 1)$ (i.e., $-2 + 3 = 1$). There is also a second solution $(3, -2)$ (i.e., $3 \cdot 2 - 2 \cdot 3 = 0$) that represents the terminating condition. Thus these two solutions are a pair and the solution exists only if (a, b) are coprime $(a \perp b)$.

Subtraction method: This method is more complicated than the division algorithm because at each stage we must check whether $a < b$. Define

$$\begin{bmatrix}a_0\\b_0\end{bmatrix} = \begin{bmatrix}a\\b\end{bmatrix}, \qquad Q = \begin{bmatrix}1 & -1\\0 & 1\end{bmatrix}, \qquad S = \begin{bmatrix}0 & 1\\1 & 0\end{bmatrix},$$

where Q sets $a_{i+1} = a_i - b_i$ and $b_{i+1} = b_i$ assuming $a_i > b_i$, and S is a swap matrix that swaps a_i and b_i if $a_i < b_i$. Using these matrices, we implement the algorithm by assigning

$$\begin{bmatrix}a_{i+1}\\b_{i+1}\end{bmatrix} = Q\begin{bmatrix}a_i\\b_i\end{bmatrix} \text{ for } a_i > b_i, \qquad \begin{bmatrix}a_{i+1}\\b_{i+1}\end{bmatrix} = QS\begin{bmatrix}a_i\\b_i\end{bmatrix} \text{ for } a_i < b_i.$$

The result of this method is a cascade of Q and S matrices. For $(a, b) = (2, 3)$, the result is

$$\begin{bmatrix}1\\1\end{bmatrix} = \underbrace{\begin{bmatrix}1 & -1\\0 & 1\end{bmatrix}}_{Q}\underbrace{\begin{bmatrix}0 & 1\\1 & 0\end{bmatrix}}_{S}\underbrace{\begin{bmatrix}1 & -1\\0 & 1\end{bmatrix}}_{Q}\underbrace{\begin{bmatrix}0 & 1\\1 & 0\end{bmatrix}}_{S}\begin{bmatrix}2\\3\end{bmatrix} = \underbrace{\begin{bmatrix}2 & -1\\-1 & 1\end{bmatrix}}_{m \quad n}\begin{bmatrix}2\\3\end{bmatrix}.$$

Thus we find two solutions (m, n) to the integer equation $2m + 3n = \gcd(2, 3) = 1$.

–6.1: By inspection, find at least one integer pair (m, n) that satisfies $12m + 15n = 3$.

–6.2: Using matrix methods for the Euclidean algorithm, find integer pairs (m, n) that satisfy $12m + 15n = 3$ and $12m + 15n = 0$. Show your work!!!

–6.3: Does the equation $12m + 15n = 1$ have integer solutions for n and m? Why or why not?

Problem #7 *Matrix approach:*
It can be difficult to keep track of the a's and b's when the algorithm has many steps. We need an alternative way to run the Euclidean algorithm using matrix algebra. Matrix methods provide a more transparent approach to the operations on (a, b). Thus the Euclidean algorithm can be classified in terms of standard matrix operations. Write out the indirect matrix approach discussed at the end of Sect. 2.4.3 (Eq. 2.4.3).

2.5.5 Continued Fractions

Problem #8 *Here we explore the continued fraction algorithm (CFA), discussed in Sect. 2.4.4.*

In its simplest form, the CFA starts with a real number, which we denote as $\alpha \in \mathbb{R}$. Let us work with an irrational real number, $\pi \in \mathbb{I}$, as an example because its CFA representation will be infinitely long. We can represent the CFA coefficients α as a vector of integers n_k, $k = 1, 2, \ldots, \infty$:

$$\begin{aligned}\alpha &= [n_1; n_2, n_3, n_4, \ldots] \\ &= n_1 + \cfrac{1}{n_2 + \frac{1}{n_3 + \frac{1}{n_4 + \cdots}}}.\end{aligned}$$

As discussed in Sect. 2.4.3, the CFA is recursive, with three steps per iteration. For $\alpha_1 = \pi$, $n_1 = 3$, $r_1 = \pi - 3$, and $\alpha_2 \equiv 1/r_1$.

$$\alpha_2 = 1/0.1416 = 7.0625\ldots$$

$$\alpha_1 = n_1 + \frac{1}{\alpha_2} = n_1 + \frac{1}{n_2 + \frac{1}{\alpha_3}} = \cdots.$$

In terms of a MATLAB/Octave script,

```
alpha0 = pi;
K=10;
n=zeros(1,K); alpha=zeros(1,K);
alpha(1)=alpha0;

for k=2:K  %k=1 to K
```

```
n(k)=round(alpha(k-1));
%n(k)=fix(alpha(k-1));
alpha(k)= 1/(alpha(k-1)-n(k));
%disp([fix(k), round(n(k)), alpha(k)]); pause(1)
end
disp([n; alpha]);
%Now compare this to MATLAB'S rat() function
rat(alpha0,1e-20)
```

–8.1: By hand (you may use MATLAB/Octave as a calculator), find the first three values of n_k for $\alpha = e^{\pi}$.

–8.2: For the preceding question, what is the error (remainder) when you truncate the continued fraction after $n_1, \ldots, n_3$? Give the absolute value of the error and the percentage error relative to the original α.

–8.3: Use the MATLAB/Octave program provided to find the first 10 values of n_k for $\alpha = e^{\pi}$, and verify your result using the MATLAB/Octave command `rat()`.

–8.4: Discuss the similarities and differences between the Euclidean algorithm and the CFA.

–8.5: Extra Credit: Show that the CFA is the inverse operation of the GCD (i.e., the CFA is the GCD run in reverse). (Hint: see Sect. 2.4.3.)

2.5.6 Continued Fraction Algorithm (CFA)

Problem #9 *CFA of ratios of large primes –9.1:Starting from the primes below* 10^6, *form the CFA of* π_j/π_k *with* $j = 78498$ *and* $k < j$.

–9.2: Look at other ratios of prime numbers and look for a pattern in the CFA of the ratios of large primes. What is the most obvious conclusion?

–9.3: ***(4pts)*** *Expand 23/7 as a continued fraction. Express your answer in bracket notation (e.g.,* $\pi = [3., 7, 16, \cdots]$*). Show your work.*

–9.4: ***(2pts)*** *Can* $\sqrt{2}$ *be represented as a finite continued fraction? Why or why not?*

–9.5: ***(2pts)*** *What is the CFA for* $\sqrt{2} - 1$*?*

Hint: $$\sqrt{2} + 1 = \frac{1}{\sqrt{2} - 1} = [2; 2, 2, 2, \cdots].$$

–9.6: Find the CFA for $1 + \sqrt{3\jmath}$

–9.7: Show that

$$\frac{1}{1 - \sqrt{a}} = a^{\frac{11}{2}} + a^{\frac{9}{2}} + a^{\frac{7}{2}} + a^{\frac{5}{2}} + a^{\frac{3}{2}} + \sqrt{a} + a^5 + a^4 + a^3 + a^2 + a + 1 = 1 - a^6$$

```
syms a,b
b= taylor(1/( 1-sqrt(a) ))
simplify((1-sqrt(a))*b) = 1-a^6
```

Use symbolic analysis to show this, then explain.

2.6 Number Theory Applications

2.6.1 *Pythagorean Triplets (Euclid's Formula)*

Euclid's formula is a method for finding three integer lengths $[a, b, c] \in \mathbb{N}$ that satisfy Eq. 1.1. It is important to ask Which set are the lengths $[a, b, c]$ drawn from? There is a huge difference, both practical and theoretical, whether they are from the real numbers $\mathbb{R}$ or from the counting numbers $\mathbb{N}$. Given $p, q \in \mathbb{N}$ with $p > q$, the three lengths $[a, b, c] \in \mathbb{N}$ of Eq. 1.1 are given by

$$a = p^2 - q^2, \qquad b = 2pq, \qquad c = p^2 + q^2. \tag{2.6.6}$$

This result may be directly verified, since

$$[p^2 + q^2]^2 = [p^2 - q^2]^2 + [2pq]^2$$

or

$$p^4 + q^4 + \cancel{2p^2q^2} = p^4 + q^4 - \cancel{2p^2q^2} + \cancel{4p^2q^2}.$$

Thus, Eqs. 2.6.6 are easily proved once given. Deriving Euclid's formula (see AE-2, problem #1) is obviously much more difficult and is similar to the proof of Pell's equation as shown in Table 2.1.

A well-known example is the right triangle depicted in Fig. 2.7, defined by the integer lengths [3, 4, 5] that have angles $[0.54, 0.65, \pi/2]$ [rad], which satisfies Eq. 1.1. As quantified by Euclid's formula (Eq. 2.6.6), there are an infinite number of Pythagorean triplets (PTs). Furthermore, the seemingly simple triangle that has angles of $[30, 60, 90] \in N$ [deg] (i.e., $[\pi/6, \pi/3, \pi/2] \in \mathbb{I}$ [rad]) has one irrational ($\mathbb{I}$) length ($[1, \sqrt{3}, 2]$).

The set from which the lengths $[a, b, c]$ are drawn was not missed by the early Asians and was documented by the Greeks. Any equation whose solution is based on integers is called a *Diophantine equation*, named for the Greek mathematician Diophantus of Alexandria (ca. 250 CE) (see Fig. 1.1).

Table 2.1 Table of Pythagorean triplets computed from Euclid's formula, Eq. 2.6.6, for various $[p, q]$. The last three columns are the first, fourth, and penultimate values of Plimpton-322, along with their corresponding $[p, q]$. In all cases $c^2 = a^2 + b^2$ and $p = q + l$, where $l = \sqrt{c - b} \in \mathbb{N}$

q	1	1	1	2	2	2	3	3	3	5	54	27
l	1	2	3	1	2	3	1	2	3	7	71	23
p	2	3	4	3	4	5	4	5	6	12	125	50
a	3	8	15	5	12	21	7	16	27	119	12709	1771
b	4	6	8	12	16	20	24	30	36	120	13500	2700
c	5	10	17	13	20	29	25	34	45	169	18541	3229

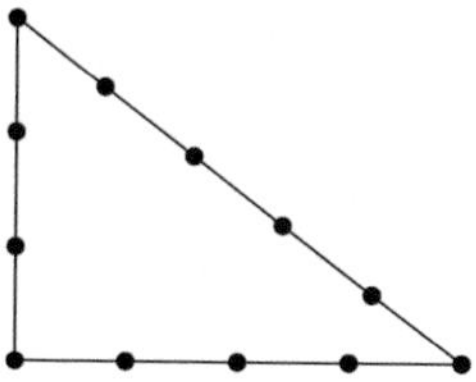

Fig. 2.7 Beads on a string form perfect right triangles when the number of unit lengths between beads on each side satisfies Eq. 1.1. For example, when $p = 2, q = 1$, the sides are $[3, 4, 5]$

A clay tablet from the nineteenth century BCE with the numbers engraved on it, as shown in Fig. 2.8, was discovered in Mesopotamia, and cataloged in 1922 by George A. Plimpton.[22] These numbers are a and c pairs from PTs $[a, b, c]$. Given this discovery, it is clear that the Pythagoreans were following those who came long before them. Recently a second tablet, dating between 350 and 50 BCE, has been reported, that indicates calculations of the apparent motion of Jupiter based on a trapezoidal graph of the rate.[23]

It is of interest that PTs play a role on atomic physics, as discussed in Appendix I.

2.6.2 *Pell's Equation*

Pell's equation,

$$x_n^2 - Ny_n^2 = (x_n - \sqrt{N}y_n)(x_n + \sqrt{N}y_n) = 1, \tag{2.6.7}$$

with non-square $N \in \mathbb{N}$ specified and $x, y \in \mathbb{N}$ unknown, has a venerable history in both physics (see Sect. 2.1) and mathematics. Given its factored form, it is obvious that every solution x_n, y_n has the asymptotic property

$$\left.\frac{x_n}{y_n}\right|_{n\to\infty} \to \pm\sqrt{N}. \tag{2.6.8}$$

It is believed that Pell's equation is directly related to the Pythagorean theorem, since both are simple binomials that have integer coefficients (Stillwell 2010, p. 48), with Pell's equation being the hyperbolic version of Eq. 1.1. For example, with $N = 2$, a solution is $x = 17, y = 12$ (i.e., $17^2 - 2 \cdot 12^2 = 1$).

[22]https://www.nytimes.com/2010/11/27/arts/design/27tablets.html.

[23]https://www.nytimes.com/2016/01/29/science/babylonians-clay-tablets-geometry-astronomy-jupiter.html.

a	c
119	169
3367	4825
4601	6649
12709	18541
65	97
319	481
2291	3541
799	1249
481	769
4961	8161
45	75
1679	2929
161	289
1771	3229
56	106

Fig. 2.8 Plimpton-322, a clay tablet from 1800 BCE that displays a and c values of the Pythagorean triplets $[a, b, c]$, with the property $b = \sqrt{c^2 - a^2} \in \mathbb{N}$. Several of the c values are primes, but not the a values. The clay is item 322 (item 3 from 1922) from the collection of George A. Plimpton

A 2×2 matrix recursion algorithm, likely due to the Chinese and used by the Pythagoreans to investigate $\sqrt{N}$, is

$$\begin{bmatrix} x \\ y \end{bmatrix}_{n+1} = \begin{bmatrix} 1 & N \\ 1 & 1 \end{bmatrix} \begin{bmatrix} x \\ y \end{bmatrix}_n, \tag{2.6.9}$$

where we indicate the index outside the vectors.

Starting with the trivial solution $[x_o, y_o]^T = [1, 0]^T$ (i.e., $x_o^2 - Ny_o^2 = 1$), additional solutions of Pell's equations are determined, having the property $\boldsymbol{x}_n/y_n \to \sqrt{N} \in \mathbb{F}$, motivated by Euclid's formula for Pythagorean triplets (Stillwell 2010, p. 44).

Note that Eq. 2.6.9 is a 2×2 linear matrix composition method (see Sect. 3.4.2), since the output of one matrix multiplication is the input to the next.

Asian solutions: The first solution of Pell's equation was published in about 628 CE by Brahmagupta, who first discovered the equation (Stillwell 2010, p. 46). Brahmagupta's novel solution also used the composition method, but in a different way from Eq. 2.6.9. Then in 1150 CE, Bhaskara II independently obtained solutions using Eq. 2.6.9 (Stillwell 2010, p.69). This is the composition method we shall explore here, as summarized in Appendix B, Table B.1.

The best way to see how this recursion results in solutions to Pell's equation is by example. Initializing the recursion with the trivial solution $[x_0, y_0]^T = [1, 0]^T$ gives

$$\begin{bmatrix} x_1 \\ y_1 \end{bmatrix} = \begin{bmatrix} 1 \\ 1 \end{bmatrix}_1 = \begin{bmatrix} 1 & 2 \\ 1 & 1 \end{bmatrix} \begin{bmatrix} 1 \\ 0 \end{bmatrix}_0 \qquad 1^2 - 2 \cdot 1^2 = -1$$

$$\begin{bmatrix} x \\ y \end{bmatrix}_2 = \begin{bmatrix} 3 \\ 2 \end{bmatrix} = \begin{bmatrix} 1 & 2 \\ 1 & 1 \end{bmatrix} \begin{bmatrix} 1 \\ 1 \end{bmatrix}_1 \qquad 3^2 - 2 \cdot 2^2 = 1$$

$$\begin{bmatrix} x \\ y \end{bmatrix}_3 = \begin{bmatrix} 7 \\ 5 \end{bmatrix} = \begin{bmatrix} 1 & 2 \\ 1 & 1 \end{bmatrix} \begin{bmatrix} 3 \\ 2 \end{bmatrix}_2 \qquad (7)^2 - 2 \cdot (5)^2 = -1$$

$$\begin{bmatrix} x \\ y \end{bmatrix}_4 = \begin{bmatrix} 17 \\ 12 \end{bmatrix} = \begin{bmatrix} 1 & 2 \\ 1 & 1 \end{bmatrix} \begin{bmatrix} 7 \\ 5 \end{bmatrix}_3 \qquad 17^2 - 2 \cdot 12^2 = 1$$

$$\begin{bmatrix} x \\ y \end{bmatrix}_5 = \begin{bmatrix} 41 \\ 29 \end{bmatrix} = \begin{bmatrix} 1 & 2 \\ 1 & 1 \end{bmatrix} \begin{bmatrix} 17 \\ 12 \end{bmatrix}_4 \qquad (41)^2 - 2 \cdot (29)^2 = -1.$$

Thus the recursion results in a modified version of Pell's equation,

$$x_n^2 - 2y_n^2 = (-1)^n, \tag{2.6.10}$$

where only even values of n are solutions. This sign change had no effect on the Pythagoreans' goal, since they cared about only the ratio $y_n/x_n \rightarrow \pm\sqrt{2}$.

Modified recursion: We may restore the solution of Pell's equation for $N = 2$ using a slightly modified linear matrix recursion. To fix the $(-1)^n$ problem, we multiply the 2×2 matrix by $1\jmath = \sqrt{-1}$, which gives

$$\begin{bmatrix} x \\ y \end{bmatrix}_1 = \jmath \begin{bmatrix} 1 \\ 1 \end{bmatrix}_1 = \jmath \begin{bmatrix} 1 & 2 \\ 1 & 1 \end{bmatrix} \begin{bmatrix} 1 \\ 0 \end{bmatrix}_0 \qquad \jmath^2 - 2 \cdot \jmath^2 = 1$$

$$\begin{bmatrix} x \\ y \end{bmatrix}_2 = \jmath^2 \begin{bmatrix} 3 \\ 2 \end{bmatrix}_2 = \jmath \begin{bmatrix} 1 & 2 \\ 1 & 1 \end{bmatrix} \jmath \begin{bmatrix} 1 \\ 1 \end{bmatrix}_1 \qquad 3^2 - 2 \cdot 2^2 = 1$$

$$\begin{bmatrix} x \\ y \end{bmatrix}_3 = \jmath^3 \begin{bmatrix} 7 \\ 5 \end{bmatrix}_3 = \jmath \begin{bmatrix} 1 & 2 \\ 1 & 1 \end{bmatrix} \jmath^2 \begin{bmatrix} 3 \\ 2 \end{bmatrix}_2 \qquad (7\jmath)^2 - 2 \cdot (5\jmath)^2 = 1$$

$$\begin{bmatrix} x \\ y \end{bmatrix}_4 = \begin{bmatrix} 17 \\ 12 \end{bmatrix}_4 = \jmath \begin{bmatrix} 1 & 2 \\ 1 & 1 \end{bmatrix} \jmath^3 \begin{bmatrix} 7 \\ 5 \end{bmatrix}_3 \qquad 17^2 - 2 \cdot 12^2 = 1$$

$$\begin{bmatrix} x \\ y \end{bmatrix}_5 = \jmath \begin{bmatrix} 41 \\ 29 \end{bmatrix}_5 = \jmath \begin{bmatrix} 1 & 2 \\ 1 & 1 \end{bmatrix} \begin{bmatrix} 17 \\ 12 \end{bmatrix}_4 \qquad (41\jmath)^2 - 2 \cdot (29\jmath)^2 = 1.$$

Solution to Pell's equation: By multiplying the matrix by $1\jmath$, all the solutions ($\boldsymbol{x}_k \in \mathbb{C}$) to Pell's equation are determined. The $1\jmath$ factor corrects the alternation in sign, so every iteration yields a solution. For $N = 2, n = 0$ (the initial solution), $[x_0, y_0]$ is $[1, 0]_0$, $[x_1, y_1] = \jmath[1, 1]_1$, and $[x_2, y_2] = -[3, 2]_2$. These are easily checked using this recursion.

The solution for $N = 3$ is given in Appendix B.2.2. Table B.1 shows that every output of this slightly modified matrix recursion gives solutions to Pell's equation: $[1, 0], [1, 1], [4, 2], [10, 6], \ldots, [76, 44], \ldots$.

At each iteration, the ratio x_n/y_n approaches $\sqrt{2}$ with increasing accuracy, coupling it to the CFA, which may also be used to find approximations to $\sqrt{N}$. The value of $41/29 \approx \sqrt{2}$, with a relative error of <0.03%.

2.6.3 *Fibonacci Sequence*

Another classic problem, also formulated by the Chinese, is the Fibonacci sequence, generated by the relationship

$$f_{n+1} = f_n + f_{n-1}. \tag{2.6.11}$$

Here the next number $f_{n+1} \in \mathbb{N}$ is the sum of the previous two. If we start from [0, 1], this linear recursion equation leads to the Fibonacci sequence $f_n = [0, 1, 1, 2, 3, 5, 8, 13, 21, 34, \ldots]$. Alternatively, if we define $y_{n+1} = x_n$, then Eq. 2.6.11 may be compactly represented by a 2×2 companion matrix recursion (see the Fibonacci exercises in NS-3).

$$\begin{bmatrix} x \\ y \end{bmatrix}_{n+1} = \begin{bmatrix} 1 & 1 \\ 1 & 0 \end{bmatrix} \begin{bmatrix} x \\ y \end{bmatrix}_n, \tag{2.6.12}$$

which has eigenvalues $(1 \pm \sqrt{5})/2$.

The correspondence of Eqs. 2.6.11 and 2.6.12 is easily verified. Starting with $[x, y]_0^T = [0, 1]^T$, we obtain for the first few steps:

$$\begin{bmatrix} 1 \\ 0 \end{bmatrix}_1 = \begin{bmatrix} 1 & 1 \\ 1 & 0 \end{bmatrix} \begin{bmatrix} 0 \\ 1 \end{bmatrix}_0, \quad \begin{bmatrix} 1 \\ 1 \end{bmatrix}_2 = \begin{bmatrix} 1 & 1 \\ 1 & 0 \end{bmatrix} \begin{bmatrix} 1 \\ 0 \end{bmatrix}_1, \quad \begin{bmatrix} 2 \\ 1 \end{bmatrix}_3 = \begin{bmatrix} 1 & 1 \\ 1 & 0 \end{bmatrix} \begin{bmatrix} 1 \\ 1 \end{bmatrix}_2, \quad \begin{bmatrix} 3 \\ 2 \end{bmatrix}_4 = \begin{bmatrix} 1 & 1 \\ 1 & 0 \end{bmatrix} \begin{bmatrix} 2 \\ 1 \end{bmatrix}_3, \quad \ldots.$$

From the above, $x_n = [0, 1, 1, 2, 3, 5, \ldots]$ is the Fibonacci sequence, since the next x_n is the sum of the previous two, and the next y_n is x_n, as shown in Fig. 2.9.

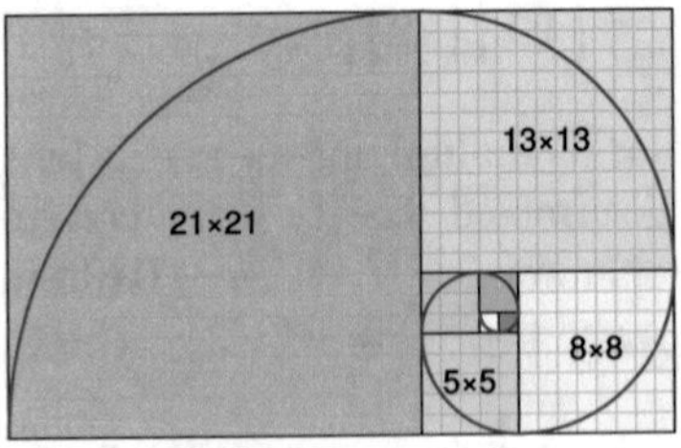

Fig. 2.9 A construction called the Fibonacci spiral. Note how it is constructed of squares that have areas given by the squares of the Fibonacci numbers. In this way, the spiral is smooth and the radius increases as the Fibonacci numbers (e.g., $8 = 3 + 5$, $13 = 5 + 8$, etc.) (Adapted from https://en.wikipedia.org/wiki/Golden_spiral)

Exercise #24 Use the Octave/MATLAB command `compan(c)` to find the companion matrix of the polynomial coefficients defined by Eq. 2.6.11.
Sol: Using MATLAB/Octave: `f=[1, -1, -1]; C=compan(f);`

$$C = \begin{bmatrix} 1 & 1 \\ 1 & 0 \end{bmatrix} \tag{2.6.13}$$

■

Exercise #25 Find the eigenvalues of matrix C.
Sol: The characteristic equation is

$$\det \begin{bmatrix} 1-\lambda & 1 \\ 1 & -\lambda \end{bmatrix} = 0$$

or $\lambda^2 - \lambda - 1 = (\lambda - 1/2)^2 - 1/4 - 1 = 0$, which has roots $\lambda_\pm = (1 \pm \sqrt{5})/2 \approx \{1.618, -0.618\}$.

■

The mean-Fibonacci sequence: Suppose that the Fibonacci sequence recursion is replaced by the mean of the last two values—namely, let

$$f_{n+1} = \frac{f_n + f_{n-1}}{2}. \tag{2.6.14}$$

This seems like a small change. But how does the solution differ? To answer this question it is helpful to look at the corresponding 2×2 matrix.

Exercise #26 Find the 2×2 matrix corresponding to Eq. 2.6.14. The 2×2 matrix may be found using the *companion matrix* method (see Sect. 3.1.3).
Sol: Using MATLAB/Octave code, we have

```
f=[1, -1/2, -1/2];
C=compan(f);
```

which returns

$$C = \frac{1}{2}\begin{bmatrix} 1 & 1 \\ 2 & 0 \end{bmatrix}. \tag{2.6.15}$$

■

Exercise #27 Find the steady-state solution for the mean-Fibonacci, starting from $[1, 0]_0$. State the nature of both solutions.
Sol: By inspection one steady-state solution is $[1, 1]^T_\infty$ or $f_n = 1$. To find the full solution, we need to find the two eigenvalues, defined by

$$\det \begin{bmatrix} 1/2-\lambda & 1/2 \\ 1 & -\lambda \end{bmatrix} = \lambda^2 - \lambda/2 - 1/2 = (\lambda - 1/4)^2 - (1/4)^2 - 1/2 = 0.$$

Thus $\lambda_\pm = (1 \pm 3)/4 = [1, -0.5]$. The first solution converges to 1 while the second solution is $(-1/2)^n$, which changes sign at each time step and quickly converges to zero. The full solution is given by $\boldsymbol{E}\Lambda^n\boldsymbol{E}^{-1}[1,0]^T$ (see Appendix B). ■

Relationships to digital signal processing: Today we recognize Eq. 2.6.11 as a discrete difference equation, which is a pre-limit (pre–Stream 3) recursive form of a differential equation. The 2×2 matrix form of Eq. 2.6.11 is an early precursor to seventeenth- and eighteenth-century developments in linear algebra. Thus the Greeks' recursive solution for the $\sqrt{2}$ and Bhaskara's solution of Pell's equation are early precursors to discrete-time signal processing as well as to calculus.

There are strong similarities between Pell's equation and the Pythagorean theorem. As we shall see, Pell's equation is related to the geometry of a hyperbola, just as the Pythagorean equation is related to the geometry of a circle. We shall show, as one might assume, that there is a counterpart to Euclid's formula for the case of Pell's equations, since these are all conic sections with closely related conic geometry. As we have seen, the solutions involve $\sqrt{-1}$. The derivation is a trivial extension of that for Euclid's formula for Pythagorean triplets. The early solution of Brahmagupta was not related to this simple formula.

2.6.4 Diagonalization of a Matrix (Eigenvalue/Eigenvector Decomposition)

As derived in Appendix B, the most efficient way to compute A^n is to diagonalize the matrix A by finding its eigenvalues and eigenvectors.

The eigenvalues λ_k and eigenvectors $\vec{e}_k$ of a square matrix A are related by

$$A\vec{e}_k = \lambda_k \vec{e}_k, \tag{2.6.16}$$

such that multiplying an eigenvector $\vec{e}_k$ of A by the matrix A is the same as multiplying by a scalar, $\lambda_k \in \mathbb{C}$ (the corresponding eigenvalue). The complete eigenvalue problem may be written as

$$AE = E\Lambda.$$

If A is a 2×2 matrix,[24] the matrices E and Λ (of eigenvectors and eigenvalues, respectively) are

$$E = \begin{bmatrix}\vec{e}_1 & \vec{e}_2\end{bmatrix} \qquad \text{and} \qquad \Lambda = \begin{bmatrix}\lambda_1 & 0 \\ 0 & \lambda_2\end{bmatrix}.$$

Thus the matrix equation $AE = \begin{bmatrix}A\vec{e}_1 & A\vec{e}_2\end{bmatrix} = \begin{bmatrix}\lambda_1\vec{e}_1 & \lambda_2\vec{e}_2\end{bmatrix} = E\Lambda$ contains Eq. 2.6.16 for each eigenvalue–eigenvector pair.

---

[24] These concepts may be easily extended to higher dimensions.

The diagonalization of the matrix A refers to the fact that the matrix of eigenvalues, Λ, has nonzero elements only on the diagonal. The key result is found by post-multiplication of the eigenvalue matrix by E^{-1}, giving

$$AEE^{-1} = A = E\Lambda E^{-1}. \tag{2.6.17}$$

If we now take powers of A, the nth power of A is

$$\begin{aligned} A^n &= (E\Lambda E^{-1})^n \\ &= E\Lambda E^{-1} E\Lambda E^{-1} \cdots E\Lambda E^{-1} \\ &= E\Lambda^n E^{-1}. \end{aligned} \tag{2.6.18}$$

This is a very powerful result because the nth power of a diagonal matrix is extremely easy to calculate:

$$\Lambda^n = \begin{bmatrix} \lambda_1^n & 0 \\ 0 & \lambda_2^n \end{bmatrix}.$$

Thus, from Eq. 2.6.18 we can calculate A^n using only two matrix multiplications:

$$A^n = E\Lambda^n E^{-1}.$$

2.6.5 *Finding the Eigenvalues*

The eigenvalues λ_k are determined from Eq. 2.6.16, by factoring out $\vec{e}_k$:

$$\begin{aligned} A\vec{e}_k &= \lambda_k \vec{e} \\ (A - \lambda_k I)\vec{e}_k &= \vec{0}. \end{aligned}$$

Matrix $I = [1, 0; 0, 1]^T$ is the identity matrix, having the dimensions of A, with elements δ_{ij} (i.e., diagonal elements $\delta_{11,22} = 1$ and off-diagonal elements $\delta_{12,21} = 0$).

The vector $\vec{e}_k$ is not zero, yet when operated on by $A - \lambda_k I$, the result must be zero. The only way this can happen is if the operator is degenerate (has no solution)—that is,

$$\det(A - \lambda I) = \det \begin{bmatrix} (a_{11} - \lambda) & a_{12} \\ a_{21} & (a_{22} - \lambda) \end{bmatrix} = 0. \tag{2.6.19}$$

This means that the two equations have the same roots (the equation is degenerate).

This determinant equation results in a second-degree polynomial in λ:

$$(a_{11} - \lambda)(a_{22} - \lambda) - a_{12}a_{21} = 0,$$

the roots of which are the eigenvalues of the matrix A.

2.6.6 *Finding the Eigenvectors*

An eigenvector $\vec{e}_k$ can be found for each eigenvalue λ_k from Eq. 2.6.16,

$$(A - \lambda_k I)\vec{e}_k = \vec{0}.$$

The left side of the above equation becomes a column-vector, where each element is an equation in the elements of $\vec{e}_k$, set equal to 0 on the right side. These equations are always degenerate, since the determinant is zero. Thus the two equations have the same slope.

Solving for the eigenvectors is often confusing because they have arbitrary magnitudes, $||\vec{e}_k|| = \sqrt{\vec{e}_k \cdot \vec{e}_k} = \sqrt{e_{k,1}^2 + e_{k,2}^2} = d$. From Eq. 2.6.16, we can determine only the relative magnitudes and signs of the elements of $\vec{e}_k$, so we have to choose a magnitude d. It is common practice to normalize each eigenvector to have unit magnitude ($d = 1$).

2.7 Problems NS-3

2.7.1 *Topic of This Homework*

Pythagorean triplets, Pell's equation, Fibonacci sequence.

2.7.2 *Pythagorean Triplets*

Problem #1 *Euclid's formula for the Pythagorean triplets a, b, c is $a = p^2 - q^2$, $b = 2pq$, and $c = p^2 + q^2$.*

–1.1: What condition(s) must hold for p and q such that a, b, and c are always positive and nonzero?

–1.2: Solve for p and q in terms of a, b, and c.

Problem #2 *The ancient Babylonians (ca. 2000 BCE) cryptically recorded (a, c) pairs of numbers on a clay tablet, archeologically denoted* Plimpton-322 *(see 2.8).*

–2.1: Find p and q for the first five pairs of a and c shown here from Plimpton-322.

a	c
119	169
3367	4825
4601	6649
12709	18541
65	97

Find a formula for a in terms of p and q.
–2.2: Based on Euclid's formula, show that $c > (a, b)$.
–2.3: What happens when $c = a$?
–2.4: Is $b + c$ *a perfect square? Discuss.*

2.7.3 Pell's Equation

Problem #3 *Pell's equation is one of the most historic (i.e., important) equations of Greek number theory because it was used to show that* $\sqrt{2} \in \mathbb{I}$. *We seek integer solutions of*

$$x^2 - Ny^2 = 1.$$

As shown on Sect. 2.6.2, the solutions x_n, y_n *for the case of* $N = 2$ *are given by the linear* 2×2 *matrix recursion*

$$\begin{bmatrix} x_{n+1} \\ y_{n+1} \end{bmatrix} = 1_J \begin{bmatrix} 1 & 2 \\ 1 & 1 \end{bmatrix} \begin{bmatrix} x_n \\ y_n \end{bmatrix}$$

with $[x_0, y_0]^T = [1, 0]^T$ *and* $1_J = \sqrt{-1} = e^{j\pi/2}$. *It follows that the general solution to Pell's equation for* $N = 2$ *is*

$$\begin{bmatrix} x_n \\ y_n \end{bmatrix} = (e^{J\pi/2})^n \begin{bmatrix} 1 & 2 \\ 1 & 1 \end{bmatrix}^n \begin{bmatrix} x_0 \\ y_0 \end{bmatrix}.$$

To calculate solutions to Pell's equation using the matrix equation above, we must calculate

$$A^n = e^{J\pi n/2} \begin{bmatrix} 1 & 2 \\ 1 & 1 \end{bmatrix}^n = e^{J\pi n/2} \begin{bmatrix} 1 & 2 \\ 1 & 1 \end{bmatrix} \begin{bmatrix} 1 & 2 \\ 1 & 1 \end{bmatrix} \begin{bmatrix} 1 & 2 \\ 1 & 1 \end{bmatrix} \cdots \begin{bmatrix} 1 & 2 \\ 1 & 1 \end{bmatrix},$$

which becomes tedious for $n > 2$.

–3.1: Find the companion matrix and thus the matrix A that has the same eigenvalues as Pell's equation. *Hint: Use MATLAB'S function* `[E,Lambda] = eig(A)` *to check your results!*

–3.2: Solutions to Pell's equation were used by the Pythagoreans to explore the value of $\sqrt{2}$. Explain why Pell's equation is relevant to $\sqrt{2}$.

–3.3: Find the first three values of $(x_n, y_n)^T$ by hand and show that they satisfy Pell's equation for $N = 2$.

By hand, find the eigenvalues $\lambda_\pm$ of the 2×2 Pell's equation matrix

$$A = \begin{bmatrix} 1 & 2 \\ 1 & 1 \end{bmatrix}.$$

–3.4: By hand, show that the matrix of eigenvectors, E, is

$$E = \begin{bmatrix} \vec{e}_+ & \vec{e}_- \end{bmatrix} = \frac{1}{\sqrt{3}} \begin{bmatrix} -\sqrt{2} & \sqrt{2} \\ 1 & 1 \end{bmatrix}.$$

–3.5: Using the eigenvalues and eigenvectors you found for A, verify that

$$E^{-1}AE = \Lambda \equiv \begin{bmatrix} \lambda_+ & 0 \\ 0 & \lambda_- \end{bmatrix}$$

–3.6: Now that you have diagonalized A (Equation 2.6.18), use your results for E and Λ to solve for the $n = 10$ solution $(x_{10}, y_{10})^T$ to Pell's equation with $N = 2$.

2.7.4 The Fibonacci Sequence

Problem #4 *Here we seek the Fibonacci sequence for x_n. Like Pell's equation, Eq. 2.6.11 has a recursive, eigenanalysis solution. To find it we must recast x_n as a 2×2 matrix relationship and then proceed as we did for the Pell case.*

–4.1: By example, show that the Fibonacci sequence x_n as described above may be generated by

$$\begin{bmatrix} x_n \\ y_n \end{bmatrix} = \begin{bmatrix} 1 & 1 \\ 1 & 0 \end{bmatrix}^n \begin{bmatrix} x_0 \\ y_0 \end{bmatrix}, \qquad \begin{bmatrix} x_0 \\ y_0 \end{bmatrix} = \begin{bmatrix} 1 \\ 0 \end{bmatrix}. \tag{NS-3.1}$$

–4.2: What is the relationship between y_n and x_n?

–4.3: Write a MATLAB/Octave program to compute x_n using the matrix equation above. Test your code using the first few values of the sequence. Using your program, what is x_{40}? Note: Consider using the eigenanalysis of A, described by Eq. *2.6.18.*

–4.4: Using the eigenanalysis of the matrix A (and a lot of algebra) shows that it is possible to obtain the general formula for the Fibonacci sequence

$$x_n = \frac{1}{\sqrt{5}}\left[\left(\frac{1+\sqrt{5}}{2}\right)^{n+1} - \left(\frac{1-\sqrt{5}}{2}\right)^{n+1}\right]. \tag{NS-3.2}$$

–4.5: What are the eigenvalues $\lambda_\pm$ of the matrix A?

–4.6: How is the formula for x_n related to these eigenvalues? Hint: Find the eigenvectors.

–4.7: What happens to each of the two terms

$$[(1 \pm \sqrt{5})/2]^{n+1}?$$

–4.8: What happens to the ratio x_{n+1}/x_n?

Problem #5 *Replace the Fibonacci sequence with*

$$x_n = \frac{x_{n-1} + x_{n-2}}{2},$$

such that the value x_n is the average of the previous two values in the sequence.

–5.1: What matrix A is used to calculate this sequence?

–5.2: Modify your computer program to calculate the new sequence x_n. What happens as $n \to \infty$?

–5.3: What are the eigenvalues of your new A? How do they relate to the behavior of x_n as $n \to \infty$? Hint: You can expect the closed-form expression for x_n to be similar to Eq. *NS-3.2*.

–5.4: What matrix A is used to calculate this sequence?

–5.5: Modify your computer program to calculate the new sequence x_n. What happens as $n \to \infty$?

–5.6: What are the eigenvalues of your new A? How do they relate to the behavior of x_n as $n \to \infty$? Hint: You can expect the closed-form expression for x_n to be similar to Eq. *NS-3.2*.

Problem #6 *Consider the expression*

$$\sum_1^N f_n^2 = f_N f_{N+1}.$$

–6.1: Find a formula for f_n that satisfies this relationship. Hint: It holds for only the Fibonacci recursion formula.

2.7.5 CFA as a Matrix Recursion

Problem #7 *The CFA may be writen as a matrix recursion. For this we adopt a special notation, unlike other matrix notations,*[25] *with* $k \in \mathbb{N}$*:*

$$\begin{bmatrix} n \\ x \end{bmatrix}_{k+1} = \begin{bmatrix} 0 & \lfloor x_k \rfloor \\ 0 & \frac{1}{x_k - \lfloor x_k \rfloor} \end{bmatrix} \begin{bmatrix} n \\ x \end{bmatrix}_k .$$

This equation says that $n_{k+1} = \lfloor x_k \rfloor$ *and* $x_{k+1} = 1/(x_k - \lfloor x_k \rfloor)$. *It does* not *mean that* $n_{k+1} = \lfloor x_k \rfloor x_k$, *as would be implied by standard matrix notation. The lower equation says that* $r_k = x_k - \lfloor x_k \rfloor$ *is the* remainder*—namely,* $x_k = \lfloor x - k \rfloor + r_k$ *(Octave/MATLAB'S* `rem(x, floor(x))` *function), also known as* `mod(x,y)`.

–7.1: Start with $n_0 = 0 \in \mathbb{N}$, $x_0 \in \mathbb{I}$, $n_1 = \lfloor x_0 \rfloor \in \mathbb{N}$, $r_1 = x - \lfloor x \rfloor \in \mathbb{I}$, *and* $x_1 = 1/r_1 \in \mathbb{I}$, $r_n \neq 0$. *For* $k = 1$ *this generates on the left the next CFA parameter* $n_2 = \lfloor x_1 \rfloor$ *and* $x_2 = 1/r_2 = 1/(x_0 - \lfloor x_0 \rfloor)$ *from* n_0 *and* x_0. *Find* $[n, x]^T_{k+1}$ *for* $k = 2, 3, 4, 5$.

[25] This notation is highly nonstandard due to the nonlinear operations. The matrix elements are *derived* from the vector rather than multiplying them. These calculation may be done with the help of MATLAB/Octave.

Chapter 3
Algebraic Equations: Stream 2

3.1 Algebra and Geometry as Physics

Stream 2 is geometry, which led to the merging of Euclid's geometrical methods and the development of algebra by al-Khwarizmi in 830 CE (Fig. 1.1). This migration of ideas led Descartes and Fermat to develop analytic geometry (Fig. 1.2).

The mathematics upto the time of the Greeks, documented and formalized by Euclid, served students of mathematics for more than two thousand years. Algebra and geometry were, at first, independent lines of thought. When merged, the focus returned to the Pythagorean theorem. Algebra generalized the analytic conic section into the complex plane, greatly extending the geometrical approach as taught in Euclid's *Elements*. With the introduction of algebra, numbers, rather than lines, could be used to reproduce geometrical lengths in the complex plane. Thus the appreciation for geometry grew after the addition of rigorous analysis using numbers as shown as a time-line in Fig. 3.1.

History of Mathematics after the 15th Century

16th	Bombelli 1526–1572; Galileo 1564–1642; Kepler 1571–1630; Mersenne 1588–1648;
17th	Huygens 1629–1695; Newton 1642–1727[a], *Principia* 1687; Bernoulli, Jakob 1655–1705; Bernoulli, Johann 1667–1748; Fermat, Pierre de 1607–1665; Pascal, Blaise 1623–1662; Descartes, René 1596–1648
18th	Bernoulli, Daniel 1700–1782; Euler, Leonhard 1707–1783; d'Alembert, Jean le Rond 1717–1783; Lagrange, Joseph-Louis 1736–1833; Laplace 1749–1827; Fourier 1768–1830; Gauss 1777–1855; Cauchy 1789–1857
19th	Helmholtz 1821–1894; Kelvin 1824–1907; Kirchhoff 1824–1887; Riemann 1826–1866; Maxwell 1831–1879; Rayleigh 1842–1919; Heaviside 1850–1925; Poincare 1854–1912; Hilbert 1862–1942; Einstein 1879–1955; Fletcher 1884–1981; Sommerfeld 1886–1951; Brillouin 1889–1969; Nyquist 1889–1976
20th	Bode 1905–1982

[a]Born Dec 25, 1942, Julian calendar

Physics inspires algebraic mathematics: The Chinese used music, art, and navigation to drive mathematics. Much of their knowledge has been handed down as either artifacts, such as musical bells and tools, or mathematical relationships documented, but not created by scholars such as Euclid, Archimedes, Diophantus, and perhaps Brahmagupta. With the invention of algebra in 830 CE by al-Khwarizmi, mathematics became more powerful and blossomed. During the sixteenth and seven-

J. Allen, *An Invitation to Mathematical Physics and Its History*,
https://doi.org/10.1007/978-3-030-53759-3_3

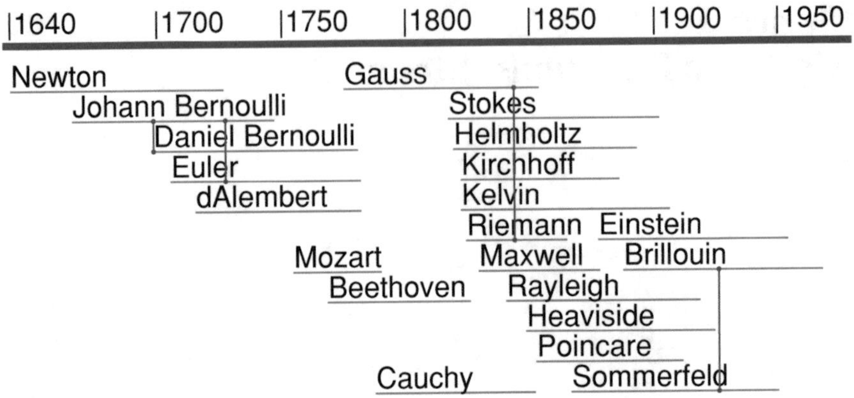

Fig. 3.1 Timeline of the 300 years from the mid-seventeenth to the mid-twentieth centuries CE, an important era in mathematics that produced a continuous stream of fundamental theorems because mathematicians were sharing information. A few of the individuals who played notable roles in this development, in chronological (birth) order, are Galileo, Mersenne, Newton, d'Alembert, Fermat, Huygens, Descartes, Helmholtz, and Kirchhoff. These were some of the first to develop the basic ideas, in various forms, that were later reworked into the proofs that today we recognize as *the fundamental theorems of mathematics*. Mozart and Beethoven are included for orientation

teenth centuries, it became clear that differential equations (DEs), such as the wave equation, can characterize a law of nature at a single point in space and time. This principle was not obvious. A desire to understand the motions of planets other objects precipitated many new discoveries. This period, centered around Galileo, Newton, and Euler, is shown on the timeline in Fig. 1.2.

As we have described, the law of gravity was first formulated by Galileo using the concept of conservation of energy, which determines how masses are accelerated when friction is not considered and the mass is constant. Tycho Brahe investigated the motion of the planets. Starting in early 1600, Kepler (1571–1630) began an extended visit with Brahe. In 1604, working with Brahe's data Kepler described the inverse square law of light and studied the workings of the human eye. Kepler was also the first to predict that the orbits of planets are described by ellipses. It seems he under-appreciated the significance of his finding, as he continued working on his five polyhedra nested model of planetary motion (Stillwell 2010, p. 23).

Building on Galileo (1638) (see the discussion on "Preface"), Newton demonstrated that there must be a gravitational potential between two masses (m_1, m_2) of the form

$$\phi_{\text{New}}(r(t)) \propto \frac{m_1 m_2}{r(t)}, \tag{3.1.1}$$

where $r = |x_1 - x_2|$ is the Euclidean distance between the two point masses at locations x_1 and x_2. Using algebra and his calculus, Newton formalized the equations of gravity, forces, and motion (Newton's three laws), the most important being

$$f(t) = \frac{d}{dt} M(t)v(t), \tag{3.1.2}$$

and showed that Kepler's discovery of planetary elliptical motion naturally follows from these laws (see Sect. 3.10.2). With the discovery of Uranus in 1781, "Kepler's theory was ruined." (Stillwell 2010, p. 23).

In Fig. 3.1 we see the important 1640–1950 timeline, from Newton to Einstein. There are two clear developments: Newton to Cauchy and Stokes to Einstein.

> Possibly the first measurement of the speed of sound, 1380 Paris feet per second, was made by Marin Mersenne in 1630.[a]
>
> ---
>
> [a] 1 English foot is 1.06575 Paris feet.

Newton and the speed of sound: After Newton proposed the basic laws of gravity and explained the elliptical motion of the planets, he proposed the first model of the speed of sound.

In 1630 Mersenne showed that the speed of sound was approximately 1000 [ft/s]. This may have been done by finding the difference between the time of the flash of an explosion and the time it is heard. For example, if the explosion is 1 [mi] away, the delay is about 5 [s]. Thus with a simple clock, such as a pendulum, and a explosive, the speed may be accurately measured. If we say the speed of sound is c_o, then the equation for the wavefront is $f(x, t) = u(x - c_o t)$, where the function $u(t) = 0$ for $t < 0$ and 1 for $t > 0$. If the wave is traveling in the opposite direction, then the formula is $u(x + c_o t)$. If one also assumes that sounds add in an independent manner (superposition holds) (see Postulate P2 on Sect. 3.10.2), then the general solution for the acoustic wave is

$$f(x, t) = Au(x - c_o t) + Bu(x + c_o t),$$

where A and B are the amplitudes of the two waves. This is the solution proposed by d'Alembert in 1747 for the acoustic wave equation

$$\frac{\partial^2}{\partial x^2} \varrho(x, t) = \frac{1}{c_o^2} \frac{\partial^2}{\partial t^2} \varrho(x, t), \tag{3.1.3}$$

one of the most important equations of mathematical physics (see Eq. 4.4.1), 20 years after Newton's death.

It was well established, at least by the time of Galileo, that the wavelength λ and frequency f of a pure tone sound wave obey the relationship

$$f\lambda = c_o. \tag{3.1.4}$$

Given what we know today, the general solution to the wave equation may be written in terms of a sum over the complex exponentials, famously credited to Euler, as

$$\varrho(x,t) = Ae^{\jmath 2\pi(ft-x/\lambda)} + Be^{\jmath 2\pi(ft+x/\lambda)}, \tag{3.1.5}$$

where t is time, x is position, and ft and x/λ are dimensionless. This equation describes only the steady-state solution, with no onsets or dispersion. Thus this solution must be generalized to include these important effects.

The basics of sound propagation were within Newton's grasp and were finally published in *Principia* in 1687. The general solution to Newton's wave equation [i.e., $p(x,t) = G(t \pm x/c)$], where G is any function, was first published 60 years later by d'Alembert. Newton's value for the speed of sound in air c_o was incorrect by the thermodynamic constant $\sqrt{\eta_o} = \sqrt{1.4}$, a problem that would take well over a century to rectify, by Laplace in 1816, and experimentally by Rankine in 1850 (Rayleigh 1896, p. 19–23, Vol. II). What was needed was the adiabatic process (the concept of constant-heat energy). For audio frequencies (0.02–20 [kHz]), the temperature gradients cannot diffuse the distance of a wavelength in one cycle, so the heat energy is trapped in the wave (Tisza 1966; Pierce 1981; Boyer and Merzbach 2011).[1] To repair Newton's formula for the sound speed it was necessary to define the adiabatic stiffness of air $\eta_o P_o$, where P_o is 1 [atm] or 10^5 [Pa] (1 [Pa] is 1 [N/m^2]). This required replacing Boyle's law (PV/T = constant) with the adiabatic expansion law (PV^{η_o} = constant). But this fix still ignores the important viscous and thermal losses, as discussed in Appendix D (Kirchhoff 1868; Rayleigh 1896; Mason 1927; Pierce 1981).

Today we know that when ignoring viscous and thermal losses, the speed of sound is given by

$$c_o = \sqrt{\frac{\eta_o P_o}{\rho_o}} = 343 \quad \text{[m/s]},$$

which is a function of the density $\rho_o = 1.12$ [kg/m^3], $P_o = 10^5$ [Pa], and the dynamic stiffness $\eta_o P_o$ of air.[2] The speed of sound stated in other units is 343 [m/s], 1234.8 [km/h], 153.33 [m/s], 15.334 [cm/ms], 1.125 [ft/ms], 1125.3 [ft/s], 4.692 [s/mi], 12.78 [mi/min], 0.213 [mi/s], and 767.27 [mi/h]. A slightly useful approximation is the time between the lightning flash and the thunder ($\approx$5 [sec/mi]).

Newton's success was important because it quantified the physics behind the speed of sound and demonstrated that momentum (mv), not mass m, is transported by the wave. His concept was correct, and his formulation using algebra and calculus represented a milestone in science.

In periodic structures, again the wave number becomes complex due to diffraction, as commonly observed in optics (e.g., diffraction gratings) and acoustics (creeping surface waves). Thus Eq. 3.1.4 holds for only the simplest cases. In general, the complex analytic (thus causal and dispersive) function *propagation vector* $\boldsymbol{\kappa}(\boldsymbol{x}, s)$ must be considered (see Eq. 3.1.6).

[1] There were other physical enigmas, such as the observation that sound disappears in a vacuum, or Pascal's observation that a vacuum cannot draw water up a column by more than 34 ft..

[2] $\eta_o = c_p/c_v = 1.4$ is the ratio of two thermodynamic constants, and $P_o = 10^5$ [Pa] is the barometric pressure of air.

The corresponding discovery of the formula for the speed of light was made nearly two centuries after *Principia* by Maxwell (ca.1861). Maxwell's formulation also required great ingenuity, as it was necessary to hypothesize an experimentally unmeasured term in his equations to get the mathematics to correctly predict the speed of light (and gravity waves). This parallel with the speed of sound is notable.

It is somewhat amazing that to this day we have failed to fully understand gravity significantly better than Newton's theory, although this may too harsh given Einstein's famous work on general relativity in 1915.[3]

Case of dispersive wave propagation: This classic relationship $\lambda f = c$ is deceptively simple, yet confusing, because the *wave number*[4] $k = 2\pi/\lambda$ becomes a complex function of frequency (has both real and imaginary parts) in dispersive media when losses are considered, as discussed in Appendix D (Kirchhoff 1868; Mason 1928).

A second important example is the case of electron waves in silicon crystals, where the wave number $k(f) = 2\pi f/c$ is replaced by the complex analytic function of the Laplace frequency s, the propagation vector $\boldsymbol{\kappa}(s)$. In this case the wave becomes the eigenfunction of the vector (3D) wave equation

$$p^{\pm}(\boldsymbol{x}, t) = P_o(s)e^{st}e^{\pm\boldsymbol{\kappa}(\boldsymbol{x},s)\cdot\boldsymbol{x}}, \tag{3.1.6}$$

where $|\boldsymbol{\kappa}(\boldsymbol{x}, s)|$ is the vector eigenvalue (Brillouin 1953). In these more general cases, $\boldsymbol{\kappa}(\boldsymbol{x}, s)$ must be a vector complex analytic function of the Laplace frequency $s = \sigma + \jmath\omega$, and inverted with the Laplace transform (Brillouin 1960, with help from Sommerfeld). This is because electron "waves" in the dispersive semiconductor (e.g., silicon) are "causally filtered" in three dimensions—in magnitude, phase, and directions $\boldsymbol{x}$. These 3D dispersion relationships are known as *Brillouin zones.*

Silicon is a highly dispersive "wave-filter," forcing the wavelength to be a function of both s and direction. This view is elegantly explained by Brillouin (1953, Chap. 1) in his historic text. Although the most famous examples come from quantum mechanics (Condon and Morse 1929), modern acoustics contains a rich source of related examples (Morse 1948; Beranek 1954; Ramo et al. 1965; Fletcher and Rossing 2008).

3.1.1 The First Algebra

Prior to the invention of algebra, people worked out mathematical problems as sentences, using an obtuse description of the problem (Stillwell 2010, p. 93). Algebra changed this approach and led to a compact language of mathematics, where numbers are represented as symbols (e.g., x and α). The problem to be solved could be formu-

[3] Gravity waves were first observed experimentally while I was writing this chapter.

[4] This term is a misnomer, since the wave number is a complex function of the complex Laplace frequency $s = \sigma + \jmath\omega$, thus not a number in the common sense. Much worse, $\kappa(s) = s/c_o$ must be complex analytic in s, which an even stronger condition. The term *wave number* is so well established, that there is little hope for recovery at this point.

lated in terms of sums of powers of smaller terms, the most common being powers of some independent variable (i.e., time or frequency). If we define $M_N(z) = P_N(z)/a_n$ and $m_k = a_k/a_n$, then

$$M_N(z) \equiv z^N + m_{N-1}z^{N-1} + \cdots + m_0 z^0 = z^N + \sum_{k=0}^{N-1} m_k z^k = \prod_{k=1}^{N}(z - z_k) \tag{3.1.7}$$

is called a *monic polynomial* or simply a monic. The coefficient a_N cannot be zero, or the polynomial would not be of degree N. The resolution is to force $a_N = 1$, since this simplifies the expression and does not change the roots.

The key question is What values of $z = z_k$ result in $M_N(z_k) = 0$? In other words, what are the roots z_k of the polynomial? Answering this question consumed thousands of years, with intense efforts by many aspiring mathematicians. In the earliest attempts, it was a competition to demonstrate mathematical acumen. Results were held as a secret to the death bed. It would be fair to view this effort as an obsession. Today the roots of any polynomial may be found, to high accuracy, by numerical methods. Finding roots is limited by the numerical limits of the representation—namely, by IEEE 754 (see Sect. 2.1.2). There are also a number of important theorems.

Of particular interest is the problem of drawing a circle and a line and finding the intersection (root). There was no solution to this venerable problem using geometry. The resolution is addressed in the solution of Euclid's formula (Problem #2 of Assignment Sect. 3.7.8).

3.1.2 Finding Roots of Polynomials

The problem of factoring polynomials has a history more than a millennium in the making.[5] While the quadratic (degree $N = 2$) was solved by the time of the Babylonians (i.e., the earliest recorded history of mathematics), the cubic solution was finally published by Cardano in 1545. The same year, Cardano's student Ferrari solved the quartic ($N = 4$). In 1826 (281 years later) it was proved that the quintic ($N = 5$) could not be factored by analytic methods.

As a concrete example we begin with the trivial case of the quadratic

$$P_2(s) = as^2 + bs + c. \tag{3.1.8}$$

First note that if $a = 0$, the quadratic reduces to the monomial $P_1(s) = bs + c$. Thus we have the necessary condition that $a \neq 0$. The best way to proceed is to divide a out and work directly with the monic $M_2(s) = \frac{1}{a}P_2(s)$. In this way we do not need to worry about the $a = 0$ exception.

[5] https://www.britannica.com/science/algebra/Fundamental-concepts-of-modern-algebra.

The roots are the values of $s = s_k$ such that $M_2(s_k) = 0$. One of the earliest mathematical results, recorded by the Babylonians in about 2000 BCE, was the factoring of the quadratic by completing the square. We can isolate s by rewriting Eq. 3.1.8 as

$$M_2(s) \equiv \frac{1}{a} P_2(s) = (s + b/2a)^2 - (b/2a)^2 + c/a. \tag{3.1.9}$$

The factorization may be verified by expanding the squared term and canceling $(b/2a)^2$:

$$M_2(s) = [s^2 + (b/a)s + \cancel{(b/2a)^2}] - \cancel{(b/2a)^2} + c/a.$$

Setting Eq. 3.1.9 equal to zero and solving for the two roots $s_\pm$ give the quadratic formula:

$$s_\pm = \left. \frac{-b \pm \sqrt{b^2 - 4ac}}{2a} \right|_{a=1} = -b/2 \pm \sqrt{(b/2)^2 - c}. \tag{3.1.10}$$

Role of the discriminant: Equation 3.1.10 can be further simplified. The term $(b/2)^2 - c > 0$ under the square root is called the *discriminant*. Nominally in physics and engineering problems, the discriminant is negative and $b/2 \ll \sqrt{c}$ may be ignored (the damping is small compared to the resonant frequency), leaving only $-c$ under the radical. Thus the most natural way (i.e., corresponding to the most common physical cases) of writing the roots (Eq. 3.1.10) is[6]

$$s_\pm \approx -b/2 \pm \jmath\sqrt{|c|} = -\sigma_o \pm s_o. \tag{3.1.11}$$

This form separates the real and imaginary parts of the solution in a natural way. The term $\sigma_o = b/2$ is called the damping, which accounts for losses in a resonant circuit; the term $\omega_o = \sqrt{|c|}$, for mechanical, acoustical, and electrical networks, is called the *resonant frequency*. The last approximation ignores the (typically) minor correction $b/2$ to the resonant frequency, which in engineering applications is almost always ignored. Knowing that there is a correction is highlighted by this formula, which makes us aware that the small approximation exists (thus can be ignored).

It is not required that $a, b, c \in \mathbb{R} > 0$, but for physical problems of interest, this is almost always true (>99.99% of the time).

Summary: The quadratic equation and its solution are ubiquitous in physics and engineering. It seems obvious that instead of memorizing the meaningless quadratic formula (Eq. 3.1.10), one should learn the physically meaningful solution (Eq. 3.1.11), obtained via Eq. 3.1.9 with $a = 1$. Arguably, the factored and normalized form (Eq. 3.1.9) is easier to remember as a method (completing the square) rather than as a formula to be memorized.

---

[6]This is the case for mechanical and electrical circuits that have small damping. Physically $b > 0$ is the damping coefficient and $\sqrt{c} > 0$ is the resonant frequency. One may then simplify and factor the form as $s^2 + 2bs + c^2 = (s + b + \jmath c)(s + b - \jmath c)$.

Additionally, the real ($b/2$) and imaginary ($\pm\jmath\sqrt{c}$) parts of the two roots have physical significance as the damping and resonant frequencies. Equation 3.1.10 has none (it is useless).

No insight is gained by memorizing the quadratic formula. To the contrary, an important concept is gained by learning how to complete the square, which is typically easier than identifying a, b, c and blindly substituting them into Eq. 3.1.10. Thus it's worth learning the alternative solution (Eq. 3.1.11), since it is more common in practice and requires less algebra to interpret the final answer.

Exercise #1 By direct substitution, demonstrate that Eq. 3.1.10 is the solution of Eq. 3.1.8. Hint: Work with $M_2(x)$.
Sol: Setting $a = 1$, we can write the quadratic formula as

$$s_\pm = \frac{-b \pm 1\jmath\sqrt{4c - b^2}}{2}.$$

Substituting this into $M_2(s)$ gives

$$\begin{aligned}
M_\pm(s_\pm) &= {s_\pm}^2 + bs_\pm + c = -b/2 \pm \jmath\sqrt{c^2 - (b/2)^2} \\
&= \left(\frac{-b \pm \sqrt{b^2 - 4c}}{2}\right)^2 + b\left(\frac{-b \pm \sqrt{b^2 - 4c}}{2}\right) + c \\
&= \frac{1}{4}\left(\cancel{b^2} \cancel{\mp 2b\sqrt{b^2 - 4c}} + (\cancel{b^2} - \cancel{4c})\right) + \frac{1}{4}\left(\cancel{-2b^2} \cancel{\pm 2b\sqrt{b^2 - 4c}}\right) + \cancel{c} \\
&= 0.
\end{aligned}$$

■

In third grade I learned the times-table trick for 9:

$$9 \cdot n = (n - 1) \cdot 10 + (10 - n).$$

With this simple rule I did not need to depend on my memory for the 9 times table. For example: $9 \cdot 7 = (7 - 1) \cdot 10 + (10 - 7) = 60 + 3$ and $9 \cdot 3 = (3 - 1) \cdot 10 + (10 - 3) = 20 + 7$. By expanding, one can see why it works: $9n = n10 + \cancel{(-10)} + \cancel{(+10)} - n = n(10 - 1)$. Note that the two terms $(n - 1)$ and $(10 - n)$ add to 9.

Learning an algorithm is much more powerful than memorizing the 9 times tables. How you think about a problem can have a great impact on your perception.

Newton's method for finding the roots of $P_N(s)$: Newton is well known for an approximate but efficient method to find the roots of a polynomial. Consider the polynomial s, $P_N(s) \in \mathbb{C}$:

$$P_N(s) = c_N(s - s_o)^N + c_{N-1}(s - s_o)^{N-1} + \cdots + c_1(s - s_o) + c_0, \tag{3.1.12}$$

where we use Taylor's formula (see Sect. 3.2.3) to determine the coefficients

$$c_k = \frac{1}{k!}\frac{d^k}{ds^k}P_N(s)\bigg|_{s=s_o}. \tag{3.1.13}$$

If our initial guess for the root s_1 is close to a root s_o (i.e., $s_1 - s_o$ is within the radius of convergence), then $|(s_1 - s_o)^k| \ll |(s_1 - s_o)|$ for $k \geq 2 \in \mathbb{N}$. Thus we may truncate $P_N(s_1)$ to its linear term c_1:

$$\begin{aligned} P_N(s_1) &\approx (s_1 - s_o)\,\frac{d}{ds}P_N(s)\bigg|_{s_o} + P_N(s_o) \\ &= (s_1 - s_o)P_N'(s_o) + P_N(s_o), \end{aligned}$$

where $P_N'(s_o)$ is shorthand for $dP_N(s_o)/ds$.

Newton's approach (approximation) was to define a recursion such that the next guess s_{n+1} is closer to the root s_o than the previous guess s_n. Replacing s_1 by s_{n+1} and s_o by s_n gives

$$P_N(s_{n+1}) = (s_{n+1} - s_n)P_N'(s_n) + P_N(s_n) \to 0.$$

Here we assume $P_N(s_{n+1}) \to 0$ because $s_{n+1} \to s_o$ as $n \to \infty$.

Solving for s_{n+1}, we get

$$s_{n+1} = s_n - \frac{P_N(s_n)}{P_N'(s_n)}. \tag{3.1.14}$$

Everything on the right is known; thus s_{n+1} should converge to the root s_o as $n \to \infty$.

Note that if s_n is at a root, the numerator term $P_N(s_n) \to 0$, thus and the update goes to zero, and $s_{n+1} \to s_n$. This difference is useful in detecting the stopping condition. Also note that the denominator can never be zero near a root because $P_N'(s)$ cannot share a root with $P_N(s)$.

In practice, it takes only a few steps to approach the root. In experimental trials (see Fig. 3.2) fewer than 10 steps give double-precision floating-point machine accuracy. If any value s_n is close to a root of P_N', the recursion fails, giving a large value for s_{n+1} and forcing the method to restart at s_{n+1}, far from the root. In such cases the solution typically converges to a different root. It should not be difficult to detect these large non-convergent steps by monitoring $|s_{n+1} - s_n|$, which should be monotonically decreasing.

However, if one assumes that the initial guess $s_1 \in \mathbb{R}$ and then evaluates the polynomial using real arithmetic, the estimate $s_{n+1} \in \mathbb{R}$. Thus the iteration will not converge if $s_o \in \mathbb{C}$.

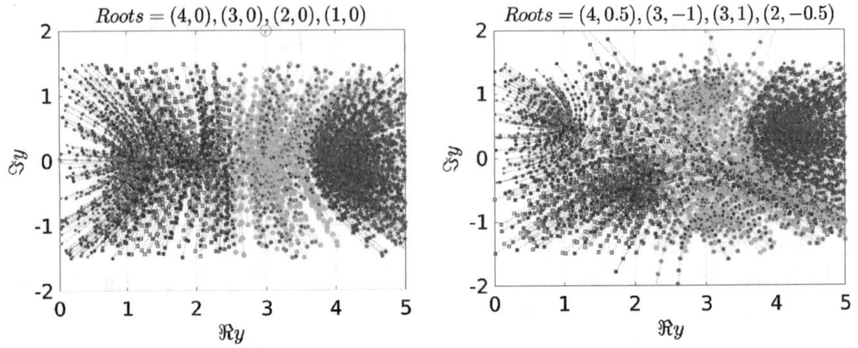

Fig. 3.2 Newton's method applied to the polynomial that has real roots [1, 2, 3, 4] (left) and 5 complex roots (right). A random starting point was chosen, and each curve shows the values of s_n as Newton's method converges to the root. Different random starting points converge to different roots. The method always results in convergence to a root. Claims to the contrary (Stewart 2012, p. 347) are a result of forcing the roots to be real. For convergence, one must assume $s_n \in \mathbb{C}$. For a related discussion see Stillwell (2010, Sect. 14.7)

Root $s_o \in \mathbb{C}$ may be found by a recursion that defines a sequence $s_n \to s_o$, $n \in \mathbb{N}$, such that $P_N(s_n) \to 0$ as $n \to \infty$. As shown in Fig. 3.2, solving for s_{n+1} using Eq. 3.1.14 always gives one of the roots, due to the analytic behavior of the complex *logarithmic derivative* P'_N/P_N.

With every step, s_{n+1} is closer to the root, finally converging to the root in the limit. As it comes closer, the linearity assumption becomes more accurate, resulting in a better approximation and thus a faster convergence.

Equation 3.1.14 depends on the log-derivative $d \log P(x)/dx = P'(x)/P(x)$. It follows that even for cases where fractional derivatives of roots are involved (see Sect. 4.5); Newton's method should converge, since the log-derivative linearizes the equation.[7]

Newton's view: Newton believed that imaginary roots and numbers have no meaning (see p. 179), thus he sought only real roots (Stillwell 2010, p. 119). In this case Newton's relationship may be explored as a graph, which puts Newton's method in the realm of analytic geometry.

Example: Given a polynomial $P_2 = 1 - x^2$ that has roots ± 1, we can use Newton's method to find the roots. Since $P_2'(x) = -2x$, Newton's iteration becomes

$$x_{n+1} = x_n + \frac{1 - x_n^2}{2x_n}.$$

From the Gauss–Lucas theorem (see Sect. 4.5), for the case of $N = 2$, the root of $P_2'(x)$ is always the average of the roots of $P_2(x)$. To start the iteration ($n = 0$) we

[7]This seems like a way to understand fractional, even irrational, roots.

need an initial guess for x_0, which is an initial random guess of where a root might be. The only place we may not start is at the roots of P'_N. For $P_2(x) = 1 - x^2$,

$$x_1 = x_0 + \frac{1 - x_0^2}{2x_0} = x_0 + \frac{1}{2}(-x_0 + 1/x_0).$$

Exercise #2 Let $P_2(x) = 1 - x^2$. Choose the expansion point as $x_0 = 1/2$. Draw a graph describing the first step of the iteration.
Sol: We start with an (x, y) coordinate system and put points at $x_0 = (1/2, 0)$ and the vertex of $P_2(x)$; that is, $(0, 1)$ ($P_2(0) = 1$). Then we draw $1 - x^2$, along with a line from x_0 to x_1. ■

Exercise #3 Calculate x_1 and x_2 of Exercise #2. What root will it converge to? What are the roots of P_2?
Sol: First we must find $P'_2(x) = -2x$. Thus the equation we will iterate is

$$x_{n+1} = x_n + \frac{1 - x_n^2}{2x_n} = \frac{x_n^2 + 1}{2x_n} = (x_n + 1/x_n)/2.$$

By hand,

$$\begin{aligned}
x_0 &= 1/2 \\
x_1 &= \frac{(1/2)^2 + 1}{2(1/2)} = \frac{1}{4} + 1 = 5/4 = 1.25 \\
x_2 &= \frac{(5/4)^2 + 1}{2(5/4)} = \frac{(25/16) + 1}{10/4} = \frac{1}{2}\left(\frac{5}{4} + \frac{4}{5}\right) = \frac{41}{40} = 1.025.
\end{aligned}$$

These estimates rapidly approach the positive-real root $x = 1$. Note that if one starts at the root of $P'(x) = 0$ (i.e., $x_0 = 0$), the first step is indeterminate. ■

Exercise #4 Write an Octave/MATLAB2 script to check your answer for part Exercise #3.
Sol:

```
x=1/2;
for n = 1:3
x = x+(1-x*x)/(2*x);
end
```

■

Exercise #5 For $n = 4$, what is the absolute difference between the root and the estimate, $|x_r - x_4|$?
Sol: 4.6E-8 (very small!) ■

Exercise #6 What happens if $x_0 = -1/2$?
Sol: The solution converges to the negative root, $x = -1$. ■

Exercise #7 Does Newton's method (Kelley 2003) work for $P_2(x) = 1 + x^2$? Hint: What are the roots in this case?
Sol: In this case $P_2'(x) = +2x$; thus the iteration gives

$$x_{n+1} = x_n - \frac{1 + x_n^2}{2x_n}.$$

The roots are purely imaginary, $x_\pm = \pm 1\jmath$. Newton's method works fine as long as you use complex arithmetic. Study Fig. 3.2, and then try Octave/MATLAB to convince yourself. ■

Exercise #8 What if you let $x_0 = 1 + \jmath$ for the case of $P_2(x) = 1 + x^2$?
Sol:

```
x=1+j;
for n = 1:4
x = x-(1+x*x)/(2*x);
end
```

After 4 steps $x_4 = -0.0000046418 + 1.0000021605i$. After 6 steps $x_6 = 8.46e - 23 + i$. On the 7th Step the result is exact.

If you use only real arithmetic, obviously Newton's method fails, because there is no way for the answer to become complex. If, like Newton, you didn't believe in complex numbers, your method would fail to converge to the complex roots (i.e., Real in = Real out). This is because Octave/MATLAB assumes $x \in \mathbb{R}$ if it is initialized as $\mathbb{R}$. By starting with a complex initial value, we fix the Real in = Real out problem. ■

3.1.2.1 Basic Properties of Polynomials

In some sense polynomials such as $P_N(z)$ are the simplest constructions used in algebra, and a summary of their most basic properties is helpful.

1. The degree of a polynomial is n.
2. Polynomials are single-valued; that is, for every z_0, there is precisely one value for $P_N(z_0)$.
3. In mathematical physics and engineering it is common to have real coefficients a_n, but complex coefficients are possible.
4. The coefficients of every polynomial are determined by its Taylor series—namely, Sect. 3.2.3.
5. If the coefficients are real and positive, then the $P_N(x)$ is positive and real if $x \geq 0$.
6. The fundamental theorem of algebra states that $P_N(z)$ has exactly N roots.
7. The number of coefficients of the monomial $M_N(x)$ is equal to N, thus the number of roots.
8. The roots of polynomials with positive and real coefficients typically have complex roots—that is, if $P_N(z_k) = 0$, then $z_k \in \mathbb{C}$.

9. The region of convergence (RoC) of every polynomial about the expansion point is infinite.
10. The roots of the derivative of a polynomial lie within the convex hull defined by the roots of $P_N(z)$, as described by the Gauss–Lucas theorem (see discussion below Eq. 3.1.15).
11. The eigenvalues of the companion matrix are identical to the roots as the monic $M_N(x)$.

Exercise #9 Find the logarithmic derivative of $f(x)g(x)$.
Sol: From the definition of the logarithmic derivative and the chain rule for the differentiation of a product, we have

$$\begin{aligned}\frac{d}{dx}\ln f(x)g(x) &= \frac{d}{dx}\ln f + \frac{d}{dx}\ln g\\ &= \frac{1}{f}\frac{d}{dx}f + \frac{1}{g}\frac{d}{dx}g.\end{aligned}$$

■

Example: If we assume that function $P_3(s) = (s-a)^2/(s-b)^\pi$, then

$$\ln P_3(s) = 2\ln(s-a) - \pi\ln(s-b)$$

and

$$\frac{d}{ds}\ln P_3(s) = \frac{2}{s-a} - \frac{\pi}{s-b}.$$

Reduction by the logarithmic derivative to simple poles: As shown for $P_3(s)$ of the previous example, a function that has poles of arbitrary degree (i.e., π in the example) may be reduced to the sum of two functions having simple poles by taking the logarithmic derivative, since

$$L_N(s) = \frac{N(s)}{D(s)} = \frac{d}{ds}\ln P_N(s) = \frac{P_N'(s)}{P_N(s)}. \tag{3.1.15}$$

Here the polynomial is the denominator $D(s) = P_N(s)$, while the numerator $N(s) = P_N'(s)$ is the derivative of $D(s)$. Thus the logarithmic derivative can play a key role in the analysis of complex analytic functions, as it reduces higher order poles, even those of irrational degree, to simple poles (those of degree 1).

The logarithmic derivative $L_N(s)$ has the following special properties:

1. $L_N(s)$ has simple poles s_p and zeros s_z.
2. The poles of $L_N(s)$ are the zeros of $P_N(s)$.
3. The zeros of $L_N(s)$ (i.e., $P_N'(s_z) = 0$) are the zeros of $P_N'(s)$.
4. $L_N(s)$ is analytic everywhere other than at its poles.
5. Since the zeros of $P_N(s)$ are simple (no second-order poles), the zeros of $L_N(s)$ always lie close to the line connecting the two poles. One may easily demonstrate

the truth of the statement numerically, and it has been quantified by the Gauss–Lucas theorem, which specifies the relationship between the roots of a polynomial and those of its derivative. Specifically, the roots of P'_{N-1} lie inside the convex hull of the roots of P_N.

6. The eigenvalues s_k of the companion matrix are equal to the roots of the monomial $M_N(s)$.
   To understand the meaning of *convex hull*, consider the following construction: If stakes are placed at each of the N roots of $P_N(x)$, and a string is then wrapped around the stakes, with all the stakes inside the string, the convex hull is then the closed set inside the string. One can begin to imagine how the $N-1$ roots of the derivative must evolve with each set inside the convex hull of the previous set. This concept may be recused to smaller values of N.
7. Newton's method may be expressed in terms of the reciprocal of the logarithmic derivative, since

$$s_{k+1} = s_k + \epsilon_o / L_N(s),$$

   where ϵ_o is called the *step size*, which is used to control the rate of convergence of the algorithm. If the step size is too large, the root-finding path may jump to a different domain of convergence and thus a different root of $P_N(s)$.
8. Not surprisingly, given all the special proprieties, $L_N(s)$ plays an key role in mathematical physics.

Euler's product formula: Counting may be written as a linear recursion simply by adding 1 to the previous value, starting from 0. The even numbers may be generated by adding 2, starting from 0. Multiples of 3 may be similarly generated by adding 3 to the previous value, starting from 0. Such recursions are fundamentally related to prime numbers $\pi_k \in \mathbb{P}$, as first investigated by Euler. This logic is the basis of the sieve (see Sect.2.4). The basic idea is both simple and important, taking almost everyone by surprise, likely even Euler. It is related to the old idea that the integers may be generated by the geometric series when viewed as a recursion.
Example: Let's look at counting modulo prime numbers. For example, if $k \in \mathbb{N}$, then

$$k \cdot \text{mod}(k, 2), \quad k \cdot \text{mod}(k, 3), \quad k \cdot \text{mod}(k, 5)$$

are all multiples of the primes $\pi_1 = 2$, $\pi_2 = 3$, and $\pi_3 = 5$.

To see this, we define the *step function* $u_n = 0$ for $n < 0$ and $u_n = 1$ for $n \geq 0$ and the *counting number function* $N_n = 0$ for $n < 0$. The counting numbers may be recursively generated from the recursion

$$N_{n+1} = N_{n-M} + u_n, \tag{3.1.16}$$

which for $M = 1$ gives $N_n = n$. For $M = 2$, $N_n = 0, 2, 4, \ldots$ gives the even numbers.

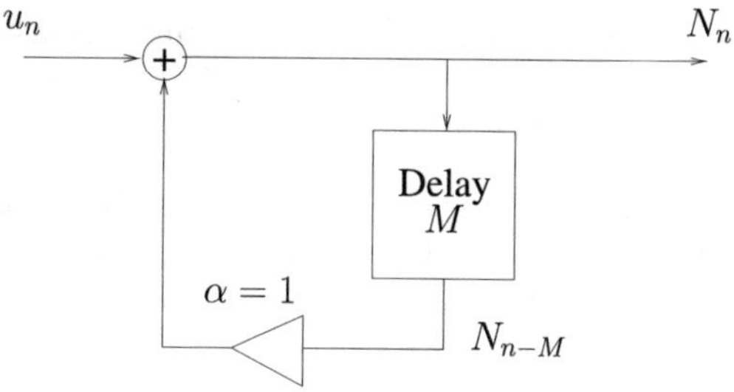

$$N_n = \alpha N_{n-M} + u_n$$

Fig. 3.3 This feedback network describes the linear discrete-time difference equation with delay M [s] given by Eq. 3.1.16. If $M = 1$ this circuit acts as an integrator. When the input is a step function, the output will be $N_n = nu_n = [0, 1, 2, 3, \ldots]$. Such discrete–time circuits are called digital filters

As was first published by Euler in 1737, one may recursively factor out the leading prime term, resulting in Euler's product formula. Based on the argument given in the discussion of the sieve on Sect. 2.2, one may automate the process and create a recursive procedure to identify multiples of the first item on the list, and then remove the multiples of that prime. The lowest number on this list is the next prime. One may then recursively generate all the multiples of this new prime and remove them from the list. Any numbers that remain are candidates for primes.

The observation that this procedure may be automated with a recursive filter, such as that shown in Fig. 3.3, implies that it may be transformed into the frequency domain and described in terms of its poles, which are related to the primes. For example, the poles of the filter shown in Fig. 3.3 may be determined by taking the z-transform of the recursion equation and solving for the roots of the resulting polynomial. The recursion equation is the time-domain equivalent to Riemann's zeta function $\zeta(s)$, which is the frequency-domain equivalent representation.

Exercise #10 Show that $N_n = n$ follows from the above recursion.
Sol: If $n = -1$, we have $N_n = 0$ and $u_n = 0$. For $n = 0$ the recursion gives $N_1 = N_0 + u_0$; thus $N_1 = 0 + 1$. When $n = 1$, we have $N_2 = N_1 + 1 = 1 + 1 = 2$. For $n = 2$, the recursion gives $N_3 = N_2 + 1 = 3$. Continuing the recursion, we find that $N_n = n$. Today we denote a recursions of this form as a *digital filter*. The state diagram for N_n is shown in Fig. 3.3. ∎

To start the recursion, we define $u_n = 0$ for $n < 0$. Thus $u_0 = u_{-1} + 1$. But since $u_{-1} = 0$, $u_0 = 1$. The counting numbers follow from this recursion. A more understandable notation is convolution of the step function with itself—namely,

$$nu_n = u_n \star u_n = \sum_{m=0}^{\infty} u_m u_{m-n} \leftrightarrow \frac{1}{(1-z)^2},$$

which says that the counting numbers $\hat{\mathbf{n}} \in \mathbb{N}$ are easily generated by convolution, which corresponds to a second-order pole at $z = 1$ in the z-transform frequency domain (see Sect. 3.4.1).

Exercise #11 Write an Octave/MATLAB program that generates the odd numbers $N_n = \{1, 0, 3, 0, 5, 0, 7, 0, 9, \ldots\}$ by removing the even numbers.
Sol:

```
M=50; N=(0:M-1);
u=ones(1,M); u(1)=0;
Dem=[1 1]; Num=[1];
n=filter(Num,Dem,u);
y2=n.*N; F1=N-y2
```

which generates: $F1 = [0, 1, 0, 3, 0, 5, 0, 7, 0, 9, 0, \ldots]$. ■

An alternative is to use the `mod(n,N)` function:

```
M=20; n=0:M; k=mod(n,2); m=(k==0).*n;
```

which generates $m = [0, 1, 0, 3, 0, 5, \ldots]$

Exercise #12 Write a program to recursively down-sample N_n by 2:1.
Sol:

```
N=[1 0 3 0 5 0 7 0 9 0 11 0 13 0 15]
M=N(2:2:end);
```

which gives: $M = [1, 3, 5, 7, 9, 11, 13, 15, \ldots]$ ■

For the next step toward a full sieve (Fig. 2.3), we generate all the multiples of 3 (the second prime) and subtract these from the list. This will either zero out these numbers from the list or create negative items, which may then be removed. Numbers are negative when the number has already been removed because it has a second factor of that number. For example, 6 is already removed because it is a multiple of 2 and thus was removed with the multiples of prime number 2.

3.1.3 Matrix Formulation of the Polynomial

There is a simple relationship between every constant coefficient differential equation, its characteristic polynomial, and the equivalent matrix form of that differential equation, defined by the *companion matrix*. The roots of the monic polynomial are the eigenvalues of the companion matrix C_N (Horn and Johnson 1988, p. 147).
The companion matrix: The $N \times N$ *companion matrix* is defined as

$$C_N = \begin{bmatrix} 0 & & & & & -c_0 \\ 1 & 0 & & 0 & & -c_1 \\ 0 & 1 & 0 & & & -c_2 \\ \vdots & 0 & 1 & 0 & \cdots & \vdots \\ & & \cdots & \ddots & 0 & \vdots \\ & 0 & & & 1 \; 0 & -c_{N-2} \\ & & & & 0 \; 1 & -c_{N-1} \end{bmatrix}_{N\times N}. \tag{3.1.17}$$

The constants c_{N-n} are from the monic polynomial of degree N,

$$\begin{aligned} P_N(s) &= s^N + c_{N-1}s^{N-1} + \cdots + c_2 s^2 + c_1 s + c_0 \\ &= s^N + \sum_{n=0}^{N-1} c_n s^n, \end{aligned}$$

which has coefficient vector

$$\mathbf{c}_N = [1, c_{N-1}, c_{N-2}, \ldots, c_1, c_0]^T.$$

Any transformation of a matrix that leaves the eigenvalues invariant (e.g., the transpose) results in an equivalent definition of C_N. Note that the Octave/MATLAB companion matrix function `C=compan(A)` returns the coefficient vector along the top row.

Example: MATLAB/Octave returns the companion matrix in a different format with the coefficients along the top row rather than on the final column. For example if $P_3(s) = [1, c_2, c_1, c_0]$ then Octave/MATLAB returns

$$P_3(s) = \begin{bmatrix} -c_2 & -c_1 & -c_0 \\ 1 & 0 & 0 \\ 0 & 1 & 0 \end{bmatrix}.$$

Exercise #13 Show that the eigenvalues of the 3×3 companion matrix are the same as the roots of $P_3(s)$.

Sol: Expanding the determinant of $C_3 - sI_3$ along the rightmost column, we get

$$P_3(s) = -\begin{vmatrix} -s & 0 & -c_0 \\ 1 & -s & -c_1 \\ 0 & 1 & -(c_2+s) \end{vmatrix} = c_0 + c_1 s + (c_2+s)s^2 = s^3 + c_2 s^2 + c_1 s + c_0.$$

Setting this to zero gives the requested result. ■

Exercise #14 Find the companion matrix for the Fibonacci sequence defined by the recursion (i.e., difference equation)

$$f_{n+1} = f_n + f_{n-1}$$

and initialized with $f_n = 0$ for $n < 0$ and $f_0 = 1$.
Sol: Taking the z-transform gives the polynomial $(z^1 - z^0 - z^{-1})F(z) = 0$, which has the coefficient vector $\mathbf{c} = [1, -1, -1]^T$, resulting in the Fibonacci companion matrix

$$C = \begin{bmatrix} 0 & 1 \\ 1 & 1 \end{bmatrix}.$$

The MATLAB/Octave companion matrix routine `compan(C)` uses an alternative definition that has the same eigenvalues (see Sect. 2.7). ■

Example: A polynomial is represented in MATLAB/Octave in terms of its coefficient vector. When the polynomial vector for the poles of a differential equation is

$$c_N = [1, c_{N-1}, c_{N-2}, \ldots, c_0]^T,$$

the coefficient $c_N = 1$. This normalization guarantees that the leading term is not zero and the number of roots (N) is equal to the degree of the monic polynomial.

3.1.4 Working with Polynomials in MATLAB/Octave

In MATLAB/Octave there are eight related functions you must become familiar with:

1. `R=roots(A)`: Vector $\mathtt{A} = [a_N, a_{N-1}, \ldots, a_0] \in \mathbb{C}$ are the complex coefficients of polynomial $P_N(z) = \sum_{n=0}^{N} a_n z^n \in \mathbb{C}$, where $N \in \mathbb{N}$ is the degree of the polynomial. It is convenient to force $a_N = 1$, corresponding to dividing the polynomial by this value, when it is not 1, thus guaranteeing it cannot be zero. Further, `R` is the vector of roots $[z_1, z_2, \ldots, z_n] \in \mathbb{C}$ such that `polyval(A,`z_k`)=0`.

   Example: `roots([1, -1])=1`
2. `y=polyval(A,x)`: This evaluates the polynomial defined by vector $\mathtt{A} \in \mathbb{C}^N$ evaluated at $\mathtt{x} \in \mathbb{C}$, returning vector $\mathtt{y(x)} \in \mathbb{C}$.

   Example: `polyval([1 -1],1)=0,polyval([1, 1],3)=4`
3. `P=poly()`: This is the inverse of `root()`, returning a vector of polynomial coefficients $\mathtt{P} \in \mathbb{C}^N$ of the corresponding characteristic polynomial, starting from either a vector of roots R or a matrix A, for example, defined with the roots on the diagonal. The characteristic polynomial is defined as the determinant of $|A - \lambda I| = 0$ that has roots R.

   Example: `poly([1])=[1, -1],poly([1,2])=[1,-3,2]`

Due to IEEE 754 scaling issues, this can give strange results that are numerically correct, but only within the limits of IEEE 754 accuracy.

4. `R=polyder(C)`: This routine takes the N coefficients of polynomial C and returns the $N-1$ coefficients of the derivative of the polynomial. This is useful when working with Newton's method, where each step is proportional to $P_N(x)/P'_{N-1}(x)$.

   Example: `polyder([1,1])= [1]`
5. `[K,R]=residue(N,D)`: Given the ratio of two polynomials `N,D`, `residue(N,D)` returns vectors `K,R` such that

$$\frac{N(s)}{D(s)} = \sum_k \frac{K_k}{s - s_k}, \tag{3.1.18}$$

   where $s_k \in \mathbb{C}$ are the roots of the denominator `D` polynomial and `K` $\in \mathbb{C}$ is a vector of residues, which characterize the roots of the numerator polynomial $N(s)$. The use of `residue(N,D)` is discussed on Sect. 4.4. This is one of the most valuable time-saving routines I know.

   Example: `residue(2, [1 0 -1])= [1 -1]`
6. `C=conv(A,B)`: Vector `C` $\in \mathbb{C}^{N+M-1}$ contains the polynomial coefficients of the convolution of the two vectors of coefficients of polynomials `A,B` $\in \mathbb{C}^N$ and `B` $\in \mathbb{C}^M$.

   Example: `[1, 2, 1]=conv([1, 1], [1, 1])`
7. `[C,R]=deconv(N,D)`: Vectors `C, N, D` $\in \mathbb{C}$. This operation uses long division of polynomials to find $C(s) = N(s)/D(s)$ with remainder $R(s)$, where `N = conv(D,C)+R`, which is

$$C = \frac{N}{D} \text{ with remainder } R. \tag{3.1.19}$$

   Example: By defining the coefficients of two polynomials as $A = [1, a_1, a_2, a_3]$ and $B = [1, b_1, b + 2]$, we can find the coefficients of the product from `C=conv(A,B)` and recover B from C with `B=deconv(C,A)`.
8. `A=compan(D)`: Vector $D = [1, d_{N-1}, d_{N-2}, \ldots, d_0]^T \in \mathbb{C}$ contains the coefficients of the monic polynomial

$$D(s) = s^N + \sum_{k=1}^{N} d_{N-k} s^k,$$

   and A is the companion matrix of vector D (Eq. 3.1.17). The eigenvalues of A are the roots of the monic polynomial $D(s)$.

Example: `compan([1 -1 -1])= [1 1; 1 0]`

Exercise #15 Practice the use of MATLAB'S/Octave's related functions that manipulate roots, polynomials, and residues: `root()`, `conv()`, `deconv()`, `poly()`, `polyval()`, `polyder()`, `residue()`, `compan()`.

Sol: We try Newton's method for various polynomials. We use `N=poly(R)` to provide the coefficients of a polynomial given the roots R. Then we use `root()` to factor the resulting polynomial. Finally, we use Newton's method and show that the iteration converges to the nearest root.[8] ■

3.2 Eigenanalysis

At this point we turn a corner in the discussion toward the important topic of eigenanalysis, which starts with the computation of the eigenvalues of a matrix, and their eigenvectors. As briefly discussed on Sect. 2.3, eigenvectors are mathematical generalizations of resonances or modes, naturally found in physical systems.

When you pluck the string of a violin or guitar or hammer a bell or tuning fork, there are natural resonances that occur. These are the eigenmodes of the instrument. The frequency of each mode is related to the eigenvalue, which in physical terms is the frequency of the mode. But this idea goes way beyond simple acoustical instruments. Wave-guides and atoms are resonant systems. The resonances of the hydrogen atom are called the Lyman series, a special case of the Rydberg series and Rydberg atom (Bohr 1954; Gallagher 2005).

Thus this stream runs deep in both physics and eventually mathematics. In some real sense, eigenanalysis was what the Pythagoreans were seeking to understand. This relationship is rarely spoken about in the literature, but once you see it, it can never be forgotten, as it colors your entire view of all aspects of modern physics.

3.2.1 Eigenvalues of a Matrix

The method for finding eigenvalues is best described with an example.[9] Starting from the matrix Eq. 2.6.15), the eigenvalues are defined by the eigenmatrix equation

$$\frac{1}{2}\begin{bmatrix}1 & 1\\ 2 & 0\end{bmatrix}\begin{bmatrix}e_1\\ e_2\end{bmatrix} = \lambda\begin{bmatrix}e_1\\ e_2\end{bmatrix}. \tag{3.2.1}$$

The unknowns here are the eigenvalue λ and the eigenvector $\boldsymbol{e} = [e_1, e_2]^T$. First we find λ by subtracting the right from the left:

[8] A MATLAB/Octave program that does this may be downloaded from https://jontalle.web.engr.illinois.edu/uploads/493/M/NewtonJPD.m.

[9] Appendix B) is an introduction to the topic of eigenanalysis for 2×2 matrices.

$$\frac{1}{2}\begin{bmatrix}1 & 1\\ 2 & 0\end{bmatrix}\begin{bmatrix}e_1\\ e_2\end{bmatrix} - \lambda\begin{bmatrix}e_1\\ e_2\end{bmatrix} = \frac{1}{2}\begin{bmatrix}1-2\lambda & 1\\ 2 & -2\lambda\end{bmatrix}\begin{bmatrix}e_1\\ e_2\end{bmatrix} = 0.$$

The only way this equation for $\boldsymbol{e}$ can have a non-trivial ($e_1 = e_2 = 0$) solution is if the matrix is singular. If it is singular, the determinant of the matrix is zero.
Example: The determinant in the above equation is the product of the diagonal elements minus the product of the off-diagonal elements, which results in the quadratic equation

$$-2\lambda(1-2\lambda) - 2 = 4\lambda^2 - 2\lambda - 2 = 0.$$

Completing the square gives

$$(\lambda - 1/4)^2 - (1/4)^2 - 1/2 = 0;$$

thus the roots (i.e., eigenvalues) are $\lambda_\pm = \frac{1\pm 3}{4} = \{1, -1/2\}$.

Exercise #16 Expand Eq. 3.2.1 and recover the quadratic equation.
Sol:

$$(\lambda - 1/4)^2 - (1/4)^2 - 1/2 = \lambda^2 - \lambda/2 + \cancel{(1/4)^2 - (1/4)^2} - 1/2 = 0.$$

Thus completing the square is the same as the original equation. ■

Exercise #17 Find the eigenvalues of the matrix of Eq. 2.6.9.

Sol: This is a minor variation on the previous example. Briefly, we have

$$\det\begin{bmatrix}1-\lambda & N\\ 1 & 1-\lambda\end{bmatrix} = (1-\lambda)^2 - N = 0.$$

Thus $\lambda_\pm = 1 \pm \sqrt{N}$. ■

Exercise #18 Starting with Eq. 3.2.1 and initial conditions $[x_1, y_1]^T = [1, 0]^T$, compute the first five values of $[x_n, y_n]^T$.
Sol: Here is a MATLAB/Octave code for computing $[x_n, y_n]^T$:

```
x=[1;0];
A=[1 1;2 0]/2;
for k=1:10; x(k+1)=A*x(:,k); end
```

which gives the rational ($x_n \in \mathbb{Q}$) sequence: $1, 1/2, 3/4, 5/8, 11/2^4, 21/2^5, 43/2^6, 85/2^7, 171/2^8, 341/2^9, 683/2^{10}, \ldots$. ■

Exercise #19 Show that the solution to the mean-Fibonacci sequence (Eq. 2.6.14) is bounded, unlike that of the Fibonacci sequence. Explain what is going on.

Sol: Because the next value is the mean of the last two, the sequence is bounded. To see this one needs to compute the eigenvalues of the matrix in Eq. 2.6.15). ■

The key to the analysis of such equations is called eigenanalysis or the modal-analysis method (see Appendix B). The eigenvalues (eigenfrequencies) are also known as resonant frequencies in engineering and eigenmodes in physics. Eigenmodes describe the naturally occurring "ringing" found in physical wave-dominated boundary value problems and in resonant circuits. Each mode's eigenvalue quantifies the mode's natural complex frequency $s_k = \sigma_k + \omega_k \jmath$.

Complex eigenvalues result in damped modes having frequencies $s_k \in \mathbb{C}$, which decay in time as $\tau_k = 1/\sigma_k$ due to energy losses, as determined by σ_k.

Two modes that have exactly the same frequency are said to be degenerate. This is a very special condition representing a very high degree of symmetry. When two modes are slightly different in frequency, one hears a beating of the modes at the difference frequency (they are not degenerate). If they have different decay, the beats will die away with the shorter of the two time constants.

Common examples include tuning forks, pendulums, bells, and the strings of musical instruments (such as guitar and fiddles), all of which (except for tuning forks) have hundreds of modes (Fletcher and Rossing 2008; Morse 1948). For those interested in musical acoustics, these books are excellent.

3.2.2 *Cauchy's Theorem and Eigenmodes*

Cauchy's residue theorem (see Sect. 4.4) is used to find the time-domain response of each frequency-domain complex eigenmode. Thus eigenanalysis and eigenmodes of physics are the same thing (see Sect. 4.4) but are described using different notional methods.[10] The eigenanalysis method is summarized in Appendix B.3.

Taking a simple example of a 2×2 matrix $\mathcal{T} \in \mathbb{C}$, we start from the definition of the two eigenequations

$$\mathcal{T}\boldsymbol{e}_\pm = \lambda_\pm \boldsymbol{e}_\pm \tag{3.2.2}$$

corresponding to two eigenvalues $\lambda_\pm \in \mathbb{C}$ and two 2×1 eigenvectors $\boldsymbol{e}_\pm \in \mathbb{C}$.

Example: Assume that $\mathcal{T}$ is the Fibonacci matrix in Eq. 2.6.12.

The eigenvalues $\lambda_\pm$ may be merged into a 2×2 diagonal eigenvalue matrix

$$\Lambda = \begin{bmatrix} \lambda_+ & 0 \\ 0 & \lambda_- \end{bmatrix},$$

while the two eigenvectors $\boldsymbol{e}_+$ and $\boldsymbol{e}_-$ are merged into a 2×2 eigenvector matrix

$$\boldsymbol{E} = [\boldsymbol{e}_+, \boldsymbol{e}_-] = \begin{bmatrix} e_1^+ & e_1^- \\ e_2^+ & e_2^- \end{bmatrix}, \tag{3.2.3}$$

[10] During the discovery or creation of quantum mechanics, two alternatives were developed: Schrödinger's differential equation method and Heisenberg's matrix method. Eventually it was realized the two were equivalent.

corresponding to the two eigenvalues. Using matrix notation, we can write this compactly as

$$\mathcal{T}\boldsymbol{E} = \boldsymbol{E}\Lambda. \tag{3.2.4}$$

Note that while $\lambda_\pm$ and $\boldsymbol{E}_\pm$ commute, $\boldsymbol{E}\Lambda \neq \Lambda\boldsymbol{E}$.

From Eq. 3.2.4 we may obtain two very important relations:

1. the diagonalization of $\mathcal{T}$

$$\Lambda = \boldsymbol{E}^{-1}\mathcal{T}\boldsymbol{E}, \tag{3.2.5}$$

and

2. the eigenexpansion of $\mathcal{T}$

$$\mathcal{T} = \boldsymbol{E}\Lambda\boldsymbol{E}^{-1}, \tag{3.2.6}$$

which is used for computing powers of $\mathcal{T}$ (i.e., $\mathcal{T}^{100} = \boldsymbol{E}\Lambda^{100}\boldsymbol{E}^{-1}$).

Example: If we take

$$\mathcal{T} = \begin{bmatrix} 1 & 1 \\ 1 & -1 \end{bmatrix},$$

then the eigenvalues are given by $(1 - \lambda_\pm)(1 + \lambda_\pm) = -1$; thus $\lambda_\pm = \pm\sqrt{2}$. This method of eigenanalysis is discussed on Sect. 3.7.8 and in Appendix B.2.

Exercise #20 Show that the geometric series formula holds for 2×2 matrices. Starting with the 2×2 identity matrix I_2 and $a \in \mathbb{C}$, with $|a| < 1$, show that

$$I_2(I_2 - aI_2)^{-1} = I_2 + aI_2 + a^2I_2^2 + a^3I_2^3 + \cdots.$$

Sol: Multiply both sides by $I_2 - aI_2^k$ results in an identity

$$\begin{aligned} I_2 &= I_2 + aI_2 + a^2I_2^2 + a^3I_2^3 + \cdots - aI_2(aI_2 + a^2I_2^2 + a^3I_2^3 + \cdots) \\ &= [1 + (a + a^2 + a^3 + \cdots) - (a + a^2 + a^3 + a^4 + \cdots)]I_2 \\ &= I_2. \end{aligned}$$

This equality requires that the two series converge, but only if $|a| < 1$. ■

When the matrix $\mathcal{T}$ is not a square matrix, Eq. 3.2.6 may be generalized as

$$\mathcal{T}_{m,n} = \mathbf{U}_{m,m}\Lambda_{m,n}\mathbf{V}^\dagger_{n,n}.$$

This useful generalization of eigenanalysis is called singlar value decomposition (SVD). To see this use the MATLAB/Octave command `[U,L,V]=svd(A)` where A is a rectangular (non-square) matrix.

Exercise #21 Verify that $\Lambda = \boldsymbol{E}^{-1}\boldsymbol{A}\boldsymbol{E}$.

Sol: We shall work with the unnormalized eigenmatrix $c\boldsymbol{E}$, where $c = \sqrt{\sqrt{2}^2 + 1} = \sqrt{3}$. To compute the inverse of $c\boldsymbol{E}$, 1) swap the diagonal values, 2) change the sign of the off diagonals, and 3) divide by the determinant Δ:

$$(c\boldsymbol{E})^{-1} = \frac{1}{2c\sqrt{2}}\begin{bmatrix} 1 & \sqrt{2} \\ -1 & \sqrt{2} \end{bmatrix} = \frac{1}{2c}\begin{bmatrix} 0.707 & 1 \\ -0.707 & 1 \end{bmatrix}.$$

We wish to show that $\Lambda = \boldsymbol{E}^{-1}\boldsymbol{A}\boldsymbol{E}$

$$\frac{1}{2c}\begin{bmatrix} 0.707 & 1 \\ -0.707 & 1 \end{bmatrix}\begin{bmatrix} 1 & 2 \\ 1 & 1 \end{bmatrix} c \begin{bmatrix} \sqrt{2} & -\sqrt{2} \\ 1 & 1 \end{bmatrix} = \begin{bmatrix} 1+\sqrt{2} & 0 \\ 0 & 1-\sqrt{2} \end{bmatrix}$$

which is best verified with MATLAB/Octave. ■

Exercise #22 Verify that $\boldsymbol{A} = \boldsymbol{E}\Lambda\boldsymbol{E}^{-1}$.
Sol: We wish to show that

$$\begin{bmatrix} 1 & 2 \\ 1 & 1 \end{bmatrix} = \frac{1}{\sqrt{3}}\begin{bmatrix} \sqrt{2} & -\sqrt{2} \\ 1 & 1 \end{bmatrix} \cdot \begin{bmatrix} 1+\sqrt{2} & 0 \\ 0 & 1-\sqrt{2} \end{bmatrix} \cdot \frac{\sqrt{3}}{2\sqrt{2}}\begin{bmatrix} 1 & \sqrt{2} \\ -1 & \sqrt{2} \end{bmatrix}.$$

All the above solutions have been verified with Octave.

Eigenmatrix diagonalization is helpful in generating solutions for finding the solutions of Pell's and Fibonacci's equations using transmission matrices. ■

Example: If the matrix corresponds to a transmission line, the eigenvalues have units of seconds [s]

$$\begin{bmatrix} V^+ \\ V^- \end{bmatrix}_n = \begin{bmatrix} e^{-sT_o} & 0 \\ 0 & e^{sT_o} \end{bmatrix}\begin{bmatrix} V^+ \\ V^- \end{bmatrix}_{n+1}. \tag{3.2.7}$$

In the time domain the forward traveling wave $v^+_{n+1}(t-(n+1)T_o) = v^+_n(t-nT_o)$ is delayed by T_o. Two applications of the matrix delay the signal by $2T_o$.

Summary: The GCD (Euclidean algorithm), Pell's equation, and the Fibonacci sequence may all be written as compositions of 2×2 matrices. Thus Pell's equation and the Fibonacci sequence are special cases of the 2×2 matrix composition

$$\begin{bmatrix} x \\ y \end{bmatrix}_{n+1} = \begin{bmatrix} a & b \\ c & d \end{bmatrix}\begin{bmatrix} x \\ y \end{bmatrix}_n.$$

This is an important and common thread of these early mathematical findings. This 2×2 linearized matrix recursion plays a special role in physics, mathematics, and engineering because one-dimensional system equations are solved using the 2×2 eigenanalysis method. More than several thousand years of mathematical trial and error set the stage for this breakthrough. But it took even longer to be fully appreciated.

The key idea of the 2×2 matrix solution, widely used in modern engineering, can be traced back to Brahmagupta's solution of Pell's equation for arbitrary N. Brahmagupta's recursion, identical to that of the Pythagoreans' $N = 2$ case (see Eq. 2.6.9), eventually led to the concept of linear algebra, defined by the simultaneous solutions of many linear equations. The recursion by the Pythagoreans (sixth century BCE) predated the creation of algebra by al-Khwarizmi (ninth century CE), as seen in Fig. 1.1.

3.2.3 Taylor Series

An analytic function is one that meets these criteria:

1. It may be expanded in a Taylor series:

$$P(x) = \sum_{n=0}^{\infty} c_n (x - x_o)^n. \tag{3.2.8}$$

2. It converges for $|x - x_o| < 1$, called the region of convergence (RoC), with coefficients c_n.
3. The Taylor series coefficients c_n are defined by taking derivatives of $P(x)$ and evaluating them at the expansion point x_o—namely,

$$c_n = \frac{1}{n!} \frac{d^n}{dx^n} P(x)\bigg|_{x=x_o}. \tag{3.2.9}$$

4. Although $P(x)$ may be multivalued, the Taylor series is always single-valued.

Exercise #23 Verify that c_0 and c_1 of Eq. 3.2.8 follow from Eq. 3.2.9.
Sol: To obtain c_0, for $n = 0$, there is no derivative (d^0/dx^0 indicates no derivative is taken), so we must simply evaluate $P(x - x_o) = c_0 + c_1(x - x_o) + \cdots$ at $x = x_o$, leaving c_0. To find c_1, we take one derivative, which results in $P'(x) = c_1 + 2c_2(x - x_o)) + \cdots$. Evaluating this at $x = x_o$ leaves c_1. Each time we take a derivative we reduce the degree of the series by 1, leaving the next constant term. ■

Exercise #24 Suppose we truncate the Taylor series expansion to N terms. What is the name of such functions?
Sol: When an infinite series is truncated, the resulting function is an Nth-degree polynomial:

$$P_N(x) = \sum_{n=0}^{N} = c_0 + c_1(x - x_o) + c_2(x - x_o)^2 + \cdots + c_N(x - x_o)^N.$$

We can find c_0 by evaluating $P_N(x)$ at the expansion point x_o, since from the above formula $P_N(x_o) = c_0$. From the Taylor formula, $c_1 = P'_N(x)\big|_{x_o}$. ■

Exercise #25 How many roots do $P_N(x)$ and $P'_N(x)$ have?
Sol: According to the fundamental theorem of algebra, $P_N(x)$ has N roots and $P'_N(x)$ has $N-1$ roots. The Gauss–Lucas theorem states that the $N-1$ roots of $P'_N(x)$ lie inside the convex hull of the N roots of $P_N(x)$ (see Sect. 3.1.3). ■

Exercise #26 Would it be possible to find an inverse Gauss–Lucas theorem, that states where the roots of the integral of a polynomial might be?
Sol: To the best of my knowledge this problem has not been addressed. With each integral there is a new degree of freedom that must be accommodated, expanding the convex hull. ■

Properties: The Taylor formula is a prescription for how to uniquely define the coefficients c_n. Without the Taylor series formula, we would have no way of determining c_n. The proof of the Taylor formula is transparent; The coefficients may be determined by simply taking successive derivatives of Eq. 3.2.8 and then evaluating the result at the expansion point. If $P(x)$ is analytic, then this procedure will always be successful. If $P(x)$ fails to have a derivative of any order, then the function is not analytic and Eq. 3.2.8 is not valid.

The Taylor series representation of $P(x)$ has special applications for solving differential equations for these reasons:

1. It is single-valued.
2. The series if valid inside the RoC (an open set).
3. All its derivatives and integrals are uniquely defined.
4. It may be continued into the complex plane by extending $x \in \mathbb{C}$. This extension is necessary because the eigenvalues are typically in $\mathbb{C}$. In fact the only reasonable eigenvalues must be complex, having negative real parts. If the real part of λ_k is zero, the solution is lossless, thus never dies away, which is non-physical in the macroscopic world.[11] If it is positive, the solution is unstable (blows up). Typically this involves expanding the series about a different expansion point.

Analytic continuation: A limitation of the Taylor series expansion is that it is not valid outside of its RoC. One method for working with this limitation is to move the expansion point. This is called analytic continuation. However, analytic continuation is a non-trivial operation because: (1) It requires manipulating an infinite number of derivatives of $P(x)$, (2) at the new expansion point x_o, where (3) $P(x - x_o)$ may not have derivatives, due to possible singularities. (4) Thus one needs to know where the singularities of $P(s)$ are in the complex s plane. Due to these many problems analytic continuation is rarely used, other than as an important theoretical concept.

Every Taylor series is a single-valued representation because powers of the variable are single-valued. Single valuedness is key feature to the series representation. However functions have regions where the series is not valid. This is best seen with a simple example using the geometric series

$$f(s) = \frac{1}{1-as} = \sum_{n=0}^{\infty} (as)^n, \quad |as| < 1$$

[11] Quantum eigenstates are lossless.

with $a, s \in \mathbb{C}$. Note that $f(s)$ has a pole at $s_o = 1/a$ and residue $1/a$.

However $f(s)$ is perfectly well defined for $|as| > 1$, for which it has a different series expansion. If we let $s = 1/z$ we find

$$f(z) = \frac{1}{1 - a/z} = \frac{-z/a}{1 - z/a} = -\frac{z}{a}\sum_{n=0}^{\infty}\left(\frac{z}{a}\right)^n, \quad |z/a| < 1.$$

Expressed in terms of $z = 1/s$ we have that $|sa| > 1$.

Thus the first expansion is good inside the circle $|s| < 1/a$ while the second is valid outside the circle. While each series is single valued within its RoC, $f(s)$ is valid everywhere, except at the pole $s = 1/a$, where it is singular.

Example: The similar case is the geometric series $P(x) = 1/(\jmath - x)$ about the expansion point $x = 1$. The function $P(x)$ is defined everywhere, except at the singular point $x = \jmath$, whereas the geometric series is valid for $|x| < 1$. However $P(x)$ is valid for $|x| > 1$. For example $P(10) = 1/(\jmath - 10) = (\jmath + 10)/(\jmath + 10)(\jmath - 10) = -(\jmath + 10)/101$.

Role of the Taylor series: The Taylor series plays a key role in the mathematics of differential equations and their solution, as the coefficients of the series uniquely determine the analytic series representation via its derivatives. The implications and limitations of the power series representation are very specific: If the series fails to converge (i.e., outside the RoC), it is meaningless.

Every differential equation has as many independent solutions as it has eigenvalues. To obtain these solutions we must use the Taylor series, with its single-value property, to uniquely represent each solution. The general solution is then the weighted sum over the independent solutions. This theory trivially follows from the Cauchy residue theorem CT-3 (Eq. 4.5.3).

Starting from a differential equation, it may be transformed to a matrix equation using the companion matrix (Sect. 3.1.3) having K eigenvalues $\lambda_1, \cdots, \lambda_k, \cdots, \lambda_K$, with a general solution

$$f(t; C_k) = \sum_{k=1}^{K} C_k e^{s_k t}.$$

The constants $C_k \in \mathbb{C}$ are determined using the initial conditions.

A very important fact about the RoC: It is relevant to only the series, not the function being expanded. Typically the function has a pole at the radius of the RoC, beyond which the series fails to converge. However, the function being expanded is valid everywhere (other than at its poles). This point has been inadequately explained in many text books. In addition, the RoC is the region of divergence (RoD), which is the RoC's complement.

The Taylor series does not need to be infinite to converge to the function it represents, since it obviously works for any polynomial $P_N(x)$ of degree N. But in the finite case ($N < \infty$), the RoC is infinite and the series is the function $P_N(x)$ exactly, everywhere. Of course, $P_N(x)$ is a polynomial of degree N. When $N \to \infty$, the

Taylor series is valid only within the RoC, and it is (typically) the representation of the reciprocal of a polynomial.

These properties are both the curse and the blessing of the analytic function. On the positive side, analytic functions are the ideal starting point for solving differential equations, which is exactly how they were used by Newton and others. Analytic functions are "smooth," since they are infinitely differentiable, with coefficients given by Eq. 3.2.9. They are single-valued, so there can be no ambiguity in their interpretation. On the negative side, they only represent the function within the RoC, which depends on the expansion point.

Two well-known analytic functions are the geometric series ($|x| < 1$)

$$\frac{1}{1-x} = 1 + x + x^2 + x^3 + \cdots = \sum_{n=0}^{\infty} x^n \tag{3.2.10}$$

and the exponential series ($|x| < \infty$)

$$e^x = 1 + x + \frac{1}{2}x^2 + \frac{1}{3 \cdot 2}x^3 + \frac{1}{4 \cdot 3 \cdot 2}x^4 + \cdots = \sum_{n=0}^{\infty} \frac{1}{n!}x^n. \tag{3.2.11}$$

Exercise #27 Provide the Taylor series expression for the following functions:

$$F_1(x) = \int^x \frac{1}{1-x} dx \tag{3.2.12}$$

Sol: $F_1(x) = x + \frac{1}{2}x^2 + \frac{1}{3}x^3 + \cdots$ ■

$$F_2(x) = \frac{d}{dx}\frac{1}{1-x} \tag{3.2.13}$$

Sol: $F_2(x) = 1 + 2x + 3x^2 + \cdots$ ■

$$F_3(x) = \ln \frac{1}{1-x} \tag{3.2.14}$$

Sol: $F_3(x) = 1 + \frac{1}{2}x + \frac{1}{3}x^2 + \cdots$ ■

$$F_4(x) = \frac{d}{dx} \ln \frac{1}{1-x} \tag{3.2.15}$$

Sol: $F_4(x) = 1 + x + x^2 + x^3 + \cdots$ ■

Exercise #28 Using symbolic manipulation (MATLAB, Octave, Mathematica), expand the function $F(s)$ in a Taylor series and find the recurrence relationships among the Taylor coefficients c_n, c_{n-1}, c_{n-2}. Assume $a \in \mathbb{C}$ and $T \in \mathbb{R}$.

$$F(s) = e^{as}$$

Sol: A Google search on *octave syms taylor* is useful. The MATLAB/Octave code to expand this in a Taylor series is

```
syms s
taylor(exp(s),s,0,'order',10)
```

■

Exercise #29 Find the coefficients of the following functions by the method of Eq. 3.2.9 and give the RoC.

1. $w(x) = \frac{1}{1-x\jmath}$.

   Sol: From a straightforward expansion we know the coefficients are

$$\frac{1}{1-x\jmath} = 1 + x\jmath + (x\jmath)^2 + (x\jmath)^3 + \cdots = 1 + x\jmath - x^2 + -\jmath x^3 + \cdots.$$

   Working this out using Eq. 3.2.9 is more work:

$$c_0 = \frac{1}{0!}w\Big|_0 = 1; c_1 = \frac{1}{1!}\frac{dw}{dx}\Big|_0 = -\frac{-\jmath}{(1-x\jmath)^2}\Big|_{x=0} = \jmath; \quad c_2 = \frac{1}{2!}\frac{d^2w}{dx^2}\Big|_0 = \frac{1}{2!}\frac{-2}{(1-x\jmath)^3}\Big|_0 = -1;$$

$$c_3 = \frac{1}{3!}\frac{d^3w}{dx^3}\Big|_0 = \frac{-\jmath}{(1-x\jmath)^4}\Big|_0 = -j.$$

   However, if we take derivatives of the series expansion, it is much easier and we can even figure out the term for c_n:

$$c_0 = 1; c_1 = \frac{d}{dx}\sum(\jmath x)^n\Big|_0 = j; c_2 = \frac{1}{2!}\frac{d^2}{dx^2}\sum(jx)^n\Big|_0 = 2(\jmath)^2;$$
$$c_3 = \frac{1}{3!}\frac{d^3}{dx^3}\sum(jx)^n\Big|_0 = (\jmath)^3 = -\jmath;$$
$$\cdots, c_n = \frac{1}{n!}\jmath^n n! = \jmath^n.$$

   The RoC is $|xj| = |x| < 1$. ■

2. $w(x) = e^{x\jmath}$.

   Sol: $c_n = \frac{1}{n!}\jmath^n$. The RoC is $|x| < \infty$. Functions with an RoC of ∞ are called *entire*. Thus $c_n = \jmath c_{n-1}/n$.

   ■

Exercise #30 Show that $Z(s) = 1/\sqrt{s}$ is positive-real but not a Brune impedance.
Sol: Since it may not be written as the ratio of two polynomials, it is not in the Brune impedance class. If we write $Z(s) = |Z(s)|e^{\phi\jmath}$ in polar coordinates, since $-\pi/4 \le \phi \le \pi/4$ when $|\angle s| < \pi/2$, $Z(s)$ satisfies the Brune condition and thus is positive-real. ■

Determining the region of convergence (RoC): Determining the RoC for a given analytic function is quite important and may not always be obvious. In general the RoC is a circle whose radius extends from the expansion point out to the nearest pole. Thus when the expansion point is moved, the RoC changes, since the location of the pole is fixed.

Example: For the geometric series (Eq. 3.2.10), the expansion point is $x_o = 0$ and the RoC is $|x| < 1$, since $1/(1-x)$ has a pole at $x = 1$. We may move the expansion point by a linear transformation—for example, by replacing x with $z+3$. Then the series becomes $1/((z+3)-1) = 1/(z+2)$, so the RoC becomes 3 because in the z plane the pole has moved to -2.

Example: A second important example is the function $1/(x^2+1)$, which has the same RoC as the geometric series, since it may be expressed in terms of its residue expansion (also called its partial fraction expansion)

$$\frac{1}{x^2+1} = \frac{1}{(x+1\jmath)(x-1\jmath)} = \frac{1}{2\jmath}\left(\frac{1}{x-1\jmath} - \frac{1}{x+1\jmath}\right).$$

Each term has an RoC of $|x| < |1\jmath| = 1$. The amplitude of each pole is called the *residue*, defined in Eq. 4.5.4. The residue for the pole at $1\jmath$ is $1/2\jmath$.

The roots must be found by factoring the polynomial (e.g., Newton's method). Once the roots are known, the residues are best found via complex linear algebra.

In summary, the function $1/(x^2+1)$ is the sum of two geometric series, with poles at $\pm 1\jmath$, which is not initially obvious because the roots are complex and conjugate. Only when the function is factored does it become clear what is going on.

Exercise #31 Verify that the above expression is correct and show that the residues are $\pm 1/2\jmath$.

Sol: We cross-multiply and cancel, leaving 1, as required. The RoC is the coefficient on the pole. Thus the residue of the pole at $x\jmath$ is $\jmath/2$. ■

Exercise #32 Find the residue of $\frac{d}{dz}z^\pi$.

Sol: Taking the derivative gives $\pi z^{\pi-1}$, which has a pole at $z = 0$. Applying the formula for the residue (Eq. 4.5.4), we find

$$c^{-1} = \pi \lim_{z\to 0} z z^{\pi-1} = \pi \lim_{z\to 0} z^\pi = 0.$$

Thus the residue is zero. ■

3.2.4 Analytic Functions

Any function that has a Taylor series expansion is called an *analytic function*. Within the RoC, the series expansion defines a single-valued function. Polynomials $1/(1-x)$ and e^x are examples of analytic functions that are real functions of their real argument x.

Every analytic function has a corresponding differential equation, which is determined by the coefficients a_k of the analytic power series. An example is the exponential, which has the property that it is the eigenfunction of the derivative operation

$$\frac{d}{dx}e^{ax} = ae^{ax},$$

which may be verified using Eq. 3.2.11. This relationship is a common definition of the exponential function, which is special because it is the eigenfunction of the derivative.

The complex analytic power series (i.e., complex analytic functions) may also be integrated term by term, since

$$\int^x f(x)dx = \sum \frac{a_k}{k+1}x^{k+1}. \qquad (3.2.16)$$

Newton took full advantage of this property of the analytic function and used the analytic series (Taylor series) to solve analytic problems, especially for working out integrals. This enabled him to solve differential equations. To fully understand the theory of differential equations, one must master single-valued analytic functions and their analytic power series.

Single- versus multivalued functions: Polynomials and their ∞-degree extensions (analytic functions) are single-valued: For each x there is a single value for $P_N(x)$. The roles of the domain and codomain may be swapped to obtain an inverse function with properties that can be very different from those of the function. For example, $y(x) = x^2 + 1$ has the inverse $x = \pm\sqrt{y-1}$, which is double-valued and complex when $y < 1$. Periodic functions such as $y(x) = \sin(x)$ are even more "exotic," since $x(y) = \arcsin(x) = \sin^{-1}(x)$ has an infinite number of $x(y)$ values for each y. This problem was first addressed in Bernhard Riemann's 1851 Ph.D. thesis, written while he was working with Gauss.

Exercise #33 Let $y(x) = \sin(x)$. Then $dy/dx = \cos(x)$. Show that $dx/dy = \pm 1/\sqrt{1-y^2}$.

Sol: Since $\sin^2 x + \cos^2 x = 1$, it follows that $y^2(x) + (dy/dx)^2 = 1$. Thus $dy/dx = \pm\sqrt{1-y^2}$. Taking the reciprocal gives the result.

To fully understand this, Google "implicit function theorem" (D'Angelo 2017, p. 104). ■

Exercise #34 Evaluate the integral

$$I(y) = \int^y \frac{dy}{\sqrt{1-y^2}}.$$

Sol: From the previous Exercise we know that

$$x(y) = \int^x dx = \int^y \frac{dy}{\sqrt{1-y^2}}.$$

But since $y(x) = \sin(x)$, it follows that $x(y) = \sin^{-1} y = \arcsin(y)$. ■

Exercise #35 Find the Taylor series coefficients of $y = \sin(x)$ and $x = \sin^{-1}(y)$. Note that $\log e^s = s$ and

$$\sin(\sin^{-1}(s)) = \sin^{-1}(\sin(s)) = 1.$$

Hint: Use symbolic Octave. Note $\sin^{-1}(y) = \arcsin(y)$.
Sol: `syms s;taylor(sin(s),'order',10);`

$$\sin(s) = s - s^3/3! + s^5/5! - s^7/7! + \cdots$$

and `syms s;taylor(asin(s),'order',15);`

$$\begin{aligned}\arcsin(s) &= s + \frac{1}{6}s^2 + \frac{3}{40}s^5 + \frac{5}{112}s^7 + \frac{35}{1152}s^9 + \frac{63}{2816}s^{11} + \frac{231}{13312}s^{13} + \cdots \\ &= s + \frac{1}{3 \cdot 2^1}s^3 + \frac{3}{5 \cdot 2^3}s^5 + \frac{5}{7 \cdot 2^4}s^7 + \frac{7 \cdot 5}{9 \cdot 2^7}s^9 + \frac{7 \cdot 3^2}{11 \cdot 2^8}s^{11} + \frac{3 \cdot 7 \cdot 11}{13 \cdot 2^{10}}s^{13} + \cdots\end{aligned}$$

Note that every complex analytic function may be expanded in a Taylor series, within its RoC. It follows that the inverse is also complex analytic, as demonstrated in this case using symbolic algebra. ■

Exercise #36 What is the necessary condition such that if $dy/dx = F(x)$, then $dx/dy = 1/F(x)$?
Sol: This will be true when $df(x)/dx = F(x)$ is complex analytic because the Fundamental Theorem of Complex Calculus (FTCC) (see Sect. 4.2.2) defines the antiderivative. In this case $dy/dx = (dx/dy)^{-1}$ (except at singular points, where it is not analytic). ■

3.2.5 Brune Impedances

A special family of functions is formed from ratios of two polynomials $Z(s) = N(s)/D(s)$ commonly used to define an impedance $Z(s)$, called a *Brune impedance*. Impedance functions are a special class of complex analytic functions because they must have a nonnegative real part

$$\Re Z(s) = \Re \frac{N(s)}{D(s)} \geq 0$$

so as to obey conservation of energy. A physical Brune impedance cannot have a negative resistance (the real part); otherwise, it would act like a power source, violating conservation of energy. Most impedances used in engineering applications are in the class of Brune impedances, defined by the ratio of two polynomials of degrees M and N:

$$Z_{\text{Brune}}(s) = \frac{P_M(s)}{P_N(s)} = \frac{s^M + a_1 S^{M-1} + \cdots + a_0}{s^N + b_1 S^{N-1} + \cdots + b_0}, \tag{3.2.17}$$

where $M = N \pm 1$ (i.e., $N = M \pm 1$). This fraction of polynomials is sometimes known as a *Padé approximation*, with poles and zeros, defined as the complex roots of the two polynomials. The key property of the Brune impedance is that the real part of the impedance is nonnegative (positive or zero) in the right s half-plane:

$$\Re Z(s) = \Re\left[R(\sigma,\omega) + jX(\sigma,\omega)\right] = R(\sigma,\omega) \geq 0 \qquad \text{for} \quad \Re s = \sigma \geq 0. \tag{3.2.18}$$

Since $s = \sigma + \omega\jmath$, the complex frequency (s) right half-plane (RHP) corresponds to $\Re s = \sigma \geq 0$. This condition defines the class of positive-real functions, also known as the *Brune condition*, which is frequently written in the abbreviated form

$$\Re Z(\Re s \geq 0) \geq 0. \tag{3.2.19}$$

As a result of this positive-real (PR) constraint, the subset of Brune impedances (those given by Eq. 3.2.17 and satisfying Eq. 3.2.18) must be complex analytic in the entire right s half-plane. This is a powerful constraint that places strict limitations on the locations of both the poles and the zeros of every positive-real Brune impedance. **A little history:** The key idea that every impedance $Z(s)$ must be complex analytic and its real part be nonnegative ($\Re Z(s) \geq 0$) for $\sigma = \Re s > 0$, as first proposed by Otto Brune in his Ph.D. thesis at MIT. His supervised was Ernst A. Guillemin, an MIT electrical engineering professor who played an important role in the development of circuit theory and likely was a student of Arnold Sommerfeld.[12] Other MIT advisers were Norbert Wiener and Vannevar Bush. Brune's primary, but non-MIT advisor was W. Cauer, who was trained in nineteenth-century German mathematics, perhaps under Sommerfeld (Brune 1931b).

3.2.6 Complex Analytic Functions

We are given that the argument of an analytic function $F(x)$ is complex; that is, $x \in \mathbb{R}$ is replaced by $s = \sigma + \omega\jmath \in \mathbb{C}$. Recall that $\mathbb{R} \subset \mathbb{C}$. Thus

$$F(s) = \sum_{n=0}^{\infty} c_n (s - s_o)^n, \tag{3.2.20}$$

with $c_n \in \mathbb{C}$. In this case, that function is said to be a *complex analytic*.

An important example is when the exponential becomes complex, since

[12] It must be noted that University of Illinois Professor "Mac" Van Valkenburg was arguably more influential in circuit theory during the same period. Mac's books are certainly more accessible, but perhaps less widely cited.

$$e^{st} = e^{(\sigma+\omega\jmath)t} = e^{\sigma t} e^{\jmath\omega t} = e^{\sigma t}\left[\cos(\omega t) + \jmath \sin(\omega t)\right]. \tag{3.2.21}$$

Taking the real part gives

$$\Re\{e^{st}\} = e^{\sigma t}\frac{e^{\omega\jmath t} + e^{-\omega\jmath t}}{2} = e^{\sigma t}\cos(\omega t)$$

and $\Im\{e^{st}\} = e^{\sigma t}\sin(\omega t)$. Once the argument is allowed to be complex, it becomes obvious that the exponential and circular functions are fundamentally related. This exposes the family of entire circular functions [i.e., e^s, $\sin(s)$, $\cos(s)$, $\tan(s)$, $\cosh(s)$, $\sinh(s)$] and their inverses [$\ln(s)$, $\arcsin(s)$, $\arccos(s)$, $\arctan(s)$, $\cosh^{-1}(s)$, $\sinh^{-1}(s)$], first fully elucidated by Euler in about 1750 (Stillwell 2010, p. 315).

Note that because $\sin(\omega t)$ is periodic, its inverse must be multivalued. What was needed is some systematic way to account for this multivalued property. This extension to multivalued functions is called a *branch cut*, invented by Riemann in his 1851 Ph.D. thesis, supervised by Gauss in the final years of Gauss's long life.

The Taylor series of a complex analytic function: However, there is a fundamental problem: We cannot formally define the Taylor series for the coefficients c_k until we have defined the derivative with respect to the complex variable $dF(s)/ds$, with $s \in \mathbb{C}$. Thus simply substituting s for x in an analytic function leaves a major hole in one's understanding of the complex analytic function.

It was Cauchy in 1814 (Fig. 3.1) who uncovered the much deeper relationships within complex analytic functions (see Sect. 3.11) by defining differentiation and integration in the complex plane, leading to several fundamental theorems of complex calculus, including the fundamental theorem of complex calculus and Cauchy's formula.

There seems to be some disagreement as to the status of multivalued functions: Are they functions, or is a function strictly single-valued? If so, then we are missing out on a host of interesting possibilities, including all the inverses of nearly every complex analytic function. For example, the inverse of a complex analytic function is a complex analytic function (e.g., e^s and $\log(s)$).

Impact of complex analytic mathematics on physics: It seems likely, if not obvious, that the success of Newton was his ability to describe physics using mathematics. He was inventing new mathematics at the same time he was explaining new physics. The same might be said for Galileo. It seems likely that Newton was extending the successful techniques and results of Galileo's work on gravity (Galileo 1638). Galileo died on January 8, 1642, and Newton was born January 4, 1643, just short of one year later. Obviously Newton was well aware of Galileo's great success and naturally would have been influenced by him (see p. 9).

The application of complex analytic functions to physics was dramatic, as may be seen in the six volumes on physics written by Arnold Sommerfeld (1868–1951), and from the productivity of his many (36) students (e.g., Debye, Lenz, Ewald, Pauli, Guillemin, Bethe, Heisenberg, Morse, and Seebach, to name a few), notable coworkers (Leon Brillouin), and others (John Bardeen) upon whom Sommerfeld had a strong influence. Sommerfeld is famous for training many students who were awarded the Nobel Prize in Physics, yet he never won a Nobel Prize (the prize is not awarded in mathematics). Sommerfeld brought mathematical physics (the merging of physical and experimental principles via mathematics) to a new level with the use of complex integration of analytic functions to solve otherwise difficult problems, thus following the lead of Newton, who used real integration of Taylor series to solve differential equations (Brillouin 1960, Chap. 3 by Sommerfeld).

3.3 Problems AE-1

Topics of this homework: Fundamental theorem of algebra, polynomials, analytic functions and their inverse, convolution, Newton's root-finding method, Riemann zeta function. Deliverables: Answers to problems.

Note: The term $\underline{analytic}$ is used in two different ways. (1) An $\underline{analytic function}$ is a function that may be expressed as a locally convergent power series; (2) $\underline{analytic geometry}$ refers to geometry using a coordinate system.

3.3.1 Polynomials and the Fundamental Theorem of Algebra (FTA)

Problem #1 A polynomial of degree N is defined as

$$P_N(x) = a_0 + a_1x + a_2x^2 + \cdots + a_Nx^N.$$

–1.1: How many coefficients a_n does a polynomial of degree N have?
–1.2: How many roots does $P_N(x)$ have?

Problem #2 The *fundamental theorem of algebra* (FTA)
–2.1: State and then explain the FTA.
–2.2: Using the FTA, *prove* your answer to question 1.2. *Hint: Apply the FTA to prove how many roots a polynomial $P_N(x)$ of order N has.*

Problem #3 Consider the polynomial function $P_2(x) = 1 + x^2$ of degree $N = 2$ and the related function $F(x) = 1/P_2(x)$. What are the roots (e.g., zeros) $x_\pm$ of $P_2(x)$? *Hint: Complete the square on the polynomial $P_2(x) = 1 + x^2$ of degree 2, and find the roots.*

Problem #4 $F(x)$ may be expressed as $(A, B, x_\pm \in \mathbb{C})$

$$F(x) = \frac{A}{x - x_+} + \frac{B}{x - x_-}, \qquad \text{(AE-1.1)}$$

where $x_\pm$ are the roots (zeros) of $P_2(x)$, which become the *poles* of $F(x)$; A and B are the *residues*. The expression for $F(x)$ is sometimes called a *partial fraction expansion* or *residue expansion*, and it appears in many engineering applications.

–4.1: Find $A, B \in \mathbb{C}$ in terms of the roots $x_\pm$ of $P_2(x)$.

–4.2: Verify your answers for A and B by showing that this expression for $F(x)$ is indeed equal to $1/P_2(x)$.]

–4.3: Give the values of the poles and zeros of $P_2(x)$.

–4.4: Give the values of the poles and zeros of $F(x) = 1/P_2(x)$.

3.3.2 Analytic Functions

Overview: Analytic functions are defined by infinite (power) series. The function $f(x)$ is said to be *analytic* at any value of constant $x = x_o$, where there exists a convergent power series

$$P(x) = \sum_{n=0}^{\infty} a_n (x - x_o)^n$$

such that $P(x_o) = f(x_o)$. The point $x = x_o$ is called the *expansion point*. The region around x_o such that $|x - x_o| < 1$ is called the *radius of convergence* or region of convergence (RoC). The local power series for $f(x)$ about $x = x_o$ is defined by the Taylor series:

$$\begin{aligned} f(x) &\approx f(x_o) + \left.\frac{df}{dx}\right|_{x=x_o} (x - x_o) + \frac{1}{2!}\left.\frac{d^2 f}{dx^2}\right|_{x=x_o} (x - x_o)^2 + \cdots \\ &= \sum_{n=0}^{\infty} \frac{1}{n!} \left.\frac{d^n}{dx^n} f(x)\right|_{x=x_o} (x - x_o)^n. \end{aligned}$$

Two classic examples are the geometric series[13] where $a_n = 1$,

$$\frac{1}{1-x} = 1 + x + x^2 + x^3 + \cdots = \sum_{n=0}^{\infty} x^n, \qquad \text{(AE-1.2)}$$

and the exponential function where $a_n = 1/n!$, Eq. 3.2.11. The coefficients for both series may be derived from the Taylor formula.

Problem #5 The geometric series

–5.1: What is the *region of convergence* (RoC) for the power series Eq. AE-1.2 of $1/(1-x)$ given above—for example, where does the power series $P(x)$ converge to the function value $f(x)$? State your answer as a condition on x.

Hint: What happens to the power series when $x > 1$?

–5.2: In terms of the pole, what is the RoC for the geometric series in Eq. AE-1.2?

–5.3: How does the RoC relate to the location of the pole of $1/(1-x)$?

[13] The geometric series is *not* defined as the function $1/(1-x)$, it is defined as the series $1 + x + x^2 + x^3 + \cdots$, such that the ratio of consecutive terms is x.

–5.4: Where are the zeros, if any, in Eq. AE-1.2?

–5.5: Assuming x is in the RoC, prove that the geometric series correctly represents $1/(1-x)$ by multiplying both sides of Eq. AE-1.2 by $(1-x)$.

Problem #6 Use the geometric series to study the degree N polynomial. It is very important to note that all the coefficients c_n of this polynomial are 1.

$$P_N(x) = 1 + x + x^2 + \cdots + x^N = \sum_{n=0}^{N} x^n. \tag{AE-1.3}$$

–6.1: Prove that

$$P_N(x) = \frac{1 - x^{N+1}}{1 - x}. \tag{AE-1.4}$$

–6.2: What is the RoC for Eq. AE-1.3?

–6.3: What is the RoC for Eq. AE-1.4?

–6.4: How many poles does $P_N(x)$ (Eq. AE-1.3) have? Where are they?

–6.5: How many zeros does $P_N(x)$ (Eq. AE-1.4) have? State where are they in the complex plane.

–6.6: Explain why Eqs. AE-1.3 and AE-1.4 have different numbers of poles and zeros.

–6.7: Is the function $1/(1-x)$ analytic outside of the RoC stated in the first question in Problem #5? *Hint: Can it be represented by a different power series outside this RoC?*

–6.8: Extra credit. Evaluate $P_N(x)$ at $x = 0$ and $x = 0.9$ for the case of $N = 100$, and compare the result to that from MATLAB.

```
%sum the geometric series and P_100(0.9)
clear all;close all;format long
N=100; x=0.9; S=0;
for n=0:N
S=S+x^n
end
P100=(1-x^(N+1))/(1-x);
disp(sprintf('S= %g, P100= %g, error= %g',S,P100, S-P100))
```

Problem #7 The exponential series

–7.1: What is the RoC for the exponential series Eq. 3.2.11?

–7.2: Let $x = \jmath$ in Eq. 3.2.11, and write out the series expansion of e^x in terms of its real and imaginary parts.

–7.3: Let $x = \jmath\theta$ in Eq. 3.2.11, and write out the series expansion of e^x in terms of its real and imaginary parts. How does your result relate to Euler's identity ($e^{\jmath\theta} = \cos(\theta) + \jmath\sin(\theta)$)?

3.3.3 *Inverse Analytic Functions and Composition*

Overview: It may be surprising, but every analytic function has an inverse function. Starting from the function ($x, y \in \mathbb{C}$)

$$y(x) = \frac{1}{1-x}$$

the inverse is

$$x = \frac{y-1}{y} = 1 - \frac{1}{y}.$$

Problem #8 Consider the inverse function described above
–8.1: Where are the poles and zeros of $x(y)$?
–8.2: Where (for what condition on y) is $x(y)$ analytic?

Problem #9 Consider the exponential function $z(x) = e^x$ ($x, z \in \mathbb{C}$).
–9.1: Find the inverse $x(z)$.
–9.2: Where are the poles and zeros of $x(z)$?
–9.3: If $y(s) = 1/(1-s)$ and $z(s) = e^s$, compose these two functions to obtain $(y \circ z)(s)$.
Give the expression for $(y \circ z)(s) = y(z(s))$.
–9.4: Where are the poles and zeros of $(y \circ z)(s)$?
–9.5: Where (for what condition on x) is $(y \circ z)(x)$ analytic?

3.3.4 *Convolution*

Multiplying two short or simple polynomials is not demanding. However, if the polynomials have many terms, it can become tedious. For example, multiplying two 10th-degree polynomials is not something one would want to do every day.

An alternative is a method called *convolution*, as described in Sect. 3.4.

Problem #10 Convolution of sequences. Practice convolution (by hand!!) using a few simple examples. Show your work!!! Check your solution using MATLAB.
–10.1: Convolve the sequence $\{0\ 1\ 1\ 1\ 1\}$ with itself.
–10.2: Calculate $\{1, 1\} \star \{1, 1\} \star \{1, 1\}$.

Problem #11 Multiplying two polynomials is the same as convolving their coefficients.

$$f(x) = x^3 + 3x^2 + 3x + 1$$
$$g(x) = x^3 + 2x^2 + x + 2.$$

–11.1: In Octave/MATLAB, compute $h(x) = f(x) \cdot g(x)$ in two ways: (1) use the commands `roots` and `poly`, and (2) use the convolution command `conv`. Confirm that both methods give the same result.

–11.2: What is $h(x)$?

3.3.5 *Newton's Root-Finding Method*

Problem #12 Use Newton's iteration to find the roots of the polynomial

$$P_3(x) = 1 - x^3.$$

–12.1: Draw a graph describing the first step of the iteration starting with $x_0 = (1/2, 0)$.

–12.2: Calculate x_1 and x_2. What number is the algorithm approaching?

–12.3: Here is an Octave/MATLAB script for the $P_2(x)$ case. Modify it to find $P_3(x)$:

```
x(1)=1/2; %x(1)=0.9; %x(1)=-10
y(1)=x(1);
 for n=2:10
x(n) = x(n-1) + (1-x(n-1)^2)/(2*x(n-1));
y(n) = (1+y(n-1)^2)/(2*x(n-1));
 end
semilogy(abs(x)-1); hold on
semilogy(abs(7)-1,'or'); hold off
```

–12.4: For $n = 4$, what is the absolute difference between the root and the estimate, $|x_r - x_4|$?

–12.5: Does Newton's method work for $P_2(x) = 1 + x^2$? If so, why? *Hint: What are the roots in this case?*

–12.6: What if we let $x_0 = (1 + \jmath)/2$ for the case of $P_2(x) = 1 + x^2$?

3.3.6 *Riemann Zeta Function $\zeta(s)$*

Definitions and preliminary analysis: The zeta function $\zeta(s)$ is defined by the complex analytic power series

$$\zeta(s) \equiv \sum_{n=1}^{\infty} \frac{1}{n^s} = \frac{1}{1^s} + \frac{1}{2^s} + \frac{1}{3^s} + \frac{1}{4^s} + \cdots .$$

This series converges, and thus is valid, only in the RoC given by $\Re s = \sigma > 1$, since there $|n^{-\sigma}| \leq 1$. To determine its formula in other regions of the s plane, one must extend the series via analytic continuation (see Sect. 3.2.3).

Euler product formula: As Euler first published in 1737, one may recursively factor out the leading prime term, which results in Euler's product formula.[14] Multiplying $\zeta(s)$ by the factor $1/2^s$ and subtracting from $\zeta(s)$ remove all the terms $1/(2n)^s$ (e.g., $1/2^s + 1/4^s + 1/6^s + 1/8^s + \cdots$)

$$\left(1 - \frac{1}{2^s}\right)\zeta(s) = 1 + \frac{1}{2^s} + \frac{1}{3^s} + \frac{1}{4^s} + \frac{1}{5^s} + \cdots - \left(\frac{1}{2^s} + \frac{1}{4^s} + \frac{1}{6^s} + \frac{1}{8^s} + \frac{1}{10^s} + \cdots\right), \tag{AE-1.5}$$

which results in

$$\left(1 - \frac{1}{2^s}\right)\zeta(s) = 1 + \frac{1}{3^s} + \frac{1}{5^s} + \frac{1}{7^s} + \frac{1}{9^s} + \frac{1}{11^s} + \frac{1}{13^s} + \cdots. \tag{AE-1.6}$$

Problem #13 Questions about the Riemann zeta function.

–13.1: What is the RoC for Eq. AE-1.6?

–13.2: Repeat the algebra of Eq. AE-1.5 using the lead factor of $1/3^s$.

–13.3: What is the RoC for Eq. AE-1.6?

–13.4: Repeat the algebra of Eq. AE-1.5 for all prime scale factors (i.e., $1/5^s$, $1/7^s, \ldots, 1/\pi_k^s, \ldots$) to show that

$$\zeta(s) = \prod_{\pi_k \in \mathbb{P}} \frac{1}{1 - \pi_k^{-s}} = \prod_{\pi_k \in \mathbb{P}} \zeta_k(s), \tag{AE-1.7}$$

where π_p represents the pth prime.

–13.5: Given the product formula, identify the poles of $\zeta_p(s)$ $(p \in \mathbb{Z})$, which is important for defining the RoC of each factor. For example, the pth factor of Eq. AE-1.7, expressed as an exponential, is

$$\zeta_p(s) \equiv \frac{1}{1 - \pi_p^{-s}} = \frac{1}{1 - e^{-sT_p}}, \tag{AE-1.8}$$

where $T_p \equiv \ln \pi_p$.

–13.6: Plot Eq. AE-1.8 using `zviz` for $p = 1$. Describe what you see.

3.4 Root Classification by Convolution

Following the exploration of algebraic relationships by Fermat and Descartes, the first theorem was being formulated by d'Alembert. The idea behind this theorem is that every polynomial of degree N (Eq. 3.1.7) has at least one root. Every polynomial

[14] This is known as *Euler's sieve*, as distinguished from the Eratosthenes sieve.

may be written as the product of a monomial root and a second polynomial of degree of $N-1$. By the recursive application of this concept, it is clear that every polynomial of degree N has N roots. Today this result is known as the *fundamental theorem of algebra:*

> *Every polynomial equation $P(z)=0$ has a solution in the complex numbers. As Descartes observed, a solution $z=a$ implies that $P(z)$ has a factor $z-a$. The quotient*
>
> $$Q(z)=\frac{P(z)}{z-a}=\frac{P(z)}{a}\left[1+\frac{z}{a}+\left(\frac{z}{a}\right)^2+\left(\frac{z}{a}\right)^3+\cdots\right] \tag{3.4.1}$$
>
> *is then a polynomial of one lower degree. ... We can go on to factorize $P(z)$ into n linear factors.*
>
> — *Stillwell (2010, p. 285).*

The ultimate expression of this theorem is given by Eq. 3.1.7, which indirectly states that an nth degree polynomial has n roots. We shall use the term *degree* when speaking of polynomials and the term *order* when speaking of differential equations. A general rule is that *order* applies to the time domain and *degree* to the frequency domain, since the Laplace transform of a differential equation, having constant coefficients, of order N, is a polynomial of degree N in Laplace frequency s.

Today this theorem is so widely accepted we fail to appreciate it. Certainly at about the time you learned the quadratic formula, you were prepared to understand the concept of polynomials having roots. The simple quadratic case may be extended to a higher degree polynomial. The Octave/MATLAB command `roots`$([1, a_2, a_1, a_0])$ provides the roots $[s_1, s_2, s_3]$ of the cubic equation, defined by the coefficient vector $[1, a_2, a_1, a_0]$. The command `poly`$([s_1, s_2, s_3])$ returns the coefficient vector. I don't know the largest degree that can be accurately factored numerically by MATLAB/Octave, but I'm sure it's well over $N=10^3$. Today, finding the roots numerically is a solved problem.

The best way to gain insight into the polynomial factorization problem is through the inverse operation, multiplication of monomials. Given the roots x_k, there is a simple algorithm for computing the coefficients a_k of $P_N(x)$ for any n, no matter how large. This method is called *convolution*. Convolution is said to be a *trapdoor function*, since it is easy, while the inverse, factoring (deconvolution), is hard and analytically intractable for degree $N \geq 5$ (Stillwell 2010, p. 102).

3.4.1 Convolution of Monomials

As outlined by Eq. 3.1.7, a polynomial has two equivalent descriptions, first as a series with coefficients a_n and second as a product of monomial roots x_r. The question is What is the relationship between the coefficients and the roots? The simple answer is that they are related by convolution.

Let us start with the quadratic

$$(x + a)(x + b) = x^2 + (a + b)x + ab, \tag{3.4.2}$$

where in vector notation $[-a, -b]$ are the roots and $[1, a + b, ab]$ are the coefficients.

To see how the result generalizes, we may work out the coefficients for the cubic ($N = 3$). Multiplying the following three factors gives

$$(x-1)(x-2)(x-3) = (x^2-3x+2)(x-3) = x(x^2-3x+2) - 3(x^2-3x+2) = x^3 - 6x^2 + 11x - 6. \tag{3.4.3}$$

When the roots are $[1, 2, 3]$, the coefficients of the polynomial are $[1, -6, 11, -6]$. To verify, we can substitute the roots into the polynomial and show that they give zero. For example, $r_1 = 1$ is a root, since $P_3(1) = 1 - 6 + 11 - 6 = 0$.

As the degree increases, the algebra becomes more difficult. Imagine trying to work out the coefficients for $N = 100$. What is needed is a simple way of finding the coefficients from the roots. Fortunately, convolution keeps track of the bookkeeping, formalizing the procedure, along with Newton's deconvolution method for finding the roots of polynomials (see Sect. 3.4.1).

Convolution of two vectors: To obtain the coefficients by convolution we may write the monomial roots as vectors $[1, a]$ and $[1, b]$. Convolution is a recursive operation described by $[1, a] \star [1, b]$, where $\star$ denotes convolution. The convolution of $[1, a] \star [1, b]$ is done as follows: Reverse one of the two monomials, padding unused elements with zeros. Next slide one monomial against the other, forming the local scalar product (element-wise multiply and add):

$$\begin{array}{ccccc}
a\ 1\ 0\ 0 & a\ \ 1\ \ 0 & a\quad 1\quad 0 & 0\quad a\quad 1 & 0\ 0\ a\ 1 \\
0\ 0\ 1\ b & 0\ \ 1\ \ b & 1\quad b\quad 0 & 1\quad b\quad 0 & 1\ b\ 0\ 0, \\
= 0 & = x^2 & = (a+b)x & = abx^0 & = 0
\end{array}$$

resulting in coefficients $[\ldots, 0, 0, 1, a + b, ab, 0, 0, \ldots]$.

If we reverse one of the polynomials and then take successive scalar products, all the terms in the sum of the scalar product correspond to the same power of x. This explains why the convolution of the coefficients gives the same answer as the product of the polynomials.

As seen from the above example, the positions of the first monomial coefficients are reversed and then slid across the second set of coefficients, the scalar product is computed, and the result is placed in the output vector. Outside the range shown, all the elements are zero. In summary,

$$[1, -1] \star [1, -2] = [1, -1 - 2, 2] = [1, -3, 2].$$

In general,

$$[a, b] \star [c, d] = [ac, bc + ad, bd].$$

Convolving a third term $[1, -3]$ with $[1, -3, 2]$ gives (Eq. 3.4.3)

$$[1, -3] \star [1, -3, 2] = [1, -3 - 3, 9 + 2, -6] = [1, -6, 11, -6],$$

which is identical to the cubic example found by the algebraic method.

When we convolve one monomial factor at a time, the overlap is always two elements; thus it is never necessary to compute more than two multiplications and one addition for each output coefficient. This greatly simplifies the operations (i.e., they are easily done in your head). Thus the final result is more likely to be correct. Comparing this to the algebraic method, we see that convolution has the clear advantage.

Exercise #37 What three nonlinear equations would we need to solve to find the roots of a cubic? ■

Sol: From our formula for the convolution of three monomials, we may find the nonlinear deconvolution relationships between the roots $[-a, -b, -c]$ and the cubic's coefficients $[1, \alpha, \beta, \gamma]$[15]:

$$\begin{aligned}(x+a) \star (x+b) \star (x+c) &= (x+c) \star (x^2 + (a+b)x + ab)\\ &= x \cdot (x^2 + (a+b)x + ab) + c \cdot (x^2 + (a+b)x + ab)\\ &= x^3 + (a+b+c)x^2 + (ab+ac+cb)x + abc\\ &= [1, a+b+c, ab+ac+cb, abc].\end{aligned}$$

It follows that the nonlinear equations must be

$$\begin{aligned}\alpha &= a+b+c\\ \beta &= ab+ac+bc\\ \gamma &= abc.\end{aligned}$$

These equations may be solved by the classic cubic solution, which therefore is a deconvolution problem, also known as *long division of polynomials*. Therefore the following long division of polynomials must be true:

$$\frac{x^3 + (a+b+c)x^2 + (ab+ac+bc)x + abc}{x+a} = x^2 + (b+c)x + bc.$$

The product of a monomial $P_1(x)$ and a polynomial $P_N(x)$ gives $P_{N+1}(x)$: This is another way of stating the fundamental theorem of algebra. Each time we convolve a monomial with a polynomial of degree N, we obtain a polynomial of degree $N + 1$. The convolution of two monomials results in a quadratic (degree 2 polynomial). The convolution of three monomials gives a cubic (degree 3). In general, the degree k of the product of two polynomials of degree n, m is the sum of the degrees ($k = n + m$). For example, if the degrees are each 5 ($n = m = 5$), then the resulting degree is 10.

While we all know this theorem from high school algebra class, it is important to explicitly identify the fundamental theorem of algebra.

---

[15]By working with the negative roots, we may avoid an unnecessary and messy alternating sign problem.

Note that the degree of a polynomial is one less than the length of the vector of coefficients. Since the leading term of the polynomial cannot be zero, or else the polynomial would not have degree N, when we look for roots, the coefficient can (and should always) be normalized to 1.

In summary, the product of two polynomials of degree m, n having m and n roots is a polynomial of degree $m + n$. This is an analysis process of merging polynomials by coefficient convolution. Multiplying polynomials is a merging process into a single polynomial.

Composition of polynomials: Convolution is not the only important operation between two polynomials. Another is composition $c(z) = f(z) \circ g(z) = f(g(z))$ which is defined for analytic functions $f(z)$ and $g(z)$. For example suppose $f(z) = 1 + z + z^2$ and $g(z) = e^{2z}$. Thus

$$f(z) \circ g(z) = 1 + e^{2z} + (e^{2z})^2 = 1 + e^{2z} + e^{4z}.$$

Note that $f(z) \circ g(z) \neq g(z) \circ f(z)$.

Exercise #38 Find $g(z) \circ f(z)$.
Sol: $e^{2f(z)} = e^{2(1+z+z^2)} = e^2 e^{(1+z+z^2)} = e^3 e^z e^{z^2}$ ■

3.4.2 *Residue Expansions of Rational Functions*

As we discussed on Sect. 3.1.4, there are eight important MATLAB/Octave routines that are closely related: `conv()`, `deconv()`, `poly()`, `polyder()`, `polyval()`, `residue()` and `root()`. Several of these are complements of each other or do a similar operation in a slightly different way. The routines `conv()` and `poly()` build polynomials from the roots, while `root()` solves for the roots given the polynomial coefficients. The operation `residue()` expands the ratio of two polynomials in a partial fraction expansion, as poles and residues.

When lines and planes are defined, the equations are said to be *linear* in the independent variables. In keeping with this definition of *linear*, we say that the equations are *nonlinear* when the equations have degree greater than 1 in the independent variables. The term *bilinear* has a special meaning: Both the domain and codomain are linearly related by lines (or planes). As an example, impedance is defined in frequency as the ratio of the voltage over the current, but it often has a representation as the ratio of two polynomials, $N(s)$ and $D(s)$:

$$Z(s) = \frac{N(s)}{D(s)} = sL_o + R_o + \sum_{k=0}^{K} \frac{K_k}{s - s_k}. \tag{3.4.4}$$

Here $Z(s)$ is the impedance, V and I are the voltage and current at radian frequency ω, and K_k, s_k are the residues and eigenvalues.[16]

[16]Note that the relationship between the impedance and the residues K_k is a linear one, ideally solved by formulating a linear system of equations in the unknown residues.

Such an impedance is typically specified as a rational or bilinear function—namely, the ratio of two polynomials, $P_N(s) = N(s) = [a_N, a_{n-1}, \ldots, a_o]$ and $P_K(s) = D(s) = [b_K, b_{K-1}, \ldots, b_o]$ of degrees $N, K \in \mathbb{N}$, as functions of complex Laplace frequency $s = \sigma + \jmath\omega$ with simple roots. Most impedances are rational functions, since they may be written as $D(s)V = N(s)I$. Since $D(s)$ and $N(s)$ are both polynomials in s, a rational function is also called a *bilinear transformation*, or in the mathematical literature a *Möbius transformation*, which comes from a corresponding scalar differential equation of the form

$$\sum_{k=0}^{K} b_k \frac{d^k}{dt^k} i(t) = \sum_{n=0}^{N} a_n \frac{d^n}{dt^n} v(t) \quad \leftrightarrow \quad I(\omega) \sum_{k=0}^{K} b_k s^k = V(\omega) \sum_{n=0}^{N} a_n s^n. \quad (3.4.5)$$

This construction is also known as the ABCD method in the engineering literature (Eq. 3.8.1). This equation, as well as Eq. 3.4.4, follows from the Laplace transform (see Sect. 3.10) of the differential equation (on left) by forming the impedance $Z(s) = V/I = A(s)/B(s)$. This form of the differential equation follows from Kirchhoff's voltage and current laws (KCL, KVL) or from Newton's laws (for the case of mechanics).

Impedance is a very important and general concept. It is typically defined as the ratio of the change in voltage across a device, over the current through the device, which is known as *Ohm's law*. However it applies to many more physical variables than just electricity (see Table 3.2, p. 147), which leads to the concept of a *generalized impedance*.
It began as the real ratio of the voltage drop over the current through, but by at least 1893 it was realized that complex numbers could be used to represent the complex impedance of inductors (mass) and capacitors (springs) (Heaviside 1892; Kennelly 1893). As we explore more deeply it is is likely that Maxwell understood this concept as well, since he first formulated his famous equations of electricity using complex analysis (Maxwell 1865).
Since impedance is the ratio of a force over a flow, it does not directly depend on either the force or the flow. Rather it is the complex, frequency dependent proportionality factor between them:

$$\text{force} = Z(s) \cdot \text{flow} \quad \text{with } s, Z(s) \in \mathbb{C},$$

where $s = \sigma + \omega\jmath$ is the Laplace frequency.

The physical properties of an impedance: Based on d'Alembert's observation that the solution to the wave equation is the sum of forward and backward traveling waves, the impedance may be rewritten in terms of forward and backward traveling waves (see Sect. 4.4):

$$Z(s) = \frac{V}{I} = \frac{V^+ + V^-}{I^+ - I^-} = r_o \frac{1 + \Gamma(s)}{1 - \Gamma(s)}, \tag{3.4.6}$$

where $r_o = V^+/I^+$ is called the *characteristic impedance* of the transmission line (e.g., wire) connected to the load impedance $Z(s)$, and $\Gamma(s) = V^-/V_+ = I^-/I^+$ is the reflection coefficient corresponding to $Z(s)$. Any impedance of this type is called a *Brune impedance* due to its special properties (Brune 1931a; Van Valkenburg 1964a). Like $Z(s)$, $\Gamma(s)$ is causal and complex analytic. The impedance and the reflectance function $\Gamma(s)$ must both be complex analytic, since they are related to the bilinear transformation, which ensures the mutual complex analytic properties.

Due to the bilinear transformation, the physical properties of $Z(s)$ and $\Gamma(s)$ are very different. Specifically, the real part of the load impedance is nonnegative ($\Re\{Z(\omega\jmath)\} \geq 0$) if and only if $|\Gamma(s)| \leq 1$. In the time domain, the impedance $z(t) \leftrightarrow Z(s)$ must have a value of r_o at $t = 0$. Correspondingly, the time-domain reflectance $\gamma(t) \leftrightarrow \Gamma(s)$ must be zero at $t = 0$.

This is the basis of conservation of energy, which may be traced back to the properties of the reflectance $\Gamma(s)$.

Exercise #39 Show that if $\Re\{Z(s)\} \geq 0$, then $|\Gamma(s)| \leq 1$.
Sol: Taking the real part of Eq. 3.4.6, which must be ≥ 0, we find

$$\Re\{Z(s)\} = \frac{r_o}{2}\left[\frac{1 + \Gamma(s)}{1 - \Gamma(s)} + \frac{1 + \Gamma^*(s)}{1 - \Gamma^*(s)}\right] = r_o \frac{1 - |\Gamma(s)|^2}{|1 + \Gamma(s)|^2} \geq 0.$$

Thus $|\Gamma| \leq 1$. ∎

3.5 Introduction to Analytic Geometry

Analytic geometry came about as Euclid's geometry merged with algebra. The combination of Euclid's (323 BCE) geometry and al-Khwarizmi's (830 CE) algebra resulted in a totally new and powerful tool, analytic geometry, independently worked out by Descartes and Fermat (Stillwell 2010). The development of matrix algebra during the eighteenth century enabled an analysis in more than three dimensions. Due to modern computation, today this is one of the most powerful tools used in artificial intelligence, data science, and machine learning. The utility and importance of these new tools cannot be overstated. The timeline for this period of development in mathematics is shown in Fig. 1.2.

There are many important relationships between Euclidean geometry and sixteenth-century algebra. Table 3.1 is an attempt at a detailed comparison. Important similarities include vectors, their Pythagorean lengths $[a, b, c]$,

Table 3.1 An ad hoc comparison of Euclidean geometry and analytic geometry. I am uncertain of the classification of the items in the third column

Euclidean geomerty: $\mathbb{R}^3$	Analytic geometry: $\mathbb{R}^n$	Uncertain
Proof	Numbers	Recursion
Line length	Algebra	Iteration $\in \mathbb{C}^2$,
Line intersection	Power series	Newton's method
Point	Analytic functions	Appximation: (Least-squares)
Projection (scalar product)	Complex analytic functions:	
Line direction	$\sin\theta$, $\cos\theta$, $e^{\theta J}$, $\log z$	
Vector (sort of)	Scalar product	
Conic section	Wedge (scalar) product $\boldsymbol{A} \curlywedge \boldsymbol{B}$	
Square roots (spiral of	Generalized scalar product	
Theodorus)	(Eq. 3.5.5)	
	Normed vector spaces	
	Composition	
	Elimination	
	Integration	
	Derivatives	
	Calculus	
	Polynomial $\in \mathbb{C}$	
	Fundamental theorem of algebra	

$$c = \sqrt{(x_2 - x_1)^2 + (y_2 - y_1)^2}, \tag{3.5.1}$$

$a = x_2 - x_1$, and $b = y_2 - y_1$, and the angles. Euclid's geometry had length and angles but no concept of coordinates or thus of vectors. One of the main innovations of analytic geometry is that we could compute with real, and soon after, complex numbers, first observed in the completion of squares, Eq. 3.1.9.

3.5.1 Merging the Concepts

Several new concepts came with the development of analytic geometry:

1. Composition of functions: If $y = f(x)$ and $z = g(y)$, then the composition of functions f and g is denoted $z(x) = g \circ f(x) = g(f(x))$.
2. Elimination: Given two functions $f(x, y)$ and $g(x, y)$, elimination removes either x or y. This procedure, well known to the Chinese, is now known as *Gaussian elimination*.
3. Intersection: One may speak of the intersection of two lines to define a point or two planes to define a line. This is a special case of elimination when the functions $f(x, y)$ and $g(x, y)$ are linear in their arguments. The term *intersection* is also an important but very different from the meaning of the term as used in set theory.
4. Vectors: Analytic geometry provides the concept of a vector (see Appendix A.3.1) as a line with length and orientation (i.e., direction). Analytic geometry defines vectors in any number of dimensions as ordered sets of points.

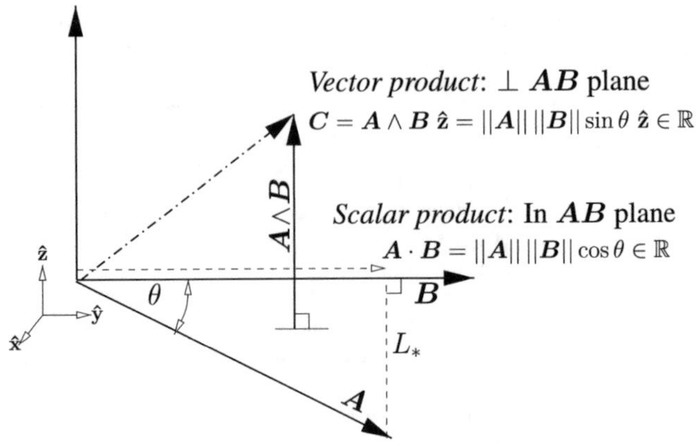

Fig. 3.4 Vectors $A, B, C \in \mathbb{C}$ are used to define the scalar product $A \cdot B \in \mathbb{C}$ and the scalar wedge-product $A \wedge B \in \mathbb{C}$. The scalar wedge-product is the same as the vector cross-product except the output is a scalar. As shown, the scalar dot and wedge-products complement each other, since one is proportional to the sine of the angle θ between them, and the other to the cosine. The dot product computes the projection of one vector on the other (the length of the base of the triangle formed by the two vectors), while the vector wedge-product $A \wedge B$ computes the area of the right triangle ($area = base \cdot height = A \cdot B\ L_*$) formed by the two vectors. Thus $|A \cdot B|^2 + |A \wedge B|^2 = ||A||^2 ||B||^2$. The *scalar triple product* $C \cdot (A \times B)$ represents the volume of the parallelepiped (i.e., prism) defined by the three vectors A, B, and C. When all the angles are 90°, the volume becomes a *cuboid*

5. Scalar products of vectors: Analytic geometry extends the ideas of Euclidean geometry with the introduction of the scalar (dot) product of two vectors $\mathbf{f} \cdot \mathbf{g}$ and the scalar wedge-product $\mathbf{f} \wedge \mathbf{g}$. The vector wedge-product adds a unit vector $\perp$ the plane of the two vectors

$$\mathbf{f} \wedge \mathbf{g}\hat{\mathbf{z}},$$

as described in the caption of Fig. 3.4.

What algebra also added to geometry was the ability to compute with complex (polar) numbers. For example, in geometry the length of a line (Eq. 3.5.1) was measured with a compass; numbers played no role. Once algebra was available, the line's Euclidean length could be computed numerically, directly from the coordinates of the two ends, defined by the 3-vector

$$\mathbf{e} = x\hat{\mathbf{x}} + y\hat{\mathbf{y}} + z\hat{\mathbf{z}} = [x, y, z]^T,$$

which represents a point at $(x, y, z) \in \mathbb{R}^3 \subset \mathbb{C}^3$ in three dimensions, having direction from the origin $(0, 0, 0)$ to (x, y, z). An alternative matrix notation is $\mathbf{e} = [x, y, z]^T$, a column-vector of three numbers. These two notations are different ways of representing a vector $\mathbf{e}$.

By defining the vector, analytic geometry allows Euclidean geometry to become quantitative, beyond the physical drawing of an object (e.g., a sphere, triangle, or line). With analytic geometry we have the Euclidean concept of a vector, a line that has a magnitude (length) and direction (angle), but analytic, defined in terms of physical coordinates (i.e., numbers). The difference between two vectors $(\mathbf{x} - \mathbf{y})$ defines a third vector form, a concept already present in Euclidean geometry. For the first time, complex numbers were allowed into geometry (but rarely used before Cauchy and Riemann).

Scalar product of two vectors: When we use algebra, many concepts that are obvious with Euclid's geometry may be made precise. There are many examples of how algebra extends Euclidean geometry, the most basic being the scalar product (also known as the dot product) between vectors $\mathbf{x} \in \mathbb{R}^3$ and $\boldsymbol{\kappa} \in \mathbb{C}^3$:

$$\begin{aligned}\mathbf{x} \cdot \boldsymbol{\kappa} &= (x\hat{\mathbf{x}} + y\hat{\mathbf{y}} + z\hat{\mathbf{z}}) \cdot (\alpha\hat{\mathbf{x}} + \beta\hat{\mathbf{y}} + \gamma\hat{\mathbf{z}}) \qquad \in \mathbb{C} \\ &= \alpha x + \beta y + \gamma z.\end{aligned}$$

Scalar products play an important role in vector algebra and calculus.

In vector notation the scalar product is written as (see Appendix A.3).

$$\mathbf{x} \cdot \boldsymbol{\kappa} = \begin{bmatrix} x \\ y \\ z \end{bmatrix}^T \begin{bmatrix} \alpha \\ \beta \\ \gamma \end{bmatrix} = \begin{bmatrix} x, y, z \end{bmatrix} \begin{bmatrix} \alpha \\ \beta \\ \gamma \end{bmatrix} = \alpha x + \beta y + \gamma z. \tag{3.5.2}$$

If $\boldsymbol{\kappa}(s) \in \mathbb{C}^3$ is a complex function of frequency s, then the scalar product is a complex function of s.

Norm (length) of a vector: The norm of a vector

$$||\mathbf{e}|| \equiv +\sqrt{\mathbf{e} \cdot \mathbf{e}} \geq 0$$

is defined as the positive square root of the scalar product of the vector with itself (see Appendix A.3). This is a generalization of the length, in any number of dimensions, that forces the sign of the square root to be nonnegative. The length is a concept of Euclidean geometry, and it must always be positive and real. A complex (or negative) length is not physically meaningful. More generally, the Euclidean length of a line is given as the norm of the difference between two real vectors $\mathbf{e}_1, \mathbf{e}_2 \in \mathbb{R}$:

$$\begin{aligned}||\mathbf{e}_1 - \mathbf{e}_2||^2 &= (\mathbf{e}_1 - \mathbf{e}_2) \cdot (\mathbf{e}_1 - \mathbf{e}_2) \\ &= (x_1 - x_2)^2 + (y_1 - y_2)^2 + (z_1 - z_2)^2 \geq 0.\end{aligned} \tag{3.5.3}$$

From this formula we see that the norm of the difference of two vectors is a compact expression for the Euclidean length. A zero-length vector, such as a point, is the result of the fact that

$$||\boldsymbol{x} - \boldsymbol{x}||^2 = (\boldsymbol{x} - \boldsymbol{x}) \cdot (\boldsymbol{x} - \boldsymbol{x}) = 0.$$

Integral definition of a scalar product: Following Euclid, we only considered a vector to be a set of elements $\{x_n\} \in \mathbb{R}$, index over $n \in \mathbb{N}$. Starting with Fig. 3.4 we assume the vectors are in $\mathbb{C}$.

An obvious question presents itself: Can we extend our definition of vectors to differentiable functions (i.e., $f(t)$ and $g(t)$) indexed over $t \in \mathbb{R}$ with coefficients labeled by $t \in \mathbb{R}$ rather than by $n \in \mathbb{N}$? Clearly, if the functions are analytic, there is no obvious reason that this should be a problem, since analytic functions may be represented by a convergent series that has Taylor coefficients and thus are integrable term by term.

Specifically, under certain conditions, the function $f(t)$ may be thought of as a vector, defining a normed vector space called a Hilbert space. This intuitive and somewhat obvious idea is powerful. In this case the scalar product can be defined in terms of the integral

$$\begin{aligned} f(t) \cdot g(t) &= \int_t f(t)g(t)dt \\ &= ||f(t)||\,||g(t)|| \cos\theta \end{aligned}$$

summed over $t \in \mathbb{R}$, rather than a sum over $n \in \mathbb{N}$.

This definition of the vector–scalar product allows for a significant but straightforward generalization of our vector space, which will turn out to be both useful and an important extension of the concept of a normed vector space. In this space we can define the derivative of a norm with respect to t, which is not possible for the discrete case, indexed over n. The distinction introduces the concept of analytic continuity in the index t, which also fails to exist for the discrete index $n \in \mathbb{N}$.

Pythagorean theorem and the Schwarz inequality: Regarding Fig. 3.4, suppose we compute the difference between vector $\boldsymbol{A} \in \mathbb{R}$ and $\alpha\boldsymbol{B} \in \mathbb{R}$ as $L = ||\boldsymbol{A} - \alpha\boldsymbol{B}|| \in \mathbb{R}$, where $\alpha \in \mathbb{R}$ is a scalar that modifies the length of $\boldsymbol{B}$. We seek the value of α, which we denote as α^*, that minimizes the length of L. From simple geometrical considerations, $L(\alpha)$ will be minimum when the difference vector is perpendicular to $\boldsymbol{B}$, as shown in the figure by the dashed line from the tip of $\boldsymbol{A} \perp \boldsymbol{B}$.

To show this algebraically, we write the expression for $L(\alpha)$, take the derivative with respect to α, and set it to zero, which gives the formula for α^*. The argument does not change, but the algebra greatly simplifies if we normalize $\boldsymbol{A}$ and $\boldsymbol{B}$ to be unit vectors $\boldsymbol{a} = \boldsymbol{A}/||\boldsymbol{A}||$ and $\boldsymbol{b} = \boldsymbol{B}/||\boldsymbol{B}||$, which each have norm = 1:

$$L^2 = (\boldsymbol{a} - \alpha\boldsymbol{b}) \cdot (\boldsymbol{a} - \alpha\boldsymbol{b}) = 1 - 2\alpha\boldsymbol{a} \cdot \boldsymbol{b} + \alpha^2. \tag{3.5.4}$$

Thus the length is shortest ($L = L_*$, as shown in Fig. 3.4) when

$$\frac{d}{d\alpha} L_*^2 = -2\boldsymbol{a} \cdot \boldsymbol{b} + 2\alpha^* = 0.$$

Solving for $\alpha^* \in \mathbb{R}$, we find $\alpha^* = \boldsymbol{a} \cdot \boldsymbol{b}$. Since $L_* > 0$ ($\boldsymbol{a} \neq \boldsymbol{b}$), Eq. 3.5.4 becomes

$$1 - 2|\boldsymbol{a} \cdot \boldsymbol{b}|^2 + |\boldsymbol{a} \cdot \boldsymbol{b}|^2 = 1 - |\boldsymbol{a} \cdot \boldsymbol{b}|^2 > 0.$$

In terms of A and B this is $|A \cdot B| < ||A||\,||B|| \cos\theta$, as shown adjacent to B in Fig. 3.4.

In conclusion, $\cos\theta \equiv |a \cdot b| < 1$. Thus the scalar product between two vectors is their direction-cosine. Furthermore, since this forms a right triangle, the Pythagorean theorem must hold. The triangle inequality says that the sum of the lengths of the two sides must be greater than the length of the hypotenuse. Note that $\Theta \in \mathbb{R} \notin \mathbb{C}$. Equality cannot be obtained because in Fourier space the scalar product defines an open set, which gives rise to Gibbs ringing in the time domain (Greenberg 1988, p. 854). This derivation is an abbreviated version of a related discussion on Sect. 3.5.5.

Vector cross ($\times$) and wedge ($\wedge$) products of two vectors: The vector product (cross-product) $A \times B$ and the exterior product (wedge-product) $A \wedge B$ are the second and third types of vector products. As shown in Fig. 3.4,

$$C = A \times B = (a_1\hat{\mathbf{x}} + a_2\hat{\mathbf{y}} + a_3\hat{\mathbf{z}}) \times (b_1\hat{\mathbf{x}} + b_2\hat{\mathbf{y}} + b_3\hat{\mathbf{z}}) = \begin{vmatrix} \hat{\mathbf{x}} & \hat{\mathbf{y}} & \hat{\mathbf{z}} \\ a_1 & a_2 & a_3 \\ b_1 & b_2 & b_3 \end{vmatrix}$$

is $\perp$ to the plane defined by A and B. The cross-product is strictly limited to two input vectors A and $B \in \mathbb{R}^2$ taken from three real dimensions (i.e., $\mathbb{R}^3$).

The exterior (wedge) product generalizes the cross-product, since it may be defined in terms of any two vectors $A, B \in \mathbb{C}^2$ taken from n dimensions ($\mathbb{C}^n$) with output in $\mathbb{C}^1$. Thus the cross-product is composed of three wedge-products.

From this specific example we see that the absolute value of the wedge-product $|\mathbf{a} \wedge \mathbf{b}| = ||\mathbf{a} \times \mathbf{b}||$, namely,

$$|(a_2\hat{\mathbf{y}} + a_3\hat{\mathbf{z}}) \wedge (b_2\hat{\mathbf{y}} + b_3\hat{\mathbf{z}})| = ||\mathbf{a} \times \mathbf{b}|| = ||\mathbf{a}||\,||\mathbf{b}||\,|\sin\theta|.$$

The wedge-product is especially useful because it is zero when the two vectors are colinear: that is, $\hat{\mathbf{x}} \wedge \hat{\mathbf{x}} = 0$ and $\hat{\mathbf{x}} \wedge \hat{\mathbf{y}} = 1$, where $\hat{\mathbf{x}}$ and $\hat{\mathbf{y}}$ are unit vectors.

Since

$$\mathbf{a} \cdot \mathbf{b} = ||\mathbf{a}||\,||\mathbf{b}|| \cos\theta \quad \text{and} \quad \mathbf{a} \wedge \mathbf{b} = ||\mathbf{a}||\,||\mathbf{b}||\,\sin\theta,$$

it follows that

$$\mathbf{a} \cdot \mathbf{b} + \jmath\, \mathbf{a} \wedge \mathbf{b} = ||\mathbf{a}||\,||\mathbf{b}||\, e^{\jmath\theta},$$

which may be viewed as a generalized complex scalar product $\in \mathbb{C}$, with the right-hand side the polar form.

The main advantage of the wedge-product is that it is valid in $n \geq 3$ dimensions since it is defined for any two vectors in any number of dimensions.

Scalar triple product: The triple of a third vector C with the vector product $A \times B \in \mathbb{R}$ is

$$C \cdot (A \times B) = \begin{vmatrix} c_1 & c_2 & c_3 \\ a_1 & a_2 & a_3 \\ b_1 & b_2 & b_3 \end{vmatrix} \qquad \in \mathbb{R}^3,$$

which equals the volume of a parallelepiped.

3.5.2 Generalized Scalar Product

As shown in Fig. 3.4, any two vectors $\boldsymbol{A}, \boldsymbol{B} \in \{\hat{\mathbf{x}}, \hat{\mathbf{y}}\}$ define a plane. There are two types of scalar products[17]: the scalar dot product

$$\boldsymbol{A} \cdot \boldsymbol{B} = ||\boldsymbol{A}||\ ||\boldsymbol{B}|| \cos\theta \in \mathbb{R},$$

and the scalar wedge-product[18]

$$\boldsymbol{A} \wedge \boldsymbol{B} = ||\boldsymbol{A}||\ ||\boldsymbol{B}|| \sin\theta \in \mathbb{R}.$$

As shown in the figure, these two products form a right triangle and thus may be naturally merged, defining the complex analytic scalar product

$$\boldsymbol{A} \curlywedge \boldsymbol{B} = \boldsymbol{A} \cdot \boldsymbol{B} + \jmath \boldsymbol{A} \wedge \boldsymbol{B} = ||\boldsymbol{A}||\ ||\boldsymbol{B}|| e^{\jmath\theta} \quad \in \mathbb{C}. \tag{3.5.5}$$

Important examples, based on the Poynting theorem, come from Maxwell's equations

$$\mathcal{P} = \boldsymbol{E} \curlywedge \boldsymbol{H} = \boldsymbol{E} \cdot \boldsymbol{H} + \jmath \boldsymbol{E} \wedge \boldsymbol{H} \qquad [\mathrm{W/m^2}]$$

(Sommerfeld 1952, p. 26), and the corresponding momentum equation (Johnson et al. 1994)

$$\mathcal{M} = \boldsymbol{D} \curlywedge \boldsymbol{B} = \boldsymbol{D} \cdot \boldsymbol{B} + \jmath \boldsymbol{D} \wedge \boldsymbol{B} = \frac{1}{c_o^2}\mathcal{P} \qquad [\mathrm{J\ s/m^4}].$$

While the *solar constant* $\mathcal{P} = 1.3$ [kW/m^2] is large, solar gravity, which follows from Maxwell's equations, is $1/c_o^2 = 1.1 \times 10^{-17}$ smaller.

Example: If we define $\boldsymbol{A} = 3\jmath\hat{\mathbf{x}} - 2\hat{\mathbf{y}} + 0\hat{\mathbf{z}}$ and $\boldsymbol{B} = 1\hat{\mathbf{x}} + 1\hat{\mathbf{y}} + 0\hat{\mathbf{z}}$, then the cross-product is

$$\boldsymbol{A} \times \boldsymbol{B} = \begin{vmatrix} \hat{\mathbf{x}} & \hat{\mathbf{y}} & \hat{\mathbf{z}} \\ 3\jmath & -2 & 0 \\ 1 & 1 & 0 \end{vmatrix} = (3\jmath + 2)\hat{\mathbf{z}}.$$

Since $a_1 \in \mathbb{C}$, this example violates the common assumption that $\boldsymbol{A} \in \mathbb{R}^3$. The wedge-product $\boldsymbol{A} \wedge \boldsymbol{B}$ takes two vectors and returns a scalar, which is the magnitude of a vector $\perp$ to the plane defined by the two input vectors (see Fig. 3.4). It is defined as

---

[17] https://en.wikipedia.org/wiki/Bivector.

[18] In some texts the wedge-product is called the vector exterior-product.

$$
\begin{aligned}
\boldsymbol{A} \wedge \boldsymbol{B} &= \begin{vmatrix} a_1 & b_1 \\ a_2 & b_2 \end{vmatrix} = \begin{vmatrix} 3\jmath & 1 \\ -2 & 1 \end{vmatrix} \\
&= (3\jmath\hat{\mathbf{x}} - 2\hat{\mathbf{y}}) \wedge (\hat{\mathbf{x}} + \hat{\mathbf{y}}) \\
&= 3 \cdot 0\, \cancelto{0}{\boldsymbol{x} \wedge \boldsymbol{x}} - 2\hat{\mathbf{y}} \wedge \hat{\mathbf{x}} + 3\jmath\hat{\mathbf{x}} \wedge \hat{\mathbf{y}} - 2\, \cancelto{0}{\hat{\mathbf{y}} \wedge \hat{\mathbf{y}}} \\
&= (3\jmath + 2)||\hat{\mathbf{x}} \wedge \hat{\mathbf{y}}|| \\
&= (3\jmath + 2)||\hat{\mathbf{y}}|| = 3\jmath + 2.
\end{aligned}
$$

This defines a compact and useful algebra (Hestenes 2003).

Impact of Analytic Geometry: The most obvious impact of analytic geometry was its detailed analysis of the conic sections using algebra rather than drawings with a compass and ruler. An important example is the composition of the line and circle, a venerable pre-Euclid construction.

Once algebra was invented, analysis could be done using formulas. With analysis came complex numbers.

The first two mathematicians to appreciate this mixture of Euclid's geometry and the new algebra were Fermat and Descartes (Fig. 1.5). Soon Newton contributed to this effort by adding physics (e.g., calculations in acoustics, orbits of the planets, and the theory of gravity and light, significant concepts for 1687) (Stillwell 2010, p. 115–117)).

Given these new methods, many new solutions to problems emerged. The complex roots of polynomials continued to appear, without any obvious physical meaning. Newton called them *imaginary*. Complex numbers seem to have been viewed as an inconvenience. Newton's solution to this dilemma was to simply ignore the "imaginary" cases (Stillwell 2010, p. 115–19).

3.5.3 Development of Analytic Geometry

The first "algebra" (*al-jabr*) is credited to al-Khwarizmi (830 CE). Its invention advanced the theory of polynomial equations in one variable, Taylor series, and composition versus intersections of curves. The solution of the quadratic equation had been worked out thousands of years earlier, but with algebra a general solution could be defined. The Chinese had found the way to solve several equations in several unknowns—for example, finding the values of the intersections of two circles. With the invention of algebra by al-Khwarizmi, a powerful tool became available to solve more difficult problems.

In algebra there are two contrasting operations on functions: composition and elimination (e.g., intersection).

3.5.3.1 Composition:

Composition is the merging of functions by feeding one into the other. If the two functions are f and g, then their composition is indicated by $f \circ g$, meaning the function $y = f(x)$ is substituted into the function $z = g(y)$, giving $z = g(f(x))$.

Composition is not limited to linear equations, even though that is where it is most frequently applied. That requires solving for that substitution variable, which is not always possible in the case of nonlinear equations. However, many tricks are available that may work around this restriction. For example, if one equation is in x^2 and the other in x^3 or $\sqrt{x}$, it may be possible to multiply the first by x or square the second. The point is that one of the variables must be isolated so that when it is substituted into the other equation, the variable is removed from the mix.
Example: Let $y = f(x) = x^2 - 2$ and $z = g(y) = y + 1$. Then

$$g \circ f = g(f(x)) = (x^2 - 2) + 1 = x^2 - 1. \tag{3.5.6}$$

In general, composition does not commute (i.e., $f \circ g \neq g \circ f$), as is easily demonstrated. Swapping the order of composition for our example gives

$$f \circ g = f(g(y)) = z^2 - 2 = (y + 1)^2 - 2 = y^2 + 2y - 1. \tag{3.5.7}$$

3.5.3.2 Intersection:

Complementary to composition is intersection (i.e., decomposition). For example, the intersection of two lines is defined as the point where they meet. This is not to be confused with finding roots. A polynomial of degree N has N roots, but the points where two polynomials intersect has nothing to do with the roots of the polynomials. The intersection is a function (equation) of lower degree, implemented by Gaussian elimination.

A system of linear equations $Ax = y$ has many interpretations, and one should not be biased by the notation. As engineers, we are trained to view x as the input and y as the output. Then $y = Ax$ seems natural, much like the functional relationship $y = f(x)$. But what does the linear relationship $x = Ay$ mean, when x is the input? The answer is $y = A^{-1}x$.

But when we work with systems of equations, there are many uses of equations, and we need to be more flexible in our interpretation. For example, $y = A^2x$ has a useful meaning, and in fact we saw this type of relationship we worked with Pell's equation (Sect. 2.6.2) and the Fibonacci sequence (Sect. 2.6.3). As another example, consider

$$\begin{bmatrix} z_1 \\ z_2 \end{bmatrix} = \begin{bmatrix} a_{1x} & a_{1y} \\ a_{2x} & a_{2y} \end{bmatrix} \begin{bmatrix} x \\ y \end{bmatrix},$$

which is reminiscent of a two-dimensional surface $\mathbf{z} = \mathbf{f}(x, y)$. We shall find that such generalizations are much more than a curiosity.

Intersection of two lines: Unless they are parallel, two lines meet at a point. In terms of linear algebra, this may be written as two linear equations[19] (on the left) along with the intersection point $[x_1, x_2]^T$ given by the inverse of the 2×2 set of equations (on the right):

$$\begin{bmatrix} a & b \\ c & d \end{bmatrix} \begin{bmatrix} x_1 \\ x_2 \end{bmatrix} = \begin{bmatrix} y_1 \\ y_2 \end{bmatrix} \qquad \begin{bmatrix} x_1 \\ x_2 \end{bmatrix} = \frac{1}{\Delta} \begin{bmatrix} d & -b \\ -c & a \end{bmatrix} \begin{bmatrix} y_1 \\ y_2 \end{bmatrix}. \tag{3.5.8}$$

By substituting the expression for the intersection point $[\boldsymbol{x}_1, \boldsymbol{x}_2]^T$ into the original equation, we see that it satisfies the equations. Thus the equation on the right is the solution to the equation on the left.

3.5.3.3 Elimination:

Note the structure of the inverse: (1) The diagonal values (a, d) are swapped, (2) the off-diagonal values (b, c) are negated, and (3) the 2×2 matrix is divided by the determinant $\Delta = ad - bc$. If $\Delta = 0$, there is no solution. When the determinant is zero ($\Delta = 0$), the slopes of the two lines

$$\text{slope} = \frac{b}{a} = \frac{d}{c}$$

are equal; thus the lines are parallel. Only if the slopes differ can there be a unique solution.

Algebra can give the solution when geometry cannot. When the two curves fail to intersect on the real plane, the solution still exists, but it is complex-valued. In such cases, geometry, which considers only the real solutions, fails. For example, when the coefficients $[a, b, c, d]$ are complex, the solution exists but the determinant can be complex. Thus algebra is much more general than geometry. Geometry fails when the solution has a complex intersection.

3.5.4 *Applications of Scalar Products*

Another important example of algebraic expressions in mathematics is Hilbert's generalization of the Pythagorean theorem (Eq. 1.1), known as the *Schwarz* inequality and shown in Fig. 3.5. What is special about this generalization is that it proves that when the vertex is 90°, the Euclidean length of the leg is minimum.

---

[19]When we write the equation $Ax = y$ in matrix format, the two equations are $ax_1 + bx_2 = y_1$ and $dx_1 + ex_2 = y_2$ with unknowns (x_1, x_2), whereas in the original equations $ay + bx = c$ and $dy + ex = f$, the unknowns are y and x. Thus in matrix format, the names are changed. The first time you see this scrambling of variables, it can be confusing.

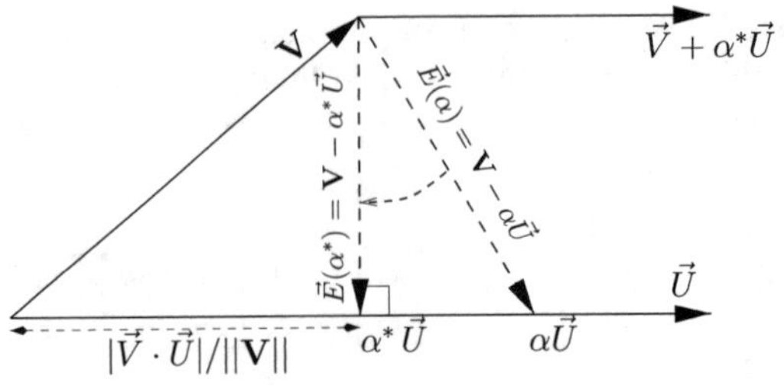

Fig. 3.5 The *Schwarz inequality* is related to the shortest distance (length of a line) between the ends of the two vectors. $||U|| = \sqrt{U \cdot U}$ is the scalar product of that vector with itself

Vectors may be generalized to have ∞ dimensions. For example $\mathbf{U} = [u_1, u_2, \ldots, u_\infty]$, $\mathbf{V} = [v_1, v_2, \ldots, v_\infty]$. The Euclidean inner product (i.e., scalar product) between two such vectors generalizes the finite-dimensional case

$$\mathbf{U} \cdot \mathbf{V} = \sum_{k=1}^{\infty} u_k v_k = ||\mathbf{U}|| \, ||\mathbf{V}|| \ \cos\theta$$

where $\theta \in \mathbb{R}$ is the multivalued angle between the two normalized (unit) vectors

$$\theta = \cos^{-1}\left(\frac{\mathbf{U}}{||\mathbf{U}||} \cdot \frac{\mathbf{V}}{||\mathbf{U}||}\right).$$

As with the finite case the norm $||\mathbf{U}|| = \sqrt{\mathbf{U}\cdot\mathbf{U}} = \sqrt{\sum u_k^2}$, the scalar product of the vector with itself, defines the length of the infinite component vector. There is an issue of convergence when the norm of the vectors is zero.

It is a somewhat arbitrary requirement that $a, b, c \in \mathbb{R}$ for (Eq. 1.1). This seems natural enough, since the sides are lengths. But what if these lengths are taken from high-dimensionality complex vectors, as for the lossy vector wave equation or the lengths of vectors in the Fourier transform $(\mathcal{FT}) \in \mathbb{C}^n$? Then the equation generalizes to $K \to \infty$ dimensions

$$\mathbf{c} \cdot \mathbf{c} = ||\mathbf{c}||^2 = \sum_{k=1}^{K\to\infty} |c_k|^2.$$

As before, $||\mathbf{c}|| = \sqrt{||\mathbf{c}||^2}$ is the *norm* of vector $\mathbf{c}$, akin to a length, which must be finite (converge). This is simply the important case of complex analytic functions, which also must converge.

Schwarz inequality: The Schwarz inequality says that the magnitude of the inner product of two vectors is less than or equal to the product of their lengths:

$$|U \cdot V| \leq ||U|| \, ||V||.$$

This may be simplified by normalizing the vectors to have unit length ($\widehat{U} = U/||U||$, $\widehat{V} = V/||V||$), in which case $-1 < \widehat{U} \cdot \widehat{V} \leq 1$. Another simplification is to define the scalar product in terms of the direction-cosine

$$\cos\theta = |\widehat{U} \cdot \widehat{V}| \leq 1.$$

From these definitions we may define the minimum difference between the two vectors as the perpendicular from the end of the first to the intersection with the second. As shown in Fig. 3.5, $U \perp V$ may be found by minimizing the length of the vector difference:

$$\begin{aligned}
\min_{\alpha} ||V - \alpha U||^2 &= ||V||^2 + 2\alpha V \cdot U + \alpha^2 ||U||^2 > 0 \\
0 &= \partial_\alpha \, (V - \alpha U) \cdot (V - \alpha U) \\
&= V \cdot U - \alpha^* ||U||^2 \\
\therefore \quad \alpha^* &= V \cdot U / ||U||^2.
\end{aligned}$$

The Schwarz inequality follows:

$$I_{\min} = ||V - \alpha^* U||^2 = ||V||^2 - \frac{|U \cdot V|^2}{||U||^2} > 0 \tag{3.5.9}$$

$$0 \leq |U \cdot V| \leq ||U|| \, ||V||.$$

An important example of such a vector space includes the definition of the $\mathcal{FT}$ where we may set

$$U(\omega) = e^{-\omega_0 \jmath t} \qquad V(\omega) = e^{\omega \jmath t} \qquad U \cdot V = \int_\omega e^{\jmath \omega t} e^{-\jmath \omega_0 t} \frac{d\omega}{2\pi} = \delta(\omega - \omega_0).$$

It seems that the Fourier transform is a result that follows from a minimization, unlike the Laplace transform, which follows from causality. This explains the important differences between the two in terms of their properties (unlike the $\mathcal{LT}$ the $\mathcal{FT}$ is not complex analytic). Recall that

$$U \cdot V + \jmath U \wedge V = ||U|| \, ||V|| e^{\jmath \theta}.$$

3.5.5 *Gaussian Elimination*

The method for finding the intersection of equations is based on the recursive elimination of all the variables but one. This method, known as *Gaussian elimination* (Appendix 5), works across a broad range of cases but may be defined as a systematic algorithm when the equations are linear in the variables (Strang et al. 1993). Rarely do we even attempt to solve problems in several variables of degree greater than 1. But Gaussian elimination may still work in such cases (Stillwell 2010, p. 90). In Appendix 3.6.6 we derive the inverse of a 2×2 linear system of equations. Even for a 2×2 case, the general solution requires a great deal of algebra. Working out a numeric example of Gaussian elimination is more instructive. For example, suppose we wish to find the intersection of the two equations

$$\begin{aligned} x - y &= 3 \\ 2x + y &= 2. \end{aligned}$$

This 2×2 system of equations is so simple that you may immediately visualize the solution: By adding the two equations, y is eliminated, leaving $3x = 5$. But doing it this way takes advantage of the specific example, and we need a method for larger systems of equations. We need a generalized (algorithmic) approach. This general approach is called Gaussian elimination.

We start by writing the equations in matrix form (note this is not in the form $Ax = y$):

$$\begin{bmatrix} 1 & -1 \\ 2 & 1 \end{bmatrix} \begin{bmatrix} x \\ y \end{bmatrix} = \begin{bmatrix} 3 \\ 2 \end{bmatrix}. \tag{3.5.10}$$

Next, we eliminate the lower left term ($2x$) using a scaled version of the upper left term (x). Specifically, we multiply the first equation by -2 and add it to the second equation, replacing the second equation with the result. This gives

$$\begin{bmatrix} 1 & -1 \\ 0 & 3 \end{bmatrix} \begin{bmatrix} x \\ y \end{bmatrix} = \begin{bmatrix} 3 \\ 2 - 3 \cdot 2 \end{bmatrix} = \begin{bmatrix} 3 \\ -4 \end{bmatrix}. \tag{3.5.11}$$

Note that the top equation did not change. Once the matrix is "upper triangular" (zero below the diagonal), we have the solution. If we start from the bottom equation, $y = -4/3$. Then the upper equation gives $x - (-4/3) = 3$ or $x = 3 - 4/3 = 5/3$.

In principle, Gaussian elimination is easy, but if you make a calculation mistake along the way, it is very difficult to find your error. The method requires a lot of mental labor, and you have a high probability of making a mistake. Thus you do not want to apply this method every time. For example, suppose the elements are complex numbers or polynomials in some other variable such as frequency. Once the coefficients become more complicated, the seemingly trivial problem becomes corrosive. There is a much better way that is easily verified; it puts all the numerics at the end in a single step.

The above operations may be automated by finding a carefully chosen upper-diagonalized matrix G. For example, we can define the *Gaussian matrix* that zeros the element 2 in the matrix in Eq. 3.5.10. More generally let

$$G = \begin{bmatrix} 1 & 0 \\ a & 1 \end{bmatrix}. \tag{3.5.12}$$

Multiplying Eq. 3.5.10 by G, we find

$$\begin{bmatrix} 1 & 0 \\ a & 1 \end{bmatrix} \begin{bmatrix} 1 & -1 \\ 2 & 1 \end{bmatrix} \begin{bmatrix} x \\ y \end{bmatrix} = \begin{bmatrix} 1 & -1 \\ a+2 & 1-a \end{bmatrix} \begin{bmatrix} x \\ y \end{bmatrix} = \begin{bmatrix} 3 \\ 3a+2 \end{bmatrix}. \tag{3.5.13}$$

Thus we obtain Eq. 3.5.11 if we let $a = -2$ (we choose a to force the lower left to be zero). At this point we can either back-substitute and obtain the solution, as we did above, or find a matrix L that finishes the job by removing elements above the diagonal. Note that the determinant of matrix G is 1, thus it will always have an inverse.

Exercise #40 Using G and A from the discussion above show that $\det(G) = \det(GA) = 3$.
Sol: A common convention is to denote $\det(A) = |A|$. The two sides of the identity are

$$|A| = \det \begin{bmatrix} 1 & -1 \\ 2 & 1 \end{bmatrix} = 1 + 2 = 3, \qquad |GA| = \det \begin{bmatrix} 1 & -1 \\ 0 & 3 \end{bmatrix} = 3,$$

and $|G| = 1$. Thus $|GA| = |G||A| = 3$. ■

Matrix inverse: In Appendix 3.6.6, finding the inverse of a general 2×2 matrix takes three steps: (1) swap the diagonal elements, (2) reverse the signs of the off-diagonal elements, and (3) divide by the determinant $\Delta = ab - cd$. Specifically,

$$\begin{bmatrix} a & b \\ c & d \end{bmatrix}^{-1} = \frac{1}{\Delta} \begin{bmatrix} d & -b \\ -c & a \end{bmatrix}. \tag{3.5.14}$$

There are very few things that you must memorize, but the inverse of a 2×2 matrix is one of them. It needs to be in your mental toolkit, like completing the square (see Sect. 3.1.1).

While it is difficult to compute the inverse matrix from scratch (see Appendix 5), it takes only a few seconds (four dot products) to verify it (steps 1 and 2):

$$\begin{bmatrix} a & b \\ c & d \end{bmatrix} \begin{bmatrix} d & -b \\ -c & a \end{bmatrix} = \begin{bmatrix} ad - bc & -ab + ab \\ cd - cd & -bc + ad \end{bmatrix} = \begin{bmatrix} \Delta & 0 \\ 0 & \Delta \end{bmatrix}. \tag{3.5.15}$$

Thus dividing by the determinant gives the 2×2 identity matrix. A good strategy (don't trust your memory) is to write down the inverse as best you recall and then verify.

Using the 2×2 matrix inverse on our example (Eq. 3.5.10), we find

$$\begin{bmatrix} x \\ y \end{bmatrix} = \frac{1}{1+2} \begin{bmatrix} 1 & 1 \\ -2 & 1 \end{bmatrix} \begin{bmatrix} 3 \\ 2 \end{bmatrix} = \frac{1}{3} \begin{bmatrix} 5 \\ -6+2 \end{bmatrix} = \begin{bmatrix} 5/3 \\ -4/3 \end{bmatrix}. \tag{3.5.16}$$

If you use this method, you will rarely (never) make a mistake and the solution is easily verified.

Augmented matrix: There is one minor notational improvement. Rather than writing the matrix equation as Eq. 3.5.10 ($Ax = y$), we place the y vector next to the elements of A to remove the equal sign, which is cumbersome. In this case we write GA_{aug}:

$$GA_{\text{aug}} = \begin{bmatrix} 1 & 0 \\ -2 & 1 \end{bmatrix} \left[\begin{array}{cc|c} 1 & -1 & 3 \\ 2 & 1 & 2 \end{array}\right] = \left[\begin{array}{cc|c} 1 & -1 & 3 \\ 0 & 3 & -4 \end{array}\right].$$

In summary, this is the same as Eq. 3.5.12, where $a = -2$. Thus the lower-left 2 is removed (becomes 0).

3.6 Matrix Algebra: Systems

3.6.1 Vectors

Vectors as columns of ordered sets of scalars $\in \mathbb{C}$. When we write them out in text, we typically use row notation, with the *transpose* symbol:

$$[a, b, c]^T = \begin{bmatrix} a \\ b \\ c \end{bmatrix}.$$

This is strictly to save space on the page. The notation for *conjugate transpose is* $\dagger$, for example,

$$\begin{bmatrix} a \\ b \\ c \end{bmatrix}^\dagger = \begin{bmatrix} a^* & b^* & c^* \end{bmatrix}.$$

The above example is said to be a *3-dimensional* vector because it has three components.

Row versus column-vectors: With rare exceptions, vectors are columns, denoted *column-major*.[20] To avoid confusion, it is a good rule to make your mental default column-major, in keeping with most signal processing (vectorized) software.[21] Column-vectors are the unstated default of MATLAB/Octave, only revealed when matrix operations are performed. The need for the column (or row) major is revealed as a consequence of efficiency when accessing long sequences of numbers from computer memory. For example, when forming the sum of many numbers using the MATLAB/Octave command `sum(A)`, where `A` is a matrix, MATLAB/Octave operates on the columns, returning a row vector of column sums:

$$\texttt{sum} \begin{bmatrix} 1 & 2 \\ 3 & 4 \end{bmatrix} = [4, 6].$$

[20] https://en.wikipedia.org/wiki/Row-_and_column-major_order.

[21] In contrast, reading words in English is "row-major".

If the data were stored in row-major order, the answer would be the Column-vector $\begin{bmatrix}3\\7\end{bmatrix}$. Thus MATLAB/Octave is column-major by default.

3.6.2 *Vector Products*

A *scalar product* (aka dot product) is defined to "weight" vector elements before summing them, resulting in a scalar. The transpose of a vector (a *row-vector*) is typically used as a *scale factor* (i.e., weights) on the elements of a vector. For example,

$$\begin{bmatrix}1\\2\\-1\end{bmatrix}\cdot\begin{bmatrix}1\\2\\3\end{bmatrix}=\begin{bmatrix}1\\2\\-1\end{bmatrix}^T\begin{bmatrix}1\\2\\3\end{bmatrix}=\begin{bmatrix}1 & 2 & -1\end{bmatrix}\begin{bmatrix}1\\2\\3\end{bmatrix}=1+2\cdot 2-3=2.$$

A more interesting example defines a polynomial

$$\begin{bmatrix}1\\2\\4\end{bmatrix}\cdot\begin{bmatrix}1\\s\\s^2\end{bmatrix}=\begin{bmatrix}1\\2\\4\end{bmatrix}^T\begin{bmatrix}1\\s\\s^2\end{bmatrix}=\begin{bmatrix}1 & 2 & 4\end{bmatrix}\begin{bmatrix}1\\s\\s^2\end{bmatrix}=1+2s+4s^2.$$

Polar scalar product: The vector–scalar product in polar coordinates is (Fig. 3.4)

$$\boldsymbol{B}\cdot\boldsymbol{C}=\|\boldsymbol{B}\|\;\|\boldsymbol{C}\|\cos\theta\in\mathbb{R},$$

where $\cos\theta\in\mathbb{R}$ is called the *direction-cosine* between $\boldsymbol{B}$ and $\boldsymbol{C}$.
Polar wedge-product: The vector wedge-product in polar coordinates is (Fig. 3.4)

$$\boldsymbol{B}\wedge\boldsymbol{C}=\|\boldsymbol{B}\|\;\|\boldsymbol{C}\|\sin\theta\in\mathbb{R},$$

where $\sin\theta\in\mathbb{R}$ is therefore the *direction-sine* between $\boldsymbol{B}$ and $\boldsymbol{C}$.
Complex polar scalar product: From these two polar definitions and $e^{\jmath\theta}=\cos\theta+\jmath\sin\theta$,

$$\boldsymbol{B}\cdot\boldsymbol{C}+\jmath\boldsymbol{B}\wedge\boldsymbol{C}=||\boldsymbol{B}||||\boldsymbol{C}||\,e^{s}.$$

Hence

$$|\boldsymbol{B}\cdot\boldsymbol{C}|^2+|\boldsymbol{B}\wedge\boldsymbol{C}|^2=|\|\boldsymbol{B}\|^2\;\|\boldsymbol{C}\|^2\cos^2\theta|+|\|\boldsymbol{B}\|^2\;\|\boldsymbol{C}\|^2\sin^2\theta|=||\boldsymbol{B}||^2||\boldsymbol{C}||^2.$$

This relationship holds true in any vector space, of any number of dimensions, containing vectors $\boldsymbol{B}$ and $\boldsymbol{C}$. In this case $s=\sigma+\omega\jmath\in\mathbb{C}$ can be the Laplace frequency. Jaynes (1991) has an relevant discussion about this type of scalar product.

3.6.3 Norms of Vectors

The norm of a vector is the scalar product of the vector with itself

$$\|\boldsymbol{A}\| = \sqrt{\boldsymbol{A} \cdot \boldsymbol{A}} \geq 0,$$

forming the Euclidean length of the vector.[22]
Euclidean distance between two points in $\mathbb{R}^3$: The scalar product of the difference between two vectors $(\boldsymbol{A} - \boldsymbol{B}) \cdot (\boldsymbol{A} - \boldsymbol{B})$ is the Euclidean distance between the points they define

$$\|\boldsymbol{A} - \boldsymbol{B}\| = \sqrt{(a_1 - b_1)^2 + (a_2 - b_2)^2 + (a_3 - b_3)^2}.$$

Triangle inequality

$$\|\boldsymbol{A} + \boldsymbol{B}\| = \sqrt{(a_1 + b_1)^2 + (a_2 + b_2)^2 + (a_3 + b_3)^2} \leq \|\boldsymbol{A}\| + \|\boldsymbol{B}\|.$$

In terms of a right triangle this says the sum of the lengths of the two sides is greater to the length of the hypotenuse, and equal when the triangle degenerates into a line.
Vector cross-product: The *vector product* (aka cross-product) $\boldsymbol{A} \times \boldsymbol{B} = \|\boldsymbol{A}\| \, \|\boldsymbol{B}\| \sin\theta$ is defined between the two vectors $\boldsymbol{A}$ and $\boldsymbol{B}$. In Cartesian coordinates

$$\boldsymbol{A} \times \boldsymbol{B} = \det \begin{vmatrix} \hat{\mathbf{x}} & \hat{\mathbf{y}} & \hat{\mathbf{z}} \\ a_1 & a_2 & a_3 \\ b_1 & b_2 & b_3 \end{vmatrix}.$$

The triple product: This is defined between three vectors as

$$\boldsymbol{A} \cdot (\boldsymbol{B} \times \boldsymbol{C}) = \det \begin{vmatrix} a_1 & a_2 & a_3 \\ b_1 & b_2 & b_3 \\ c_1 & c_2 & c_3 \end{vmatrix}.$$

This may be indicated without the use of parentheses, since there can be no other meaningful interpretation. However for clarity, parentheses should be used. The triple product is the volume of the parallelepiped (3D-crystal shape) outlined by the three vectors, as shown in Fig. 3.4.
Dialects of vector notation: Physical fields are, by definition, functions of space $\boldsymbol{x}$ [m], and in the most general case, time t [s]. When Laplace transformed, the fields become functions of space and complex frequency (e.g., $\boldsymbol{E}(\boldsymbol{x}, t) \leftrightarrow \boldsymbol{E}(\boldsymbol{x}, s)$). As before, there are several equivalent vector notations. For example, $\boldsymbol{E}(\boldsymbol{x}, t) = \left[E_x, E_y, E_z\right]^T = E_x(\boldsymbol{x}, t)\hat{\mathbf{x}} + E_y(\boldsymbol{x}, t)\hat{\mathbf{y}} + E_z(\boldsymbol{x}, t)\hat{\mathbf{z}}$ is "in-line," to save space. The same equation may written in "displayed" notation as

[22]This leaves open the interpretation of the curly norm $\|\boldsymbol{A}\| = \sqrt{\boldsymbol{A} \curlywedge \boldsymbol{A}} \geq 0$. The obvious answer is in the properties of the Schwarz inequality, which has a complex analytic angle $\theta \in \mathbb{C}(\notin \mathbb{R})$.

$$\boldsymbol{E}(\boldsymbol{x},t) = \begin{bmatrix} E_x(\boldsymbol{x},t) \\ E_y(\boldsymbol{x},t) \\ E_z(\boldsymbol{x},t) \end{bmatrix} = \begin{bmatrix} E_x \\ E_y \\ E_z \end{bmatrix} (\boldsymbol{x},t) = \left[E_x,\ E_y,\ E_z\right]^T \equiv E_x\hat{\mathbf{x}} + E_y\hat{\mathbf{y}} + E_z\hat{\mathbf{z}}.$$

Note the four notations for vectors, bold font, element-wise columns, element-wise transposed rows, and dyadic format. These are all shorthand notations for expressing the vector. Such usage is similar to a dialect in a language.

Complex elements: When the elements are complex ($\in \mathbb{C}$), the transpose is defined as the complex conjugate of the elements. In such complex cases the transpose conjugate may be denoted with a † rather than T

$$\begin{bmatrix} -2\jmath \\ 3\jmath \\ 1 \end{bmatrix}^\dagger = \left[-2\jmath,\ 3\jmath,\ 1\right] \in \mathbb{C}.$$

For this case when the elements are complex, the dot product is a real number (like a length)

$$\boldsymbol{a}\cdot\boldsymbol{b} = \boldsymbol{a}^\dagger\boldsymbol{b} = \left[a_1^*\ a_2^*\ a_3^*\right]\begin{bmatrix} b_1 \\ b_2 \\ b_3 \end{bmatrix} = a_1^*b_1 + a_2^*b_2 + a_3^*b_3 \in \mathbb{R}.$$

Norm of a complex vector: The dot product of a vector with itself is called the *norm* of $\boldsymbol{a}$

$$||\boldsymbol{a}|| = \sqrt{\boldsymbol{a}^\dagger\boldsymbol{a}} \geq 0.$$

which is always nonnegative, and real.

Such a construction is useful when $\boldsymbol{a}$ and $\boldsymbol{b}$ are related by an impedance matrix

$$\boldsymbol{V}(s) = \boldsymbol{Z}(s)\boldsymbol{I}(s)$$

and we wish to compute the power. For example, the impedance of a mass is ms and a capacitor is $1/sC$. When given a system of equations (a mechanical or electrical circuit) one may define an impedance matrix.

Complex power: In this special case, the *complex power* $\mathcal{P}(s) \in \mathbb{C}(s)$ is defined, in the complex frequency domain (s), as

$$\mathcal{P}(s) = \boldsymbol{I}^\dagger(s)\boldsymbol{V}(s) = \boldsymbol{I}^\dagger(s)\boldsymbol{Z}(s)\boldsymbol{I}(s) \leftrightarrow p(t) \quad [\mathrm{W}].$$

The real part of the complex power must be positive. The imaginary part corresponds to available stored energy.

3.6.4 Matrices

When working with matrices, the role of the weights and vectors can change, depending on the context. A useful way to view a matrix is as a set of column-vectors, weighted by the elements of the column-vector of weights multiplied from the right. For example,

$$\begin{bmatrix} a_{11} & a_{12} & a_{13} & \cdots & a_{1M} \\ a_{21} & a_{22} & a_{23} & \cdots & a_{2M} \\ & & \ddots & & \\ a_{N1} & a_{N2} & a_{N3} & \cdots & a_{NM} \end{bmatrix} \begin{bmatrix} w_1 \\ w_2 \\ w_3 \\ \vdots \\ w_M \end{bmatrix} = w_1 \begin{bmatrix} a_{11} \\ a_{21} \\ a_{31} \\ \vdots \\ a_{N1} \end{bmatrix} + w_2 \begin{bmatrix} a_{12} \\ a_{22} \\ a_{32} \\ \vdots \\ a_{N2} \end{bmatrix} + \cdots + w_M \begin{bmatrix} a_{1M} \\ a_{2M} \\ a_{3M} \\ \vdots \\ a_{NM} \end{bmatrix},$$

where the weights are $\left[w_1, w_2, \ldots, w_M\right]^T$. Note that a_{23} is in row 2, column 3, thus is $a_{\text{row,col}}$. Rows are index vertically, according to the column definition of a vector. Think of the matrix as M column vectors with a_{n1} being the first vector.

Alternatively, the matrix is a set of row vectors of weights, each of which is applied to the column-vector on the right ($[w_1, w_2, \ldots, W_M]^T$). Both views are important (and correct). Don't think of a matrix as being just one or the other. It is both, but not at the same time.

The determinant of a matrix is denoted as either det $\boldsymbol{A}$ or simply $|\boldsymbol{A}|$ (as in the absolute value). The inverse of a square matrix is $\boldsymbol{A}^{-1}$ or inv$\boldsymbol{A}$. If $|A| = 0$, the inverse does not exist. If it does then $\boldsymbol{A}\boldsymbol{A}^{-1} = \boldsymbol{A}^{-1}\boldsymbol{A}$.

MATLAB/Octave's notional convention for a row-vector is $[a, b, c]$ and a column-vector is $[a; b; c]$. A prime on a vector takes the complex conjugate transpose. To suppress the conjugation, place a period before the prime. The argument converts the array into a column-vector, without conjugation. A tacit notation in MATLAB is that *vectors* are columns and the index to a vector is a row vector. MATLAB defines the notation 1:4 as the "row-vector" $[1, 2, 3, 4]$, which is unfortunate as it leads users to assume that the default vector is a row. This can lead to serious confusion later, as MATLAB'S default vector is a column. I have not found the above convention explicitly stated, and it took me years to figure this out for myself.

3.6.5 N x M Complex Matrices

Here are some definitions to learn:

1. *Scalar*: A number—for example, $\{a, b, c, \alpha, \beta, \ldots\} \in \{\mathbb{Z}, \mathbb{Q}, \mathbb{I}, \mathbb{R}, \mathbb{C}\}$
2. *Vector*: A quantity having direction as well as magnitude, often denoted by a bold letter **x**, or with an arrow over the top **x**. In matrix notation, this is typically represented as a single row $[x_1, x_2, x_3, \ldots]$ or single column $[x_1, x_2, x_3 \ldots]^T$ (where T indicates the transpose). In this text we will typically use column-vectors. The

vector may also be written out using unit vector notation to indicate direction. For example, $\mathbf{x}_{3,1} = x_1\hat{\mathbf{x}} + x_2\hat{\mathbf{y}} + x_3\hat{\mathbf{z}} = [x_1, x_2, x_3]^T$, where $\hat{\mathbf{x}}, \hat{\mathbf{y}}, \hat{\mathbf{z}}$ are unit vectors in the x, y, z Cartesian directions (here the vector's subscript 3, 1 indicates its dimensions). The type of notation used frequently depends on the engineering problem you are solving.

3. *Matrix:* $A = \left[\mathbf{a}_1, \mathbf{a}_2, \mathbf{a}_3, \ldots, \mathbf{a}_M\right]_{N,M} = \{a_{n,m}\}_{N,M}$ can be a non-square matrix if the number of elements in each of the vectors (N) is not equal to the number of vectors (M). When $M = N$, the matrix is square. It may be inverted if its determinant $|A| = \prod \lambda_k \neq 0$ (where λ_k are the eigenvalues). In this text we mainly work only with 2×2 and 3×3 square matrices.
4. *Linear system of equations*: $A\mathbf{x} = \mathbf{b}$ where $\mathbf{x}$ and $\mathbf{b}$ are vectors and matrix A is a square.

   (a) *Inverse:* The solution of this system of equations may be found by finding the inverse $\mathbf{x} = A^{-1}\mathbf{b}$.
   (b) *Equivalence:* If two systems of equations $A_0\mathbf{x} = \mathbf{b}_0$ and $A_1\mathbf{x} = \mathbf{b}_1$ have the same solution (i.e., $\mathbf{x} = A_0^{-1}\mathbf{b}_0 = A_1^{-1}\mathbf{b}_1$), they are said to be equivalent.
   (c) *Augmented matrix:* The first type of augmented matrix is defined by combining the matrix with the right-hand side. For example, given the linear system of equations of the form $A\mathbf{x} = \mathbf{y}$

$$\begin{bmatrix} a & b \\ c & d \end{bmatrix} \begin{bmatrix} x_1 \\ x_2 \end{bmatrix} = \begin{bmatrix} y_1 \\ y_2 \end{bmatrix},$$

   the augmented matrix is

$$[A|y] = \begin{bmatrix} a & b & y_1 \\ c & d & y_2 \end{bmatrix}.$$

   A second type of augmented matrix may be used for finding the inverse of a matrix (rather than solving a specific instance of linear equations $Ax = b$). In this case the augmented matrix is

$$[A|I] = \begin{bmatrix} a & b & 1 & 0 \\ c & d & 0 & 1 \end{bmatrix}.$$

   Performing Gaussian elimination on this matrix, until the left side becomes the identity matrix, yields A^{-1}. This is because multiplying both sides by A^{-1} gives $A^{-1}A|A^{-1}I = I|A^{-1}$.

5. *Permutation matrix* (P): A matrix that is equivalent to the identity matrix, but with scrambled rows (or columns). Such a matrix has the properties $\det(P) = \pm 1$ and $P^2 = I$. For the 2×2 case, there is only one permutation matrix:

$$P = \begin{bmatrix} 0 & 1 \\ 1 & 0 \end{bmatrix} \qquad P^2 = \begin{bmatrix} 0 & 1 \\ 1 & 0 \end{bmatrix} \begin{bmatrix} 0 & 1 \\ 1 & 0 \end{bmatrix} = \begin{bmatrix} 1 & 0 \\ 0 & 1 \end{bmatrix}.$$

A permutation matrix P swaps rows or columns of the matrix it operates on. For example, in the 2×2 case, pre-multiplication swaps the rows,

$$PA = \begin{bmatrix} 0 & 1 \\ 1 & 0 \end{bmatrix} \begin{bmatrix} a & b \\ \alpha & \beta \end{bmatrix} = \begin{bmatrix} \alpha & \beta \\ a & b \end{bmatrix},$$

whereas post-multiplication swaps the columns,

$$AP = \begin{bmatrix} a & b \\ \alpha & \beta \end{bmatrix} \begin{bmatrix} 0 & 1 \\ 1 & 0 \end{bmatrix} = \begin{bmatrix} b & a \\ \beta & \alpha \end{bmatrix}.$$

For the 3×2 case there are $3 \cdot 2/2 = 3$ such matrices (swap a row with the other 2, then swap the remaining two rows).

6. *Gaussian elimination (GE) operations* G_k: There are three types of elementary row operations, which may be performed without fundamentally altering a system of equations (e.g., the resulting system of equations is *equivalent*). These operations are (1) swap rows (e.g., using a permutation matrix), (2) scale rows, or (3) perform addition/subtraction of two scaled rows. All such operations can be performed using matrices.
   For lack of a better term, we'll describe these as "Gaussian elimination" or "GE" matrices.[23] We will categorize any matrix that performs only elementary row operations (but any number of them) as a "GE" matrix. Therefore, a cascade of GE matrices is also a GE matrix.
   Consider the GE matrix
   $$G = \begin{bmatrix} 1 & 0 \\ 1 & -1 \end{bmatrix}.$$
   (a) This pre-multiplication scales and subtracts row (2) from (1) and returns it to row (2).
   $$GA = \begin{bmatrix} 1 & 0 \\ 1 & -1 \end{bmatrix} \begin{bmatrix} a & b \\ \alpha & \beta \end{bmatrix} = \begin{bmatrix} a & b \\ a - \alpha & b - \beta \end{bmatrix}.$$
   The shorthand for this Gaussian elimination operation is $(1) \leftarrow (1)$ and $(2) \leftarrow (1) - (2)$.
   (b) Post-multiplication adds and scales *columns*.
   $$AG = \begin{bmatrix} a & b \\ \alpha & \beta \end{bmatrix} \begin{bmatrix} 1 & 0 \\ -1 & 1 \end{bmatrix} = \begin{bmatrix} a - b & b \\ \alpha - \beta & \beta \end{bmatrix}.$$
   Here the second column is subtracted from the first, and placed in the first. The second column is untouched. This operation is *not* a Gaussian elimination.

[23] The term "elementary matrix" may also be used to refer to a matrix that performs an elementary row operation. Typically, each elementary matrix differs from the identity matrix by a single row operation. A cascade of elementary matrices could be used to perform Gaussian elimination.

Therefore, to put Gaussian elimination operations in matrix form, we form a cascade of pre-multiplication matrices.
Here $\det(G) = 1, G^2 = I$, which won't always be true if we scale by a number greater than 1. For instance, if $G = \begin{bmatrix} 1 & 0 \\ m & 1 \end{bmatrix}$ (scale and add), then we have $\det(G) = 1,\ G^n = \begin{bmatrix} 1 & 0 \\ n \cdot m & 1 \end{bmatrix}$.

3.6.6 *Inverse of the 2 × 2 Matrix*

We shall now apply Gaussian elimination to find the solution $[x_1, x_2]$ for the 2×2 matrix equation $Ax = y$ (Eq. 3.5.8, left). We assume to know $[a, b, c, d]$ and $[y_1, y_2]$.

Here we wish to prove that the left equation (i) has an inverse given by the right equation (ii):

$$\begin{bmatrix} a & b \\ c & d \end{bmatrix} \begin{bmatrix} x_1 \\ x_2 \end{bmatrix} = \begin{bmatrix} y_1 \\ y_2 \end{bmatrix} \quad (i); \qquad \begin{bmatrix} x_1 \\ x_2 \end{bmatrix} = \frac{1}{\Delta} \begin{bmatrix} d & -b \\ -c & a \end{bmatrix} \begin{bmatrix} y_1 \\ y_2 \end{bmatrix} \quad (ii).$$

To take the inverse:
(1) swap the diagonal, (2) change the off-diagonal signs, and (3) normalize by the determinant Δ. We wish to show that the intersection (solution) is given by the equation on the right.

Exercise #1 Show that the equation on the right is the solution of the equation on the left.
Sol: By direct substitution (composition) of the right equation into the left equation, we have

$$\begin{bmatrix} a & b \\ c & d \end{bmatrix} \cdot \frac{1}{\Delta} \begin{bmatrix} d & -b \\ -c & a \end{bmatrix} \begin{bmatrix} y_1 \\ y_2 \end{bmatrix} = \frac{1}{\Delta} \begin{bmatrix} ad - bc & -ab + ab \\ cd - cd & -cb + ad \end{bmatrix} = \frac{1}{\Delta} \begin{bmatrix} \Delta & 0 \\ 0 & \Delta \end{bmatrix}, \qquad (3.6.1)$$

which gives the identity matrix. ■

3.7 Problems AE-2

3.7.1 *Topics of This Homework*

Linear versus nonlinear systems of equations, Euclid's formula, Gaussian elimination, matrix permutations, Ohm's law, two-port networks,

Deliverables: Answers to problems

3.7.2 Gaussian Elimination

Problem #1 Gaussian elimination

–1.1: Find the inverse of

$$A = \begin{bmatrix} 1 & 2 \\ 4 & 3 \end{bmatrix}.$$

–1.2: Verify that $A^{-1}A = AA^{-1} = \begin{bmatrix} 1 & 0 \\ 0 & 1 \end{bmatrix}$.

Problem #2 Find the solution to the following 3×3 matrix equation $Ax = b$ by GE. Show your intermediate steps. You can check your work at each step using Octave/MATLAB.

$$\begin{bmatrix} 1 & 1 & -1 \\ 3 & 1 & 1 \\ 1 & -1 & 4 \end{bmatrix} \begin{bmatrix} x_1 \\ x_2 \\ x_3 \end{bmatrix} = \begin{bmatrix} 1 \\ 9 \\ 8 \end{bmatrix}.$$

–2.1 Show (i.e., verify) that the first GE matrix G_1, which zeros out all entries in the first column is given by

$$G_1 = \begin{bmatrix} 1 & 0 & 0 \\ -3 & 1 & 0 \\ -1 & 0 & 1 \end{bmatrix}.$$

Identify the elementary row operations that this matrix performs.

–2.2 Find a second GE matrix, G_2, to put $G_1 A$ in upper triangular form. Identify the elementary row operations that this matrix performs.

–2.3 Find a third GE matrix G_3 that scales each row so that its leading term is 1. Identify the elementary row operations that this matrix performs.

–2.4: Find the last GE matrix, G_4, which subtracts a scaled version of row 3 from row 2, and scaled versions of rows 2 and 3 from row 1, such that you are left with the identity matrix ($G_4G_3G_2G_1A = I$).

–2.5: Solve for $\{x_1, x_2, x_3\}^T$ using the augmented matrix format $G_4G_3G_2G_1\{A|b\}$ (where $\{A|b\}$ is the augmented matrix). Note that if you've performed the preceding steps correctly, $x = G_4G_3G_2G_1b$.

–2.6: Find the pivot matrix G that rescales the second row of the augmented matrix $A|b$ by 1/3.

3.7.3 Two Linear Equations

Problem #3 In this problem we transition from a general pair of equations

$$f(x, y) = 0$$
$$g(x, y) = 0$$

to the important case of two linear equations

$$y = ax + b$$
$$y = \alpha x + \beta.$$

Note that to help keep track of the variables, roman coefficients (a, b) are used for the first equation and Greek (α, β) for the second.

–3.1: What does it mean, graphically, if these two linear equations have (1) a unique solution, (2) a nonunique solution, or (3) no solution?

–3.2: Assuming the two equations have a unique solution, find the solution for x and y.

–3.3: When will this solution fail to exist (for what conditions on a, b, α, and β)?

–3.4: Write the equations as a 2×2 matrix equation of the form $A\mathbf{x} = \mathbf{b}$, where $\mathbf{x} = \{x, y\}^T$.

–3.5: Find the inverse of the 2×2 matrix, and solve the matrix equation for x and y.

–3.6: Discuss the properties of the determinant of the matrix (Δ) in terms of the slopes of the two equations (a and α).

Problem #4 The application of *linear functional relationships* between *two variables* We use 2×2 matrices to describe two-port networks, as discussed in Sect. 3.8. Transmission lines are a great example: Both voltage and current must be tracked as they travel along the line. Figure 3.10 shows an example segment of a transmission line.

Suppose you are given the following pair of linear relationships between the input (source) variables V_1 and I_1 and the output (load) variables V_2 and I_2 of the transmission line:

$$\begin{bmatrix} V_1 \\ I_1 \end{bmatrix} = \begin{bmatrix} J & 1 \\ 1 & -1 \end{bmatrix} \begin{bmatrix} V_2 \\ I_2 \end{bmatrix}.$$

–4.1: Let the output (the load) be $V_2 = 1$ and $I_2 = 2$ (i.e., $V_2/I_2 = 1/2$ $\{\Omega\}$). Find the input voltage and current, V_1 and I_1.

–4.2: Let the input (source) be $V_1 = 1$ and $I_1 = 2$. Find the output voltage and current, V_2 and I_2.

3.7.4 Integer Equations: Applications and Solutions

Any equation for which we seek only integer solutions is called a *Diophantine* equation.

Problem #5 A practical example of using a Diophantine equation:

"A merchant had a 40-pound weight that broke into four pieces. When the pieces were weighed, it was found that each piece was a whole number of pounds and that the four pieces could be used to weigh every integral weight between 1 and 40 pounds. What were the weights of the pieces?" - *Bachet de Bèziriac (1623).*[24]

Here, weighing is performed using a balance scale that has two pans, with weights on either pan. Thus, given weights of 1 and 3 pounds, one can weigh a 2-pound weight by putting the 1-pound weight in the same pan with the 2-pound weight, and the 3-pound weight in the other pan. Then the scale will be balanced. A solution to the four weights for Bachet's problem is $1 + 3 + 9 + 27 = 40$ pounds.

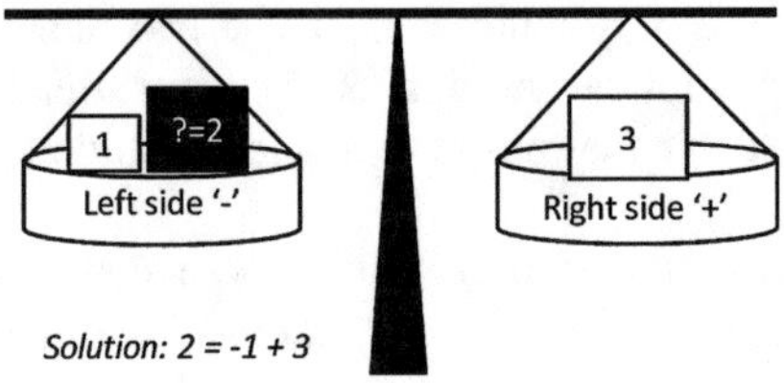

–5.1: Show how the combination of 1-, 3-, 9-, and 27-pound weights can be used to weigh $1, 2, 3, \dots, 8, 28$, and 40 pounds of milk (or something else, such as flour). Assuming that the milk is in the *left* pan, provide the position of the weights using a negative sign $-$ to indicate the left pan and a positive sign $+$ to indicate the right pan. For example, if the left pan has 1-pound of milk, then 1-pound of milk in the right pan, $+1$, will balance the scales.

Hint: It is helpful to write the answer in matrix form. Set the vector of values to be weighed equal to a matrix indicating the pan assignments, multiplied by a vector of the weights $[1, 3, 9, 27]^T$. The pan assignments matrix should contain only the values -1 (left pan), $+1$ (right pan), and 0 (leave out). You can indicate these using $-$, $+$, and blanks.

3.7.5 Vector Algebra in $\mathbb{R}^3$

Definitions of the scalar (also called a dot product) $\boldsymbol{A} \cdot \boldsymbol{B}$, cross $\boldsymbol{A} \times \boldsymbol{B}$ and triple product $\boldsymbol{A} \cdot (\boldsymbol{B} \times \boldsymbol{C})$, may be found in Appendix A, where $\boldsymbol{A}, \boldsymbol{B}, \boldsymbol{C}$ in $\mathbb{R}^3 \subset \mathbb{C}^3$), as shown in Fig. 3.4. A fourth "double-cross" (☠) vector product is:[25]

$$\boldsymbol{A} \times (\boldsymbol{B} \times \boldsymbol{C}) = \alpha_o \boldsymbol{B} - \beta_o \boldsymbol{C},$$

where $\alpha_o = \boldsymbol{A} \cdot \boldsymbol{C}$ and $\beta_o = \boldsymbol{A} \cdot \boldsymbol{B}$ (Note: $\boldsymbol{A} \times (\boldsymbol{B} \times \boldsymbol{C}) \neq (\boldsymbol{A} \times \boldsymbol{B}) \times \boldsymbol{C}$).

Problem #6 Scalar product A · B

–6.1: If $\boldsymbol{A} = a_x\hat{\mathbf{x}} + a_y\hat{\mathbf{y}} + a_z\hat{\mathbf{z}}$ and $\boldsymbol{B} = b_x\hat{\mathbf{x}} + b_y\hat{\mathbf{y}} + b_z\hat{\mathbf{z}}$, write out the definition of $\mathbf{A} \cdot \mathbf{B}$.

---

[24]Taken from Rotman (1996, p. 50).

[25]Greenberg p. 694, Eq. 8.

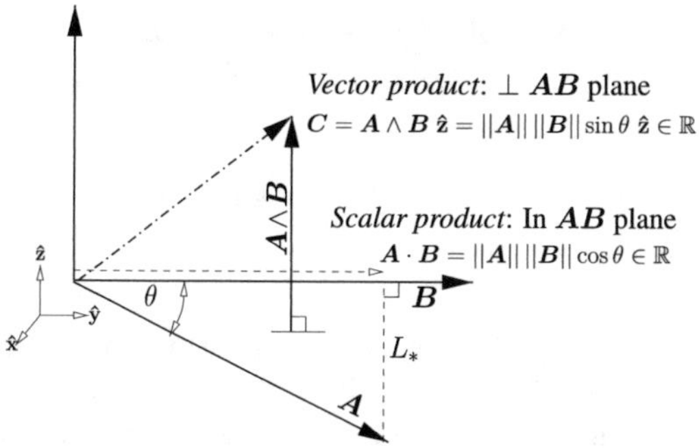

Fig. 3.6 This is figure is identical to Fig. 3.4, Sect.3.5. Definitions of vectors $\boldsymbol{A}, \boldsymbol{B}, \boldsymbol{C}$ (vectors in $\mathbb{R}^3$) used in the definition of $\boldsymbol{A} \cdot \boldsymbol{B}$, $\boldsymbol{A} \times \boldsymbol{B}$ and $\boldsymbol{A} \cdot (\boldsymbol{B} \times \boldsymbol{C})$. There are two algebraic vector products, the *scalar (dot) product* $\boldsymbol{A} \cdot \boldsymbol{B} \in \mathbb{R}$ and the *vector (cross) product* $\boldsymbol{A} \times \boldsymbol{B} \in \mathbb{R}^3$. Note that the result of the dot product is a scalar, while the vector product yields a vector, which is $\perp$ to the plane containing $\boldsymbol{A}, \boldsymbol{B}$

–6.2: The dot product is often defined as $||\boldsymbol{A}||\,||\boldsymbol{B}|| \cos(\theta)$, where $||\boldsymbol{A}|| = \sqrt{\boldsymbol{A} \cdot \boldsymbol{A}}$ and θ is the angle between $\boldsymbol{A}, \boldsymbol{B}$. If $||A|| = 1$, describe how the dot product relates to the vector **B**.

Problem #7 Vector (cross) product A × B

–7.1: If $\boldsymbol{A} = a_x\hat{\mathbf{x}} + a_y\hat{\mathbf{y}} + a_z\hat{\mathbf{z}}$ and $\boldsymbol{B} = b_x\hat{\mathbf{x}} + b_y\hat{\mathbf{y}} + b_z\hat{\mathbf{z}}$, write out the definition of **A** × **B**.

–7.2: Show that the cross-product is equal to the area of the parallelogram formed by $\boldsymbol{A}, \boldsymbol{B}$, namely, $||\boldsymbol{A}||\,||\boldsymbol{B}|| \sin(\theta)$, where $||\boldsymbol{A}|| = \sqrt{\boldsymbol{A} \cdot \boldsymbol{A}}$ and θ is the angle between $\boldsymbol{A}$ and $\boldsymbol{B}$.

Problem #8 Triple product $\boldsymbol{A} \cdot (\boldsymbol{B} \times \boldsymbol{C})$

Let $\boldsymbol{A} = [a_1, a_2, a_3]^T, \boldsymbol{B} = [b_1, b_2, b_3]^T, \boldsymbol{C} = [c_1, c_2, c_3]^T$ be three vectors in $\mathbb{R}^3$.

–8.1: Starting from the definition of the dot and cross-product, explain using a diagram and/or words, how one shows that: $\boldsymbol{A} \cdot (\boldsymbol{B} \times \boldsymbol{C}) = \begin{vmatrix} a_1 & a_2 & a_3 \\ b_1 & b_2 & b_3 \\ c_1 & c_2 & c_3 \end{vmatrix}$.

–8.2: Describe why $|\boldsymbol{A} \cdot (\boldsymbol{B} \times \boldsymbol{C})|$ is the volume of parallelepiped generated by $\boldsymbol{A}$,$\boldsymbol{B}$, and $\boldsymbol{C}$.

–8.3: Explain why three vectors **A**, **B**, **C** are in one plane if and only if the triple product $\mathbf{A} \cdot (\mathbf{B} \times \mathbf{C}) = 0$.

Problem #9 Given two vectors $\boldsymbol{A}, \boldsymbol{B}$ in the $\hat{\mathbf{x}}, \hat{\mathbf{y}}$ plane shown in Fig. 3.6 (same as Fig. 3.4), with $\boldsymbol{B} = \hat{\mathbf{y}}$ (i.e., $||\boldsymbol{B}|| = 1$).

–9.1: Show that $\boldsymbol{A}$ may be split into two orthogonal parts, one in the direction of $\boldsymbol{B}$ and the other perpendicular ($\perp$) to $\boldsymbol{B}$. Hint: Express the vector products of $\boldsymbol{A}$ and $\boldsymbol{B}$ (dot and cross) in polar coordinates (Greenberg 1988).

$$\begin{aligned} A &= (A \cdot B)B + B \times (A \times B) \\ &= A_{\parallel} + A_{\perp}. \end{aligned}$$

3.7.6 Ohm's Law

In general, impedance is defined as the ratio of a force to a flow. For electrical circuits, the voltage is the force and the current is the flow. Ohm's law states that the voltage across and the current through a circuit element are related by the impedance of that element (which may be a function of frequency). For resistors, the voltage over the current is called the *resistance* and is a constant (e.g., the simplest case is $V/I = R$). For inductors and capacitors, the voltage over the current is a frequency-dependent impedance (e.g., $V/I = Z(s)$, where s is the complex frequency $s \in \mathbb{C}$).

As shown in Table 3.2, the impedance concept also holds in mechanics and acoustics. In mechanics, the force is equal to the mechanical force on an element (e.g., a mass, dashpot, or spring) and the flow is the velocity. In acoustics, the force is pressure and the flow is the volume velocity or particle velocity of air molecules.

Problem #10 The resistance of an incandescent (filament) lightbulb, measured cold, is about 100 ohms. As the bulb lights up, the resistance of the metal filament increases. Ohm's law says that the current

$$\frac{V}{I} = R(T),$$

where T is the temperature. In the United States, the voltage is 120 volts (RMS) at 60 [Hz].

Problem #11 The power in watts is the product of the force and the flow. What is the power of the lightbulb of Problem #10?

Problem #12 State the impedance $Z(s)$ of each of the following circuit elements: (1) a resistor with resistance R, (2) an inductor with inductance L, and (3) a capacitor with capacitance C.

Problem #13 Consider what happens at the triple point of water. As water freezes or thaws, the temperature remains constant at 0 (C°). Once all the water is frozen and more heat is removed, the temperature drops below 0°. As heat is added, water thaws but the temperature remains at 0°. Once all the ice has melted, what is the temperature as more heat is added?

Model the triple point using a Zener diode, a resistor, and a capacitor. A Zener diode holds the voltage constant independent of current. For the case of water's triple point, the voltage represents the temperature of water at the triple point, clamped at 0 [C°]. The current represents the heat flux. The latent heat of water at the triple point is 32 Cal/gm. Thus as the temperature rises from below freezing, the water is clamped at 0° once the triple point is reached. At that point, adding more heat flux has no effect on the temperature until all the ice melts. Once the ice has melted, the temperature again begins to rise until it hits the boiling point, where it again stays at 100° until all the water has evaporated.

3.7.7 *Nonlinear (Quadratic) to Linear Equations*

In the following problems we deal with algebraic equations in more than one variable that are not linear equations. For example, the circle $x^2 + y^2 = 1$ may be solved for $y(x) = \pm\sqrt{1 - x^2}$. If we let $z_+ = x + y\jmath = x + \jmath\sqrt{1 - x^2} = e^{\theta\jmath}$, we obtain the equation for half a circle ($y > 0$). The entire circle is described by the magnitude of z as $|z|^2 = (x + y\jmath)(x - y\jmath) = 1$.

Problem #14 Give the curve defined by the equation:

$$x^2 + xy + y^2 = 1$$

–14.1: Find the function $y(x)$.
–14.2: Using MATLAB/Octave, plot $y(x)$ and describe the graph.
–14.3: What is the name of this curve?
–14.4: Find the solution (in x, p, and q) to these equations:

$$\begin{aligned} x + y &= p \\ xy &= q. \end{aligned}$$

–14.5: Find an equation that is linear in y starting from equations that are quadratic (second-degree) in the two unknowns x and y:

$$x^2 + xy + y^2 = 1 \tag{AE-2.1}$$
$$4x^2 + 3xy + 2y^2 = 3. \tag{AE-2.2}$$

–14.6: Compose the following two quadratic equations and describe the results.

$$\begin{aligned} x^2 + xy + y^2 &= 1 \\ 2x^2 + xy \quad\quad &= 1 \end{aligned}$$

3.7.8 *Nonlinear Intersection in Analytic Geometry*

Euclid's formula for Pythagorean triplets (Eq. 2.6.6) can be derived by intersecting a circle and a secant line. Consider the nonlinear equation of a unit circle having radius 1, centered at $(x, y) = (0, 0)$,

Euclidean Proof:

1) $2\phi + \eta = \pi$

2) $\eta + \Theta = \pi$

3) $\therefore \phi = \Theta/2$

Diophantus's Proof:

1) $c^2 = a^2 + b^2$

2) $b(a) = t(a + c)$

3) $\zeta(t) \equiv a + {\jmath}b = \dfrac{1 - t^2 + {\jmath}2t}{1 + t^2}$

4) $\zeta = |c|e^{i\theta} = |c|\dfrac{1 + it}{1 - it} = |c|(\cos(\theta) + i\sin(\theta))$

Pythagorean triplets:

1) $t = p/q \in \mathbb{Q}$

2) $a = p^2 - q^2$

3) $b = 2pq$

4) $c = p^2 + q^2$

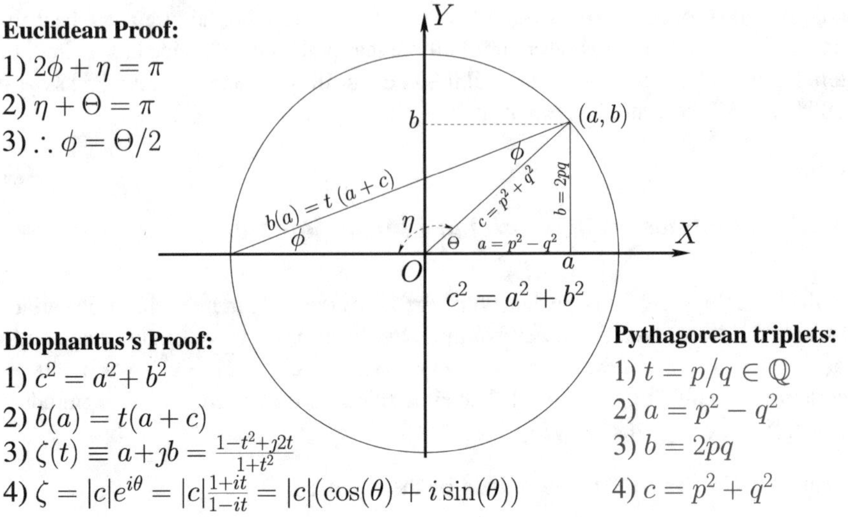

Fig. 3.7   Derivation of Euclid's formula for the Pythagorean triplets (PT) $[a, b, c]$, based on a composition of a line, having a rational slope $t = p/q \in \mathbb{F}$, and a circle $c^2 = a^2 + b^2$, $[a, b, c] \in \mathbb{N}$. This analysis is attributed to Diophantus (Di·o·phan′·tus) (250 CE), and today such equations are called Diophantine (Di·o·phan′·tine) *equations*. PTs have applications in architecture and scheduling, and many other practical problems. Most interesting is their relation to Rydberg's formula for the eigenstates of the hydrogen atom (Appendix I)

$$x^2 + y^2 = 1,$$

and the secant line through $(-1, 0)$,

$$y = t(x + 1),$$

a linear equation having slope t and intercept $x = -1$. If the slope $0 < t < 1$, the line intersects the circle at a second point (a, b) in the positive x, y quadrant. The goal is to find $a, b \in \mathbb{N}$ and then show that $c^2 = a^2 + b^2$. Since the construction gives a right triangle with short sides $a, b \in \mathbb{N}$, then it follows that $c \in \mathbb{N}$.

Problem #15 Derive Euclid's formula, following the arguments provided in Fig. 3.7.

    –15.1: Draw the circle and the line, given a positive slope $0 < t < 1$.

Problem #16 Next define $y = t(x + 1)$ (the line equation) into the equation for the circle, and solve for $x(t)$.

Hint: Because the line intersects the circle at two points, you will get two solutions for x. One of these solutions is the trivial solution $x = -1$.

    –16.1: Substitute the $x(t)$ you found back into the line equation, and solve for $y(t)$.

    –16.2: Let $t = q/p$ be a rational number, where p and q are integers. Find $x(p, q)$ and $y(p, q)$.]

–16.3: Substitute $x(p, q)$ and $y(p, q)$ into the equation for the circle, and show how Euclid's formula for the Pythagorean triples is generated.

For full points you must show that you understand the argument. Explain the meaning of the comment "magic happens" when t^4 cancels.

3.8 Transmission (ABCD) Matrix Composition Method

Matrix composition: Matrix multiplication represents a composition of 2×2 matrices because the input to the second matrix is the output of the first [this follows from the definition of composition: $f(x) \circ g(x) = f(g(x))$]. Thus the ABCD matrix is also known as the *transmission matrix* or occasionally the *chain matrix*. The general expression for the transmission matrix $\mathcal{T}(s)$ is

$$\begin{bmatrix} V_1 \\ I_1 \end{bmatrix} = \begin{bmatrix} \mathcal{A}(s) & \mathcal{B}(s) \\ \mathcal{C}(s) & \mathcal{D}(s) \end{bmatrix} \begin{bmatrix} V_2 \\ -I_2 \end{bmatrix} = \mathcal{T}(s) \begin{bmatrix} V_2 \\ -I_2 \end{bmatrix}. \tag{3.8.1}$$

The four coefficients $\mathcal{A}(s), \mathcal{B}(s), \mathcal{C}(s), \mathcal{D}(s)$ are all complex analytic functions of the Laplace frequency $s = \sigma + \jmath\omega$ (see Sects. 3.1–3.4). Typically they are polynomials in s. For example, $\mathcal{C}(s) = s^2 + 1$. A sum and parallel combination of inductors (masses), capacitors (springs), and resistors (dashpots) results in an Brune impedance, defined as the ratio of two polynomials. Thus such methods are called lumped-element networks. A symbolic eigenanalysis of 2×2 matrices may be found in Appendix B.3.

It is a standard convention to always define the current (flow) into the node. Since the input current on the left of Eq. 3.8.1 is the same as the output current on the right (I_2), we need the negative sign on I_2 to match the sign convention of current into every node. When we use this construction, all the currents will all agree.

We have already used 2×2 matrix composition for: (1) representing complex numbers (see Sect. 2.2), (2) computing the $\gcd(m, n)$ of $m, n \in \mathbb{N}$ (see Sect. F), (3) computing Pell's equation (see Sect. 2.6.2), and (4) computing the Fibonacci sequence (see Sect. 2.6.3). Thus it appears that 2×2 complex analytic matrices have high utility.

Definitions of $\mathcal{A}, \mathcal{B}, \mathcal{C}, \mathcal{D}$: By writing the equations that correspond to Eq. 3.8.1, we see that

$$\mathcal{A}(s) = \left.\frac{V_1}{V_2}\right|_{I_2=0}, \quad \mathcal{B}(s) = -\left.\frac{V_1}{I_2}\right|_{V_2=0}, \quad \mathcal{C}(s) = \left.\frac{I_1}{V_2}\right|_{I_2=0}, \quad \mathcal{D}(s) = -\left.\frac{I_1}{I_2}\right|_{V_2=0}. \tag{3.8.2}$$

Example: Figure 3.8 shows two examples of networks that may be analyzed using the ABCD transmission matrix method.

Each equation has a physical interpretation and a corresponding name. Functions $\mathcal{A}$ and $\mathcal{C}$ are said to be *blocked* because the output current I_2 is zero. Functions $\mathcal{B}$ and $\mathcal{D}$ are said to be *short-circuited* because the output voltage V_2 is zero. These two

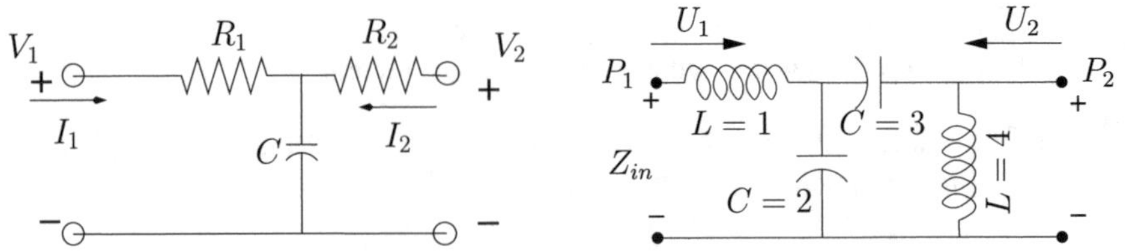

Fig. 3.8 Left: A low-pass RC electrical filter. The circuit elements R_1, R_2, and C are defined. **Right: A band-pass acoustic filter.** Here, the pressure P is analogous to voltage, and the velocity U is analogous to current. The circuit elements are labeled with their L and C values as integers, to make the algebra simple

terms (blocked vs. short-circuited) are electrical engineering–centric, arbitrary, and fail to generalize to other cases; thus we should avoid these terms.

For example, in a mechanical system *blocked* would correspond to an output isometric (no length change) velocity of zero. In mechanics the *isometric force* is defined as the maximum applied force conditioned on zero velocity (the blocked force). Thus the *short-circuited* force ($\mathcal{B}$) would correspond to zero force, which is nonsense. These engineering-centric terms do not gracefully generalize, so better terminology is needed. Much of this was sorted out by Thévenin in about 1883 (Van Valkenburg 1964a; Johnson 2003; Kennelly 1893).

$\mathcal{A}$ and $\mathcal{D}$ are called voltage (force) and current (velocity) *transfer functions*, since they are ratios of voltages and currents, whereas $\mathcal{B}$ and $\mathcal{C}$ are known as the *transfer impedance* and *transfer admittance*. For example, the unloaded (blocked) ($I_2 = 0$) output voltage $V_2 = I_1/\mathcal{C}$ corresponds to the isometric force in mechanics. In this way each term expresses an output (port 2) in terms of an input (port 1) for a given load condition.

Exercise #2 Derive the formula for $\mathcal{C}$ in terms of the input and output currents and voltages. Hint: See Eq. 3.8.2.
Solution Writing out the lower equation gives $I_1 = \mathcal{C}V_2 - \mathcal{D}I_2$ and setting $I_2 = 0$, we may obtain the equation for $\mathcal{C} = I_1/V_2|_{I_2=0}$. ■

Exercise #3 Can $\mathcal{C} = 0$?
Solution: Yes, if $I_2 = 0$ and $I_1 = I_2$, then $\mathcal{C} = 0$. In such cases the 2-port is ill-conditioned as shown in Appendix B.3, Eq. B.3.1. For $\mathcal{C} \neq 0$, there needs to be a finite shunt impedance across V_1, so that $I_1 \neq I_2 = 0$. ■

3.8.1 *Thévenin Parameters of a Source*

An important concept in circuit theory is that of the Thévenin parameters: the blocked force (zero flow) and the blocked flow (zero force). Their ratio defines the Thévenin impedance (Johnson 2003). The open-circuit voltage is defined as the voltage V_2 when the load current is zero ($I_2 = 0$), which was shown in Eq. 3.8.2 to be $V_2/I_1|_{I_2=0} = 1/\mathcal{C}$.

Thévenin Voltage: From Eq. 3.8.1 there are two definitions for the Thévenin voltage $V_{\text{Thev}} = V_2$, conditioned on the source on the left:

$$\left.\frac{V_{\text{Thev}}}{I_1}\right|_{I_2=0} = \frac{1}{\mathcal{C}} \quad \text{and} \quad \left.\frac{V_{\text{Thev}}}{V_1}\right|_{I_2=0} = \frac{1}{\mathcal{A}}. \tag{3.8.3}$$

A more general expression is needed when the source impedance is mixed.

Thévenin impedance The Thévenin impedance is the impedance looking into port 2 with $V_1 = 0$; thus

$$Z_{\text{Thev}} = \left.\frac{V_2}{I_2}\right|_{V_1=0}. \tag{3.8.4}$$

From the upper equation of Eq. 3.8.1, with $V_1 = 0$, we obtain $\mathcal{A}V_2 = \mathcal{B}I_2$; thus

$$Z_{\text{Thev}} = \frac{\mathcal{B}}{\mathcal{A}}. \tag{3.8.5}$$

3.8.2 The Impedance Matrix

With a bit of algebra, we can find the impedance matrix in terms of $\mathcal{A}, \mathcal{B}, \mathcal{C}, \mathcal{D}$ (Van Valkenburg 1964a, p. 310):

$$\begin{bmatrix} V_1 \\ V_2 \end{bmatrix} = \begin{bmatrix} z_{11} & z_{12} \\ z_{21} & z_{22} \end{bmatrix} \begin{bmatrix} I_1 \\ I_2 \end{bmatrix} = \mathcal{Z}(s) \begin{bmatrix} I_1 \\ I_2 \end{bmatrix} = \frac{1}{\mathcal{C}} \begin{bmatrix} \mathcal{A} & \Delta_{\mathcal{T}} \\ 1 & \mathcal{D} \end{bmatrix} \begin{bmatrix} I_1 \\ -I_2 \end{bmatrix}. \tag{3.8.6}$$

The determinate of the transmission matrix is $\Delta_{\mathcal{T}} = \pm 1$, and if $\mathcal{C} = 0$, the impedance matrix does not exist (see Exercise #42).

Definitions of $z_{11}(s), z_{12}(s), z_{21}(s), z_{22}(s)$: The definitions of the matrix elements are easily read off of the equation as

$$z_{11} \equiv \left.\frac{V_1}{I_1}\right|_{I_2=0}, \quad z_{12} \equiv -\left.\frac{V_1}{I_2}\right|_{I_1=0}, \quad z_{21} \equiv \left.\frac{V_2}{I_1}\right|_{I_2=0}, \quad z_{22} \equiv -\left.\frac{V_2}{I_2}\right|_{I_1=0}. \tag{3.8.7}$$

These definitions follow trivially from Eq. 3.8.6 and each element has a physical interpretation. For example, the unloaded ($I_2 = 0$, also called blocked or isometric) input impedance is $z_{11}(s) = \mathcal{A}(s)/\mathcal{C}(s)$, while the unloaded transfer impedance is $z_{21}(s) = 1/\mathcal{C}(s)$. For reciprocal systems (Postulate P6, Sect. 3.10.2), $z_{12} = z_{21}$, since $\Delta_{\mathcal{T}} = 1$. For antireciprocal systems, such as dynamic (also called magnetic) loudspeakers and microphones (Kim and Allen 2013), $\Delta_{\mathcal{T}} = -1$; thus $z_{21} = -z_{12} = 1/\mathcal{C}$. Finally z_{22} is the impedance looking into port 2 with port 1 open/blocked ($I_1 = 0$).

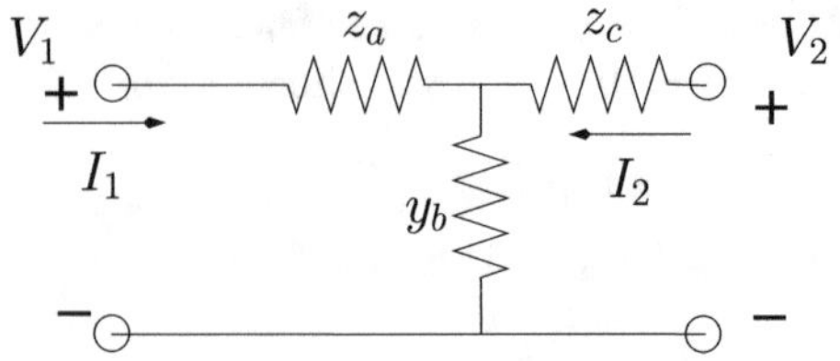

Fig. 3.9 Equivalent circuit for a transmission matrix. This allows us to better visualize the matrix elements in terms of complex impedances $z_a(s)$, $z_c(s)$, $y_b(s)$, as defined in this figure

To understand the meaning of the four impedance variables we analyze the transmission matrix of Fig. 3.9

$$\begin{bmatrix} V_1 \\ I_1 \end{bmatrix} = \begin{bmatrix} 1 + z_a y_b & z_c(1 + z_a y_b) + z_a \\ y_b & 1 + y_b z_c \end{bmatrix} \begin{bmatrix} V_2 \\ -I_2 \end{bmatrix}. \tag{3.8.8}$$

Note that it is easy to invert the $\mathcal{T}(s)$ matrix because $\Delta_{\mathcal{T}} = \pm 1$.

From the circuit elements defined in Fig. 3.9 (i.e., z_a, z_c, y_b) we can compute the impedance matrix elements of Eq. 3.8.6 (i.e., $z_{11}, z_{12}, z_{21}, z_{22}$). For example, the impedance matrix element z_{11}, in terms of z_a and y_b, is easily read off of Fig. 3.9 as the sum of the series and shunt impedances:

$$z_{11}(s)|_{I_2=0} = z_a + 1/y_b = \frac{\mathcal{A}}{\mathcal{C}}.$$

Given the impedance matrix, we can then compute transmission matrix $\mathcal{T}(s)$—namely, from Eq. 3.8.6,

$$\frac{1}{\mathcal{C}(s)} = z_{21}, \qquad \frac{\mathcal{A}(s)}{\mathcal{C}(s)} = z_{11}.$$

The theory is best modeled using the transmission matrix (Eq. 3.8.1), while experimental data are best modeled using the impedance matrix (Eq. 3.8.6).

Rayleigh reciprocity: Figure 3.9 is particularly helpful in understanding the Rayleigh reciprocity Postulate P6 ($\mathcal{B}(s) = \pm\mathcal{C}(s)$, Sect. 3.10.2, G):

$$\left.\frac{V_2}{I_1}\right|_{I_2=0} = \left.\frac{V_1}{I_2}\right|_{I_1=0}.$$

This says that the unloaded output voltage over the input current is symmetric, which is obvious from Fig. 3.9.

Table 3.2 The *generalized impedance* is defined as the ratio of a force to a flow, a concept that also holds in mechanics and acoustics. In mechanics, the force is the mechanical force on an element (e.g., a mass, dashpot, or spring) and the flow is the velocity. In acoustics, the force is the gradient of the pressure, and the flow is the volume velocity or particle velocity of air molecules

Case	Potential	Flow	Impedance	Units *ohms* [Ω]
Electrical	Voltage (V)	Current (I)	$Z = -\nabla V/I$	[Ω]
Mechanics	Force (F)	Velocity (U)	$Z = -\nabla F/U$	Mechanical [Ω]
Acoustics	Pressure (P)	Particle velocity (V)	$Z = -\nabla P/V$	Specific [Ω]
Acoustics	Mean pressure ($\mathcal{P}$)	Volume velocity ($\mathcal{V}$)	$Z = -\nabla \mathcal{P}/\mathcal{V}$	Acoustic [Ω]
Thermal	Temperature (T)	Entropy ($\mathcal{S}$)	$Z = -\nabla T/\mathcal{S}$	Thermal [Ω]

3.8.3 Network Power Relationships

Impedance is a general concept, closely tied to the definition of power $\mathcal{P}(t)$ (and energy). *Power* is defined as the product of the effort (force) and the flow (current). As described in Table 3.2, these concepts are very general, applying to mechanics, electrical circuits, acoustics, thermal circuits, and any other case where conservation of energy applies. Two basic variables are defined, generalized force and generalized flow, also called *conjugate variables*. The product of the conjugate variables is the power, and the ratio is the impedance. For example, for the case of voltage and current,

$$\mathcal{P}(t) \equiv v(t)i(t), \qquad v(t) = z(t) \star i(t), \qquad i(t) = y(t) \star v(t)$$

where $\star$ represents convolution (Sect. 4.7.4)

$$v(t) = z(t) \star i(t) \equiv \int_{t=0}^{\infty} z(\tau)i(t-\tau)d\tau \leftrightarrow Z(\omega)I(\omega).$$

Power versus power series, linear versus nonlinear Another place where second-degree equations appear in physical applications is in energy and power calculations. The electrical power is given by the product of the voltage $v(t)$ and current $i(t)$ (or in mechanics as the force times the velocity). For example, if we define $\mathcal{P} = v(t)i(t)$ to be the power $\mathcal{P}$ [watts], then the total energy [joules] at time t is (Van Valkenburg 1964a, Sect. 14)

$$\mathcal{E}(t) = \int_0^t v(t)i(t)dt.$$

From this observe that the power is the rate of change of the total energy

$$\mathcal{P}(t) = \frac{d}{dt}\mathcal{E}(t),$$

reminiscent of the fundamental theorem of calculus (Eq. 4.2.2).

3.8.4 Ohm's Law and Impedance

The ratio of voltage to current is called the *impedance* and it has units of [ohms]. For example, given a resistor of $R = 10$ [ohms],

$$v(t) = R\, i(t);$$

namely, 1 [amp] flowing through the resistor would give 10 [volts] across it. Merging the linear relationship due to Ohm's law with the definition of power shows that the instantaneous power in a resistor is quadratic in voltage and current:

$$\mathcal{P}(t) = v(t) \cdot i(t) = v(t)^2/R = i(t)^2 R, \qquad \mathcal{E}(t) = \int_{-\infty}^{t} \mathcal{P}(t)dt. \tag{3.8.9}$$

Note that Ohm's law is linear in its relationship between voltage and current, whereas power and energy are quadratic nonlinear functions.

Ohm's law generalizes the $I(\omega)$, $V(\omega)$ relation in a very important way, resulting in a linear complex analytic function of complex frequency $s = \sigma + \omega\jmath$ (Kennelly 1893; Brune 1931a). Impedance is a fundamental concept in many fields of engineering. For example:[26] Newton's second law $F = ma$ obeys Ohm's law, with mechanical impedance $Z(s) = sm$. Hooke's law $F = kx$ for a spring is described by a mechanical impedance $Z(s) = k/s$. In mechanics a resistor is called a *dashpot* and its impedance is a positive-real constant.

Kirchhoff's laws: KCL and KVL: The laws of electricity and mechanics may be written using Kirchhoff's current and voltage laws (KCL and KVL), which lead to linear systems of equations in the currents and voltages (velocities and forces) of the system under study, with complex coefficients having positive-real parts.

Transfer functions (transfer matrix): The most common standard reference is a physical system that has an input $x(t)$ and an output $y(t)$. If the system is linear, then it may be represented by its impulse response $h(t)$. In such cases, the system equation is

$$y(t) = h(t) \star x(t) \leftrightarrow Y(\omega) = H(s)|_{\sigma=0}\, X(\omega),$$

namely, the convolution of the input with the impulse response gives the output. This relationship may be written in the frequency domain as a product of the Laplace

---

[26] In acoustics the pressure is a potential, like voltage. The force per unit area is given by $f = -\nabla p$; thus $F = -\int \nabla p\, dS$. Velocity is analogous to a current. In terms of the velocity potential, the velocity per unit area is $v = -\nabla\phi$.

transform of the impulse response evaluated on the $\omega \jmath$-axis and the Fourier transform of the input $X(\omega) \leftrightarrow x(t)$ and output $Y(\omega) \leftrightarrow y(t)$.

If the system is nonlinear, then the output is not given by a convolution, and the Fourier and Laplace transforms have no obvious meaning.

The question that must be addressed is why the *power* is nonlinear, whereas a *power series* of $H(s)$ is linear: Both have powers of the underlying variables. This is confusing and rarely, if ever, addressed. The quick answer is that powers of the Laplace frequency s correspond to derivatives, which are linear operations, whereas the product of the voltage $v(t)$ and current $i(t)$ is nonlinear. It is confusing because the work power has two different meanings. The important and interesting question will be addressed on Sect. 3.10.2 in terms of the system postulates of physical systems.

Ohm's law: In general, impedance is defined as the ratio of a force to a flow. For electrical circuits (Table 3.2), the voltage difference is the force and the current is the flow. Ohm's law states that the voltage across and the current through a circuit element are linearly related by the impedance of that element (which is typically a complex function of the complex Laplace frequency $s = \sigma + \omega \jmath$). For resistors, the voltage over the current is called the *resistance* and is a constant (e.g., the simplest case is $V/I = R \in \mathbb{R}$). For inductors and capacitors, the impedance depends on the Laplace frequency s (e.g., $V/I = Z(s) \in \mathbb{C}$).

As shown in Table 3.2, the impedance concept also holds for mechanics and acoustics. In mechanics, the force is equal to the mechanical force on an element (e.g., a mass, dashpot, or spring) and the flow is the velocity. In acoustics, the force density is the negative of the gradient of the pressure, and the flow is the volume velocity (or particle velocity) of air molecules.

In this section we shall derive the method of the linear composition of systems, also known as the *ABCD transmission matrix method* , or in the mathematical literature, the *Möbius (bilinear) transformation*. With the method of matrix composition, we can use a linear system of 2 × 2 matrices to represent a significant family of networks. By the application of Ohm's law to the circuit shown in Fig. 3.10, we can model a cascade of such cells, which characterize transmission lines (Campbell 1903).

Example of the use of the ABCD matrix composition: Figure 3.10 shows a network composed of a series inductor (mass) that has an impedance $Z_l = sL$ and a shunt capacitor (compliance) that has an admittance $Y_c = sC \in \mathbb{C}$. As determined by Ohm's law, each equation describes a linear relationship between the current and the voltage. For the inductive impedance, applying Ohm's law gives

$$Z_l(s) = (V_1 - V_2)/I_1,$$

where $Z_l(s) = Ls \in \mathbb{C}$ is the complex impedance of the inductor. For the capacitive impedance, applying Ohm's law gives

$$Y_c(s) = (I_1 + I_2)/V_2,$$

where $Y_c = sC \in \mathbb{C}$ is the complex admittance of the capacitor.

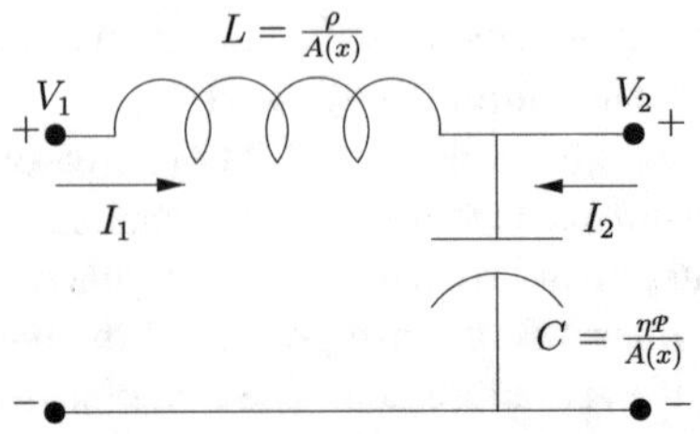

Fig. 3.10 A single LC cell of the LC transmission line. Every cell of any transmission line may be modeled by the ABCD method as the product of two matrices. For the example shown here, the inductance L of the coil and the capacitance C of the capacitor are in units of [henry/m] and [farad/m]; thus they depend on length Δ_x [m] that the cell represents. Note the flows are always defined as into the + node.

Each of these linear impedance relationships may be written in a 2×2 matrix format. The series inductor ($C = 0$) equation gives ($I_1 = -I_2$)

$$\begin{bmatrix} V_1 \\ I_1 \end{bmatrix} = \begin{bmatrix} 1 & Z_l \\ 0 & 1 \end{bmatrix} \begin{bmatrix} V_2 \\ -I_2 \end{bmatrix}, \tag{3.8.10}$$

while the shunt capacitor ($L = 0$) equation yields ($V_1 = V_2$)

$$\begin{bmatrix} V_1 \\ I_1 \end{bmatrix} = \begin{bmatrix} 1 & 0 \\ Y_c & 1 \end{bmatrix} \begin{bmatrix} V_2 \\ -I_2 \end{bmatrix}. \tag{3.8.11}$$

When the second matrix equation for the shunt admittance (Eq. 3.8.11) is substituted into the series impedance equation (Eq. 3.8.10), we find that the ABCD matrix composition ($\mathcal{T}_{12} = \mathcal{T}_1 \circ \mathcal{T}_2$) for the cell is the product of two matrices:

$$\begin{bmatrix} V_1 \\ I_1 \end{bmatrix} = \begin{bmatrix} 1 & Z_l \\ 0 & 1 \end{bmatrix} \begin{bmatrix} 1 & 0 \\ Y_c & 1 \end{bmatrix} \begin{bmatrix} V_2 \\ -I_2 \end{bmatrix} = \begin{bmatrix} 1 + Z_l Y_c & Z_l \\ Y_c & 1 \end{bmatrix} \begin{bmatrix} V_2 \\ -I_2 \end{bmatrix}. \tag{3.8.12}$$

Note that the determinant of the matrix $\Delta_\mathcal{T} = \mathcal{AD} - \mathcal{BC} = 1$. This is not an accident, since the determinants of the two matrices are each 1; thus the determinant of their product is 1. Every cascade of series and shunt elements will always have $\Delta_\mathcal{T} = \pm 1$.

For the case of Fig. 3.10, Eq. 3.8.12 has $\mathcal{A}(s) = 1 + s^2LC$, $\mathcal{B}(s) = sL$, $\mathcal{C}(s) = sC$, and $\mathcal{D} = 1$. These equations characterize the four possible relationships of the cell's input and output voltage and current. For example, the ratio of the output to input voltage, with the output unloaded, is

$$\left.\frac{V_2}{V_1}\right|_{I_2=0} = \frac{1}{\mathcal{A}(s)} = \frac{1}{1 + Z_l Y_c} = \frac{1}{1 + s^2LC}.$$

This is known as the *voltage divider relationship*. To derive the *current divider relationship*, we use the lower equation with $V_2 = 0$:

$$\left.\frac{-I_2}{I_1}\right|_{V_2=0} = 1.$$

Exercise #4 What happens if the order of Z and Y are reversed?
Solution:

$$\begin{bmatrix} V_1 \\ I_1 \end{bmatrix} = \begin{bmatrix} 1 & 0 \\ Y_c & 1 \end{bmatrix} \begin{bmatrix} 1 & Z_l \\ 0 & 1 \end{bmatrix} \begin{bmatrix} V_2 \\ -I_2 \end{bmatrix} = \begin{bmatrix} 1 & Z_l \\ Y_c & 1 + Z_l Y_c \end{bmatrix} \begin{bmatrix} V_2 \\ -I_2 \end{bmatrix} \tag{3.8.13}$$

This is the same network, reversed in direction. ■

Exercise #5 What happens if the series element is a capacitor and the shunt an inductor?
Solution:

$$\begin{bmatrix} V_1 \\ I_1 \end{bmatrix} = \begin{bmatrix} 1 & 1/Y_c \\ 0 & 1 \end{bmatrix} \begin{bmatrix} 1 & 0 \\ 1/Z_l & 1 \end{bmatrix} \begin{bmatrix} V_2 \\ -I_2 \end{bmatrix} = \begin{bmatrix} 1 + 1/Z_l Y_c & 1/Y_c \\ 1/Z_l & 1 \end{bmatrix} \begin{bmatrix} V_2 \\ -I_2 \end{bmatrix} \tag{3.8.14}$$

This circuit is a high-pass filter rather than a low-pass. ■

Properties of the transmission matrix: The transmission matrix is always constructed from the product of elemental matrices of the form

$$\begin{bmatrix} 1 & Z(s) \\ 0 & 1 \end{bmatrix} \quad \text{or} \quad \begin{bmatrix} 1 & 0 \\ Y(s) & 1 \end{bmatrix}.$$

Thus for the case of reciprocal systems (Postulate P6, Sect. 3.10),

$$\Delta_{\mathcal{T}} = \det \begin{bmatrix} \mathcal{A}(s) & \mathcal{B}(s) \\ \mathcal{C}(s) & \mathcal{D}(s) \end{bmatrix} = 1,$$

since the determinant of the product of each elemental matrix is 1 and the determinant of their product is 1. An antireciprocal system may be synthesized by the use of a gyrator, and for such cases $\Delta_{\mathcal{T}} = -1$.

The eigenvalue and vector equations for a T matrix are summarized in Appendix B and discussed in Appendix B.3. The basic postulates of network theory also apply to the matrix elements $\mathcal{A}(s)$, $\mathcal{B}(s)$, $\mathcal{C}(s)$, $\mathcal{D}(s)$, which place restrictions on their functional relationships. For example, Postulate P1 (Sect. 3.10.2) places limits on the poles and/or zeros of each function, since the time response must be causal.

3.9 Signals: Fourier Transforms

The two most fundamental tools for dealing with differential equations in engineering mathematics are the Fourier and the Laplace transforms, which deal with time-frequency analysis (Papoulis 1962).

The Fourier transform ($\mathcal{FT}$) takes a time-domain signal $f(t) \in \mathbb{R}$ and transforms it to the frequency domain by taking the scalar product (also called dot product) of $f(t)$ with the complex time vector $e^{-\jmath\omega t}$:

$$f(t) \leftrightarrow F(\omega) = f(t) \cdot e^{-\jmath\omega t},$$

where $F(\omega)$ and $e^{-\jmath\omega t} \in \mathbb{C}$ and $\omega, t \in \mathbb{R}$. Here $f(t)$ and $e^{\jmath\omega t}$ are in a Hilbert space, as discussed in Sect. 3.5.1. The scalar product between two vectors results in a scalar (number), as discussed in Appendix A.3.

Definition of the Fourier transform: The forward transform takes $f(t)$ to $F(\omega)$, while the inverse transform takes $F(\omega)$ to $\widetilde{f}(t)$. The tilde indicates that, in general, the recovered inverse transform signal can be slightly different from $f(t)$. Examples are presented in Table 3.6.

$$F(\omega) = \int_{-\infty}^{\infty} f(t)e^{-\jmath\omega t}dt \qquad \widetilde{f}(t) = \frac{1}{2\pi}\int_{-\infty}^{\infty} F(\omega)e^{\jmath\omega t}d\omega \tag{3.9.1}$$

$$F(\omega) \leftrightarrow f(t) \qquad \widetilde{f}(t) \leftrightarrow F(\omega).$$

It is accepted in the engineering and physics literature to use the case of the variable to indicate the type of argument. A time-domain function is $f(t)$, where t has units of seconds [s] and is lowercase. Its Fourier transform is uppercase $F(\omega)$ and is a function of frequency, having units of either hertz [Hz] or radians per second [2π Hz]. This case convention helps the reader parse the variable under consideration. This notation is helpful but not agree with the notation used in mathematics, where units are rarely cited.

Table 3.3 The general rule is that if a function is discrete in one domain (time or frequency) it is periodic in the other. Abbreviations: $\mathcal{FT}$: Fourier Transform; FS: Fourier Series; DTFT: Discrete-time Fourier transform; DFT: Discrete Fourier transform (the FFT is a "fast" DFT)

FREQUENCY \ TIME	Continuous t	Discrete t_k	Periodic $((t))_{T_o}$
Continuous ω	$\mathcal{FT}$	–	–
Discrete ω_k	–	DFT (FFT)	FS
Periodic $((\omega))_{\Omega_o}$	–	DTFT	DFT (FFT)

Types of Fourier transforms: As summarized in Table 3.3, each $\mathcal{FT}$ type is determined by symmetries in time and frequency. A time function $f(t)$ may be continuous in time, with $-\infty < t < \infty$, discrete in time, $f_n = f(t_n)$ with $t_k = kT_s$, where T_o is called the Nyquist sample period, or periodic in time, $f((t))_{T_p} = f(t + kT_p)$, where T_p is called the period. Here $k, n \in \mathbb{Z}$ and $T_o, T_p \in \mathbb{R}$. When time is discrete it is commonly represented as either $x[n]$ or $x(t_n)$.

A general rule is that if a function is discrete in one domain (time or frequency), it is periodic in the other domain (frequency or time). For example, the discrete-time function f_n must have a periodic frequency response—namely, $f_n \leftrightarrow F((\omega))_{T_p}$. This is the case of the discrete-time Fourier transform (DTFT). Alternatively, when the time function is periodic, the frequencies must be discrete—namely, $f((t))_{T_p} \leftrightarrow F(\omega_k)$. This is the case of the Fourier series (FS). When both the time and frequencies are discrete, both the time and frequencies must be periodic. This is the case of the discrete Fourier transform (DFT). These four cases are summarized in Table 3.3.

Table 3.4 As summarized in this table of scalar products (dot products), the four types of Fourier transforms differ in their support in time and frequency. These four are the (1) *Fourier transform*, (2) *Fourier series*, (3) *discrete-time Fourier transform*, and (4) *discrete Fourier transform*. The support in time and frequency defines the form of the inner product. In this way all the various forms of Fourier transforms may be reduced to differences in the scalar product, as dictated by the support of the signals in time and frequency. In the above $t_n = nT_s$, $f_k = k/T_s$ represent discrete-time and frequency samples, where T_s is one sample period. The signal period for the Fourier series (FS) is T [s]. For the discrete Fourier transform (DFT) the signal period is NT, where N is the length of the DFT. Typically the transform length is taken to be a power of 2, such as $N = 1024$ samples. This is done to improve the speed of the transform, known as the fast Fourier transform (FFT). The term *form* provides the mathematical form of the scalar product which depends on the signal symmetry (finite-duration, periodic, causal/one-sided, discrete-time/frequency, continuous-time/frequency, etc.). *ON* stands for *ortho-normal*. This column shows the signals that are used when taking the transform. The signal is projected onto these vectors by the scalar product

Name	Domain	Scalar product	Form	ON
(1) FT	$-\infty < t \in \mathbb{R} < \infty$	$x(t) \cdot y(t)$	$\int_{-\infty}^{\infty} x(t)y(t)dt$	$e^{-\jmath 2\pi ft}$
	$-\infty < f \in \mathbb{R} < \infty$	$X(f) \cdot Y(f)$	$\int_{-\infty}^{\infty} X(f)Y(f)\frac{d\omega}{2\pi}$	$e^{\jmath 2\pi ft}$
(2) FS	$0 \le t \in \mathbb{R} \le T$	$x((t)) \cdot y((t))$	$\frac{1}{T}\int_{t=0}^{T} x(t)y(t)dt$	$e^{-\jmath 2\pi f_k t}$
	$-\infty < f_k = \frac{k}{T} \in \mathbb{N} < \infty$	$X_k \cdot Y_k$	$\sum_{k=-\infty}^{\infty} X_k Y_k$	$e^{\jmath 2\pi f_k t}$
(3) DTFT	$-\infty < t_n < \infty$	$x_n \cdot y_n$	$\sum_{n=-\infty}^{\infty} x_n y_n$	$e^{-\jmath 2\pi t_n \Omega}$
	$-\pi < \Omega < \pi$	$X((\Omega)) \cdot Y((\Omega))$	$\int_{-\pi}^{\pi} X(e^{\jmath\Omega})Y(e^{\jmath\Omega})\frac{d\Omega}{2\pi}$	$e^{\jmath 2\pi t_n \Omega}$
(4) DFT	$0 \le t_n = nT \le (N-1)T$	$x_n y_n$	$\sum_{n=0}^{N-1} x_n y_n$	$e^{-\jmath 2\pi t_n f_k}$
	$0 \le f_k = \frac{k}{NT} \le \frac{(N-1)}{NT}$	$X_k Y_k$	$\frac{1}{N}\sum_{n=0}^{N-1} X_k Y_k$	$e^{\jmath 2\pi t_n f_k}$

3.9.1 Properties of the Fourier Transform

1. Both time t and frequency ω are real.
2. When a function is periodic in one domain (t, f), it must be discrete in the other (Table 3.3).
3. For the forward transform (time to frequency), the sign of the exponential is negative.
4. The limits on the integrals in both the forward and reverse FTs are $[-\infty, \infty]$.
5. When we take the inverse Fourier transform, the scale factor of $1/2\pi$ is required to cancel the 2π in the frequency differential $d\omega = 2\pi df$.
6. The Fourier step function is defined by the use of superposition of 1 and $\text{sgn}(t) = t/|t|$ as

$$\tilde{u}(t) \equiv \frac{1+\text{sgn}(t)}{2} = \begin{cases} 1 & t > 0 \\ 1/2 & t = 0 \\ 0 & t < 0 \end{cases}.$$

   Taking the FT of a delayed step function, we get

$$\tilde{u}(t - T_o) \leftrightarrow \frac{1}{2}\int_{-\infty}^{\infty} \left[1 - \text{sgn}(t - T_o)\right] e^{-j\omega t} dt = \pi\tilde{\delta}(\omega) + \frac{e^{-j\omega T_o}}{j\omega}.$$

   Thus the FT of the step function has the term $\pi\delta(\omega)$ due to the 1 in the definition of the Fourier step. This term introduces a serious flaw with the FT of the step function: While it appears to be causal, it is not. Compare this to the convolution $u(t) \star u(t)$ in Table 3.9.
7. The convolution $\tilde{u}(t) \star \tilde{u}(t)$ is not defined because both $1 \star 1$ and $\tilde{\delta}^2(\omega)$ are not defined.
8. The inverse $\mathcal{FT}$ has convergence issues whenever there is a discontinuity in the time response. We indicate this with a hat over the reconstructed time response. The error between the target time function and the reconstructed is zero in the root-mean sense, but not point-wise.
   Specifically, at the discontinuity point for the Fourier step function ($t = 0$) $\tilde{u}(t) \neq u(t)$, yet $\int |\tilde{u}(t) - u(t)|^2 dt = 0$. At the point of the discontinuity, the reconstructed function displays Gibbs ringing (it oscillates around the step and hence does not converge at the jump). The $\mathcal{LT}$ does not exhibit Gibbs ringing and is exact.
9. The $\mathcal{FT}$ is not always analytic in ω, as in this example of the step function. The step function cannot be expanded in a Taylor series about $\omega = 0$ because $\tilde{\delta}(\omega)$ is not analytic in ω.
10. The Fourier δ function is denoted $\tilde{\delta}(t)$ to differentiate it from the Laplace delta function $\delta(t)$. They differ because the step functions differ due to the convergence problem.

Table 3.5 Functional relationships between signals and their $\mathcal{FT}$'s

$\mathcal{FT}$	Functional properties
$\frac{d}{dt}v(t) \leftrightarrow \jmath\omega V(\omega)$	deriv
$\int_{-\infty}^{\infty} f(t-\tau)g(\tau)d\tau = f(t) \star g(t) \leftrightarrow F(\omega)G(\omega)$	conv
$f(t)g(t) \leftrightarrow \frac{1}{2\pi}F(\omega) \star G(\omega)$	mult
$f(at) \leftrightarrow \frac{1}{a}F\left(\frac{\omega}{a}\right)$	scale

11. One may define

$$\widetilde{u}(t) = \int_{-\infty}^{t} \widetilde{\delta}(t)dt$$

and the somewhat questionable notation

$$\widetilde{\delta}(t) = \frac{d}{dt}\widetilde{u}(t),$$

since the Fourier step function is not analytic.
12. The rec(t) function is defined as

$$\text{rec}(t) = \frac{\widetilde{u}(t) - \widetilde{u}(t - T_o)}{T_o} = \begin{cases} 0 & t < 0 \\ 1/T_o & 0 < t < T_o \\ 0 & t > T_0 \end{cases}.$$

It follows that $\widetilde{\delta}(t) = \lim_{T_o \to 0}$. Like $\widetilde{\delta}(t)$, the rec(t) has unit area (Table 3.5).

Periodic signals: As shown in Table 3.3 there are four variants of the $\mathcal{FT}$ that depend on the symmetry in time and frequency. For example, when the time signal is sampled (discrete in time), the frequency response becomes periodic, leading to the DTFT. When a time response is periodic, the frequency response is sampled (discrete in frequency), leading to the FS. These two symmetries may be simply characterized only as *periodic in time* $\Rightarrow$ discrete in frequency and *periodic in frequency* $\Rightarrow$ discrete in time. When a function is discrete in both time and frequency, it is necessarily periodic in time and frequency, leading to the DFT. The DFT is typically computed with an algorithm called the FFT, which can dramatically speed up the calculation when the data are a power of 2 in length.

Exercise #6 Consider the Fourier series scalar (dot) product (Eq. 3.5.2) between "vectors" $f((t))_{T_o}$ and $e^{-\jmath\omega_k t}$:

$$\begin{aligned} F(\omega_k) &= f((t))_{T_o} \cdot e^{-\jmath\omega_k t} \\ &\equiv \frac{1}{T_o}\int_0^{T_o} f(t)e^{-\jmath\omega_k t}dt, \end{aligned}$$

Table 3.6 Basic (Level I) Fourier transforms. Note that $a > 0 \in \mathbb{R}$ has units [rad/s]. To flag this necessary condition, we use $|a|$ to ensure this condition will be met. The other constant $T_o \in \mathbb{R}$ [s] has no restrictions, other than being real. Complex constants may not appear as the argument to a delta function, since complex numbers do not have the order property

$f(t) \leftrightarrow F(\omega)$	Name
$\widetilde{\delta}(t) \leftrightarrow 1(\omega) \equiv 1\ \forall\ \omega$	Delta
$1(t) \equiv 1\, \forall t \leftrightarrow 2\pi\widetilde{\delta}(\omega)$	Step
$\mathrm{sgn}(t) = \frac{t}{\lvert t\rvert} \leftrightarrow \frac{2}{\jmath\omega}$	
$\widetilde{u}(t) = \frac{1(t)+\mathrm{sgn}(t)}{2} \leftrightarrow \pi\widetilde{\delta}(\omega) + \frac{1}{\jmath\omega} \equiv \widetilde{U}(\omega)$	Step
$\widetilde{\delta}(t - T_o) \leftrightarrow e^{-\jmath\omega T_o}$	Delay
$\widetilde{\delta}(t - T_o) \star f(t) \leftrightarrow F(\omega)e^{-\jmath\omega T_o}$	Delay
$\widetilde{u}(t)e^{-\lvert a\rvert t} \leftrightarrow \frac{1}{\jmath\omega+\lvert a\rvert}$	Exp
$\mathrm{rec}(t) = \frac{1}{T_o}\left[\widetilde{u}(t) - \widetilde{u}(t - T_o)\right] \leftrightarrow$ $\frac{1}{T_o}\left(1 - e^{-\jmath\omega T_o}\right)$	Pulse
$\widetilde{u}(t) \star \widetilde{u}(t) \leftrightarrow \widetilde{\delta}^2(\omega)$ Not defined	NaN

where $\omega_0 = 2\pi/T_o$ and $f(t)$ has period T_o—that is, $f(t) = f(t + nT_o) = e^{\jmath\omega_n t}$ with $n \in \mathbb{N}$ and $\omega_k = k\omega_o$. What is the value of the Fourier series scalar product?
Sol: Evaluating the scalar product, we find

$$
\begin{aligned}
e^{\jmath\omega_n t} \cdot e^{-\jmath\omega_k t} &= \frac{1}{T_o}\int_0^{T_o} e^{\jmath\omega_n t} e^{-\jmath\omega_k t} dt \\
&= \frac{1}{T_o}\int_0^{T_o} e^{2\pi\jmath(n-k)t/T_o} dt = \begin{cases} 1 & n = k \\ 0 & n \neq k \end{cases}.
\end{aligned}
$$

The two signals (vectors) are orthogonal. ■

Exercise #7 Consider the discrete-time $\mathcal{FT}$ (DTFT) as a scalar (dot) product (Eq. 3.5.2, between "vectors" $f_n = f(t)|_{t_n}$ and $e^{-\jmath\omega t_n}$, where $t_n = nT_s$ and $T_s = 1/2F_{\max}$ is the sample period (Tables 3.4 and 3.6).
Sol: The scalar product over $n \in \mathbb{Z}$ is

$$
\begin{aligned}
F((\omega))_{2\pi} &= f_n \cdot e^{-\jmath\omega t_n} \\
&\equiv \sum_{n=-\infty}^{\infty} f_n e^{-\jmath\omega t_n},
\end{aligned}
$$

where $\omega_0 = 2\pi/T_o$ and $\omega_k = k\omega_o$ is periodic (i.e., $F(\omega) = F(\omega + k\omega_o)$). ■

3.10 Systems: Laplace Transforms

The Laplace transform $\mathcal{LT}$takes real causal signals $f(t)u(t) \in \mathbb{R}$, as a function of real time $t \in \mathbb{R}$, that are strictly zero for negative time ($f(t) = 0$ for $t < 0$), and transforms them into complex analytic functions ($F(s) \in \mathbb{C}$) of complex frequency $s = \sigma + \omega \jmath$. As we did for the Fourier transform, we use the same upper–lowercase notation: $f(t) \leftrightarrow F(s)$.

When a signal is zero for negative time $f(t < 0) = 0$, it is said to be *causal*, and the resulting transform $F(s)$ must be complex analytic over significant regions of the s plane. For a function of time to be causal, time *must* be real ($t \in \mathbb{R}$), since if it were complex, it would lose the order property (thus it could not be causal). It is helpful to emphasize the causal nature of $f(t)u(t)$ to force causality, with the Heaviside step function $u(t)$. Any restriction on a function (e.g., real, causal, periodic, positive-real part, etc.) is called a *symmetry property*. There are many forms of symmetry. The concept of symmetry is very general and widely used in both mathematics and physics, where it is more generally known as *group theory*. One-sided periodic transforms also exist, such as the system shown in Fig. 3.3.

Definition of the Laplace transform: The forward and inverse Laplace transforms are defined in Eq. 3.10.1. Here $s = \sigma + j\omega \in \mathbb{C}$ [2πHz] is the complex Laplace frequency in radians and $t \in \mathbb{R}$ [s] is the time in seconds.

Forward and inverse Laplace transforms (Table 3.7):

$$F(s) = \int_{0^-}^{\infty} f(t)e^{-st}dt \qquad f(t) = \frac{1}{2\pi\jmath}\int_{\sigma_o-\infty\jmath}^{\sigma_o+\infty\jmath} F(s)e^{st}ds \tag{3.10.1}$$

$$F(s) \leftrightarrow f(t) \qquad f(t) \leftrightarrow F(s).$$

Tables of functional properties are shown in Table 3.8, while basic transforms are provided in Appendix C, Table 3.9. Properties of more advanced $\mathcal{LT}$s are in Table C.2. When we deal with engineering problems, it is convenient to separate the *signals* we use from the *systems* that process them. We do this by treating signals, such as speech and music, differently from a system, such as a filter. In general, signals may

Table 3.7 Laplace transforms are complementary to the class of Fourier transforms $\mathcal{FT}$ because the time function must be a causal function. All $\mathcal{LT}$s are *complex analytic* in the complex frequency $s = \sigma + \omega\jmath$ domain. As an example, a causal function that is continuous but one-sided in time is the *step function* $u(t)$, which has the $\mathcal{LT}u(t) \leftrightarrow 1/s$. When a function is discrete in time but one-sided, it has a zeta-transform. The discrete-time step function is $u_n = u[n] \leftrightarrow 1/(1 - z^{-n})$

FREQUENCY \ TIME	Continuous t	Discrete $t[k]$	Causal-periodic $((t))_{T_o}$
Continuous s	$\mathcal{LT}$	–	–
Discrete $\omega[k]$	–	–	Unknown
Periodic $\|z\|e^{\theta\jmath}$	–	z-Transform	–

Table 3.8 Functional relationships between systems and their $\mathcal{LT}$s

$\mathcal{LT}$ Functional properties		
$\frac{d}{dt}f(t) = \delta'(t) \star f(t)$	$\leftrightarrow sF(s)$	Deriv
$f(t) \star g(t) = \int_{t=0}^{t} f(t-\tau)g(\tau)d\tau$	$\leftrightarrow F(s)G(s)$	causal convolution
$u(t) \star f(t) = \int_{0^-}^{t} f(t)dt$	$\leftrightarrow \frac{F(s)}{s}$	convolution with step
$f(at)u(at)$	$\leftrightarrow \frac{1}{a}F\left(\frac{s}{a}\right) \quad a \in \mathbb{R} \neq 0$	Scaling
$f(t)e^{-at}u(t)$	$\leftrightarrow F(s+a)$	Damped
$f(t-T)e^{-a(t-T)}u(t-T)$	$\leftrightarrow e^{-sT}F(s+a)$	Damped and delayed
$f(-t)u(-t)$	$\leftrightarrow F(-s)$	Reverse time
$f(-t)e^{-at}u(-t)$	$\leftrightarrow F(a-s)$	Time-reversed and damped
$\frac{\sin(t)u(t)}{t}$	$\leftrightarrow \tan^{-1}(1/s)$	Half-sync

start and end at any time. The concept of causality has no mathematical meaning in signal space. Systems, on the other hand, obey rigid rules (to ensure that they remain physical). These physical restrictions are described in terms of the system postulates, which we present on Sect. 3.10.2. There is a question as to why postulates are needed and which ones are the best choices. These questions are discussed in lectures by Feynman (1968, 1970a). The original video is also available online in many places, including YouTube.[27] There may be no definitive answers to these questions, but having a set of postulates is a useful way of thinking about physics.
Types of Laplace transforms: As shown in Table 3.7 there are three types of $\mathcal{LT}$s. The function may be continuous in time, in which it is also continuous in the Laplace frequency s. It may be discrete in time and therefore periodic in frequency θ, which is called the z-transform. Or it may be causal-periodic in time and therefore discrete in frequency. This transform has no name ("unknown," as best I know). An example is the Riemann zeta function (Fig. C.1).

3.10.1 *Properties of the Laplace Transform*

The following is a summary description of the $\mathcal{LT}$:

1. Time $t \in \mathbb{R}$ [s] and the Laplace frequency [rad] are defined as $s = \sigma + \omega\jmath \in \mathbb{C}$.
2. Given a Laplace transform ($\mathcal{LT}$) pair $f(t) \leftrightarrow F(s)$, in the engineering literature, the time domain is always lowercase [$f(t)$] and causal [i.e., $f(t < 0) = 0$], and the *frequency domain* is uppercase [$F(s)$]. Maxwell's venerable equations are the unfortunate exception to this otherwise universal rule.
3. The target time function $f(t < 0) = 0$ (i.e., it must be causal). The time limits are $0^- < t < \infty$. Thus the integral must start from slightly below $t = 0$ to integrate over a delta function at $t = 0$. For example, if $f(t) = \delta(t)$, the integral must

[27]https://www.youtube.com/watch?v=JXAfEBbaz_4, https://www.youtube.com/watch?v=YaUlqXRPMmY, https://www.youtube.com/watch?v=xnzB_IHGyjg.

Table 3.9 Laplace transforms of $f(t)$, $\delta(t)$, $u(t)$, rect(t), t_o, p, $e \in \mathbb{R}$ and $F(s)$, $G(s)$, s, $a \in \mathbb{C}$. Given a *Laplace transform* ($\mathcal{LT}$) pair $f(t) \leftrightarrow F(s)$, the frequency domain is always uppercase [e.g., $F(s)$] and the time-domain lowercase [$f(t)$] and causal (i.e., $f(t<0)=0$). An extended list of transforms is given in Appendix C, Table C.2

$f(t) \leftrightarrow F(s)$	$t \in \mathbb{R};\ s, F(s) \in \mathbb{C}$	Name
$\delta(t) \leftrightarrow 1$		Dirac
$\delta(\lvert a\rvert t) \leftrightarrow \frac{1}{\lvert a\rvert}$	$a \neq 0$	Time-scaled Dirac
$\delta(t-t_0) \leftrightarrow e^{-st_0}$		Delayed Dirac
$\delta(t-t_0) \star f(t) \leftrightarrow F(s)e^{-st_0}$		–
$\sum_{n=0}^{\infty} \delta(t-nt_0) = \frac{1}{1-\delta(t-t_0)} \leftrightarrow \frac{1}{1-e^{-st_0}} = \sum_{n=0}^{\infty} e^{-snt_0}$		One-sided impulse train
$u(t) \leftrightarrow \frac{1}{s}$		Heaviside step
$u(-t) \leftrightarrow -\frac{1}{s}$		Anticausal step
$u(at) \leftrightarrow \frac{a}{s}$	$a \neq 0 \in \mathbb{R}$	Dilated or reversed step
$e^{at}u(-t) \leftrightarrow \frac{1}{-s+a}$		Anticausal damped step
$e^{-at}u(t) \leftrightarrow \frac{1}{s+a}$	$a > 0 \in \mathbb{R}$	Damped step
$\cos(at)u(t) \leftrightarrow \frac{1}{2}\left(\frac{1}{s-a} + \frac{1}{s+a}\right)$	$a \in \mathbb{R}$	cos
$\sin(at)u(t) \leftrightarrow \frac{1}{2\jmath}\left(\frac{1}{s-a} - \frac{1}{s+a}\right)$	$a \in \mathbb{C}$	Damped sin
$u(t-t_0) \leftrightarrow \frac{1}{s}e^{-st_0}$	$t_0 > 0 \in \mathbb{R}$	Time delay
$\mathrm{rect}(t) = \frac{1}{t_0}[u(t) - u(t-t_0)] \leftrightarrow \frac{1}{t_0}\left(1 - e^{-st_0}\right)$		Rect-pulse
$u(t) \star u(t) = tu(t) \leftrightarrow 1/s^2$		ramp
$u(t) \star u(t) \star u(t) = \frac{1}{2}t^2u(t) \leftrightarrow 1/s^3$		Double ramp
$\frac{1}{\sqrt{t}}u(t) \leftrightarrow \sqrt{\frac{\pi}{s}}$		
$t^p u(t) \leftrightarrow \frac{\Gamma(p+1)}{s^{p+1}}$		$\Re p > -1, q \in \mathbb{C}$

include both sides of the impulse. If we want to include noncausal functions such as $\delta(t+1)$, we must extend the lower time limit. In such cases we simply set the lower limit of the integral to $-\infty$ and let the integrand ($f(t)$) determine the limits.

4. When we take the forward transform ($t \to s$), the sign of the exponential is negative. This is necessary to ensure that the integral converges when the integrand $f(t) \to \infty$ as $t \to \infty$. For example, if $f(t) = e^t u(t)$ (i.e., without the negative σ exponent), the integral does not converge.
5. The limits on the integrals of the reverse $\mathcal{LT}$s are $[\sigma_o - \infty\jmath, \sigma_o + \infty\jmath] \in \mathbb{C}$. These limits are further discussed in Sect. 4.7.4.
6. When we take the inverse Laplace transform, the normalization factor of $1/2\pi\jmath$ is required to cancel the $2\pi\jmath$ in the differential ds of the integral.
7. The frequencies for the $\mathcal{LT}$ must be complex, and in general $F(s)$ is complex analytic for $\sigma > \sigma_o$. It follows that the real and imaginary parts of $F(s)$ are related by the Cauchy–Riemann conditions. Given $\Re\{F(s)\}$, it is possible to find $\Im\{F(s)\}$ (Boas 1987). Read more on this in Sect. 4.2.3.

8. To take the inverse Laplace transform, we must learn how to integrate in the complex s plane. This is explained on Sect. 4.5–4.7.4.
9. The Laplace Heaviside step function is defined as

$$u(t) = \int_{-\infty}^{t} \delta(t)dt = \begin{cases} 1 & \text{if } t > 0 \\ \text{NaN} & \text{if } t = 0 \\ 0 & \text{if } t < 0 \end{cases}.$$

Alternatively, we can define $\delta(t) = du(t)/dt$.

10. It is easily shown that $u(t) \leftrightarrow 1/s$ by direct integration,

$$F(s) = \int_{0}^{\infty} u(t)\, e^{-st} dt = -\frac{e^{-st}}{s}\bigg|_{o}^{\infty} = \frac{1}{s}.$$

With the $\mathcal{LT}$ step ($u(t)$), there is no Gibbs ringing effect.

11. The Laplace transform of a Brune impedance takes the form of a ratio of two polynomials. In such cases, the roots of the numerator polynomial are called the *zeros* while the roots of the denominator polynomial are called the *poles*. For example, the $\mathcal{LT}$ of $u(t) \leftrightarrow 1/s$ has a pole at $s = 0$, which represents integration, since

$$u(t) \star f(t) = \int_{-\infty}^{r} f(\tau)d\tau \leftrightarrow \frac{F(s)}{s}.$$

12. The $\mathcal{LT}$ is quite different from the $\mathcal{FT}$ in terms of its analytic properties. For example, the step function $u(t) \leftrightarrow 1/s$ is complex analytic everywhere except at $s = 0$. The $\mathcal{FT}$ of $1 \leftrightarrow 2\pi\tilde{\delta}(\omega)$ is not analytic anywhere.
13. The dilated step function ($a \in \mathbb{R}$) is

$$u(at) \leftrightarrow \int_{-\infty}^{\infty} u(at)e^{-st}dt = \frac{1}{a}\int_{-\infty}^{\infty} u(\tau)e^{-(s/a)\tau}d\tau = \frac{a}{|a|}\frac{1}{s} = \pm\frac{1}{s},$$

where we have made the change of variables $\tau = at$. The only effect that a has on $u(at)$ is the sign of t, since $u(t) = u(2t)$. However, $u(-t) \neq u(t)$, since $u(t) \cdot u(-t) = 0$, and $u(t) + u(-t) = 1$, except at $t = 0$, where it is not defined. Once complex integration in the complex plane has been defined (see Sect. 4.2.3), we can justify the definition of the inverse $\mathcal{LT}$ (Eq. 3.10.1).[28]

Causal-periodic signals: This is a special symmetry that occurs due to functions that are causal *and* periodic in frequency. The best example is the z-transform, which applies to causal (one-sided in time) discrete-time signals. The harmonic series (Eq. 3.2.10) is the z-transform of the discrete-time step function and is thus, due to symmetry, analytic within the RoC in the complex frequency (z) domain.

[28] https://en.wikipedia.org/wiki/Laplace_transform#Table_of_selected_Laplace_transforms.

The double brackets on $f((t))_{T_o}$ indicate that $f(t)$ is periodic in t with period T_o—that is, $f(t) = f(t + kT_o)$ for all $k \in \mathbb{N}$. Averaging over one period and dividing by T_o give the average value.

3.10.1.1 Inverse $\mathcal{LT}$

As we will discuss on Sect. 3.10, to invert the $\mathcal{LT}$ one must use the Cauchy residue theorem (CT-3), which requires closure of the contour $\mathcal{C}$ at $\omega_J \to \pm j\infty$,

$$\oint_{\mathcal{C}} = \int_{\sigma_0 - j\infty}^{\sigma_0 + j\infty} + \int_{\subset_\infty},$$

where the path represented by $\subset_\infty$ is a semicircle of infinite radius. For a causal, stable (e.g., doesn't "blow up" in time) signal, all of the poles of $F(s)$ *must* be inside of the Laplace contour, in the left half s-plane.

Example: Hooke's law for a spring states that the force $f(t)$ is proportional to the displacement $x(t)$—that is, $f(t) = Kx(t)$. The formula for a dashpot is $f(t) = Rv(t)$, and Newton's famous formula for mass is $f(t) = d[Mv(t)]/dt$, which for a constant mass M_o is $f(t) = M_o dv/dt$.

The equation of motion for the mechanical oscillator in Fig. 3.11 is given by Newton's second law; the sum of the forces must balance to zero:

$$M_o \frac{d^2}{dt^2} x(t) + R_o \frac{d}{dt} x(t) + K_o x(t) = f(t) \leftrightarrow (M_o s^2 + R_o s + K_o) X(s) = F(s). \tag{3.10.2}$$

These three constants—mass M_o, resistance R_o, and stiffness K_o ($\in \mathbb{R} \geq 0$)—are real and nonnegative. The dynamical variables are the driving force $f(t) \leftrightarrow F(s)$, the position of the mass $x(t) \leftrightarrow X(s)$, and its velocity $v(t) \leftrightarrow V(s)$, with $v(t) = dx(t)/dt \leftrightarrow V(s) = sX(s)$.

Newton's second law (ca.1650) is the mechanical equivalent of Kirchhoff's (ca.1850) voltage law (KVL), which states that the sum of the voltages around a loop must be zero. The gradient of the voltage results in a force on a charge (i.e., $F = q_o E$). The current may be thought of as the flow of charge.

Equation 3.10.2 may be re-expressed in the frequency domain in terms of an impedance (i.e., Ohm's law), defined as the ratio of the force $F(s)$ to velocity $V(s) = sX(s)$, and the sum of three impedances:

$$Z(s) = \frac{F(s)}{V(s)} = \frac{Ms^2 + Rs + K}{s} = Ms + R + \frac{K}{s}. \tag{3.10.3}$$

Example: The divergent series

$$e^t u(t) = \sum_0^\infty \frac{1}{n!} t^n \leftrightarrow \frac{1}{s-1}$$

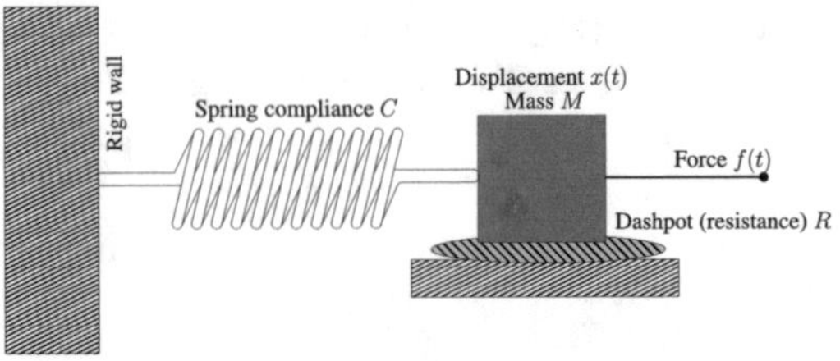

Fig. 3.11 Three-element mechanical resonant circuit consisting of a spring, mass, and dashpot (e.g., viscous fluid)

is a valid description of $e^t u(t)$, with an unstable pole at $s = 1$. For values of $|x - x_o| < 1$ ($x \in \mathbb{R}$), the analytic function $P(x)$ is said to have a *region of convergence* (RoC). For cases where the argument is complex ($s \in \mathbb{C}$), this is called the *radius of convergence* (RoC). We will call the region $|s - s_o| > 1$ the *region of divergence* (RoD) and $|s - s_o| = 0$ the *singular circle*. Typically the underlying function $P(s)$, defined by the series, has a pole on the singular circle.
Summary: While the definitions of the $\mathcal{FT}$ and $\mathcal{LT}$may appear similar, they are not. The key difference is that the time response of the Laplace transform is causal, leading to a complex analytic frequency response. The frequency response of the Fourier transform is complex but not complex analytic, since the frequency ω is real. Fourier transforms do not have poles.

The concept of symmetry is helpful in understanding the many different types of time-frequency transforms. The two most fundamental types of symmetry are causality and periodicity.

The $\mathcal{FT}$ characterizes the steady-state response, while the $\mathcal{LT}$characterizes both the transient and steady-state responses. Given a causal system force response (Eq. 3.10.3), $F(s) \leftrightarrow f(t)$ with input velocity $V(s) \leftrightarrow v(t)$, the response is

$$f(t) = z(t) \star v(t) \leftrightarrow Z(\omega) = F(s)\Big|_{s=\jmath\omega} V(\omega),$$

which says that the force is the convolution of the mechanical impedance $z(t)$ with the input velocity $v(t)$.

3.10.2 System Postulates

Solutions of differential equations, such as the wave equation, are conveniently described in terms of mathematical properties, which we present here in 11 system postulates (see Appendix 3.10, for greater detail):

(P1) *Causality* (noncausal/acausal): Causal systems respond when acted upon. All physical systems obey causality. An example of a causal system is an integrator, which has a response of a step function. Filters are also examples of causal systems. Signals represent acausal responses. They do not have a clear beginning or end, such as the sound of the wind or traffic noise. A causal linear system is typically complex analytic and is naturally represented in the complex s plane via Laplace transforms. A nonlinear system may be causal but not complex analytic.

(P2) *Linearity* (nonlinear): Linear systems obey superposition. Let two signals $x(t)$ and $y(t)$ be the inputs to a linear system, producing outputs $x'(t)$ and $y'(t)$. When the inputs are presented together as $ax(t) + by(t)$ with constant weights $a, b \in \mathbb{C}$, the output is $ax'(t) + by'(t)$. If either a or b is zero, the corresponding signal is removed from the output.

Nonlinear systems mix the two inputs, thereby producing signals that are not present in the input. For example, if the inputs to a nonlinear system are two sine waves, the output contains distortion components that have frequencies not present at the input. One example of a nonlinear system is one that multiplies the two inputs. A second is a diode, which rectifies a signal, letting current flow in only one direction. Most physical systems have some degree of nonlinear response, but this is not always desired. Other systems are designed to be nonlinear, such as the diode example.

(P3) *Passive* (active): An active system has a power source, such as a battery, while a passive system has no power source. Although you may consider a transistor amplifier to be active, it is so only when connected to a power source. Brune impedances satisfy the positive-real condition (Eq. 3.2.19).

(P4) *Real* (complex) time response: All physical systems are Real in = Real out. They do not naturally have complex responses (real and imaginary parts). While a Fourier transform takes real inputs and produces complex outputs, this is not an example of a complex time response. This postulate is a characterization of the input signal, not its Fourier transform.

(P5) *Time-invariant* (time varying): For a system to be a time-varying system, the output must depend on when the input signal starts or stops. If the output, relative to the input, is independent of the starting time, then the system is said to be *time-invariant*.

(P6) *Reciprocal* (non- or antireciprocal): In many ways this is the most difficult postulate to understand. It is best characterized by the ABCD matrix (see Sect. 3.8). If $\Delta_T = 1$, the system is said to be *reciprocal*. If $\Delta_T = -1$, it is said to be *antireciprocal*. The impedance matrix is reciprocal when $z_{12} = z_{21}$ and antireciprocal when $z_{12} = -z_{21}$. Dynamic loudspeakers are antireciprocal and must be modeled by a gyrator, which may be thought of as a transformer that swaps the force and flow variables (Kim and Allen 2013). For example, the input impedance of a gyrator terminated by an inductor is a capacitor. This property is best explained by Fig. 3.9. For an extended discussion on reciprocity, see Appendix G.

(P7) *Reversibility* (nonreversible): If swapping the input and output of a system leaves the system invariant, it is said to be reversible. When $A = D$, the system is reversible. Note the distinction between reversible and reciprocal.

(P8) *Space-invariant* (space-variant): If a system operates independently as a function of where it physically is in space, then it is space-invariant. When the parameters that characterize the system depend on position, it is space-variant.

(P9) *Deterministic* (random): Given the wave equation along with the boundary conditions, the system's solution may be deterministic, or not, depending on its extent. Consider a radar or sonar wave propagating out into uncharted territory. When the wave hits an object, the reflection can return waves that are not predicted due to unknown objects. This is an example where the boundary condition is not known in advance.

(P-10) *Quasistatic* ($ka < 1$): Quasistatics follows the Nyquist sampling theorem for systems that have dimensions that are small compared to the local wavelength (Nyquist 1924). This assumption fails when the frequency is raised (the wavelength becomes short). Thus this is also known as the *long-wavelength* approximation. Quasistatics is typically stated as $ka < 1$, where $k = 2\pi/\lambda = \omega/c_o$ and a is the smallest dimension of the system. See Sect. 3.10.2 for a method on how to integrate the transmission matrix and Nyquist sampling.

Postulate P10 is closely related to the Feynman lecture *The "underlying unity" of nature,* where Feynman asks (Feynman 1970b, Ch. 12-7): "Why do we need to treat the fields as smooth?" His answer is related to the wavelength of the probing signal relative to the dimensions of the object being probed. This raises the fundamental question: Are Maxwell's equations a band-limited approximation to reality? Today we have no definite answer to this question. The following quote seems relevant [29]:

> *The Lorentz force formula and Maxwell's equations are two distinct physical laws, yet the two methods yield the same results.*

Why the two results coincide was not known. In other words, the flux rule consists of two physically different laws in classical theories. Interestingly, this problem was also a motivation behind the development of the theory of relativity by Albert Einstein. In 1905, Einstein wrote in the opening paragraph of his first paper on relativity theory, "It is known that Maxwell's electrodynamics—as usually understood at the present time—when applied to moving bodies, leads to asymmetries which do not appear to be inherent in the phenomena." But Einstein's argument moved away from this problem and formulated special theory of relativity, thus the problem was not solved.

Richard Feynman once described this situation in his famous lecture (The Feynman Lectures on Physics, Vol. II, 1964), "we know of no other place in physics where such a simple and accurate general principle requires for its real understanding an analysis in terms of two different phenomena. Usually such a beautiful generalization is found to stem from a single deep underlying principle. ... We have to understand the "rule" as the combined effects of two quite separate phenomena."

(P11) *Periodic* $\leftrightarrow$ *discrete:* When a function is discrete in one domain (e.g., time or frequency), it is periodic in the other (frequency or time).

Summary of the 11 system postulates: Each postulate has at least two categories. For example, (P1) is causal, noncausal, or acausal, while (P2) is linear or nonlinear. (P6) and (P9) apply to only two-port algebraic networks (those that have an input and an output). The others apply to both two- and one-port networks (e.g., an impedance

[29] https://www.sciencedaily.com/releases/2017/09/170926085958.htm.

is a one-port). An important example of a two-port is the antireciprocal transmission matrix of a dynamic (EM) loudspeaker (see Sect. 3.10).

Related forms of these postulates may be found in the network theory literature (Van Valkenburg 1964a, b; Ramo et al. 1965). Postulates P1–P6 were introduced by Carlin and Giordano (1964), and Postulates P7–P9 were added by Kim et al. (2016). While linearity (P2), passivity (P3), realness (P4), and time-invariant (P5) are independent, causality (P1) is a consequence of linearity (P2) and passivity (P3) (Carlin and Giordano 1964, p. 5).

3.10.3 Probability

Many things in life follow rules we don't understand, and thus are unpredictable, yet they have structure due to some underlying poorly understood physics (e.g., quantum mechanics). Unlike mathematicians, engineers are taught to deal with uncertainty in terms of random processes using probability theory. For many this starts out as a large set of boring incomprehensible definitions, but once you begin to understand, it becomes interesting mathematics. It needs to be *in your skin*. If you don't have an intuition for it, either keep working on it or else find another job. Don't memorize a bunch of formulas, because that won't work over the long run.

A friend was once told "You're amazing in how you think outside the box." He responded "There is no box."

Some view probability as combinatorics and permutations. In my view probability is much more. Probability is about the signal processing of noise and signals (i.e., not combinatorics). The units of probability are [certainty] (Fry (1928)). An important goal in using probability is to find correlations in observations, such as the relative frequency of observations in sequential observations of events. Hamming (2004) presents an insightful discussion on probability.

3.10.3.1 Definitions:

1. An *event* is an unpredictable outcome (Papoulis and Pillai 2002). For example, measuring the temperature $T(\boldsymbol{x}, t) \in \mathbb{R}$ with $\boldsymbol{x} \in \mathbb{R}^3$ at time t [s] is an event. Measuring the temperature every hour gives 24 events per day [degrees/h]. Also, the single toss of a coin, resulting in $\{H, T\}$, is an event.

Exercise #8 What are the units of a temperature event?

Solution: Although we might think the answer is degrees, that unit is not the data that are being observed. Rather, the *relative frequency* of temperatures is the observable. For example, how many times was the event between 20° and 21° or between 22° and 27°? Events are dimensionless numbers with unit of [certainty]. ∎

2. A *trial* is N events.

3. An *experiment* $\{M, N\}$ is M trials of N events.
4. We must always keep track of the *number of events* so that we can compute the mean (i.e., average) and the uncertainty of an observable outcome.
5. The *mean* of many trials is the average.
6. A *random variable* X is the outcome from an experiment. A random variable rarely has stated units. For example, flipping a coin $N = 8$ times defines the number of trials.

Exercise #9 What are the units of coin flips?

$$X \equiv \{H, H, H, T, H, T, T, T\}$$

Solution: The random variable $\{H, T\}$ has units of [certainty], best measured in terms of odds, the ratio of the number of tails to heads. ■

Exercise #10 How do you identify the meaning of a variable that is dimensionless (has no units)?
Solution: One must get creative. We can let $H = 1$ and $T = -1$ so that the mean can be zero. ■

Exercise #11 What are the mean and standard deviation of the coin toss?
Solution: To compute the mean (or standard deviation) we assign numbers to H and T. For example, we let $H = 1$ and $T = 0$. Then we use the usual formula to compute the numerical values. ■

7. The *expected value* is the mean of N events.

Exercise #12 What is the difference between the mean, the expected value, and the average?
Solution: These terms all mean the same thing. Having several terms that mean the same thing is one of the many things that make probability theory so arbitrary. It is sloppy to have unclear terminology. ■

Exercise #13 How do you assign a numerical mean to random outcomes $\{H, T\}$?
Solution: If we let $H = 1$ and $T = 0$, then the mean is

$$\mu = (1 + 1 + 1 + 0 + 1 + 0 + 1 + 0)/8 = 5/8.$$

The odds are defined as the ratio of either P_T/P_H or P_H/P_T. ■

It is critically important to keep track of the number of events ($N = 8$ in the Exercise 13). In some sense N is more important than the actual measured sequence. It is helpful to think of N as the independent variable and X as the dependent variable; that is, think of $X(N)$, not $N(X)$.
Example: We define a trial by flipping a coin $N = 10$ times. We form an experiment by M repeated trials ($M = 1000$).

Exercise #14 A measure of the quantization in the estimate of the probability density due to the sample size N is defined as the magnitude of *sampling noise*.
Solution: When we compute the average (the mean μ_N) of N samples the error is bounded by $1/N$; thus the variance σ_N^2 from the mean is quantized to $1/N$. It follows that the root-mean-square (RMS) sample error must be bounded by $\sigma_N < \sqrt{2/N}$, independent of frequency (i.e., the $\mathcal{FT}$of the N-sample probability density function). ■

3.11 Complex Analytic Mappings (Domain-Coloring)

One of the most difficult aspects of complex functions of a complex variable is visualizing the mappings from the $z = x + y\jmath$ to $w(z) = u + v\jmath$ planes. For example, $w(z) = \sin(x)$ is trivial;

$$\sin(y\jmath) = \frac{e^{-y} - e^{y}}{2\jmath} = -\jmath \sinh(y)$$

is pure imaginary. However, the more general case

$$w(z) = \sin(z) \in \mathbb{C}$$

when $z = x + y\jmath$ is real (i.e., $y = 0$) because $\sin(x)$ is real. Likewise, the case where $x = 0$ is not easily visualized. And when $u(x, y)$ and $v(x, y)$ are less well-known functions, $w(z)$ can be even more difficult to visualize. For example, if $w(z) = J_0(z)$, then $u(x, y)$ and $v(x, y)$ are the real and imaginary parts of the Bessel function.
Visualizing complex functions: The mapping from $s = \sigma + \omega\jmath$ to $w(s) = u(\sigma, \omega\jmath) + \imath v(\sigma, \omega\jmath)$ is difficult to visualize because for each point in the domain $s = \sigma + \omega\jmath$, we would like to represent both the magnitude and phase (or real and imaginary parts) of $w(s)$. The accepted way to visualize these mappings is to use color (hue) to represent the phase and intensity (dark to light) to represent the magnitude.

Fortunately with computer software today, this mapping problem can be solved by adding color to the chart. An Octave/MATLAB script[30] `zviz.m` has been devised to make the charts shown in Fig. 3.12. Such charts are known as *domain-coloring*.

In Fig. 3.12, rather than plotting $u(x, y)$ and $v(x, y)$ separately, domain-coloring allows us to display the entire function on one color chart (i.e., colorized plot). For this visualization we see the complex polar form of $w(s) = |w|e^{\jmath\angle w}$ rather than the 2×2 (four-dimensional) Cartesian graph $w(x + y\jmath) = u(x, y) + v(x, y)\jmath$. On the left is the reference condition, the identity mapping ($w = s$), and on the right the origin has been shifted to the right and up by $\sqrt{2}$.

Mathematicians typically use the abstract (i.e., non-physical) notation $w(z)$, where $w = u + v\imath$ and $z = x + y\imath$. Engineers typically work in terms of a physical com-

[30]https://jontalle.web.engr.illinois.edu/uploads/298/zviz.zip.

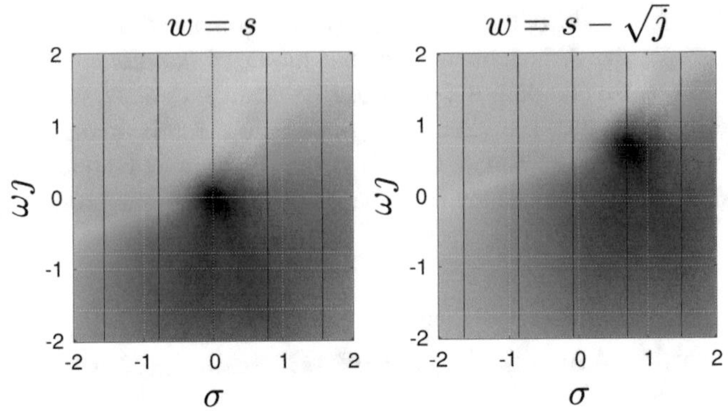

Fig. 3.12 Left: Domain-colorized map showing the complex mapping from the $s = \sigma + \omega \jmath$ plane to the $w(s) = u(\sigma, \omega) + v(\sigma, \omega)\jmath$ plane. This mapping may be visualized by the use of intensity (light/dark) to indicate magnitude, and color (hue) to indicate angle (phase) of the mapping. Right: The $w(s) = s - \sqrt{\jmath}$ plane shifted to the right and up by $\sqrt{2}/2 = 0.707$. The white and black lines are the iso-real and iso-imaginary contours of $u(\sigma, \omega)$ and $v(\sigma, \omega)$.

plex impedance $Z(s) = R(s) + \jmath X(s)$ that has resistance $R(s)$ and reactance $X(s)$ [ohms] as a function of the complex Laplace radian frequency $s = \sigma + \omega \jmath$ [rad], as used, for example, with the Laplace transform (see Sect. 3.10). In Fig. 3.12 we use a mixed notation, with $Z(s) = s$ on the left and $w(s) = s - \sqrt{\jmath}$ on the right, where we show this color code as a 2×2 dimensional domain-coloring graph. Intensity (dark to light) represents the magnitude of the function, while hue (color) represents the phase, where red is 0°, sea-green is 90°, blue-green is 135°, blue is 180°, and violet is −90° (or 270°).[31]

The function $w = s = |s|e^{\jmath\theta}$ has a dark spot (zero) at $s = 0$ and becomes brighter away from the origin. On the right is $w(s) = s - \sqrt{\jmath}$, which shifts the zero (dark spot) to $s = \sqrt{\jmath}$. Thus domain-coloring gives the full 2×2 complex analytic function mapping $w(x, y) = u(x, y) + v(x, y)\jmath$ in colorized polar coordinates.

Example: Figure 3.13 shows a colorized plot of $w(z) = \sin(\pi(s - \jmath)/2)$ resulting from the MATLAB/Octave command `zviz sin(pi*(s-j)/2)`. The abscissa (horizontal axis) is the real σ-axis and the ordinate (vertical axis) is the complex $\jmath\omega$-axis. The graph is offset along the ordinate axis by $1j$, since the argument $s - \jmath$ causes a shift of the sine function by 1 in the positive imaginary direction.

The visible zeros of $w(s)$ appear as dark regions at (−2, 1), (0, 1), (2, 1). As a function of σ, $w(\sigma + 1\jmath)$ oscillates between red (phase is zero degrees), meaning the function is positive and real, and sea-green (phase is 180°), meaning the function is negative and real.

To use the program, we use the syntax, for example, `zviz s.^2`). Note the period between s and `^2`. This will render a domain-coloring (colorized) version of the function. Examples you can render with `zviz` are given in the comments at the

[31] Hue depends on both the display medium and the eye.

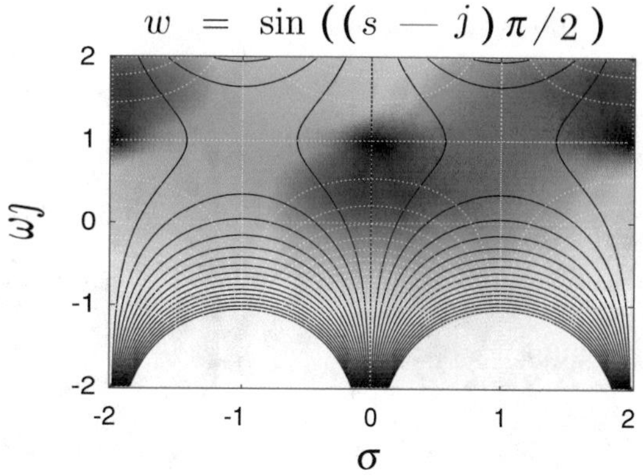

Fig. 3.13 Plot of $\sin(0.5\pi(z - \jmath))$

top of the `zviz.m` program. A good example for testing is `zviz z-sqrt(j)`, which has a dark spot (zero) at $(1 + 1\jmath)/\sqrt{2} = 0.707(1 + 1\jmath)$.

Along the vertical axis, the displayed function is either $\cosh(y)$ or $\sinh(y)$, depending on the value of x. The intensity becomes lighter as $|w|$ increases.

What is being plotted? The axes are either $s = \sigma$ and ω, or $z = x$ and y. Superimposed on the s-axis is the function $w(s) = u(\sigma, \omega) + v(\sigma, \omega)\jmath$, represented in polar coordinates by the intensity and color of $w(s)$. The density (dark vs. light) displays the magnitude $|w(s)|$, while the color (hue) displays the angle ($\angle w(s)$) as a function of s. Thus the intensity becomes darker as $|w|$ decreases and lighter as $|w(s)|$ increases. The angle $\angle(w)$ to color map is defined by Fig. 3.12. For example, 0° is red, 90° is green, −90° is purple, and 180° is blue-green.

Example: Additional examples are given in Fig. 3.14 using the notation $w(s) = u(\sigma, \omega) + v(\sigma, \omega)\jmath$. We see the two complex mappings $w = e^s$ (left) and its inverse $s = \ln(w)$. The exponential is relatively to understand because $w(s) = |e^\sigma e^{\omega\jmath}| = e^\sigma$.

The red region is where $\omega \approx 0$, in which case $w \approx e^\sigma$. As σ becomes large and negative, $w \to 0$; thus the entire field becomes dark on the left. The field is becoming light on the right where $w = e^\sigma \to \infty$. If we let $\sigma = 0$ and look along the ω-axis, we see that the function is changing phase: sea-green (90°) at the top and violet (−90°) at the bottom.

In the right panel note the zero for $s(z) = \ln(z) = \ln|z| + \omega\jmath$ at $z = 1$. The root of the $\log(z)$ function is $\log(z_r) = 0$, $w_r = 1$, $\angle z = \phi = 0$, since $\log(1) = 0$. More generally, the $\log(z)$ of $z = |z|e^{\phi\jmath}$ is $s(z) = \ln|z| + \phi\jmath$. Thus $s(w)$ can be zero only when the angle of w is zero.

The $\ln(z)$ function has a branch cut along the $\phi(z) = \angle z = 180°$ axis. As one crosses over the cut, the phase goes above 180° and the plane changes to the next sheet of the log function. The only sheet with a zero is the principal value, as shown. For all others, the log function is either increasing or decreasing monotonically, and there is no zero.

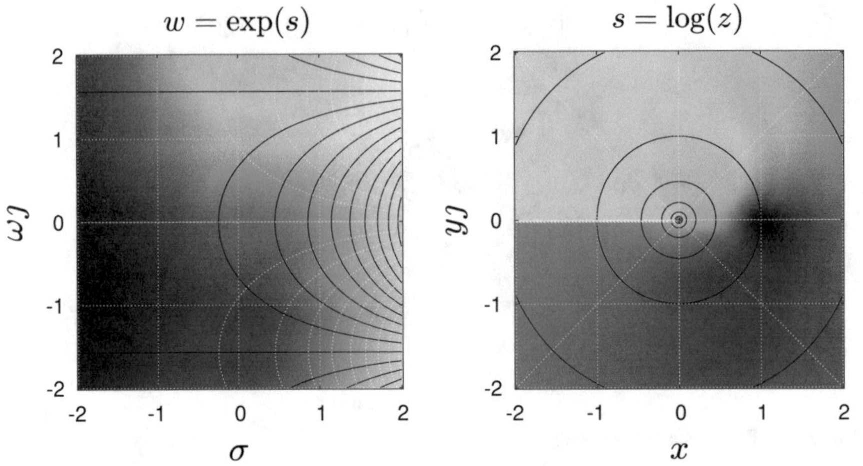

Fig. 3.14 This domain-color map allows us to visualize complex mappings by the use of intensity (light/dark) to indicate magnitude and color (hue) to indicate angle (phase). The white and black lines are the iso-real and iso-imaginary contours of the complex mapping. Left: The domain-color map for the complex mapping from the $s = \sigma + \omega\jmath$ plane to the $w(s) = u + v\jmath = e^{\sigma+\omega\jmath} = e^{\sigma}e^{\omega\jmath}$ plane, which goes to zero as $\sigma \to -\infty$, causing the domain-color map to become dark for $\sigma < -2$. The white and black lines are always perpendicular because e^s is complex analytic everywhere. Right: The *principal value* of the inverse function $s(z) = u(x, y) + v(x, y)\jmath = \log(z)$, which has a zero (dark) at $x = 1$, since there $\log(1) = 0$ and the imaginary part is zero. Note the branch cut, where the color is discontinuous, from $x = [0, -\infty\jmath)$. On branches other than the one shown, there are no zeros, since the phase ($\angle s = 2\pi n \in \mathbb{Z}$) is not zero. n is called the *branch index*. See Sect. 4.4.3 for a discussion of branch cuts and multivalued functions

3.11.1 The Riemann Sphere

Once algebra was formulated, in about 830 CE, mathematicians were able to expand beyond the limits set by geometry on the real plane and the verbose descriptions of each problem in prose (Stillwell 2010, p. 93). The geometry of Euclid's *Elements* had paved the way, but after 2000 years, the addition of the language of algebra changed everything. The analytic function was a key development, heavily used by both Newton and Euler. Also Cauchy made important headway with his investigation of complex variables. Of special note were integration and differentiation in the complex plane of complex analytic functions, which is the topic of Chap. 4. It was Riemann, working with Gauss in the final years of Gauss's life, who made the breakthrough with the concept of the extended complex plane.[32] This concept was based on the composition of a line with the sphere, similar in concept to the derivation of Euclid's formula for Pythagorean triplets (see Sect. 3.7.8). While the importance

[32]"Gauss did lecture to Riemann but he was only giving elementary courses and there is no evidence that at this time he recognized Riemann's genius." Then "In 1849 he [Riemann] returned to Göttingen and his Ph.D. thesis, supervised by Gauss, was submitted in 1851." See https://www-groups.dcs.st-and.ac.uk/~history/Biographies/Riemann.html.

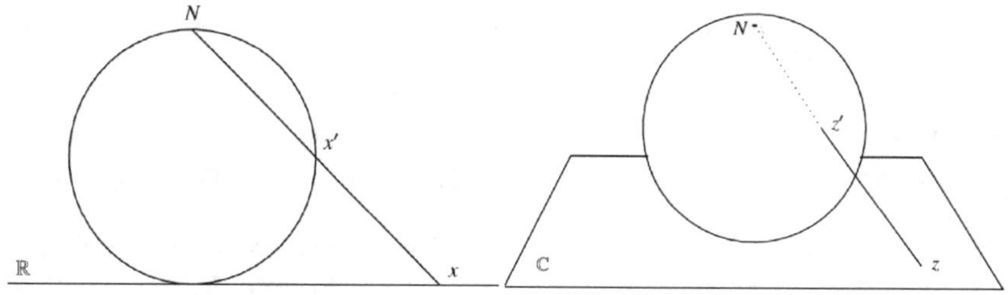

Fig. 3.15 The left panel shows how the real line may be composed with the circle. Each real x value maps to a corresponding point $x\prime$ on the unit circle. The point $x \to \infty$ maps to the north pole N. This simple idea may be extended with the composition of the complex plane with the unit sphere, thus mapping the plane onto the sphere. As with the circle, the point on the complex plane $z \to \infty$ maps onto the north pole N. This construction is important because, while the plane is open (does not include $z \to \infty$), the sphere is analytic at the north pole. Thus the sphere defines the *closed extended plane*. Figure adapted from Stillwell (2010, pp. 299–300)

of the extended complex plane was unforeseen, it changed analytic mathematics forever, along with the physics it supported. It unified and thus simplified many important integrals to the extreme. The basic idea is captured by the fundamental theorem of complex integral calculus (see Table 4.1).

The idea is outlined in Fig. 3.15. On the left is a circle and a line. The difference between this case and the derivation of the Pythagorean triplets is that the line starts at the north pole and ends on the real $x \in \mathbb{R}$ axis at point x. At point x', the line cuts through the circle. Thus the mapping from x to x' takes every point on $\mathbb{R}$ to a point on the circle. For example, the point $x = 0$ maps to the south pole (not indicated). To express x' in terms of x one must compose the line and the circle, similar to the composition used in the derivation of Euclid's formula (see Sect. 3.7.8). The points on the circle, indicated here by x', require a traditional polar coordinate system, with a unit radius and an angle defined between the radius and a vertical line passing through the north pole. When $|x| \to \infty$, the point $x' \to N$, known as the *point at infinity*. But this idea goes much further, as shown on the right of Fig. 3.15.

Here the real tangent line is replaced by a tangent complex plane $z \in \mathbb{C}$ and the complex puncture point $z' \in \mathbb{C}$—in this case on the complex sphere, called the *extended complex plane*. This is a natural extension of the chord/tangent method on the left, but with significant consequences. The main difference between the complex plane z and the extended complex plane, other than the coordinate system, is what happens at the north pole. The point at $|z| = \infty$ is not defined on the plane, whereas on the sphere, the point at the north pole is simply another point, like every other point on the sphere.

Open versus closed sets: Mathematically the plane is said to be an *open set*, since the limit $z \to \infty$ is not defined, whereas on the sphere, the point z' is a member of a *closed set*, since the north pole *is* defined. The distinction between an open and closed sets is important because the closed set allows the function to be complex analytic at the north pole, which it cannot be on the plane (since the point at infinity is not defined).

The z plane may be replaced with another tangent plane—say, the $w = F(z) \in \mathbb{C}$ plane, where w is some function F of $z \in \mathbb{C}$. For the moment we shall limit ourselves to complex analytic functions of z—namely, $w = F(z) = u(x, y) + v(x, y)\jmath = \sum_{n=0}^{\infty} c_n z^n$.

In summary, given a point $z = x + y\jmath$ on the open complex plane, we map it to $w = F(z) \in \mathbb{C}$, the complex $w = u + v\jmath$ plane, and from there to the closed extended complex plane $w'(z)$. The point of doing this is that it allows the function $w'(z)$ to be analytic at the north pole, meaning it can have a convergent Taylor series at the point at infinity $z \to \infty$.

Since we have not yet defined $dw(z)/dz$, the concept of a complex Taylor series remains undefined.

3.11.2 Bilinear Transformation

In mathematics the bilinear transformation has special importance because it is linear in its action on both the input and output variables. Since we are engineers, we shall stick with the engineering terminology. But if you wish to read about this on the Internet, be sure to also search for the mathematical term *Möbius transformation.*

When a point on the complex plane $z = x + y\jmath$ is composed with the bilinear transformation $(a, b, c, d \in \mathbb{C})$, the result is $w(z) = u(x, y) + v(x, y)\jmath$ (this is related to the Möbius transformation, Sect. 2.2):

$$w = \frac{az + b}{cz + d}. \tag{3.11.1}$$

The transformation $z \to w$ is a cascade of four independent compositions:

1. Translation ($w = z + b$: $a = 1, b \in \mathbb{C}, c = 0, d = 1$)
2. Scaling ($w = |a|z$: $a \in \mathbb{R}, b = 0, c = 0, d = 1$)
3. Rotation ($w = \frac{a}{|a|}z$: $a \in \mathbb{C}, b = 0, c = 0, d = |a|$)
4. Inversion ($w = \frac{1}{z}$: $a = 0, b = 1, c = 1, d = 0$)

Each of these transformations is a special case of Eq. 3.11.1, with inversion being the most complicated. I highly recommend a video showing the effect of the bilinear (Möbius) transformation on the plane (Arnold and Rogness 2019).[33]

The bilinear transformation is the most general way to move the expansion point in a complex analytic expansion. For example, when we start from the harmonic series, the bilinear transformation gives

[33] https://www.youtube.com/watch?v=0z1fIsUNhO4.

$$\begin{aligned}\frac{1}{1-w} &= \frac{1}{1-\frac{az+b}{cz+d}} \\ &= \frac{cz+d}{(c-a)z+(d-b)} \\ &= \frac{1}{1-\frac{a}{c}} \cdot \frac{z+\frac{d}{c}}{z-\frac{a-b}{c-a}}.\end{aligned}$$

The RoC is transformed from $|w| < 1$ to $|(az-b)/(cz-d)| < 1$. An interesting application might be to move the expansion point until it is on top of the nearest pole, so that the RoC goes to zero. This might be a useful way of finding a pole, for example.

When the extended plane (Riemann sphere) is analytic at $z = \infty$, we can take the derivatives there, defining the Taylor series with the expansion point at ∞. When the bilinear transformation rotates the Riemann sphere, the point at infinity is translated to a finite point on the complex plane, revealing the analytic nature at infinity. A second way to transform the point at infinity is by the bilinear transformation $\zeta = 1/z$, mapping a zero (or pole) at $z = \infty$ to a pole (or zero) at $\zeta = 0$. Thus this construction of the Riemann sphere and the Möbius (bilinear) transformation allows us to understand the point at infinity and treat it like any other point. If you felt that you never understood the meaning of the point at ∞ (likely), this should help.

3.12 Problems AE-3

3.12.1 Topics of This Homework

Visualizing complex functions, bilinear/Möbius transformation, Riemann sphere.

Deliverables: Answers to problems

3.12.2 Two-Port Network Analysis

Problem #1 Perform an analysis of electrical two-port networks, shown in Fig. 3.8. This can be a mechanical system if the capacitors are taken to be springs and inductors taken as mass, as in the suspension of the wheels of a car. In an acoustical circuit, the low-pass filter could be a car muffler. While the physical representations will be different, the equations and the analysis are exactly the same.

The definition of the ABCD *transmission matrix* ($\mathcal{T}$) is

$$\begin{bmatrix} V_1 \\ I_1 \end{bmatrix} = \begin{bmatrix} \mathcal{A} & \mathcal{B} \\ \mathcal{C} & \mathcal{D} \end{bmatrix} \begin{bmatrix} V_2 \\ -I_2 \end{bmatrix}. \tag{AE-3.1}$$

The *impedance matrix*, where the determinant $\Delta_{\mathcal{T}} = AD - BC$, is given by

$$\begin{bmatrix} V_1 \\ V_2 \end{bmatrix} = \frac{1}{\mathcal{C}} \begin{bmatrix} \mathcal{A} & \Delta_{\mathcal{T}} \\ 1 & \mathcal{D} \end{bmatrix} \begin{bmatrix} I_1 \\ I_2 \end{bmatrix}. \tag{AE-3.2}$$

–1.1: Derive the formula for the impedance matrix (Eq. AE-3.2) given the transmission matrix definition (Eq. AE-3.1). Show your work.

Problem #2 Consider a single circuit element with impedance $Z(s)$.
–2.1: What is the ABCD matrix for this element if it is in series?
–2.2: What is the ABCD matrix for this element if it is in shunt?

Problem #3 Find the ABCD matrix for each of the circuits of Fig. 3.8.
For each circuit, (i) show the cascade of transmission matrices in terms of the complex frequency $s \in \mathbb{C}$, then (ii) substitute $s = 1\jmath$ and calculate the total transmission matrix at this single frequency.
–3.1: Left circuit (let $R_1 = R_2 = 10$ kilo-ohms and $C = 10$ nano-farads)
–3.2: Right circuit (use L and C values given in the figure), where the pressure P is analogous to the voltage V, and the velocity U is analogous to the current I.
–3.3: Convert both transmission (ABCD) matrices to impedance matrices using Eq. AE-3.2. Do this for the specific frequency $s = 1\jmath$ as in the previous part (feel free to use MATLAB/Octave for your computation).
–3.4: Right circuit: Repeat the analysis as in Question 3.3.

3.12.3 Algebra

Problem #4 Fundamental theorem of algebra (FTA).
–4.1: State the fundamental theorem of algebra (FTA).

3.12.4 Algebra with Complex Variables

Problem #5 Order and complex numbers: One can always say that $3 < 4$—namely, that real numbers have order. One way to view this is to take the difference and compare it to zero, as in $4 - 3 > 0$. Here we will explore how complex variables may be ordered. Define the complex variable $z = x + y\jmath \in \mathbb{C}$.
–5.1: Explain the meaning of $|z_1| > |z_2|$.
–5.2: If $x_1, x_2 \in \mathbb{R}$ (are *real* numbers), define the meaning of $x_1 > x_2$. *Hint: Take the difference.*
–5.3: Explain the meaning of $z_1 > z_2$.
–5.4: If time were complex, how might the world be different?

Problem #6 It is sometimes necessary to consider a function $w(z) = u + v\jmath$ in terms of the real functions $u(x, y)$ and $v(x, y)$ (e.g., separate the real and imaginary parts). Similarly, we can consider the inverse $z(w) = x + y\jmath$, where $x(u, v)$ and $y(u, v)$ are real functions.

–6.1: Find $u(x, y)$ and $v(x, y)$ for $w(z) = 1/z$.

Problem #7 Find $u(x, y)$ and $v(x, y)$ for $w(z) = c^z$ with complex constant $c \in \mathbb{C}$ for Questions #7, #7, and #7:

–7.1: $c = e$

–7.2: $c = 1$ (recall that $1 = e^{\jmath k 2\pi k}$ for $k = 0, 1, 2, \ldots$)

–7.3: $c = \jmath$. *Hint:* $\jmath = e^{\jmath\pi/2 + \jmath 2\pi m}, \quad m \in \mathbb{Z}$.

–7.4: Find $u(x, y)$ for $w(z) = \sqrt{z}$. *Hint: Begin with the inverse function* $z = w^2$.

Problem #8 Convolution of an impedance $z(t)$ and its inverse $y(t)$: In the frequency domain a Brune impedance is defined as the ratio of a numerator polynomial $N(s)$ to a denominator polynomial $D(s)$.

–8.1: Consider a Brune impedance defined by the ratio of numerator and denominator polynomials, $Z(s) = N(s)/D(s)$. Since the admittance $Y(s)$ is defined as the reciprocal of the impedance, the product must be 1. If $z(t) \leftrightarrow Z(s)$ and $y(t) \leftrightarrow Y(s)$, it follows that $z(t) \star y(t) = \delta(t)$. What property must $n(t) \leftrightarrow N(s)$ and $d(t) \leftrightarrow D(s)$ obey for this to be true?

–8.2: The definition of a minimum phase function is that it must have a causal inverse. Show that every impedance is minimum phase.

3.12.5 *Schwarz Inequality*

Problem #9 Figure 3.16 shows three vectors for an arbitrary value of $\alpha \in \mathbb{R}$ and a specific value of $\alpha = \alpha^*$.

–9.1: Find the value of $\alpha \in \mathbb{R}$ such that the length (norm) of $\mathbf{E}$ (i.e., $||\mathbf{E}|| \geq 0$) is minimum. Show your derivation, not the answer ($\alpha = \alpha^*$).

–9.2: Find the formula for $||\boldsymbol{E}(\alpha^*)||^2 \geq 0$. *Hint: Substitute* α^* *into Eq.* 3.5.9) and show that this results in the Schwarz inequality (Fig. 3.16)

$$|\mathbf{U} \cdot \mathbf{V}| \leq ||\mathbf{U}||||\mathbf{V}||.$$

Problem #10 Geometry and scaler products

–10.1: What is the geometrical meaning of the dot product of two vectors?

–10.2: Give the formula for the dot product of two vectors. Explain the meaning based on Fig. 3.4.

–10.3: Write the formula for the dot product of two vectors $\mathbf{U} \cdot \mathbf{V}$ in $\mathbb{R}^n$ in polar form (e.g., assume the angle between the vectors is θ).

–10.4: How is the Schwarz inequality related to the Pythagorean theorem?

–10.5: Starting from $||\boldsymbol{U} + \boldsymbol{V}||$, derive the *triangle inequality*

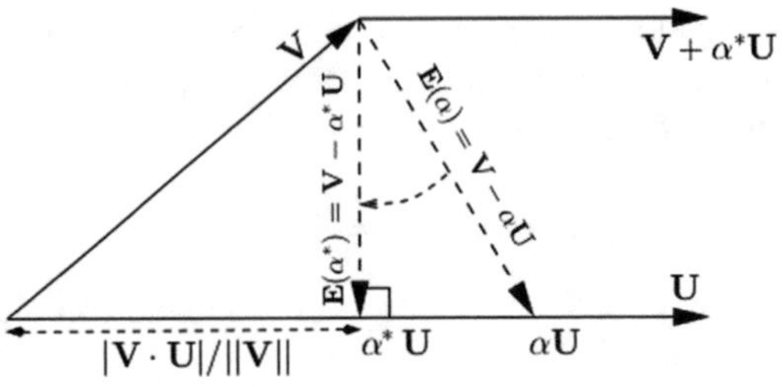

Fig. 3.16 This figure shows how to derive the Schwarz inequality, by finding the value of $\alpha = \alpha^*$ corresponding to $\min_{\alpha}[\boldsymbol{E}(\alpha)]$. It is identical to Fig. 3.5

$$||\mathbf{U} + \mathbf{V}|| \leq ||\mathbf{U}|| + ||\mathbf{V}||.$$

–10.6: The *triangle inequality* $||\mathbf{U} + \mathbf{V}|| \leq ||\mathbf{U}|| + ||\mathbf{V}||$ is true for two and three dimensions: Does it hold for five-dimensional vectors?

–10.7: Show that the wedge-product $\mathbf{U} \wedge \mathbf{V} \perp \mathbf{U} \cdot \mathbf{V}$.

3.12.6 Probability

Problem #11 Basic terminology of experiments

–11.1: What is the *mean* of a trial, and what is the average over all trials?

–11.2: What is the *expected value* of a random variable X?

–11.3: What is the *standard deviation* about the mean?

–11.4: What is the definition of *information* of a random variable?

–11.5: How do you combine events? Hint: If the event is the flip of a biased coin, the events are $H = p, T = 1 - p$, so the event is $\{p, 1 - p\}$. To solve the problem, you must find the probabilities of two independent events.

–11.6: What does the term *independent* mean in the context of question #11? Give an example.

–11.7: Define *odds*.

Chapter 4
Stream 3A: Scalar Calculus

Stream 3 is ∞, a concept that typically means unbounded (immeasurably large), but in the case of calculus, ∞ means infinitesimal (immeasurably small), since taking a limit requires small numbers. Taking a limit means you may never reach the target, a concept that the Greeks called *Zeno's paradox* (Stillwell 2010, p. 76).

When we speak of the class of *ordinary* (versus *vector*) differential equations, the term *scalar* is preferable, since the term ordinary is vague, if not meaningless. For scalar calculus, a special subset of fundamental theorems about integration is summarized in Table 4.1, starting with Leibniz's theorem.

Following our discussion of the integral theorems on scalar calculus, those on vector calculus, without which there can be no understanding of Maxwell's equations. Of these, the fundamental theorem of vector calculus (also known as Helmholtz decomposition), Gauss's law, and Stokes's theorem form the three cornerstones of modern vector field analysis. These theorems allow us to connect the differential (point) and macroscopic (integral) relationships. For example, we can write Maxwell's equations either as vector differential equations, as shown by Heaviside (along with Gibbs and Hertz), or in integral form. It is helpful to place these two forms side by side to fully

Table 4.1 The fundamental theorems of integral calculus, each of which deals with integration. At least two main theorems relate to scalar calculus, and three more to vector calculus

Name	Mapping	Section	Description
Leibniz (FTC)	$\mathbb{R}^1 \to \mathbb{R}^0$	4.2	Area under a real curve
Cauchy (FTCC)	$\mathbb{C}^1 \to \mathbb{R}^0$	4.2	Area under a complex curve
Cauchy's theorem	$\mathbb{C}^1 \to \mathbb{C}^0$	4.5	Close integral over analytic region is zero
Cauchy's integral formula	$\mathbb{C}^1 \to \mathbb{C}^0$	4.5	Fundamental theorem of complex integral calculus
Residue theorem	$\mathbb{C}^1 \to \mathbb{C}^0$	4.5	Residue integration
Helmholtz's theorem			

J. Allen, *An Invitation to Mathematical Physics and Its History*,
https://doi.org/10.1007/978-3-030-53759-3_4

appreciate their significance. To understand the differential (microscopic) view, one must fully understand the integral (macroscopic) view (see Figs. 5.5 and 5.6).

4.1 The Beginning of Modern Mathematics

As shown in Fig. 1.2, mathematics as we know it today began in the sixteenth to eighteenth centuries, arguably starting with Galileo, Descartes, Fermat, Newton, the Bernoulli family, and most important Euler. Galileo was formidable because of his fame, fortune, and "successful" stance against the powerful Catholic establishment. His creativity in scientific circles was certainly well known due to his many skills and accomplishments. Descartes and Fermat were at the forefront of merging algebra and geometry. While Fermat kept meticulous notebooks, he did not publish and tended to be secretive. Thus Descartes's contributions were more widely acknowledged, though not necessarily deeper.

Regarding the development of calculus, much was yet to be developed by Newton and Leibniz using term by term integration of functions based on Taylor series representation. This was a powerful technique but, as stated earlier, incomplete because the Taylor series can represent only single-valued functions within the RoC. More important, Newton (and others) failed to recognize (i.e., rejected) the powerful generalization to complex analytic functions. The first major breakthrough was Newton's publication of *Principia* (1687), and the second was by Riemann (1851), advised by Gauss but possibly more influenced by Cauchy.

Following both Galileo's and Newton's lead, the secretive and introverted behavior of the typical mathematician dramatically changed with the Bernoulli family (Fig. 3.1). The oldest brother Jacob taught his much younger brother Johann, who then taught his son Daniel. But Johann's star pupil was Leonhard Euler. Euler first mastered all the tools and then published with a prolifically previously unknown.

Euler and the circular functions: Euler's first major task was to understand the family of analytic circular functions (e^x, $\sin(x)$, $\cos(x)$, and $\log(x)$), a task begun by the Bernoulli family. Euler sought relationships among these many functions, some of which may not be thought of as being related, such as the log and sin functions. The connection that may "easily" be made is through their complex Taylor series representation (Eq. 3.2.9). By the manipulation of the analytic series representations, the relationship between e^x and $\sin(x)$ and $\cos(x)$ was precisely captured with the equation

$$e^{j\omega} = \cos(\omega) + \jmath \sin(\omega) \tag{4.1.1}$$

and its analytic inverse (Greenberg 1988, p. 1135)

$$\tan^{-1}(z) = \frac{1}{2\jmath} \ln\left(\frac{1\jmath - z}{1\jmath + z}\right) = \frac{\jmath}{2} \ln\left(\frac{1 - z\jmath}{1 + z\jmath}\right). \tag{4.1.2}$$

Exercise #1 Starting from Eq. 4.1.1, derive Eq. 4.1.2.
Solution: We let $z(\omega) = \tan\omega$; then

$$z(\omega) = \tan(\omega) = \frac{\sin\omega}{\cos\omega} = -\jmath\frac{e^{\omega\jmath} - e^{-\omega\jmath}}{e^{\omega\jmath} + e^{-\omega\jmath}} = -\jmath\frac{e^{2\omega\jmath} - 1}{e^{2\omega\jmath} + 1}. \tag{4.1.3}$$

Solving for $e^{-2\omega\jmath}$, we get

$$e^{-2\omega\jmath} = \frac{1 - z\jmath}{1 + z\jmath}. \tag{4.1.4}$$

Taking ln() of both sides and using the definition of $z(\omega)$ gives Eq. 4.1.2:

$$\omega = \tan^{-1}(z) = \frac{\jmath}{2}\ln\frac{1 - z\jmath}{1 + z\jmath},$$

as shown in Fig. 4.1.

These equations are the basis of transmission lines (TL). Here $z(\omega)$ of Eq. 4.1.3 is the TL's input impedance and Eq. 4.1.4 is the reflectance. ■

Although many high school students memorize Euler's relationship, it seems unlikely that they appreciate the utility of the complex analytic function.

History of complex analytic functions: Newton (ca. 1650) famously ignored imaginary numbers and called them imaginary in a disparaging (pejorative) way. Given Newton's prominence, his view must have keenly attenuated interest in complex algebra, even though it had been described by Bombelli in 1526, likely based on his serendipitous finding of Diophantus's book *Arithmetic* in the Vatican library.

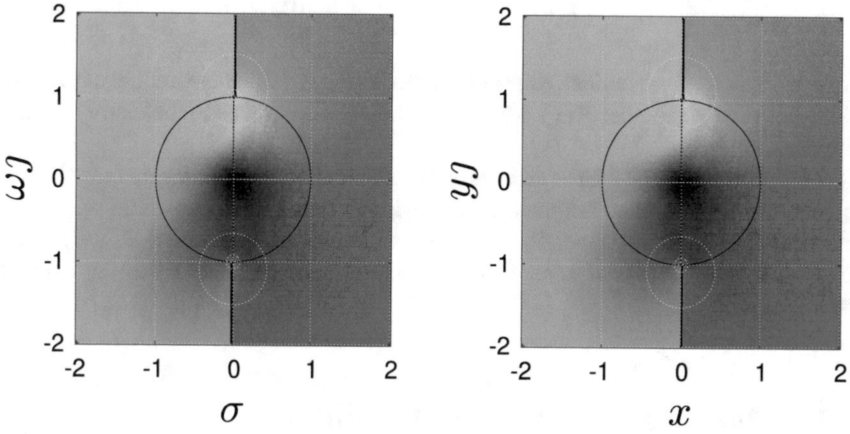

Fig. 4.1 Colorized plots of $z(s) = tan^{-1}(s)$ and $s(z) = \frac{\jmath}{2}\ln(1 - z\jmath)/(1 + z\jmath)$, verifying that they are the same complex analytic function

Euler did not appreciate the role of complex analytic functions. They were first fully appreciated well after his death in 1785 by Augustin-Louis Cauchy (1789–1857), and then extended by Riemann in 1851.

Euler derived his relationships using real-power series (i.e., real-analytic functions). While Euler was fluent with $\jmath = \sqrt{-1}$, he did not consider functions to be complex analytic. That concept was first explored by Cauchy almost a century later. The missing link to the concept of complex analytic functions is the definition of the derivative with respect to the complex argument

$$F'(s) = \frac{dF(s)}{ds}, \tag{4.1.5}$$

where $s = \sigma + \omega\jmath$, without which the complex analytic Taylor coefficients are not defined.

4.2 Fundamental Theorem of Scalar Calculus

History of scalar calculus: It some sense the story of calculus begins with the fundamental theorem of calculus (Table 4.1), also known generically as *Leibniz's formula*. The simplest integral is the length of a line $L = \int_0^L dx$. If we label a point on a line as $x = 0$ and wish to measure the distance to any other point x, we form the line integral between the two points. If the line is straight, this integral is the Euclidean length given by the difference between the two ends (Eq. 3.5.3).

If $F(\chi) \in \mathbb{R}$ describes a height above the line $\chi \in \mathbb{R}$, then

$$f(x) - f(0) = \int_{x=0}^{x} F(\chi)d\chi \tag{4.2.1}$$

may be viewed as the *antiderivative* of $F(\chi)$. Here χ is a dummy variable of integration. Thus the area under $F(\chi)$ depends on only the difference in the area evaluated at the end points.

This property of the area as an integral over an interval, depending on only the end points, has important consequences in physics in terms of conservation of energy, allowing for important generalizations. For example, as long as $\chi \in \mathbb{R}$, we can let $F(\chi) \in \mathbb{C}$ with no loss of generality, due to the linear Postulate P1 of the integral (see Sect. 3.10.2).

4.2.1 Fundamental Theorem of Real Calculus

If $f(x)$ is analytic (Eq. 3.2.8), then

$$F(x) = \frac{d}{dx} f(x) \tag{4.2.2}$$

is an exact real differential. It follows that $F(x)$ is analytic. This is known as the *fundamental theorem of (real) calculus* (FTC). Thus Eq. 4.2.2 may be viewed as an exact real differential. This is easily shown by evaluating

$$\frac{d}{dx}f(x) = \lim_{\delta \to 0} \frac{f(x+\delta) - f(x)}{\delta} = F(x)$$

starting from the antiderivative, Eq. 4.2.1. If $f(x)$ is not analytic then the limit may not exist, so this is a necessary condition.

There are many important variations on this very basic theorem (see Table 4.1). For example, the limits could depend on time. Also when we take Fourier transforms, the integrand depends on both time $t \in \mathbb{R}$ and frequency $\omega \in \mathbb{R}$ via a complex exponential "kernel" function $e^{\pm \jmath \omega t} \in \mathbb{C}$, which is analytic in both t and ω.

4.2.2 The Fundamental Theorem of Complex Calculus

The fundamental theorem of complex calculus (FTCC) states (Greenberg 1988, p. 1197) that for any complex analytic function $F(s) \in \mathbb{C}$ with $s = \sigma + \omega \jmath \in \mathbb{C}$,

$$f(s) - f(s_o) = \int_{s_o}^{s} F(\zeta) d\zeta. \tag{4.2.3}$$

Equations 4.2.1 and 4.2.3 differ because the path of the integral is complex. Thus the line integral is over $s \in \mathbb{C}$ rather than a real integral over $\chi \in \mathbb{R}$. The FTCC states that the integral depends on only the end points, since

$$F(s) = \frac{d}{ds} f(s). \tag{4.2.4}$$

Comparing exact differentials, Eq. 4.1.5 (FTCC) and Eq. 4.2.2 (FTC), we see that $f(s) \in \mathbb{C}$ must be complex analytic and have a Taylor series in powers in $s \in \mathbb{C}$. It follows that $F(s)$ is also complex analytic.

Complex analytic functions: The definition of a *complex analytic function* $F(s)$ of $s \in \mathbb{C}$ is that the function may be expanded in a Taylor series (Eq. 3.2.20) about an expansion point $s_o \in \mathbb{C}$. This definition follows the same logic as the FTC. Thus we need a definition for the coefficients $c_n \in \mathbb{C}$, which most naturally follow from Taylor's formula

$$c_n = \frac{1}{n!} \frac{d^n}{ds^n} F(s) \bigg|_{s=s_o}. \tag{4.2.5}$$

The requirement that $F(s)$ has a Taylor series naturally follows by taking derivatives with respect to s at s_o. The problem is that both integration and differentiation of functions of complex Laplace frequency $s = \sigma + \omega \jmath$ have not yet been defined.

Thus the question: What does it mean to take the derivative of a function $F(s) \in \mathbb{C}$, $s = \sigma + \omega \jmath \in \mathbb{C}$, with respect to s, where s defines a plane rather than a real line? We learned how to form the derivative on the real line. Can the same derivative concept be extended to the complex plane?

The answer is affirmative. The question may be resolved by applying the rules of the real derivative when defining the derivative in the complex plane. However, for the complex case, there is an issue regarding direction. Given any analytic function $F(s)$, is the partial derivative with respect to σ different from the partial derivative with respect to $\omega \jmath$? For complex analytic functions, the FTCC states that the integral is independent of the path in the s plane. Based on the chain rule, the derivative must also be independent of the direction at s_o. This directly follows from the FTCC. If the integral of a function of a complex variable is to be independent of the path, then the derivative of a function with respect to a complex variable must be independent of the direction. This follows from Taylor's formula for the coefficients of the complex analytic formula (Eq. 4.2.5).

The Cauchy–Riemann conditions: The FTC defines the area as an integral over a real differential ($dx \in \mathbb{R}$), while the FTCC relates an integral over a complex function $F(s) \in \mathbb{C}$ along a complex interval (i.e., path) ($ds \in \mathbb{C}$). For the FTC the area under the curve depends on only the end points of the antiderivative $f(x)$. But what is the meaning of an "area" along a complex path? The Cauchy–Riemann conditions provide the answer.

4.2.3 *Cauchy–Riemann Conditions*

For the integral of $Z(s) = R(\sigma, \omega) + \jmath X(\sigma, \omega)$ to be independent of the path, the derivative of $Z(s)$ must also be independent of the path. This requirement leads to a pair of equations known as the *Cauchy–Riemann conditions*.

To define

$$\frac{d}{ds} Z(s) = \frac{d}{ds} \left[R(\sigma, \omega) + \jmath X(\sigma, \omega) \right],$$

we take partial derivatives of $Z(s)$ with respect to σ and $\jmath\omega$, and equate them:

$$\frac{\partial Z}{\partial \sigma} = \frac{\partial R}{\partial \sigma} + \jmath \frac{\partial X}{\partial \sigma} \qquad \equiv \qquad \frac{\partial Z}{\partial \jmath\omega} = \frac{\partial R}{\partial \jmath\omega} + \jmath \frac{\partial X}{\partial \jmath\omega}.$$

This says that a horizontal derivative, with respect to σ, is equivalent to a vertical derivative, with respect to $\omega \jmath$. Taking the real and imaginary parts gives the two equations

$$\frac{\partial R(\sigma, \omega)}{\partial \sigma} = \not\jmath \frac{\partial X(\sigma, \omega)}{\partial \omega \not\jmath} \quad \text{(CR-1)} \quad \text{and} \quad \frac{\partial R(\sigma, \omega)}{\partial \omega \not\jmath} = -\not\jmath \frac{\partial X(\sigma, \omega)}{\partial \sigma} \quad \text{(CR-2)}, \tag{4.2.6}$$

known as the *Cauchy–Riemann (CR) conditions*. The $\jmath$ cancels in CR-1 but introduces a $\jmath^2 = -1$ in CR-2. They may also be written in polar coordinates ($s = re^{\theta\jmath}$) as

$$\frac{\partial R}{\partial r} = \frac{1}{r}\frac{\partial X}{\partial \theta} \quad \text{and} \quad \frac{\partial X}{\partial r} = -\frac{1}{r}\frac{\partial R}{\partial \theta}.$$

The FTCC (Eq. 4.2.3) follows from CR-1 and CR-2 (Eq. 4.2.6).

You may wonder what would happen if we took a derivative at 45°. To do this we only need to multiply the function by $e^{\jmath\pi/4}$. But doing so will not change the derivative. Thus we may take the derivative in any direction by multiplying by $e^{\theta\jmath}$, and the CR conditions will not change.

The CR conditions are necessary so that the integral of $Z(s)$, and thus its derivative, is independent of the path, expressed in terms of conditions on the real and imaginary parts of Z. This is a very strong condition on $Z(s)$, which follows assuming that $Z(s)$ may be written as a Taylor series in s:

$$Z(s) = Z_0 + Z_1 s + \frac{1}{2} Z_2 s^2 + \cdots, \tag{4.2.7}$$

where $Z_n \in \mathbb{C}$ are complex constants given by the Taylor series formula (Eq. 4.2.5). As with the real Taylor series, there is the convergence condition that $|s| < 1$, called the *radius of convergence* (RoC). This is an important generalization of the region of convergence (RoC) for real $s = x$.

Every function that may be expressed as a Taylor series in $s - s_o$ about point $s_o \in \mathbb{C}$ is said to be *complex analytic* at s_o. This series, which is single-valued, is said to converge within a radius of convergence (RoC). This highly restrictive condition has significant physical consequences. For example, every impedance function $Z(s)$ obeys the CR conditions over large regions of the s plane, including the entire *right half s plane* (RHP) ($\sigma > 0$). This condition is summarized by the Brune condition $\Re\{Z(\sigma > 0)\} \geq 0$, or alternatively $\angle Z(s) < \angle s$

When the CR conditions are generalized to volume integrals, this is called either Gauss's Law or Green's theorem, which is used in the solution of boundary value problems in engineering and physics (Kusse and Westwig 2010).

We may merge these equations into a pair of second-order equations by taking a second round of partials. Specifically, eliminating the real part $R(\sigma, \omega)$ of Eq. 4.2.6 gives

$$\frac{\partial^2 R(\sigma,\omega)}{\partial\sigma\partial\omega} = \frac{\partial^2 X(\sigma,\omega)}{\partial^2\omega} = -\frac{\partial^2 X(\sigma,\omega)}{\partial^2\sigma}, \quad \text{(CR-3)} \tag{4.2.8}$$

which may be written compactly as $\nabla^2 X(\sigma, \omega) = 0$. Eliminating the imaginary part gives

$$\frac{\partial^2 X(\sigma,\omega)}{\partial\omega\partial\sigma} = \frac{\partial^2 R(\sigma,\omega)}{\partial^2\sigma} = -\frac{\partial^2 R(\sigma,\omega)}{\partial^2\omega}, \quad \text{(CR-4)} \tag{4.2.9}$$

which may be written as $\nabla^2 R(\sigma, \omega) = 0$.

In summary, for a function $Z(s)$ to be complex analytic, the derivative dZ/ds must be independent of direction (path), which requires that the real and imaginary parts of the function obey Laplace's equation; that is,

$$\nabla^2 R(\sigma, \omega) = 0 \quad \text{and} \quad \nabla^2 X(\sigma, \omega) = 0. \tag{4.2.10}$$

Equations CR-1 and CR-2 are easy to work with because they are first order, but the intuition behind them best follows from the properties of Laplace's equation (Eq. 4.2.10). Note two facts: (1) the derivative of a complex analytic function is independent of its direction, and (2) the real and imaginary parts of the function obey Laplace's equation. Such relationships are known as *harmonic functions*.

As we shall see in the next few sections, complex analytic functions must be smooth, since every analytic function may be differentiated an infinite number of times within the RoC. The magnitude must attain its maximum and minimum on the boundary. For example, when you stretch a rubber sheet over a jagged frame, the height of the rubber sheet obeys Laplace's equation. Nowhere can the height of the sheet rise above or below its value at the boundary.

Harmonic functions define *conservative fields*, which means that energy (like a volume or area) is conserved. The work done in moving a mass from a to b in such a field is conserved. If you return the mass from b back to a, the stored energy is retrieved, thus zero network is consumed.

4.3 Problems DE-1

4.3.1 *Topics of This Homework*

Complex numbers and functions (ordering and algebra), complex power series, fundamental theorem of calculus (real and complex); Cauchy–Riemann conditions, multivalued functions (branch cuts and Riemann sheets)

4.3.2 *Complex Power Series*

Problem #1 *In each case derive (e.g., using Taylor's formula) the power series of* $w(s)$ *about* $s = 0$ *and give the RoC of your series. If the power series doesn't exist, state why!* Hint: In some cases, you can derive the series by relating the function to another function for which you already know the power series at $s = 0$.

–1.1: $1/(1 - s)$
–1.2: $1/(1 - s^2)$
–1.3: $1/(1 + s^2)$.

–1.4: $1/s$
–1.5: $1/(1-|s|^2)$

Problem #2 *Consider the function* $w(s) = 1/s$
–2.1: Expand this function as a power series about $s = 1$. Hint: Let $1/s = 1/(1 - 1 + s) = 1/(1 - (1 - s))$.

–2.2: What is the RoC?
–2.3: Expand $w(s) = 1/s$ *as a power series in* $s^{-1} = 1/s$ *about* $s^{-1} = 1$.
–2.4: What is the RoC?
–2.5: What is the residue of the pole?

Problem #3 *Consider the function* $w(s) = 1/(2-s)$
–3.1: Expand $w(s)$ *as a power series in* $s^{-1} = 1/s$. *State the RoC as a condition on* $|s^{-1}|$. Hint: Multiply top and bottom by s^{-1}.
–3.2: Find the inverse function $s(w)$. *Where are the poles and zeros of* $s(w)$, *and where is it analytic?*

Problem #4 *Summing the series*
The Taylor series of functions have more than one region of convergence.
–4.1: Given some function $f(x)$, *if* $a = 0.1$, *what is the value of*

$$f(a) = 1 + a + a^2 + a^3 + \cdots ?$$

Show your work.
–4.2: Let $a = 10$. *What is the value of*

$$f(a) = 1 + a + a^2 + a^3 + \cdots ?$$

4.3.3 *Cauchy–Riemann Equations*

Problem #5 *For this problem* $\jmath = \sqrt{-1}$, $s = \sigma + \omega\jmath$, *and* $F(s) = u(\sigma,\omega) + \jmath v(\sigma,\omega)$. *According to the fundamental theorem of complex calculus (FTCC), the integration of a complex analytic function is independent of the path. It follows that the derivative of* $F(s)$ *is defined as*

$$\frac{dF}{ds} = \frac{d}{ds}\left[u(\sigma,\omega) + \jmath v(\sigma,\omega)\right]. \qquad \text{(DE-1.1)}$$

If the integral is independent of the path, then the derivative must also be independent of the direction:

$$\frac{dF}{ds} = \frac{\partial F}{\partial \sigma} = \frac{\partial F}{\partial \jmath\omega}. \qquad \text{(DE-1.2)}$$

The Cauchy–Riemann (CR) conditions

$$\frac{\partial u(\sigma,\omega)}{\partial\sigma} = \frac{\partial v(\sigma,\omega)}{\partial\omega} \quad \textit{and} \quad \frac{\partial u(\sigma,\omega)}{\partial\omega} = -\frac{\partial v(\sigma,\omega)}{\partial\sigma}$$

may be used to show where Eq. DE-1.2 holds.
–5.1: Assuming Eq. DE-1.2 is true, use it to derive the CR equations.
–5.2: Merge the CR equations to show that u and v obey Laplace's equations

$$\nabla^2 u(\sigma,\omega) = 0 \quad \textit{and} \quad \nabla^2 v(\sigma,\omega) = 0.$$

What can you conclude?

Problem #6 *Apply the CR equations to the following functions. State for which values of $s = \sigma + i\omega$ the CR conditions do or do not hold (e.g., where the function $F(s)$ is or is not analytic).* Hint: Review where CR-1 and CR-2 hold.
–6.1: $F(s) = e^s$
–6.2: $F(s) = 1/s$

4.3.4 Branch Cuts and Riemann Sheets

Problem #7 *Consider the function $w^2(z) = z$. This function can also be written as $w_\pm(z) = \sqrt{z_\pm}$. Assume $z = re^{\phi\jmath}$ and $w(z) = \rho e^{\theta\jmath} = \sqrt{r}e^{\phi\jmath/2}$.*
–7.1: How many Riemann sheets do you need in the domain *(z) and the* range *(w) to fully represent this function as single-valued?*
–7.2: Indicate (e.g., using a sketch) how the sheet(s) in the domain map to the sheet(s) in the range.
–7.3: Use `zviz.m` *to plot the positive and negative square roots $+\sqrt{z}$ and $-\sqrt{z}$. Describe what you see.*
–7.4: Where does `zviz.m` *place the branch cut for this function?*
–7.5: Must the branch cut necessarily be in this location?

Problem #8 *Consider the function $w(z) = \log(z)$. As in Problem #7, let $z = re^{\phi\jmath}$ and $w(z) = \rho e^{\theta\jmath}$.*
–8.1: Describe with a sketch and then discuss the branch cut *for $f(z)$.*
–8.2: What is the inverse of the function $z(f)$? Does this function have a branch cut? If so, where is it?
–8.3: Using `zviz.m`*, show that*

$$\tan^{-1}(z) = -\frac{\jmath}{2}\log\frac{\jmath - z}{\jmath + z}. \qquad \text{(DE-1.3)}$$

In Fig. 4.1 these two functions are shown to be identical.
–8.4: Algebraically justify Eq. DE-1.3. Hint: Let $w(z) = \tan^{-1}(z)$ and $z(w) = \tan w = \sin w/\cos w$; then solve for $e^{w\jmath}$.

4.3.5 A Cauer Synthesis of any Brune Impedance

Problem #9 *One may synthesize a transmission line (ladder network) from a positive-real impedance $Z(s)$ by using the continued fraction method. To obtain the series and shunt impedance values, we can use a residue expansion. Here we shall explore this method.*

–9.1: Starting from the Brune impedance $Z(s) = \frac{1}{s+1}$, find the impedance network as a ladder network.

–9.2: Use a residue expansion in place of the CFA floor function (Sect. 2.4.4) for polynomial expansions. Find the residue expansion of $H(s) = s^2/(s+1)$ and express it as a ladder network.

–9.3: Discuss how the series impedance $Z(s,x)$ and shunt admittance $Y(s,x)$ determine the wave velocity $\kappa(s,x)$ and the characteristic impedance $z_o(s,x)$ when (1) $Z(s)$ and $Y(s)$ are both independent of x; (2) $Y(s)$ is independent of x, $Z(s,x)$ depends on x; (3) $Z(s)$ is independent of x, $Y(s,x)$ depends on x; and (4) both $Y(s,x)$, $Z(s,x)$ depend on x.

4.4 Complex Analytic Brune Admittance

It is rarely stated that the variable that we are integrating over, either x (space) or t (time), is real ($x, t \in R$), since that fact is implicit, due to the physical nature of the formulation of the integral. But this intuition must be refined once complex numbers are included with $s \in \mathbb{C}$, where $s = \sigma + \omega\jmath$.

That time and space are real variables and are more than an assumption: it is a requirement that follows from the *order property*. Real numbers have order. For example, if $t = 0$ is now (the present), then $t < 0$ is the past and $t > 0$ is the future. Since time and space are real ($t, \boldsymbol{x} \in \mathbb{R}$), they obey this order property. To have time travel, time and space would need to be complex (they are not), since if the space axis were complex the order property would be invalid.

Interestingly, it was shown by d'Alembert (1747) that time and space are related by the pure delay. To obtain a solution to the governing wave equation, which d'Alembert first proposed for sound waves, $x, t \in \mathbb{R}$ may be functionally combined as

$$\zeta_\pm = t \pm x/c_o,$$

where $c_o \in \mathbb{R}$ [m/s] is the wave phase velocity. The d'Alembert solution to the wave equation, describing waves on a string under tension, is

$$u(x,t) = f(t - x/c_o) + g(t + x/c_o), \tag{4.4.1}$$

which describes the transverse velocity (or displacement) of two independent waves $f(\zeta_-)$, $g(\zeta_+) \in \mathbb{R}$ on the string, which represent forward and backward traveling waves.[1] For example, starting with a string at rest, if one displaces the left end, at $x = 0$, by a step function $u(t)$, then that step displacement will propagate to the right as $u(t - x/c_o)$, arriving at location x_o [m], at time x_o/c_o [s]. Before this time, the string will not move to the right of the wavefront, at x_o [m], and after t_o [s] it will have a nonzero displacement. Since the wave equation obeys superposition (postulate P2, Sect. 3.10.2), it follows that the "plane-wave" eigenfunction of the wave equation for $\mathbf{x}, \mathbf{k} \in \mathbb{R}^3$ is given by

$$\psi_{\pm}(\mathbf{x}, t) = \delta(t \mp \mathbf{k} \cdot \mathbf{x}) \leftrightarrow e^{st \pm \jmath \mathbf{k}\cdot\mathbf{x}}, \tag{4.4.2}$$

where $|\mathbf{k}| = 2\pi/|\breve{}| = \omega/c_o$ is the *wave number*, $|\boldsymbol{\lambda}|$ is the wavelength, and $s = \sigma + \omega\jmath$, the Laplace frequency.

Complex propagation function $\boldsymbol{\kappa}(s)$: When propagation dispersion and losses are considered, we must replace the *wave number* $\jmath\mathbf{k} \in \mathbb{C}$ with a complex analytic vector wave number $\boldsymbol{\kappa}(s) = \mathbf{k}_r(s) + \jmath\mathbf{k}(s)$. This is known by several names: (1) the *complex propagation function*, (2) the *dispersion relation*, and (3) the *propagation function*. Function $\boldsymbol{\kappa}(s)$ is a subtle and important generalization of the scalar wave number $k = 2\pi/\lambda$.[2]

An interesting example is the exact solution to the acoustic wave equation, including viscous and thermal losses, as discussed in Appendix D Eq. D.1.5, where it is shown that the eigenvalues are

$$\kappa_{\pm}(s) = \frac{s \pm 2\beta_o\sqrt{s}}{c_o} = \frac{(\beta_o \pm \sqrt{s})^2 - {\beta_o}^2}{c_o},$$

with $\beta_o \in \mathbb{R} \geq 0$.

Forms of energy loss, which include viscosity and radiation, require $\boldsymbol{\kappa}(s) \in \mathbb{C}$. Physical examples include acoustic plane waves, electromagnetic wave propagation, antenna theory, and one of the most difficult cases, that of 3D electron wave propagating in crystals (e.g., silicon), where electrons and electromagnetic (EM) waves are in a state of quantum mechanical equilibrium.

Even when we cannot solve these more difficult problems, we can still appreciate their qualitative solutions. One of the principles that allows us to do that is the causal nature of $\boldsymbol{\kappa}(s)$. Namely, the $\mathcal{LT}^{-1}$ of $\boldsymbol{\kappa}(s)$ must be causal, thus Eq. 4.4.2 must be causal. The *group delay* then describes the nature of the frequency dependent causal delay. For example, if the group delay is large at some frequency, then the solutions will have the largest causal delay at that frequency (Brillouin 1953; Papoulis 1962). Qualitatively this gives a deep insight into the solutions, even when if we cannot compute them.

Electrons and photons are simply different EM states, and $\boldsymbol{\kappa}(x, s)$ describes the crystal's dispersion relations as functions of both frequency and direction, famously known as *Brillouin zones*. Dispersion is a property of the medium such that the wave

[1] d'Alembert's solution is valid for functions that are not differentiable, such as $\delta(t - x/c_o)$.

[2] Recall that for lossless plane waves $\lambda f = c_o$, and $k = 2\pi/\lambda$.

velocity is a function of frequency and direction, as in silicon.[3] Highly readable discussions on the history of this topic may be found in Brillouin (1953).

4.4.1 Generalized Admittance/Impedance

The most elementary examples of Brune admittance and impedance are those made of resistors, capacitors, and inductors. Such discrete element circuits arise not only in electrical networks but in mechanical, acoustical, and thermal networks as well (Table 3.2). These lumped-element networks can always be represented by ratios of polynomials in s. This gives them a similar structure, with easily classified properties. Such circuits are called *Brune admittances* (or impedances).[4] An example of a special symmetry is when the degrees of the numerator and denominator polynomials cannot differ by more than one. This restriction on the degrees comes about because the real part of the admittance/impedance must be positive, due to physical constraints.

But there is a much broader class of admittances that comes from transmission lines and other physical structures, which we refer to as *generalized admittances*. An interesting example is an admittance of the form $1/\sqrt{s}$, called a *semicapacitor*, and $\sqrt{s}$, called a *semiinductor*. Generalized admittance/impedance is not the ratio of two polynomials. As a result, they are more difficult to characterize.

When a generalized admittance $Y(s)$ or its impedance $Z(s) = 1/Y(s)$ is transformed into the time domain, it must have a real and positive surge admittance $\mathcal{Y}_r \in \mathbb{R}$ or surge impedance $\mathcal{Z}_r \in \mathbb{R}$, followed by the residual response $\nu(t), \zeta(t)$. We define the following notation for the admittance (both frequency and time responses),

$$Y(s) = \mathcal{Y}_r + \nu(s) \leftrightarrow y(t) = \mathcal{Y}_r\delta(t) + \nu(t), \tag{4.4.3}$$

and the impedance,

$$Z(s) = \mathcal{Z}_r + \mathcal{Z}_i(s) \leftrightarrow z(t) = \mathcal{Z}_r\delta(t) + \zeta(t). \tag{4.4.4}$$

The complexity of the notation is necessary and follows from the fact that $z(t) \leftrightarrow Z(s)$ and $y(t) \leftrightarrow Y(s)$ are positive-real and thus minimum phase.

When we are dealing with a transmission line (i.e., wave-guides), the generalized admittance is defined as the ratio of the flow to the force. For an electrical system (voltage Φ, current I), the input admittance looking to the right from location x is

$$Y_{in}^{+}(x > 0, s) = \frac{I^+(x,\omega)}{\Phi^+(x,\omega)},$$

[3]In case you missed it, what I'm suggesting is that photons (propagating waves) and electrons (evanescent waves) are different EM wave "states" (Jaynes 1991). This difference depends on the medium, which determines the dispersion relation (Papasimakis et al. 2018).

[4]Some texts prefer the term *immittance* to include both admittance and impedance.

and looking to the left is

$$Y_{in}^{-}(x < 0, s) = \frac{I^{-}(x, \omega)}{\Phi^{-}(x, \omega)}.$$

In general these two admittances $Y_{in}^{\pm}(x, s)$ are different.
Generalized reflectance: A function related to the generalized impedance is the *reflectance* $\Gamma(s)$, defined as the ratio of a reflected wave to the incident wave. For the case of acoustics (pressure $\mathcal{P}$, volume velocity $\mathcal{V}$),

$$Y_{in}(x, s) \equiv \frac{\mathcal{V}(\omega)}{\mathcal{P}(\omega)} = \frac{\mathcal{V}^{+} - \mathcal{V}^{-}}{\mathcal{P}^{+} + \mathcal{P}^{-}} \tag{4.4.5}$$

$$= \frac{\mathcal{V}^{+}}{\mathcal{P}^{+}} \frac{1 - \mathcal{V}^{-}/\mathcal{V}_{+}}{1 + \mathcal{P}^{-}/\mathcal{P}^{+}} \tag{4.4.6}$$

$$= \mathcal{Y}_r^{+} \frac{1 - \Gamma(x, s)}{1 + \Gamma(x, s)}. \tag{4.4.7}$$

When the physical system is continuous at the measurement point x, $\mathcal{Y}_r^{+}(x) = \mathcal{Y}_r^{-}(x) \in \mathbb{R}$. The reflectance $\Gamma(x, s)$ depends on the area function, the boundary conditions, or both.

There is a direct relationship between a transmission line's area function $A(x) \in \mathbb{R}$, its characteristic impedance $\mathcal{Y}_r(x) \in \mathbb{R}$, and its eigenfunctions. We shall provide specific examples as they arise in our analysis of transmission lines (e.g., Fig. 5.3).

A few papers that deal with the relationship between $Y_{in}(s)$ and the area function $A(x)$ are Youla (1964), Sondhi and Gopinath (1971), Rasetshwane et al. (2012). However, the general theory of this important and interesting problem is beyond the scope of this text (See homework DE-3, Problem # 2), as well as Appendix D.

4.4.1.1 Complex Analytic $\Gamma(s)$ and $Y_{in}(s) = Z_{in}^{-1}(s)$

When we define the complex reflectance $\Gamma(s)$, we make a key assumption: Even though $\Gamma(s)$ is defined by the ratio of two functions of real (radian) frequency ω, like the impedance, the reflectance must be causal (Postulate P1, Sect. 3.10.2). That $\gamma(t) \leftrightarrow \Gamma(s)$ and $\zeta(t) \leftrightarrow Z_{in}(s) = 1/Y_{in}(s)$ are causal is required by the physics.

4.4.2 Complex Analytic Impedance

Conservation of energy (or power) is a cornerstone of modern physics. It may first have been under consideration by Galileo Galilei (1564–1642) and Marin Mersenne (1588–1648). Today the question is not whether it is true, but why. Specifically, what is the physics behind conservation of energy? Surprisingly, the answer is straightforward, based on its definition and the properties of impedance. Recall that the power is the product of the force and the flow, and impedance is their ratio.

The power is given by the product of two variables, sometimes called *conjugate variables*, the force and the flow. In electrical terms, these are voltage (force) ($v(t) \leftrightarrow V(\omega)$) and current (flow) ($i(t) \leftrightarrow I(\omega)$); thus, the electrical power at any instant of time is[5]

$$\mathcal{P}(t) = v(t)i(t). \tag{4.4.8}$$

The total energy $\mathcal{E}(t)$ is the integral of the power, since $\mathcal{P}(t) = d\mathcal{E}/dt$. Thus if we start with all the elements at rest (no currents or voltages), then the energy as a function of time is always positive

$$\mathcal{E}(t) = \int_0^t \mathcal{P}(t)dt \geq 0 \tag{4.4.9}$$

and is simply the total energy applied to the network (Van Valkenburg 1964a, p. 376). Since the voltage and current are related by either an impedance or an admittance, conservation of energy depends on the property of impedance. From Ohm's law and Postulate P1 (every impedance is causal), and we have

$$\begin{aligned} v(t) &= z(t) \star i(t) \\ &= \int_{\tau=0}^t z(\tau)i(t-\tau)d\tau \\ &\leftrightarrow V(s) = Z(s)I(s). \end{aligned}$$

Example: Let $i(t) = \delta(t)$ (a perfect impulse). Then

$$I_{xx}(t) = \int_{\tau=0}^t z(t-\tau)\delta(\tau)d\tau = \int_0^t z(-\tau)d\tau.$$

Every Brune impedance always has the form $z(t) = r_o\delta(t) + \zeta(t)$. The *characteristic impedance* r_o (also called *surge impedance*) may be defined as (Lundberg et al. 2007)

$$r_o = \int_{0^-}^{\infty} z(t)dt.$$

This definition requires that the integral of $\zeta(t)$ is zero, a conclusion that warrants further investigation.

These ideas are perhaps easier to visualize if we work in the Laplace frequency domain. Then the total energy, equal to the integral of the real part of the power, is

[5]The voltage is sometimes called the electromotive force (EMF). However $v(t)$ is relative to a reference ground. The actual EMF is $-\nabla v(t)$.

$$\frac{1}{s}\Re VI = \frac{1}{2s}(V^*I + VI^*) = \frac{1}{2s}(Z^*I^*I + ZII^*) = \frac{1}{s}\Re Z(s)|I|^2 \geq 0.$$

Mathematically this is called a *positive-definite operator*, since the positive and real resistance is sandwiched between the current, forcing the "definiteness."

In conclusion, conservation of energy is totally dependent on the properties of the impedance. Thus one of the most important and obvious applications of complex analytic functions of a complex variable is the impedance function. This seems to be the ultimate example of the FTCC applied to $z(t)$.

Every impedance must obey conservation of energy (Postulate P3): The impedance function $Z(s)$ has resistance R and reactance X as a function of complex frequency $s = \sigma + \jmath\omega$. From the causality postulate P1, $z(t < 0) = 0$. Every impedance is defined by a Laplace transform pair

$$z(t) \leftrightarrow Z(s) = R(\sigma, \omega) + \jmath X(\sigma, \omega),$$

with $R, X \in \mathbb{R}$.

According to postulate P3 a system is passive if it does not contain a power source. Drawing power from an impedance violates conservation of energy. This property is also called *positive-real*, which was defined by Brune (1931b, a)

$$\Re\{Z(s \geq 0)\} \geq 0. \tag{4.4.10}$$

Positive-real systems cannot draw more power than is stored in the impedance. The region $\sigma \leq 0$ is called the *left half s plane* (LHP), and the complementary region $\sigma > 0$ is called the *right half s plane* (RHP). According to the Brune condition, the real part of every impedance must be nonnegative (the RHP).

It is easy to construct examples of second-order poles or zeros in the RHP such that Postulate P3 is violated. Thus Postulate P3 implies that the impedance may *not* have more than simple (first-order) poles and zeros, strictly in the LHP. But there is more: These poles and zeros in the LHP must meet a *minimum phase condition*, a condition that is easily stated:

$$\angle Z(s) < \angle s \tag{4.4.11}$$

but difficult to prove. There seems to be no proof that second-order poles and zeros (e.g., second-order roots) are not allowed.[6] However, such roots must violate a requirement that the poles and zeros must alternate on the $\sigma = 0$ axis, which follows from Postulate P3. In the complex plane the concept of "alternate" is not defined (complex numbers cannot be ordered). What *has* been proved (i.e., Foster's reactance theorem), that says that if the poles are on the real or imaginary axis, they must alternate, leading to simple poles and zeros (Van Valkenburg 1964a). The restriction on poles is sufficient but not necessary, as $Z(s) = 1/\sqrt{s}$ is a positive-real impedance

[6] As best I know, this is an open problem in network theory (Brune 1931a; Van Valkenburg 1964a).

but is less than a first-degree pole (Kim and Allen 2013). The corresponding condition in the LHP, and its proof, remains elusive (Van Valkenburg 1964a).

For example, a series resistor R_o and capacitor C_o have an impedance given by (Table 3.9)

$$Z(s) = R_o + 1/sC_o \leftrightarrow R_o\delta(t) + \frac{1}{C_o}u(t) = z(t), \tag{4.4.12}$$

with constants $R_o, C_o \in \mathbb{R} \geq 0$. In mechanics, an impedance composed of a dashpot (damper) and a spring has the same mathematical form.

A full 2d order resonant system has an inductor, resistor, and capacitor, with an impedance given by

$$Z(s) = \frac{sC_o}{1 + sC_oR_o + s^2C_oM_o} \leftrightarrow C_o\frac{d}{dt}\left(c_+e^{s_+t} + c_-e^{s_-t}\right) = z(t), \tag{4.4.13}$$

which is a second-degree polynomial with two complex resonant frequencies $s_\pm$. When $R_o > 0$, these roots are in the LHP, with $z(t) \leftrightarrow Z(s)$.

Systems (networks) that contain many elements and transmission lines can be much more complicated, yet still have a simple frequency-domain representation. This is the key to understanding how these physical systems work, as we describe next.

Poles and zeros of positive-real functions must be first-degree: The definition of *positive-real* (PR) functions requires that the poles and zeros of the impedance function be simple (only first-degree). Second-degree poles would have a reactive "secular" response of the form $h(t) = t\sin(\omega_k t + \phi)u(t)$, and these terms would not average to zero, depending on the phase, as is required of an impedance. As a result, only single-degree poles are possible.[7] I believe that no one has ever reported an impedance that has second-degree poles and zeros. Network analysis books never report second-degree poles and zeros in their impedance functions. Nor has there ever been any guidance about where the poles and zeros might lie in the LHP. Understanding the exact relationships between pairs of poles and zeros, to assure that the real part of the impedance is real, would resolve this longstanding unsolved problem (Van Valkenburg 1964b). It is the residues that determine the LHP simple pole degree.

Calculus on Complex analytic functions: To solve a differential equation or integrate a function, Newton used the Taylor series to integrate one term at a time. However, he used only real functions of a real variable due to his fundamental lack of appreciation of the complex analysis. This same method is how one finds solutions to scalar differential equations today, but using an approach that makes the solution method less obvious. Rather than working directly with the Taylor series, today we use the complex exponential, since the complex exponential is an eigenfunction of the derivative

[7] Secular terms result from second-degree poles, since $u(t) \star u(t) = tu(t) \leftrightarrow 1/s^2$.

$$\frac{d}{dt}e^{st} = se^{st}.$$

Since e^{st} may be expressed as a Taylor series that has coefficients $c_n = 1/n!$, in some real sense the modern approach is a compact way of doing what Newton did. Thus every linear constant coefficient differential equation in time may be simply transformed into a polynomial in complex Laplace frequency s, by looking for solutions of the form $F(s)e^{st}$ and transforming the differential equation into polynomial $F(s)$ in complex frequency.

For example

$$\frac{d}{dt}f(t) + af(t) = \delta(t) \leftrightarrow (s + a)F(s) = 1.$$

The pole of $F(s_r)$ is $s_r + a = 0$ is the eigenvalue of the differential equation. Thus a powerful tool for understanding the solutions of differential equations, both scalar and vector, is to work in the Laplace frequency domain using their eigenvalues (i.e., $s_r = -a$) and their eigenfunctions.

The Taylor series may be replaced by e^{st}, which transforms Newton's real Taylor series into the complex exponential eigenfunction. In some sense, these are the same methods, since

$$e^{s_r t} = \sum_{n=0}^{\infty} \frac{(s_r t)^n}{n!}. \tag{4.4.14}$$

Taking the derivative with respect to time gives

$$\frac{d}{dt}e^{st} = se^{st} = s\sum_{n=0}^{\infty} \frac{(st)^n}{n!}, \tag{4.4.15}$$

which is also complex analytic. Thus if the series for $F(s)$ is valid (i.e., it converges), then its derivative is also valid. This was a very powerful concept, exploited by Newton for real functions of a real variable, and later by Cauchy and Riemann for complex functions of a complex variable. The key question is "Where does the series fail to converge?" This is the main concept behind the FTCC (Eq. 4.2.3).

The FTCC (Eq. 4.2.3) is formally the same as the FTC (Eq. 4.2.2) (Leibniz's formula), the key (and significant) difference being that the argument of the integrand $s \in \mathbb{C}$. Thus this integration is a line integral in the complex plane. One would naturally assume that the value of the integral depends on the path of integration. And it does, but in a subtle way, as quantified by Cauchy's various theorems. If the path stays in the same RoC region, then the integral is independent of that path. If a path includes a different pole, then the integral depends on the path, as quantified by the Cauchy residue theorem. The test is to deform the path from the first to the second. If in that deformation the path crosses a pole, then the integral will change (namely, it will depend on the path). All of this follows from causality.

The FTC and FTCC are clearly distinguishable yet the reasoning is the same. If $F(s) = df(s)/ds$ is complex analytic [i.e., has a power series $f(s) = \sum_k c_k s^k$, with $f(s), c_k, s \in \mathbb{C}$], then it may be integrated, term by term, and yet the integral does not depend on the path. At first blush, this is sort of amazing. The key is that $F(s)$ and $f(s)$ must be complex analytic, which means they are differentiable. This all follows from the Taylor series formula Eq. 4.2.5 for the coefficients of the complex analytic series. For Eq. 4.2.3 to hold, the derivatives must be independent of the direction (the path), as discussed in Sect. 4.2.2. The concept of a complex analytic function, therefore, has eminent consequences in the form of several key theorems on complex integration, as first discovered by Cauchy in about 1820.

Role of the Complex Taylor series: The complex Taylor series generalizes the functions it describes, with unpredictable consequences, as nicely shown by the domain-coloring diagrams in Fig. 3.12, where a simple translation of the s plane by a complex number can void the positive-real property ($s - \sqrt{j}$) cannot be a physical impedance).

Cauchy's of complex integration tools were first exploited in physics by Sommerfeld (1952), to explain the onset (e.g., causal) transients in waves, as he explained in detail in Brillouin (1960, Chap. 3).

The importance of causality: Up to 1910, when Sommerfeld first published his results using complex analytic signals and saddle point integration in the complex plane, the implications of the causal wavefront were poorly understood. It would be reasonable to say that Sommerfeld's insight changed our understanding of wave propagation for both light and sound. Unfortunately, in my view, his insight has never been fully appreciated, perhaps even to this day. If you question this summary, please read Brillouin (1960, Chap. 1).

The full power of the complex analytic function was first appreciated by Bernard Riemann (1826–1866) in his University of Göttingen PhD thesis of 1851, under the tutelage of Gauss (1777–1855), which drew heavily on the work of Cauchy (Fig. 3.1).

The key definition of a complex analytic function is that it has a Taylor series representation, over a region of the complex frequency plane $s = \sigma + j\omega$, that converges in its RoC about the expansion point, with a radius determined by the nearest pole of the function. A further surprising feature of all analytic functions is that within the RoC, the inverse of that function also has a complex analytic expansion. Thus given $w(s)$, in theory, we can determine $s(w)$ to any desired accuracy, critically depending on the RoC. As an example if $w(s) = e^s$ then its inverse is $s(w) = log(w)$. Given the right software (e.g., `zviz.m`), this relationship may be made precise for every complex analytic function.

4.4.3 *Multivalued Functions*

In the field of mathematics there seems to have been a tug-of-war regarding the basic definition of a function. The accepted definition today is a single-valued (i.e.,

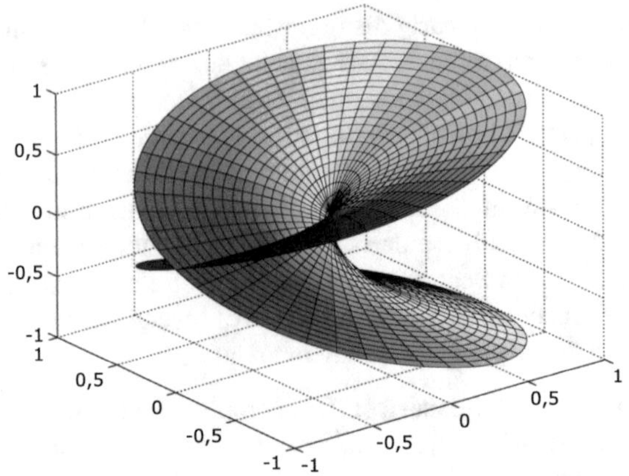

Fig. 4.2 The mapping for the square root function $z = \pm\sqrt{x}$, the inverse of $x = z^2$. This function has two single-valued sheets of the x plane corresponding to the two signs of the square root. The best way to view this function is in polar coordinates, with $x = |x|e^{\jmath\phi}$ and $z = \sqrt{|x|}e^{\jmath\phi/2}$. The color code in this figure is unrelated to that of the colorized plots. Note that the axes are not labeled. Figure from: https://en.wikipedia.org/wiki/Riemann_surface

complex analytic) mapping from the domain to the codomain (or range). This makes the discussion of complex analytic multivalued functions somewhat awkward. In 1851 Riemann (working with Gauss) resolved this problem for the complex analytic set of multivalued functions by introducing the geometric concept of single-valued sheets, delineated by branch cuts.

Two simple yet important examples of multivalued functions are the circle $z^2 = x^2 + y^2$ and $w = \log(z)$. For example, if we assume $\rho = |z|$ is the radius of the circle, then solving for $y(x)$ gives the double-valued function

$$y(x) = \pm\sqrt{\rho^2 - x^2}.$$

The related function $z = \pm\sqrt{x}$, with $x \in \mathbb{C}$, is shown in Fig. 4.2 as a three-dimensional display in polar coordinates, with $z(r)$ as the vertical axis, as a function of the angle and radius of $x \in \mathbb{C}$.

If we accept the modern injective definition of a function, as the mapping from one set to a second set, then $y(x)$ is not a function, or even two functions. For example, what if $x > z$? Or worse, what if $z = 2\jmath$ with $|x| < 1$? Riemann's construction, using branch cuts for multivalued function, resolves all these difficulties (as best I know).

To proceed, we need definitions and classifications of the various types of complex singularities:

1. Poles of degree 1 are called *simple poles*. Their amplitude is called the *residue* (e.g., α/s has residue α). Simple poles are special (see Eq. 4.5.3); they play a key role in mathematical physics, since their inverse Laplace transform defines a causal eigenfunction.

2. When the numerator and denominator of a rational function (i.e., ratio of two polynomials) have a common root (i.e., factor), that root is said to be *removable*.
3. A singularity that is not removable, a pole, or a branch point is called *essential*.
4. A complex analytic function (except for isolated poles) is called *meromorphic* (Boas 1987). Meromorphic functions can have any number of poles, even an infinite number. The poles need not be simple.
5. When the first derivative of a function $Z(s)$ has a simple pole at a, then a is said to be a *branch point* of $Z(s)$. An important example is the logarithmic derivative:

$$d\ln(s-a)^{\alpha}/ds = \alpha/(s-a), \quad \alpha \in \mathbb{I}.$$

However, the converse does not necessarily hold.
6. I am not clear about the interesting case of an irrational pole ($\alpha \in \mathbb{I}$). In some cases (e.g., $\alpha \in \mathbb{F}$) this may be simplified with the logarithmic derivative operation, as mentioned in Sect. 3.1.2.

More complex topologies are being researched today, and progress is expected to accelerate due to modern computing technology.[8] It is helpful to identify the physical meaning of these more complex surfaces, to guide us in their interpretation and possible applications.[9]

Branch cuts: Up to this point we have considered only poles of degree $\alpha \in \mathbb{N}$ of the form $1/s^{\alpha}$. The concept of a branch cut allows us to manipulate (and visualize) multivalued functions for which $\alpha \in \mathbb{F}$. This is done by breaking each region into single-valued sheets, as shown in Fig. 4.3 (right). The branch cut is a curve $\in \mathbb{C}$ that separates the various single-valued sheets of a multivalued function. The concepts of branch cuts, sheets, and the extended plane were first devised by Riemann, as described in his thesis of 1851. It was these three mathematical and geometrical constructions that provided deep insight into complex analytic functions, greatly extending the important earlier work of Cauchy (1789–1857) on the calculus of complex analytic functions. For an alternative helpful discussion of Riemann sheets and branch cuts, see Boas (1987, pp. 221–25), Kusse and Westwig (2010), and Greenberg (1988).

To study the properties of multivalued functions and branch cuts, we look at $w(s) = \sqrt{s}$ and $w(s) = \log(s)$, along with their inverse functions $w(s) = s^2$ and $w(s) = e^s$. For uniformity we refer to the *complex abscissa* ($s = \sigma + \omega \jmath$) and the *complex ordinate* ($w(s) = u + v\jmath$). When the complex domain and range are swapped, by taking the inverse of a function, multivalued functions are a common consequence. For example, $f(t) = \sin(t)$ is single-valued and analytic in t and thus has a Taylor series. The inverse function $t(f)$ is multivalued.

The best way to explore the complex mapping from the complex planes $s \rightarrow w(s)$ is to master the single-valued function $s = w^2(s)$ and its double-valued inverse $w(s) = \sqrt{s}$.

---

[8] https://www.maths.ox.ac.uk/about-us/departmental-art/theory.

[9] https://www.quantamagazine.org/secret-link-uncovered-between-pure-math-and-physics-20171201.

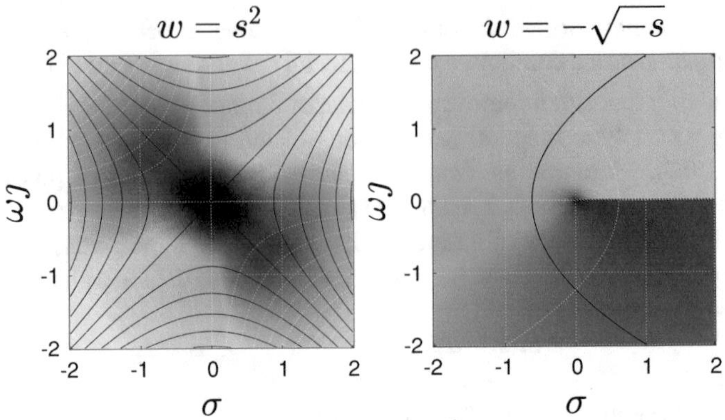

Fig. 4.3 Here we use Cartesian coordinates in the domain $s = \sigma + \omega j$ and polar coordinates for the range $w(\sigma, \omega) = |w|e^{\psi j}$. The color intensity indicates the magnitude $|s|$, with black being $|s = 0|$ and bright (eventually white) indicating $|s| \to \infty$. **Left:** Mapping: $w(s) = s^2$. **Right:** Mapping of the principal branch from $-s$ to $w(s) = -\sqrt{-s}$ (i.e., the rotated inverse of s^2). This sheet was generated by rotating w by 180°. The branch cut is on the $\psi = 0$ axis, with branch points at $|w| = 0$ and ∞. Neither of these functions are Brune impedances, since they violate the positive-real condition (Eq. 3.2.19)

Figure 4.3 (left) shows the single-valued function $w(s) = s^2$, and (right) one sheet of its inverse, the double-valued mapping of $w(s) = -\sqrt{-s}$. Single-valued functions such as $w(s) = s^2$ are relatively straightforward. Multivalued functions require the concept of a branch cut, defined in the image plane (also called the codomain or range). This is a technique to render the multiple values as single-valued on each of several sheets. Each sheet is labeled in the domain (s) plane by a sheet index $k \in \mathbb{Z}$, while branch points and cuts are defined in the image (w) plane.

The sheets are indexed by a sheet index, and separated by the branch cut. It is important to understand that the path of every branch cut is not unique and may be moved. However, branch points are unique and thus not movable.

A function may be multivalued in both the domain and image planes. As an example consider $w(s) = s^{3/2}$.

The multivalued nature of $w(s) = \sqrt{s}$ is best understood by working with the function in polar coordinates. We let

$$s_k = re^{\theta j}e^{2\pi k j}, \tag{4.4.16}$$

where $r = |s|$, $\theta = \angle s, \in \mathbb{R}$, and $k \in \mathbb{Z}$ is the sheet index.

This concept of analytic inverses becomes important only when the function is multivalued. For example, since $w(s) = s^2$ has a period of 2, $s(w) = \pm\sqrt{w}$ is multivalued. Riemann dealt with such extensions using the concept of a branch cut with multiple sheets labeled by sheet numbers. Each sheet describes an analytic function (Taylor series) that converges within some RoC that has a radius out to the nearest pole. Thus Riemann's branch cuts and sheets explicitly deal with the need to

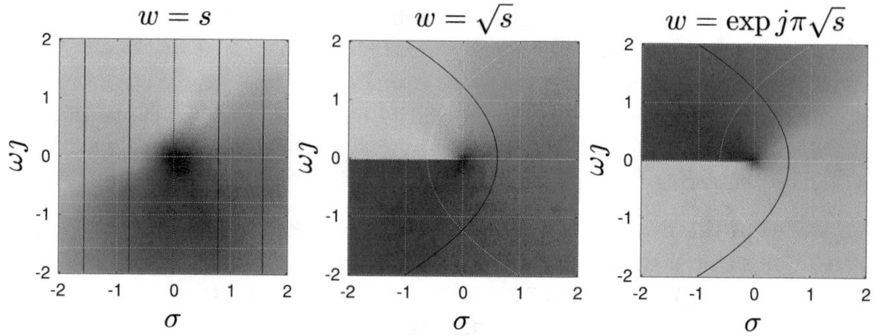

Fig. 4.4 Colorized plots of $s_k = |s|e^{\theta\jmath}e^{2\pi k\jmath}$ and $w_k(s) = \sqrt{s_k} = \sqrt{|s|}e^{\theta\jmath/2}e^{\pi k\jmath}$, as defined in polar coordinates by Eqs. 4.4.16 and 4.4.17. **Left:** Color map hue reference plane $s = |s|e^{\theta\jmath}e^{2\pi k\jmath}$. This function is a Brune impedance (it represents an inductor). **Center:** Sheet index $k = 0$, $w(s) = \sqrt{|s|}e^{\jmath\theta/2}$ for $-\pi < \theta < +\pi$ and $\psi = \angle w(s) = \theta/2$ between $\pm\pi/2$ [rad]. This function is positive-real (Eq. 3.2.19) but is not a Brune impedance, since it is not the ratio of polynomials in s. Thus it is the *generalized impedance*, known as a semiinductor. **Right:** Sheet index $k = 1$, $w(s) = e^{\jmath\theta/2}e^{\jmath\pi}\sqrt{|s|}$ for $\pi < \theta < 3\pi$ and $\pi/2 < \psi < 3\pi/2$ [rad]. The branch cut is at $\theta = \angle s = \pm 180°$ (π [rad]) where the hue of w changes abruptly from green to purple (center) and from purple back to green (right). Note how the hue matches between the center and right panels at the branch cut: in the center panel purple runs along -180° and along +180° on the right. Likewise, green runs along +180° in the : $\pm\sqrt{s}$, $e^{\pi\jmath}\sqrt{s}$ center and along -180° on the right. Thus $w(s) = \sqrt{s}$ is analytic on the branch cut connecting the two sheets ($k = 0 \to k = 1$). This function is not an impedance, since it is negative in the RHP

define unique single-valued inverses of multivalued functions. Since the square root function has two overlapping regions corresponding to the $\pm$ due to the radical, there must be two connected region—sort of like mathematical Siamese twins: distinct, yet the same.

Hue: By studying Fig. 4.4, we can appreciate domain-coloring. The angle-to-hue map is shown in the left panel of Fig. 4.4. The domain angles $\angle s$ go from $-90° < \theta < 90°$ (purple to red to green). For $w(s) = \sqrt{s}$, the $\angle s$ is *compressed* by a factor of 2 ($\psi = \theta/2$) with purple being $-180°$, and green being $+180°$.

Thus for Fig. 4.4 the principal ($k = 0$) angle $\angle s$ ($-180° < \theta < 180°$) maps to $w(s) = \sqrt{s}$ (middle panel) to half the w plane ($-90° < \psi < 90°$ (from purple to red to green).

The $k = 1$ branch, of $\angle s$ ($+180° < \theta < 180 + 360 = 520°$) maps to $\angle w$ (on the right) to green to blue to purple ($\psi = \theta/2$). Note how the panel on the left matches the right half of s (green = $+90°$, purple = $-90°$). The center panel $\angle w$ is green where $\angle s = 180°$. Thus $\angle w = \frac{1}{2}\angle s$. Going around the s plane one more time gives the right most figure. $w(s)$ is analytic everywhere except at the branch points $s = 0$ and $s = \infty$.

Moving the branch cut: Furthermore we can change $\angle s$ by 180° to move the branch cut

$$w = \rho e^{\jmath\psi} = \sqrt{r}e^{\jmath\theta/2}e^{\jmath\pi k}, \tag{4.4.17}$$

where $\rho = |w|, \psi \angle w, \in \mathbb{R}$. The generic Cartesian coordinates are $s = \sigma + \omega \jmath$ and $w(s) = u(\sigma, \omega) + v(\sigma, \omega)\jmath$. For single-valued functions such as $w(s) = s^2$ on the left in Fig. 4.3 there is no branch cut, since $\psi = 2\theta$. Note how the red color ($\theta = 0°$) appears twice in this mapping. For multivalued functions, a branch cut is required, typically along the negative $v(\sigma, \omega)$ axis (i.e., $\psi = \pi$), but it may be freely distorted, as seen by comparing the right panel of Fig. 4.3 with the right panel of Fig. 4.4.

Properties of the branch cut: It is important to understand that the function is analytic on the branch cut but not at the branch points. One is free to move the branch cut (at will). It does not need to be on a line: it could be cut in almost any connected manner, such as a spiral. The only rule is that it must start and stop at the matching branch points, or at ∞, which must have the same degree.

The location of the branch cut may be moved by rotating the s-coordinate system of Fig. 4.2. For example, $w(s) = \pm\jmath\sqrt{s}$ and $w(s) = \pm\sqrt{-s}$ have different branch cuts, as may be verified using the Matlab/Octave commands `j*zviz(s)` and `zviz(-s)`. Every function is analytic on the branch cut (since moving it does not change the function). If a Taylor series is formed on the branch cut, it will describe the function on the two different sheets. Thus the complex analytic series (i.e., the Taylor formula, Eq. 4.2.5) does not depend on the location of a branch cut, as it only describes the function uniquely (as a single-valued function), valid in its local region of convergence.

The second sheet ($k = 1$) in Fig. 4.4 picks up at $\theta = \pi$ [rads] and continues on to $\pi + 2\pi = 3\pi$. The first sheet maps the angle of w (i.e., $\phi = \angle w = \theta/2$) from $-\pi/2 < \phi < \pi/2$ ($w = \sqrt{r}e^{\jmath\theta/2}$). This corresponds to $u = \Re\{w(s)\} > 0$. The second sheet maps $\pi/2 < \psi < 3\pi/2$ (i.e., 90° to 270°), which is $\Re\{w\} = u < 0$. In summary, twice around the s plane is once around the $w(s)$ plane because the angle is half due to the $\sqrt{s}$.

Branch cuts emanate and terminate at *branch points*, defined as singularities (poles) that can even have fractional degree, as, for example, $1/\sqrt{s}$, and terminate at one of the matching roots, which includes the possibility of ∞.[10] For example, suppose that in the neighborhood of the pole, at s_o the function is

$$f(s) = \frac{w(s)}{(s - s_o)^k}, \tag{4.4.18}$$

where $w, s, s_o \in \mathbb{C}$ and $k \in \mathbb{Q}$. When $k = 1$, $s_o = \sigma_o + \omega_o\jmath$ is a first-degree "simple pole," having degree 1 in the s plane, with residue $w(s_o)$. Typically the order and degree are positive integers, but fractional degrees and orders are common in modern engineering applications (Kirchhoff 1868; Lighthill 1978). Here we allow both the degree and the order to be fractional ($\in \mathbb{F}$). When $k \in \mathbb{F} \subset \mathbb{R}$, $k = n/m$ is a real reduced fraction—namely, when `GCD(n,m)=1`, $n \perp m$). This defines the *degree* of a fractional pole. In such cases there must be two sets of branch cuts of degrees n and

[10] This presumes that poles and zeros appear in pairs, one of which may be at ∞.

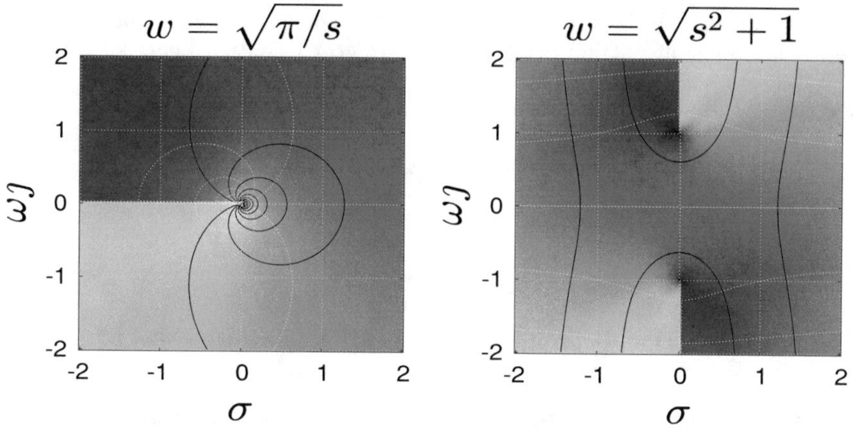

Fig. 4.5 Colorized plots of two $\mathcal{LT}$pairs: Left: $\sqrt{\pi/s} \leftrightarrow u(t)/\sqrt{t}$. Right: $\sqrt{s^2+1} \leftrightarrow \delta(t) + \frac{1}{t}J_1(t)u(t)$

m. For example, if $k = 1/2$, the singularity (branch cut) is of degree 1/2 and there are two Riemann sheets, as shown in Fig. 4.3.

Fractional-order Bessel function: An important example is the Bessel function and its Laplace transform ($\mathcal{LT}$)

$$\delta(t) + \frac{1}{t}J_1(t)u(t) \leftrightarrow \sqrt{s^2+1},$$

as shown in Fig. 4.5, which is related to the solution to the wave equation in two-dimensional cylindrical coordinates (see Table C.2). Bessel functions are the solutions (i.e., eigenfunctions) of guided acoustic waves in round pipes, or surface waves on the Earth (seismic waves), or waves on the surface of a pond (Table 5.2).

There are a limited number of possibilities for the degree, $k \in \mathbb{Z}$ or $\in \mathbb{F}$ of Eq. 4.4.18. If the degree is drawn from $\mathbb{R} \notin \mathbb{F}$, the pole cannot have a residue. According to the definition of the residue, $k \in \mathbb{F}$ has no residue. But there remains open the possibility of generalizing the concept of the Riemann integral theorem to include $k \in \mathbb{F}$. One way to do this is to use the logarithmic derivative, which transforms fractional poles to simple poles with fractional residues. (See Ex.# 9, Sect. 3.1.2).

If the singularity has an irrational degree ($k \in \mathbb{I}$), the branch point has the same irrational degree. Accordingly there are an infinite number of Riemann sheets, as in the case of the log function. An example is $k = \pi$, for which

$$F(s) = \frac{1}{s^\pi} = e^{-\log(s^\pi)} = e^{-\pi\log(s)} = e^{-\pi\log(\rho)}e^{-\pi\theta\jmath},$$

where the domain is expressed in polar coordinates $s = \rho e^{\theta\jmath}$. When $k \in \mathbb{F}$, it may be close (e.g., $k = \pi_{152}/\pi_{153} = \pi_{152}/(\pi_{152}+2) = 881/883 \approx 0.99883$, or its reciprocal ≈ 1.0023). The branch cut could be subtle (it could even go unnoticed), but it would have a significant impact on the function and on its inverse Laplace transform.

Exercise #2 Find the poles, zeros, and residues of $F(s)$.

1.
$$F(s) = \frac{d}{ds}\ln\frac{s+e}{s+\pi}$$

Solution:

$$F(s) = \frac{d}{ds}\left[\ln(s+e) - \ln(s+\pi)\right] = \left(\frac{1}{s+e} - \frac{1}{s+\pi}\right)$$

The poles are at $s_1 = -e$ and $s_2 = -\pi$ with respective residues of ± 1. ∎

2.
$$F(s) = \frac{d}{ds}\ln\frac{(s+3)^e}{(s+\jmath)^{-\pi}}$$

Solution:

$$F(s) = \frac{d}{ds}\left(e\ln(s+3) + \pi\ln(s+\jmath)\right) = \frac{e}{s+3} + \frac{\pi}{s+\jmath}.$$

There is a very important take-home message here regarding the utility of the logarithmic derivative, which "linearizes" the fractional pole. ∎

3.
$$F(s) = e^{\pi \ln s}$$

Solution: To simplify this expression take the log

$$\ln F(s) = \ln e^{\pi \ln s} = \pi \ln s = \ln s^{\pi}.$$

Thus $F(s) = s^{\pi}$. The only pole is $s \to \infty$. Thus the definition of the residue is to multiply by the pole and take the limit as $s \to \infty$

$$c_{-1} = \lim_{s\to\infty}\frac{s^{\pi}}{s} = 0^{\pi-1} = 0.$$

∎

4.
$$F(s) = \pi^{-s}$$

Solution: Converting to exponential format: $F(s) = e^{-s \ln \pi} \leftrightarrow \delta(t - \ln \pi)$. I don't think the pure time delay has poles. I'm not sure what this tells us about the residue.■

Log function: Next we discuss the multivalued nature of the log function. In this case there are an infinite number of Riemann sheets, not captured by Fig. 3.14, which displays only the principal sheet. However, if we look at the formula for the log function, the nature is easily discerned. The abscissa s may be defined as multivalued, since

$$s_k = re^{2\pi k \jmath} e^{\theta \jmath}.$$

Here we have extended the angle of s by $2\pi k$, where k is the sheet index $\in \mathbb{Z}$. Now we take the log:

$$\log(s) = \log(r) + (\theta + 2\pi k)\jmath.$$

When $k = 0$, we have the principal value sheet, which is zero when $s = 1$. For any other value of k, $w(s) \neq 0$, even when $r = 1$, since the angle is not zero, except for the $k = 0$ sheet.

4.5 Three Cauchy Integral Theorems

4.5.1 *Cauchy's Theorems for Integration in the Complex Plane*

There are three basic definitions related to Cauchy's integral formula. They are closely related and can greatly simplify integration in the complex plane. The choice of names is unfortunate, if not totally confusing. Hence I call them CT-1, CT-2, and CT-3.

1. **Cauchy's (integral) theorem (CT-1):**

$$\oint_C F(s)ds = 0 \tag{4.5.1}$$

   if and only if $F(s)$ is complex analytic inside of a simple closed curve C (Stillwell 2010; Boas 1987, p. 45). The FTCC (Eq. 4.2.3) says that the integral depends on only the end points if $F(s)$ is complex analytic. With the path (contour C) closed, the end points are the same and thus the integral must be zero as long as $F(s)$ is complex analytic.
2. **Cauchy's integral formula (CT-2):**

$$\frac{1}{2\pi j} \oint_{\mathcal{B}} \frac{F(s)}{s - s_o} ds = \begin{cases} F(s_o), \; s_o \in \mathbb{C} < \mathcal{B} \text{ (inside)} \\ 0 \quad , s_o \in \mathbb{C} > \mathcal{B} \text{ (outside)}. \end{cases} \tag{4.5.2}$$

Here $F(s)$ is required to be analytic everywhere within (and on) the boundary $\mathcal{B}$ of integration (Greenberg 1988, p. 1200); (Boas 1987, p. 51); (Stillwell 2010, p. 220). When the point $s_o \in \mathbb{C}$ is within the boundary, the value $F(s_o) \in \mathbb{C}$ is the residue of the pole s_o of $F(s)/(s - s_o)$. When the point s_o lies outside the boundary, the integral is zero.

3. **The (Cauchy) residue theorem (CT-3):** (Greenberg 1988, p. 1241); (Boas 1987, p. 73)

$$\oint_C f(s)ds = 2\pi j \sum_{k=1}^{K} c_k = \sum_{k=1}^{K} \oint \frac{F(s)}{s - s_k} ds, \tag{4.5.3}$$

where the residues $c_k \in \mathbb{C}$ correspond to the kth pole of $f(s)$ enclosed by the contour C (Greenberg 1988, p. 1241); (Boas 1987, p. 73). Cauchy's integral formula (CT-2) is a special case of the residue theorem (CT-3).

How to calculate the residue: The case of first-degree poles has special significance because the Brune impedance allows only simple poles and zeros, thus increasing its utility. The residues for simple poles are $F(s_k)$, which is complex analytic in the neighborhood of the pole, but not at the pole.

Consider the function $f(s) = F(s)/(s - s_k)$, where we have factored $f(s)$ to isolate the first-order pole at $s = s_k$, with $F(s)$ analytic at s_k. Then the residue of the poles at $c_k = F(s_k)$. This coefficient is computed by removing the singularity, placing a zero at the pole frequency, and taking the limit as $s \to s_k$—namely,

$$c_k = \lim_{s \to s_k} [(s - s_k) f(s)] \tag{4.5.4}$$

(Greenberg 1988; Boas 1987, p. 72).

When the pole is an Nth degree, the procedure is much more complicated and requires taking $N - 1$ order derivatives of $f(s)$ followed by the limit process (Greenberg 1988, p. 1242). Higher degree poles are rarely encountered; thus it is good to know that this formula exists, but perhaps it is not worth the effort to learn (i.e., memorize) it.

Some open questions: Without the use of CT-3 it is hard to evaluate the inverse Laplace transform of $1/s$ directly. For example, how do we show that the integral (Eq. 4.5.2) is zero for negative time (or 1 for positive time)? CT-3 resolves this difficult problem by the convergence of the integral for negative and positive times. Clearly the convergence of the integral at $\omega \to \infty$ plays an important role.

4.5.2 Cauchy Integral Formula and Residue Theorem

CT-2 is an important extension of CT-1, in that a pole has been explicitly represented in the integrand at $s = s_o$. If the pole location is outside the curve C, the result of the integral is zero, in keeping with CT-1. When the pole is inside C, the integrand is no longer complex analytic at the enclosed pole. When this pole is simple, the residue theorem applies. For the related CT-3 the same result holds, except it is assumed that there are K simple poles in the function $F(s)$. This requires K repeated applications of CT-2. Thus it represents a minor extension of CT-2. When the integrand is $f(s)/P_K(s)$, where $f(s)$ is analytic in $\mathcal{C}$ and $P_K(s)$ is a polynomial of degree K, with all of its roots $s_k \in \mathcal{C}$, then CT-3 applies.

Nonintegral degree singularities: A key point is that this theorem applies when $k \in \mathbb{I}$, including fractionals $k \in \mathbb{F}$, but for these cases the residue is always zero, since by definition the residue is the amplitude of the $1/s$ term (Boas 1987, p. 73). Below are examples:

1. When $k \in \mathbb{F}$ (e.g., $k = 2/3$), the residue of s^k is zero, by definition.
2. The function $1/\sqrt{s}$ has a zero residue (we apply the definition of the residue, Eq. 4.5.4).
3. When $k \neq 1 \in \mathbb{I}$, the residue is, by definition, zero.
4. When $k = 1$, the residue is given by Eq. 4.5.4.
5. CT-1, CT-2, and CT-3 are essential when computing the inverse Laplace transform.

Summary and examples: These three CT theorems, all attributed to Cauchy, collectively are related to the fundamental theorems of complex calculus. The general principles are

1. In general it makes no sense (nor is there any need) to integrate *through* a pole; thus the poles (or other singularities) must not lie on $\mathcal{C}$.
2. CT-1 (Eq. 4.5.1) follows trivially from the fundamental theorem of complex calculus (Eq. 4.2.3), since if the integral is independent of the path and the path returns to the starting point, the closed integral must be zero. Thus Eq. 4.5.1 holds when $F(s)$ is complex analytic within C.
3. Since the real and imaginary parts of every complex analytic function obey Laplace's equation (Eq. 4.2.8), it follows that every closed integral over a Laplace field—that is, those defined by Laplace's equation—must be zero. In fact, this is the property of a conservative system, corresponding to many physical systems. If a closed box has fixed potentials on the walls, with any distribution whatsoever, and a point charge (i.e., an electron) is placed in the box, then a force equal to $F = qE$ is required to move that charge, and thus work is done. However, if the point is returned to its starting location, the network done is zero.
4. Work is done in charging a capacitor, and energy is stored. However, when the capacitor is discharged, all of the energy is returned to the load.
5. Soap bubbles and rubber sheets on a wire frame obey Laplace's equation.
6. These are all cases where the fields are Laplacian, thus closed line integrals must be zero. Laplacian fields are commonly seen because they are so basic.
7. We have presented the impedance as the primary example of a complex analytic function. Physically, every impedance has an associated stored energy, and every system having stored energy has an associated impedance. This impedance is usually defined in the frequency s domain, as a force *over* a flow (i.e., voltage

over current). The power $\mathcal{P}(t)$ is defined as the force *times* the flow and the energy $\mathcal{E}(t)$ as the time integral of the power

$$\mathcal{E}(t) = \int_{-\infty}^{t} \mathcal{P}(t)dt, \tag{4.5.5}$$

which is similar to Eq. 4.2.1 [see Sect. 3.8.3, Eq. 3.8.9]. In summary, impedance and power and energy are all fundamentally related.

4.6 Problems DE-2

4.6.1 Topics of This Homework

Integration of complex functions, Cauchy's theorem, integral formula, residue theorem, power series, Riemann sheets and branch cuts, inverse Laplace transforms, Quadratic forms.

4.6.2 Two Fundamental Theorems of Calculus

4.6.2.1 Fundamental Theorem of Calculus (Leibniz)

According to the fundamental theorem of (real) calculus (FTC),

$$f(x) = f(a) + \int_{a}^{x} F(\xi)d\xi, \tag{DE-2.1}$$

where $x, a, \xi, F, f \in \mathbb{R}$. This is an indefinite integral (since the upper limit is unspecified). It follows that

$$\frac{df(x)}{dx} = \frac{d}{dx}\int_{a}^{x} F(x)dx = F(x).$$

This justifies also calling the indefinite integral the *antiderivative*.

For a closed interval $[a, b]$, the FTC is

$$\int_{a}^{b} F(x)dx = f(b) - f(a), \tag{DE-2.2}$$

thus the integral is independent of the path from $x = a$ to $x = b$.

4.6.2.2 Fundamental Theorem of Complex Calculus

According to the fundamental theorem of complex calculus (FTCC),

$$f(z) = f(z_0) + \int_{z_0}^{z} F(\zeta)d\zeta, \quad \text{(DE-2.3)}$$

where $z_0, z, \zeta, f, F \in \mathbb{C}$. It follows that

$$\frac{df(z)}{dz} = \frac{d}{dz}\int_{z_0}^{z} F(\zeta)d\zeta = F(z). \quad \text{(DE-2.4)}$$

For a closed interval $[s, s_o]$, the FTCC is

$$\int_{s_o}^{s} F(\zeta)d\zeta = f(s) - f(s_o), \quad \text{(DE-2.5)}$$

thus the integral is independent of the path from $x = a$ to $x = b$.

Problem #1 *–1.1: Consider Eq. DE-2.1. What is the condition on $F(x)$ for which this formula is true?*

–1.2: Consider Eq. DE-2.3. What is the condition on $F(z)$ for which this formula is true?

–1.3: Let $F(z) = \sum_{k=0}^{\infty} c_k z^k$.

–1.4: Let

$$F(z) = \frac{\sum_{k=0}^{\infty} c_k z^k}{z - \jmath}.$$

Problem #2 *In the following problems, solve the integral*

$$I = \int_{\mathcal{C}} F(z)dz$$

for a given path $\mathcal{C} \in \mathbb{C}$.

–2.1: Perform the following integrals ($z = x + iy \in \mathbb{C}$):

1. $I = \int_0^{1+\jmath} zdz$
2. $I = \int_0^{1+\jmath} zdz$, *but this time make the path explicit: from 0 to 1, with $y = 0$, and then to $y = 1$, with $x = 1$.*
3. *Discuss whether your results agree with Eq. DE-2.4?*

–2.2: Perform the following integrals on the closed path $\mathcal{C}$, which we define to be the unit circle. You should substitute $z = e^{i\theta}$ and $dz = ie^{i\theta}d\theta$, and integrate from $\{-\pi, \pi\}$ to go once around the unit circle.

Discuss whether your results agree with Eq. DE-2.4?

1. $\int_C z dz$
2. $\int_C \frac{1}{z} dz$
3. $\int_C \frac{1}{z^2} dz$
4. $I = \int_C \frac{1}{(z+2\jmath)^2} dz.$
   Recall that the path of integration is the unit circle, starting and ending at -1.

Problem #3 *FTCC and integration in the complex plane*
Let the function $F(z) = c^z$, where $c \in \mathbb{C}$ is given for each question. Hint: Can you apply the FTCC?

–*3.1: For the function $f(z) = c^z$, where $c \in \mathbb{C}$ is an arbitrary complex constant, use the Cauchy–Riemann (CR) equations to show that $f(z)$ is analytic for all $z \in \mathbb{C}$.*

–*3.2: Find the antiderivative of $F(z)$.*

–*3.3: $c = 1/e = 1/2.7183, \ldots$ where $\mathcal{C}$ is $\zeta = 0 \to i \to z$*

–*3.4: $c = 2$, where $\mathcal{C}$ is $\zeta = 0 \to (1+i) \to z$*

–*3.5: $c = i$, where the path $\mathcal{C}$ is an inward spiral described by $z(t) = 0.99^t e^{i2\pi t}$ for $t = 0 \to t_0 \to \infty$*

–*3.6: $c = e^{t-\tau_0}$, where $\tau_0 > 0$ is a real number and $\mathcal{C}$ is $z = (1 - i\infty) \to (1 + i\infty)$.* Hint: Do you recognize this integral? If you do not, please do not spend a lot of time trying to solve it via the "brute force" method.

4.6.3 Cauchy's Theorems CT-1, CT-2, CT-3

Problem #4 *Cauchy's theorems for integration in the complex plane*
There are three basic definitions related to Cauchy's integral formula. They are all related and can greatly simplify integration in the complex plane. When a function depends on a complex variable, we use uppercase notation, consistent with the engineering literature for the Laplace transform.

1. **Cauchy's (Integral) Theorem CT-1** (Stillwell, 2010, p. 319; Boas, 1987, p. 45)

$$\oint_{\mathcal{C}} F(z) dz = 0$$

if and only if $F(z)$ is complex analytic inside of $\mathcal{C}$. This is related to the FTCC,

$$f(z) = f(a) + \int_a^z F(z) dz,$$

where $f(z)$ is the antiderivative of $F(z)$—namely, $F(z) = df/dz$. The FTCC requires $F(z)$ to be complex analytic for all $z \in \mathbb{C}$. By closing the path (contour $\mathcal{C}$), Cauchy's theorem (and the following theorems) allows us to integrate functions that may not be complex analytic for all $z \in \mathbb{C}$.

2. **Cauchy's Integral Formula CT-2** (Boas (1987), p. 51; Stillwell, 2010, p. 220)

$$\frac{1}{2\pi j}\oint_C \frac{F(z)}{z-z_0}dz = \begin{cases} F(z_0),\ z_0 \in \mathcal{C}\ (inside) \\ 0,\ z_0 \notin \mathcal{C}\ (outside). \end{cases}$$

Here $F(z)$ is required to be analytic everywhere within (and on) the contour $\mathcal{C}$. $F(z_0)$ is called the *residue* of the pole.

3. **(Cauchy's) Residue Theorem CT-3** (Boas (1987), p. 72)

$$\oint_C F(z)dz = 2\pi j \sum_{k=1}^{K} Res_k,$$

where Res_k are the residues of all poles of $F(z)$ enclosed by the contour $\mathcal{C}$.
How to calculate the residues: The residues can be rigorously defined as

$$Res_k = \lim_{z\to z_k} [(z-z_k)f(z)].$$

This can be related to Cauchy's integral formula: Consider the function $F(z) = w(z)/(z-z_k)$, where we have factored $F(z)$ to isolate the first-order pole at $z = z_k$. If the remaining factor $w(z)$ is analytic at z_k, then the residue of the pole at $z = z_k$ *is* $w(z_k)$.

–4.1: Describe the relationships between the theorems:

1. CT-1 and CT-2
2. CT-1 and CT-3
3. CT-2 and CT-3

–4.2: Consider the function with poles at $z = \pm j$,

$$F(z) = \frac{1}{1+z^2} = \frac{1}{(z-j)(z+j)}.$$

Find the residue expansion.

–4.3: Apply Cauchy's theorems to solve the following integrals. State which theorem(s) you used and show your work.

1. $\oint_C F(z)dz$, where $\mathcal{C}$ is a circle centered at $z = 0$ with a radius of $\frac{1}{2}$
2. $\oint_C F(z)dz$, where $\mathcal{C}$ is a circle centered at $z = j$ with a radius of 1
3. $\oint_C F(z)dz$, where $\mathcal{C}$ is a circle centered at $z = 0$ with a radius of 2

4.6.4 *Integration of Analytic Functions*

Problem #5 *Integration in the complex plane*
In the following questions, you'll be asked to integrate $F(s) = u(\sigma, \omega) + iv(\sigma, \omega)$ around the contour C for complex $s = \sigma + i\omega$,

$$\oint_C F(s)ds.$$

Follow the directions carefully for each question. When asked to state where the function is and is not analytic, you are not required to use the Cauchy–Riemann equations (but you should if you can't answer the question "by inspection").

–5.1: $F(s) = \sin(s)$

–5.2: Given function $F(s) = \frac{1}{s}$

1. State where the function is and is not analytic.
2. Explicitly evaluate the integral when C is the unit circle, defined as $s = e^{i\theta}$, $0 \le \theta \le 2\pi$.
3. Evaluate the same integral using Cauchy's theorem and/or the residue theorem.

–5.3: $F(s) = \frac{1}{s^2}$

1. State where the function is and is not analytic.
2. Explicitly evaluate the integral when C is the unit circle, defined as $s = e^{i\theta}$, $0 \le \theta \le 2\pi$.
3. What does your result imply about the residue of the second-order pole at $s = 0$?

–5.4: $F(s) = e^{st}$

1. State where the function is and is not analytic.
2. Explicitly evaluate the integral when C is the square $(\sigma, \omega) = (1, 1) \rightarrow (-1, 1) \rightarrow (-1, -1) \rightarrow (1, -1) \rightarrow (1, 1)$.
3. Evaluate the same integral using Cauchy's theorem and/or the residue theorem.

–5.5: $F(s) = \frac{1}{s+2}$

1. State where the function is and is not analytic.
2. Let C be the unit circle, defined as $s = e^{i\theta}$, $0 \le \theta \le 2\pi$. Evaluate the integral using Cauchy's theorem and/or the residue theorem.
3. Let C be a circle of radius 3, defined as $s = 3e^{i\theta}$, $0 \le \theta \le 2\pi$. Evaluate the integral using Cauchy's theorem and/or the residue theorem.

–5.6: $F(s) = \frac{1}{2\pi i} \frac{e^{st}}{(s+4)}$

1. State where the function is and is not analytic.
2. Let C be a circle of radius 3, defined as $s = 3e^{i\theta}$, $0 \le \theta \le 2\pi$. Evaluate the integral using Cauchy's theorem and/or the residue theorem.

3. Let C contain the entire left half s plane. Evaluate the integral using Cauchy's theorem and/or the residue theorem. Do you recognize this integral?

–5.7: $F(s) = \pm\frac{1}{\sqrt{s}}$ (e.g., $F^2 = \frac{1}{s}$)

1. State where the function is and is not analytic.
2. This function is multivalued. How many Riemann sheets do you need in the domain (s) and the range (f) to fully represent this function? Indicate (e.g., using a sketch) how the sheet(s) in the domain map to the sheet(s) in the range.
3. Explicitly evaluate the integral

$$\int_C \frac{1}{\sqrt{z}} dz$$

when C is the unit circle, defined as $s = e^{i\theta}$, $0 \le \theta \le 2\pi$. Is this contour closed? State why or why not.

4. Explicitly evaluate the integral

$$\int_C \frac{1}{\sqrt{z}} dz$$

when C is *twice* around the unit circle, defined as $s = e^{i\theta}$, $0 \le \theta \le 4\pi$. Is this contour closed? State why or why not. Hint: Note that]

$$\sqrt{e^{i(\theta+2\pi)}} = \sqrt{e^{i2\pi}e^{i\theta}} = e^{i\pi}\sqrt{e^{i\theta}} = -1\sqrt{e^{i\theta}}.$$

5. What does your result imply about the residue of the (twice-around $\frac{1}{2}$ order) pole at $s = 0$?
6. Show that the residue is zero. Hint: Apply the definition of the residue.

4.6.5 *Laplace Transform Applications*

Problem #6 *A two-port network application for the Laplace transform*

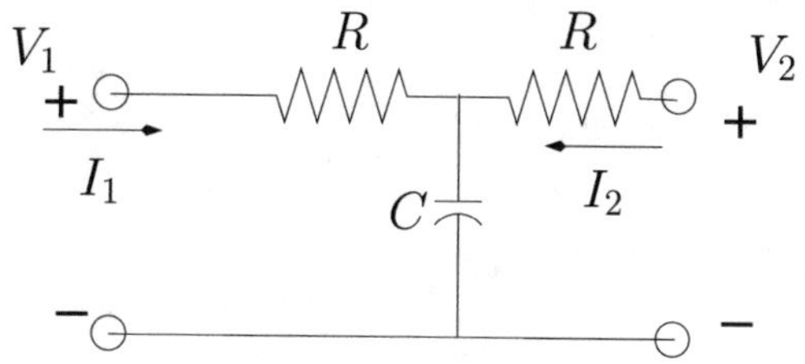

Fig. 4.6 This three-element electrical circuit is a system that acts to low-pass filter the signal voltage $V_1(\omega)$, to produce signal $V_2(\omega)$. It is convenient to define the dimensionless ratio $s/s_c = RCs$ in terms of a time constant $\tau = RC$ and cutoff frequency $s_c = 1/\tau$

–6.1: Find the 2×2 *ABCD matrix representation of Fig. 4.6. Express the results in terms of the dimensionless ratio* s/s_c*, where* $s_c = 1/\tau$ *is the cutoff frequency and* $\tau = RC$ *is the time constant.*

–6.2: Find the eigenvalues of the 2×2 *ABCD matrix. Hint: See Appendix B.3.*

–6.3: Assuming that $I_2 = 0$*, find the transfer function* $H(s) \equiv V_2/V_1$*. From the results of the ABCD matrix you determined in questions 6.1 and 6.2, show that*

$$H(s) = \frac{s_c}{s + s_c}. \qquad \text{(DE-2.6)}$$

–6.4: The transfer function $H(s)$ *has one pole. Where is the pole and residue?*

–6.5: Find $h(t)$*, the inverse Laplace transform of* $H(s)$*.*

–6.6: Assuming that $V_2 = 0$*, find* $Y_{12}(s) \equiv I_2/V_1$*.*

–6.7: Find the input impedance to the right-hand side of the system $Z_{22}(s) \equiv V_2/I_2$ *for two cases: (1)* $I_1 = 0$ *and (2)* $V_1 = 0$*.*

–6.8: Compute the determinant of the ABCD matrix. Hint: It is always ± 1.

–6.9: Given the result of the previous problem Eq. DE-2.6, compute the derivative of $H(s) = \left.\frac{V_2}{V_1}\right|_{I_2=0}$.

4.6.6 Computer Exercises with Matlab/Octave

Problem #7 *With the help of a computer*
Now we look at a few important concepts using Matlab/Octave's `syms` commands or Wolfram Alpha's symbolic math toolbox.[11]

For example, to find the Taylor series expansion about $s = 0$ of

$$F(s) = -\log(1 - s),$$

we first consider the derivative and its Taylor series (about $s = 0$)

$$F'(s) = \frac{1}{1-s} = \sum_{n=0}^{\infty} s^n.$$

Then, we integrate this series term by term:

$$F(s) = -\log(1 - s) = \int^s F'(s)ds = \sum_{n=0}^{\infty} \frac{s^n}{n}.$$

Alternatively we can use Matlab/Octave commands:

```
syms s
```

[11] https://www.wolframalpha.com/.

```
taylor(-log(1-s),'order',7)
```

–7.1: Use Octave's `taylor(-log(1-s))` *to the seventh order, as in the example above.*

1. Try the above Matlab/Octave commands. Give the first seven terms of the Taylor series (confirm that Matlab/Octave agrees with the formula derived above).
2. What is the inverse Laplace transform of this series? Consider the series term by term.

–7.2: The function $1/\sqrt{z}$ *has a branch point at* $z = 0$*; thus it is singular there.*

1. Can you apply Cauchy's integral theorem when integrating around the unit circle?
2. This Matlab/Octave code computes $\int_0^{4\pi} \frac{dz}{\sqrt{z}}$ using Matlab's/Octave's symbolic analysis package:

```
syms  z
I=int(1/sqrt(z))
J = int(1/sqrt(z),exp(-j*pi),exp(j*pi))
eval(J)
```

Run this script. What answers do you get for I and J?

3. Modify this code to integrate $f(z) = 1/z^2$ once around the unit circle. What answers do you get for I and J?

–7.3: Bessel functions can describe waves in a cylindrical geometry. The Bessel function has a Laplace transform with a branch cut

$$J_0(t)u(t) \leftrightarrow \frac{1}{\sqrt{1+s^2}}.$$

Draw a hand sketch showing the nature of the branch cut. Hint: Use `zviz`.

–7.4: Try the following Matlab/Octave commands, and then comment on your findings.

```
%Take the inverse LT of 1/sqrt(1+s^2)
syms s
I=ilaplace(1/(sqrt((1+s^2))));
disp(I)

%Find the Taylor series of the LT
T = taylor(1/sqrt(1+s^2),10); disp(T);
```

```
%Verify this
syms t
J=laplace(besselj(0,t));
disp(J);

%plot the Bessel function
t=0:0.1:10*pi;
b=besselj(0,t);
plot(t/pi,b);
grid on;
```

–7.5: When did Friedrich Bessel live?

–7.6: What did he use Bessel functions for?

–7.7: Use `zviz` *for each of the following:*

1. Describe the plot generated by `zviz S=Z`.
2. Are the functions that follow legal Brune impedances? [Do they obey $\Re Z(\sigma > 0) \geq 0$?] Hint: Consider the phase (color). Plot `zviz Z` for a reminder of the color map.

1. `zviz 1./sqrt(1+S.^2)`
2. `zviz 1./sqrt(1-S.^2)`
3. `zviz 1./(1+sqrt(S))`

4.6.7 *Inverse of Riemann $\zeta(s)$ Function*

Problem #8 *Inverse zeta function (This problem is for extra credit).*

–8.1: Find the $\mathcal{LT}^{-1}$ of one factor of the Riemann zeta function $\zeta_p(s)$, where $\zeta_p(s) \leftrightarrow z_p(t)$. Describe your results in words. Hint: See Eq. AE-1.7. Hint: Consider the geometric series representation

$$\zeta_p(s) = \frac{1}{1 - e^{-sT_p}} = \sum_{k=0}^{\infty} e^{-skT_p}, \tag{DE-2.7}$$

for which you can look up the $\mathcal{LT}^{-1}$ of each term.

Problem #9 *Inverse transform of products:*

The time-domain version of Eq. DE-2.7 may be written as the convolution of all the $z_k(t)$ *factors:*

$$z(t) \equiv z_2(t) \star z_3(t) \star z_5(t) \star z_7(t) \star \cdots \star z_p(t) \star \cdots, \tag{DE-2.8}$$

where $\star$ represents time convolution.

Explain what this means in physical terms. Start with two terms (e.g., $z_1(t) \star z_2$). Hint: The input admittance of this cascade may be interpreted as the analytic continuation *of* $\zeta(s)$ by defining a cascade of eigenfunctions with eigenvalues derived from the primes. For a discussion of this idea see Sects. 3.2.3 and C.1.1.

Physical interpretation: Such functions may be generated in the time domain, as shown in Fig. 4.7, using a feedback delay of T_p seconds, described by the two equations in the Fig. 4.7 with a unity feedback gain $\alpha = -1$. Taking the Laplace transform of the system equation, we see that the transfer function between the state variable $q(t)$ and the input $x(t)$ is given by $\zeta_p(s)$, which is an all-pole function, since

$$Q(s) = e^{-sT_n} Q(s) + V(s), \;\; or \;\; \zeta_p(s) \equiv \frac{Q(s)}{V(s)} = \frac{1}{1 - e^{-sT_p}}. \tag{DE-2.9}$$

Closing the feed-forward path gives a second transfer function $Y(s) = I(s)/V(s)$—namely,

$$q(t) = \alpha q(n - T_p) + v(t)$$
$$i(t) = q(t) - (1/\alpha)q(t - T_p)$$

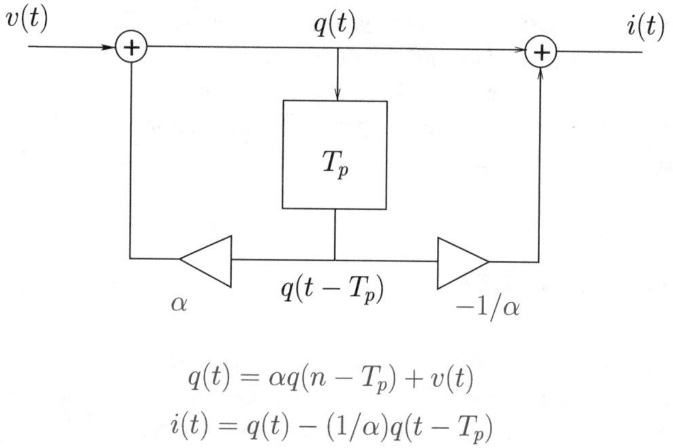

Fig. 4.7 This feedback network, described by a time-domain difference equation with delay T_p, has an all-pole transfer function $\zeta_p(s) \equiv Q(s)/I(s)$ given by Eq. DE-2.9, which physically corresponds to a stub of a transmission line, with the input at one end and the output at the other. To describe the $\zeta(s)$ function we must take $\alpha = -1$. A transfer function $Y(s) = V(s)/I(s)$ that has the same poles as $\zeta_p(s)$, but with zeros as given by Eq. DE-2.11, is the input admittance $Y(s) = I(s)/V(s)$ of the transmission line, defined as the ratio of the Laplace transform of the current $i(t) \leftrightarrow I(s)$ to the voltage $v(t) \leftrightarrow V(s)$

$$Y(s) \equiv \frac{I(s)}{V(s)} = \frac{1 - e^{-sT_p}}{1 + e^{-sT_p}}. \tag{DE-2.10}$$

If we take $i(t)$ as the current and $v(t)$ as the voltage at the input to the transmission line, then $y_p(t) \leftrightarrow \zeta_p(s)$ represents the input impedance at the input to the line. The poles and zeros of the impedance interleave along the $j\omega$ axis. By a slight modification, $\zeta_p(s)$ may alternatively be written as

$$Y_p(s) = \frac{e^{sT_p/2} + e^{-sT_p/2}}{e^{sT_p/2} - e^{-sT_p/2}} = j\tan(sT_p/2). \tag{DE-2.11}$$

Every impedance $Z(s)$ has a corresponding *reflectance* function given by a Möbius transformation, which may be read off of Eq. DE-2.11 as

$$\Gamma(s) \equiv \frac{1 + Z(s)}{1 - Z(s)} = e^{-sT_p}, \tag{DE-2.12}$$

since impedance is also related to the round-trip delay T_p on the line. The inverse Laplace transform of $\Gamma(s)$ is the round-trip delay T_p on the line

$$\gamma(t) = \delta(t - T_p) \leftrightarrow e^{-sT_p}. \tag{DE-2.13}$$

Working in the time domain provides a key insight, as it allows us to parse out the best analytic continuation of the infinity of possible continuations that are not obvious in the frequency domain (See Sect. 3.2.3). Transforming to the time domain is a form of analytic continuation of $\zeta(s)$ that depends on the assumption that $\text{Zeta}(t) \leftrightarrow \zeta(s)$ is one-sided in time (causal).

4.6.8 Quadratic Forms

A matrix that has positive eigenvalues is said to be positive-definite. The eigenvalues are real if the matrix is symmetric, so this is a necessary condition for the matrix to be positive-definite. This condition is related to conservation of energy, since the power is the voltage times the current. Given an impedance matrix

$$\mathbf{V} = \mathcal{Z}\mathbf{I},$$

the power $\mathcal{P}$ is

$$\mathcal{P} = \mathbf{I} \cdot \mathbf{V} = \mathbf{I} \cdot \mathcal{Z}\mathbf{I},$$

which must be positive-definite for the system to obey conservation of energy.

Problem #10 *In this problem, consider the* 2×2 *impedance matrix*

$$\mathcal{Z} = \begin{bmatrix} 2 & 1 \\ 1 & 4 \end{bmatrix}.$$

–10.1: Solve for the power $\mathcal{P}(i_1, i_2)$ *by multiplying out this matrix equation (which is a* quadratic form*):*

$$\mathcal{P}(i_1, i_2) = \mathbf{I}^T \begin{bmatrix} 2 & 1 \\ 1 & 4 \end{bmatrix} \mathbf{I}.$$

–10.2: Is the impedance matrix positive-definite? Show your work by finding the eigenvalues of the matrix **Z**.

–10.3: Should an impedance matrix always be positive-definite? Explain.

4.7 The Laplace Transform and Its Inverse

The Laplace transform $\mathcal{LT}$ take causal time functions into the complex analytic frequency domain s. The inverse Laplace transform $\mathcal{LT}^{-1}$ (Eq. 3.10.1) transforms a function of complex frequency $F(s)$ and returns a causal function of time $f(t)$,

$$f(t)u(t) \leftrightarrow F(s),$$

where $u(t) = 0$ for $t < 0$. Examples are provided in Sect. 3.10, Table 3.9. The forward transform is typically a relatively simple set of integrals to find $F(s)$. However the inverse transform is the key to understanding this powerful tool. Here we discuss the details of finding the inverse transform by using CT-3, and we see how the causal requirement $f(t < 0) = 0$ comes about.

As shown in Fig. 4.8, the integrand of the inverse transform is $F(s)e^{st}$ and the limits of integration are $\sigma_o \mp \omega \jmath$ with $\sigma_o > 0$. To find the inverse using CT-3 we must close the curve at $\omega \to \infty$, and specify that the integral converges. There are two ways to close the integral: to the right $\sigma > 0$ (RHP), and to the left $\sigma \leq 0$ (LHP). But there must be some logical reason for this choice. That logic is determined by the sign of t. For the integral to converge, the term $|e^{st}| = e^{\sigma t}$ must go to zero as $\omega \to \infty$. Note that both t and ω go to ∞. Thus it is the interaction between these two limits that determines how we pick the closure, RHP or LHP.

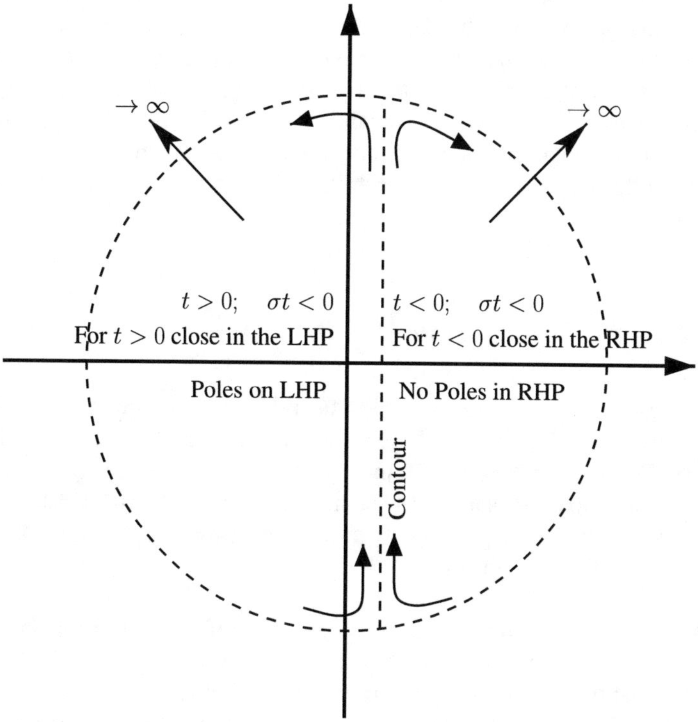

Fig. 4.8 When computing the $\mathcal{LT}^{-1}$ we must take advantage of the powerful Cauchy residue integral theorem (CT-3). To use CT-3 we must close the integral at $s \to \infty$. Furthermore this closure integral must go to zero as $\{t, \sigma\} \to \infty$. For $t < 0$, the convergence of the residue $e^{-\sigma t}$ depends on $\sigma t < 0$, since the product st must have a negative real part. For convergence for $t < 0$, $\sigma > 0$ so that $\sigma t < 0$. Thus the integral must be close in the RHP. Likewise, for $t > 0$, $\sigma < 0$, so that $\sigma t < 0$. Thus the integral must be close in the LHP. Following these guidelines based on CT-3, poles in the LHP lead to causal stable solutions, since there $\sigma < 0$ and $t \to \infty$, while poles in the RHP will lead to causal unstable solutions, each the form $e^{\sigma t}u(t)$

4.7.1 Case for Negative Time ($t < 0$) and Causality

Let us first consider negative time, including $t \to -\infty$. If we were to close $\mathcal{C}$ in the LHP ($\sigma < 0$), then the product σt is positive ($\sigma < 0, t < 0$, thus $\sigma t > 0$). In this case, as $\omega \to \infty$, the closure integral $|s| \to \infty$ will diverge. If we close in the RHP ($\sigma > 0$), then the product $\sigma t < 0$ and e^{st} will go to zero as $\omega \to \infty$. Thus we may not close in the LHP for negative time. This then justifies closing the contour, allowing for the use of Cauchy CT-3.

4.7.2 Case for Zero Time ($t = 0$)

When the time is zero, the integral does not, in general, converge, which leaves $f(t)$ undefined. This is most obvious in the case of the step function $u(t) \leftrightarrow 1/s$, where the integral may not be closed because the convergence factor $e^{st} = 1$ fails for $t = 0$.

The fact that $u(t)$ does not exist at $t = 0$ helps to explain the Gibbs phenomenon in the inverse Fourier transform. At the time where a jump occurs, the derivative of the function does not exist, and thus the time response function is not analytic. The Fourier expansion does not point-wise converge where the function is not analytic. A low-pass filter may be used to smooth the function, but at the cost of temporal resolution.

4.7.3 Case for Positive Time ($t > 0$)

Next we investigate the convergence of the integral for positive time $t > 0$. In this case we must close the integral in the LHP ($\sigma < 0$) for convergence, so that $st < 0$ ($\sigma \leq 0$ and $t > 0$). When there are poles on the $\omega\jmath = 0$ axis, $\sigma_o > 0$ assures convergence by keeping the on-axis poles inside the contour. At this point, CT-3 is relevant. If we restrict ourselves to simple poles (as required for a Brune impedance), the residue theorem may be directly applied.

Unstable poles: An important but subtle point arises: If $F(s)$ has a pole in the RHP, then the above argument still applies if we pick σ_o to be to the right of the RHP pole. This means that the inverse transform may still be applied to unstable poles (those in the RHP). This then explains the need for the σ_o in the limits. If $F(s)$ has no RHP poles in the extended RHP ($\sigma \geq 0$), we may take $\sigma_o = 0$.

The simplest example is the step function, for which $F(s) = 1/s$, thus

$$u(t) = \oint_{\text{LHP}} \frac{e^{st}}{s} \frac{ds}{2\pi\jmath} \leftrightarrow \frac{1}{s},$$

which is a direct application of CT-3. The forward transform of $u(t)$ is straightforward, as discussed in Appendix D. This is true of most of the elementary forward Laplace transforms. In these cases, causality may be built into the integral by the limits. An interesting problem is how to prove that $u(t)$ is not defined at $t = 0$.

The inverse Laplace transform of $F(s) = 1/(s+1)$ has a residue of 1 at $s = -1$, thus that is the only contribution to the integral. More demanding cases are Laplace transform pairs

$$\frac{1}{\sqrt{t}}u(t) \leftrightarrow \sqrt{\frac{\pi}{s}} \quad \text{and} \quad J_o(t)u(t) \leftrightarrow \frac{1}{\sqrt{s^2+1}},$$

as shown in Fig. 4.9 (right), and more in Table C.2. Many of these are easily proved in the forward direction but are much more difficult in the inverse direction due to the properties at $t = 0$, unless CT-3 is invoked.

Along the x-axis of Fig. 4.9, $\cos(\pi x)$ is periodic with a period of π. The dark spots are at the zeros at $\pm\pi/2, \pm 3\pi/2, \ldots$. Along the $\jmath y$-axis, the function goes to either zero (black) or ∞ (white). This behavior carries the same π periodicity as on the $x = 0$ line.

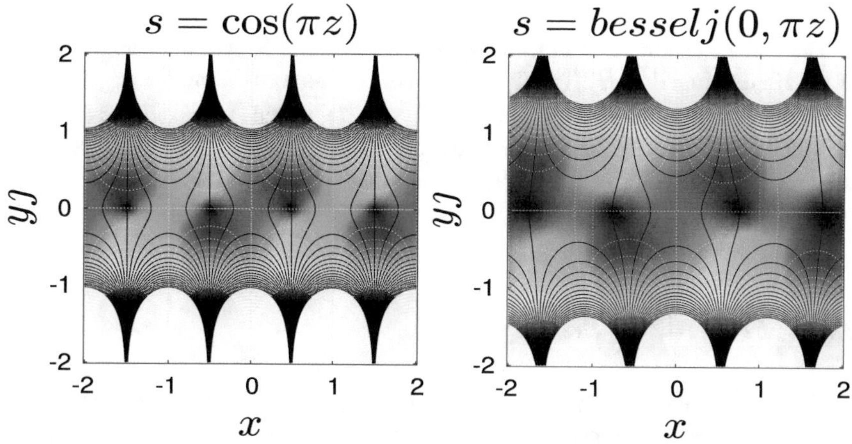

Fig. 4.9 Left: Colorized plot of $w(z) = sin(z)$. Right: Colorized plot of $w(z) = J_o(\pi z)$. Note the similarity of the two functions. The first Bessel zero is at 2.405 and thus appears at $0.7655 = 2.405/\pi$, about 1.53 times larger than the root of $\cos(\pi z)$ at $1/2$. Other than this minor distortion of the first few roots, the two functions are basically identical. It follows that their $\mathcal{LT}$s must have similar characteristics, as documented in Table C.2. These colorized plots show that these two functions become the same for $x = \Re z > 0$. The black lines indicate where the function has a constant real part

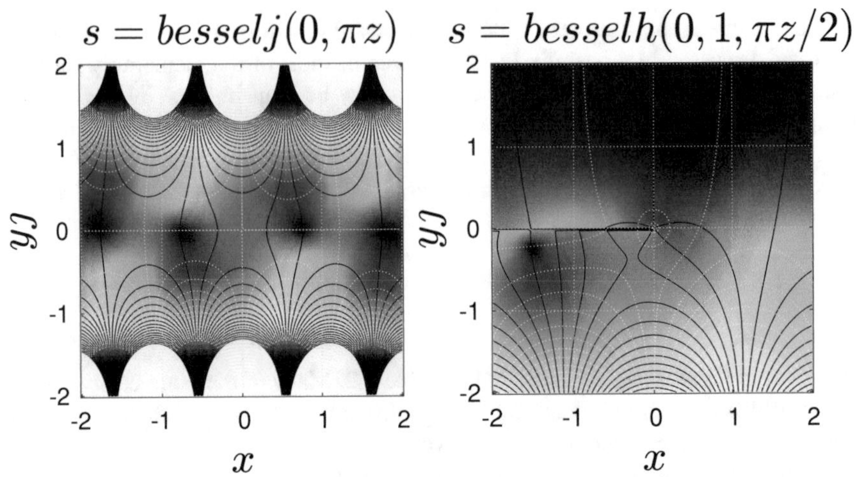

Fig. 4.10 Left: The Bessel function $J_0(\pi z)$, which is similar to $\cos(\pi z)$, except the zeros are distorted away from $s = 0$ by a small amount, due to the cylindrical geometry. Right: The related Hankel function $H_0^{(1)}(\pi z/2)$. The Hankel function $H_0^{(1)}(\pi z/2)$ has a branch cut and a complex zero at $z_{0,1} 2/\pi = -1.5 - 0.1\jmath$, as may be seen in the plot

The last $\mathcal{LT}$ example of Fig. 4.10 gives an important insight into the properties of the Hankel function $H_0^{(1)}(\pi z/2)$, which has a branch cut along the negative real axis. On the right is the Hankel function $H_0^{(1)}(\pi z/2)$, which is a mixed and distorted version of $\cos(\pi z)$ with the zeros pushed downward, and $e^{\pi z}$. Note how the white and black contour lines of the colorized maps are always perpendicular where they cross.

4.7.4 *Properties of the* $\mathcal{LT}$

As shown in Table 3.8 of Laplace transforms, there are integral (i.e., integration, not integer) relationships, or properties, that are helpful to identify. The first of these is a definition, not a property:

$$f(t) \leftrightarrow F(s).$$

Causality: When we take the $\mathcal{LT}$, the time-domain response is in lowercase (e.g., $f(t)$) and the frequency-domain transform is in uppercase (e.g., $F(s)$). It is required, but not always explicitly specified, that $f(t < 0) = 0$; that is, the time function must be causal, as stated by Postulate P1 (Sect. 3.10.2).

Linearity: The most basic property is the linearity (superposition) property of the $\mathcal{LT}$, stated by Postulate P2 (Sect. 3.10.2).
Convolution property: The product of two $\mathcal{LT}$s in frequency results in convolution in time:

$$F(s)G(s) \leftrightarrow f(t) \star g(t) = \int_0^t f(\tau)g(t-\tau)d\tau,$$

where we use $\star$ to indicate the convolution of two time functions.

A key application of convolution is filtering, which takes many forms. The most basic filter is the moving average, the moving sum of data samples, normalized by the number of samples. Such a filter has very poor performance. It also introduces a delay of half the length of the average, which may or may not constitute a problem, depending on the application. Other important examples are a low-pass filter that removes high-frequency noise and a notch filter that removes line noise (i.e., 60 [Hz] in the United States, and its second and third harmonics, 120 and 180 [Hz]). Such noise is typically a result of poor grounding and ground loops. It is better to solve the problem at its root than to remove it with a notch filter. Still, filters are very important in engineering.

By taking the $\mathcal{LT}$of the convolution we can derive this relationship:

$$\begin{aligned}\int_0^\infty [f(t) \star g(t)]e^{-st}dt &= \int_{t=0}^\infty \left[\int_0^t f(\tau)g(t-\tau)d\tau\right] e^{-st}dt \\ &= \int_0^t f(\tau)\left(\int_{t=0}^\infty g(t-\tau)e^{-st}dt\right)d\tau \\ &= \int_0^t f(\tau)\left(e^{-s\tau}\int_{t'=0}^\infty g(t')e^{-st'}dt'\right)d\tau \\ &= G(s)\int_0^t f(\tau)e^{-s\tau}d\tau \\ &= G(s)F(s).\end{aligned}$$

We first encountered this relationship in Sect. 3.4 in the context of multiplying polynomials, which is the same as convolving their coefficients. The parallel should be obvious. In the case of polynomials, the convolution is discrete in the coefficients, and here it is continuous in time. But the relationships are the same.

Time-shift property: When a function is time-shifted by time T_o, the $\mathcal{LT}$is modified by e^{-sT_o}, leading to the property

$$f(t-T_o) \leftrightarrow e^{-sT_o}F(s).$$

This is easily shown by applying the definition of the $\mathcal{LT}$to a delayed time function.

Time derivative: The key to the eigenfunction analysis provided by the $\mathcal{LT}$is the transformation of a time derivative on a time function—that is,

$$\frac{d}{dt}f(t) \leftrightarrow sF(s).$$

Here s is the eigenvalue corresponding to the time derivative of e^{st}. Given the definition of the derivative of e^{st} with respect to time, this definition seems trivial. Yet that definition was not obvious to Euler. It needed to be extended to the space of the complex analytic function e^{st}, which did not happen before Cauchy's key results.

Given a differential equation of order K, the $\mathcal{LT}$ results in a polynomial in s of degree K. It follows that this $\mathcal{LT}$ property is the cornerstone of why the $\mathcal{LT}$ is so important to scalar differential equations, as it was to the early analysis of Pell's equation and the Fibonacci sequence, presented in Chap. 2. While the relation $e^{\jmath\theta} = \cos\theta + \jmath\sin\theta$ was first uncovered by Euler. By the time of his death the formula's significance would have been clear to him. Who first coined the terms *eigenvalue* and *eigenfunction*? The word *eigen* is a German word meaning *of one*.

Initial and final value theorems: There are much more subtle relationships between $f(t)$ and $F(s)$ that characterize $f(0^+)$ and $f(t \to \infty)$, which are known as initial value theorems. If the system under investigation has potential energy at $t = 0$, then the voltage (velocity) need not be zero for negative time. An example is a charged capacitor or a moving mass. These are important concepts, but best explored in a more in-depth treatment. They are not violations of causality.

4.7.5 Solving Differential Equations

Many differential equations may be solved by assuming a power series (i.e., Taylor series) solution of the form

$$y(x) = x^r \sum_{n=0}^{\infty} c_n x^n, \tag{4.7.1}$$

with $r \in \mathbb{Z}$ and coefficients $c_n \in \mathbb{C}$. This method of Frobenius is quite general (Greenberg 1988, p. 193).

Example: When a solution of this form is substituted into the differential equation, a recursion relationship in the coefficients results. For example, if the equation is

$$y''(x) = \lambda^2 y(x),$$

the recursion is $c_n = c_{n-1}/n$. The resulting equation is

$$y(x) = e^{\lambda x} = x^0 \sum_{n=0}^{\infty} \frac{1}{n!} x^n,$$

namely, $c_n = 1/n!$, thus $nc_n = 1/(n-1)! = c_{n-1}$.

Exercise #3 Find the recursion relationship for $y(x) = J_\nu(x)$ of order ν that satisfies Bessel's equation

$$x^2 y''(x) + xy'(x) + (x^2 - \nu^2)y(x) = 0.$$

Solution: If we assume a complex analytic solution of the form of Eq. 4.7.1, we find the Bessel recursion relationship for coefficients c_k (Greenberg 1988, p. 231):

$$c_k = -\frac{1}{k(k+2\nu)} c_{k-2}.$$

■

4.8 Problems DE-3

4.8.1 *Topics of This Homework: Brune Impedance*

lattice transmission line analysis

4.8.2 *Brune Impedance*

Problem #1 *Residue form*

A Brune impedance is defined as the ratio of the force $F(s)$ to the flow $V(s)$ and may be expressed in residue form as

$$Z(s) = c_0 + \sum_{k=1}^{K} \frac{c_k}{s - s_k} = \frac{N(s)}{D(s)} \tag{DE-3.1}$$

with

$$D(s) = \prod_{k=1}^{K}(s - s_k) \quad and \quad c_k = \lim_{s \to s_k}(s - s_k)D(s) = \prod_{n'=1}^{K-1}(s - s_n).$$

The prime on the index n' means that $n = k$ is not included in the product.

–1.1: Find the Laplace transform ($\mathcal{LT}$) of a (1) spring, (2) dashpot, and (3) mass. Express these in terms of the force $F(s)$ and the velocity $V(s)$, along with the electrical equivalent impedance: (1) Hooke's law $f(t) = Kx(t)$, (2) dashpot resistance $f(t) = Rv(t)$, and (3) Newton's law for mass $f(t) = Mdv(t)/dt$.

–1.2: Take the Laplace transform ($\mathcal{LT}$) of Eq. DE-3.2 and find the total impedance $Z(s)$ of the mechanical circuit.

$$M\frac{d^2}{dt^2}x(t) + R\frac{d}{dt}x(t) + Kx(t) = f(t) \leftrightarrow (Ms^2 + Rs + K)X(s) = F(s). \quad \text{(DE-3.2)}$$

–1.3: What are $N(s)$ and $D(s)$ (see Eq. DE-3.1)?

–1.4: Assume that $M = R = K = 1$ and find the residue form of the admittance $Y(s) = 1/Z(s)$ (see Eq. DE-3.1) in terms of the roots $s_\pm$. Hint: Check your answer with Octave's/Matlab's `residue` *command.*

–1.5: By applying Eq. 4.5.3, find the inverse Laplace transform ($\mathcal{LT}^{-1}$). Use the residue form of the expression that you derived in question #1.

4.8.3 Transmission Line Analysis

Problem #2 Train-mission-line *We wish to model the dynamics of a freight train that has N such cars and study the* velocity transfer function *under various load conditions.*

As shown in Fig. 4.11, the train model consists of masses connected by springs.

Use the ABCD method (see the discussion in Appendix B.3) to find the matrix representation of the system of Fig. 4.11. Define the force on the nth train car $f_n(t) \leftrightarrow F_n(\omega)$ and the velocity $v_n(t) \leftrightarrow V_n(\omega)$.

Break the model into cells consisting of three elements: a series inductor representing half the mass ($M/2$), a shunt capacitor representing the spring ($C = 1/K$), and another series inductor representing half the mass ($L = M/2$), transforming the model into a cascade of symmetric ($\mathcal{A} = \mathcal{D}$) identical cell matrices $\mathcal{T}(s)$.

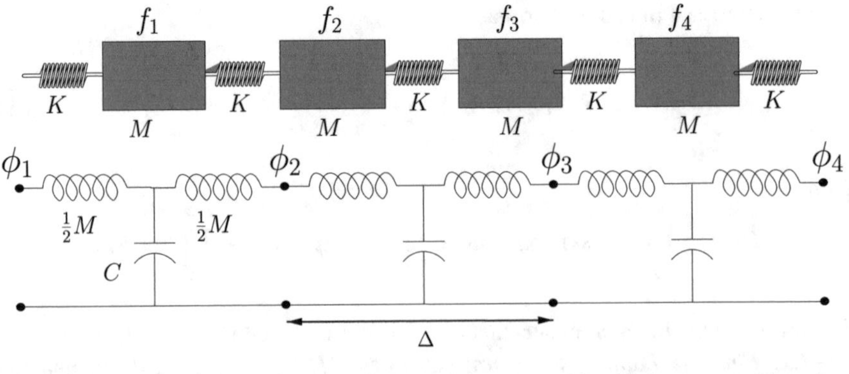

Fig. 4.11 Depiction of a train consisting of cars treated as masses M and linkages treated as springs of stiffness K or compliance $C = 1/K$. Below it is the electrical equivalent circuit for comparison. The masses are modeled as inductors and the springs as capacitors to ground. The velocity is analogous to a current and the force $f_n(t)$ to the voltage $\phi_n(t)$. The length of each cell is Δ [m]. The train may be accurately modeled as a transmission line (TL), since the equivalent electrical circuit is a lumped model of a TL. This method, called a *Cauer synthesis*, is based on the ABCD transmission line method of Sect. 3.8

–2.1: Find the elements of the ABCD matrix $\mathcal{T}$ for the single cell that relate the input node 1 to output node 2]

$$\begin{bmatrix} F \\ V \end{bmatrix}_1 = \mathcal{T} \begin{bmatrix} F(\omega) \\ -V(\omega) \end{bmatrix}_2. \tag{DE-3.3}$$

–2.2: Express each element of $\mathcal{T}(s)$ in terms of the complex Nyquist ratio $s/s_c < 1$ ($s = 2\pi jf$, $s_c = 2\pi jf_c$). The Nyquist wavelength sampling condition is $\lambda_c > 2\Delta$. It says the critical wavelength $\lambda_c > 2\Delta$. Namely it is defined in terms the minimum number of cells 2Δ, per minimum wavelength λ_c.

The Nyquist wavelength sampling theorem says that there are at least two cars per wavelength.

Proof: From the figure, the distance between cars $\Delta = c_o T_o$ [m], where

$$c_o = \frac{1}{\sqrt{MC}} \quad [m/s].$$

The cutoff frequency obeys $f_c \lambda_c = c_o$. The Nyquist critical wavelength is $\lambda_c = c_o/f_c > 2\Delta$. Therefore the Nyquist sampling condition is

$$f < f_c \equiv \frac{c_o}{\lambda_c} = \frac{c_o}{2\Delta} = \frac{1}{2\Delta\sqrt{MC}} \quad [Hz]. \tag{DE-3.4}$$

Finally, $s_c = \jmath 2\pi f_c$.

–2.3: Use the property of the Nyquist sampling frequency $\omega < \omega_c$ *(Eq. DE-3.4) to remove higher order powers of frequency*

$$1 + \left(\frac{s}{s_c}\right)^{2 \to 0} \approx 1 \tag{DE-3.5}$$

to determine a band-limited approximation of $\mathcal{T}(s)$.

Problem #3 *Now consider the cascade of N such $\mathcal{T}(s)$ matrices and perform an eigenanalysis.*

–3.1: Find the eigenvalues and eigenvectors of $\mathcal{T}(s)$ as functions of s/s_c.

Problem #4 *Find the velocity transferfunction $H_{12}(s) = V_2/V_1|_{F_2=0}$.*

–4.1: Assuming that $N = 2$ and $F_2 = 0$ (two half-mass problem), find the transfer function $H(s) \equiv V_2/V_1$. From the results of the $\mathcal{T}$ matrix, find]

$$H_{21}(s) = \left.\frac{V_2}{V_1}\right|_{F_2=0}$$

Express H_{12} in terms of a residue expansion.

–4.2: Find $h_{21}(t) \leftrightarrow H_{21}(s)$.

–4.3: What is the input impedance $Z_2 = F_2/V_2$, assuming $F_3 = -r_0 V_3$?
–4.4: Simplify the expression for Z_2 as follows:

1. Assuming the *characteristic impedance* $r_0 = \sqrt{M/C}$,
2. terminate the system in r_0: $F_2 = -r_0 V_2$ (i.e., $-V_2$ cancels).
3. Assume higher order frequency terms are less than 1 ($|s/s_c| < 1$).
4. Let the number of cells $N \to \infty$. Thus $|s/s_c|^N = 0$.

When a transmission line is terminated in its characteristic impedance r_0, the input impedance $Z_1(s) = r_0$. Thus, when we simplify the expression for $\mathcal{T}(s)$, it should be equal to r_0. Show that this is true for this setup.

–4.5: State the ABCD matrix relationship between the first and Nth nodes in terms of the cell matrix. Write out the transfer function for one cell, H_{21}.

–4.6: What is the velocity transfer function $H_{N1} = \frac{V_N}{V_1}$?

Chapter 5
Stream 3B: Vector Calculus

5.1 Properties of Fields and Potentials

Before we can define the vector operations $\nabla()$, $\nabla\cdot()$, $\nabla\times()$, and $\nabla^2()$, we must define the objects they operate on: scalar and vector fields. The word *field* has two very different meanings: a mathematical one, which defines an algebraic structure, and a physical one, discussed next.

Ultimately we wish to integrate in $\in \mathbb{R}^3, \mathbb{R}^n$, and $\mathbb{C}^n$. Integration is quantified by several fundamental theorems of calculus, each about integration (see Sects. 4.2–4.2.3).

5.1.1 *Scalar and Vector Fields*

5.1.1.1 Scalar Fields

We use the term *scalar field* interchangeably with *analytic* in a connected region of the spatial vector $\boldsymbol{x} = [x, y, z]^T \in \mathbb{R}^3$. In mathematics, functions that are piecewise differentiable are called *smooth*, which is different from *analytic*. A smooth function has at least one or more derivatives. Every analytic function is single-valued and is an infinitely differentiable power series (Sect. 3.2.4).

5.1.1.2 Vector Fields

A vector field is composed of three scalar fields. For example, the electric field used in Maxwell's equations, $\boldsymbol{E}(\boldsymbol{x}, t) = [E_x, E_y, E_z]^T$ [V/m], has three components, each a scalar field. When the magnetic flux vector $\boldsymbol{B}(\boldsymbol{x})$ is static (Postulate P5, Sect. 3.3), the potential $\phi(\boldsymbol{x})$ [V] uniquely defines $\boldsymbol{E}(\boldsymbol{x}, t)$ via the gradient,

J. Allen, *An Invitation to Mathematical Physics and Its History*,
https://doi.org/10.1007/978-3-030-53759-3_5

$$E(\boldsymbol{x}, t) = -\nabla\phi(\boldsymbol{x}, t) \quad \text{[V/m]}. \tag{5.1.1}$$

The electric force on a charge q is $\boldsymbol{F} = q\boldsymbol{E}$; thus $\boldsymbol{E}$ is proportional to the force, and when the medium is conductive, the current density (a flow) is $\boldsymbol{J_m} = \sigma_o \boldsymbol{E}$ [A/m^2]. The ratio of the potential to the flow is an impedance, so σ_o is a conductance.

Example: Suppose we are given the vector field in $\mathbb{R}^3$

$$\boldsymbol{A}(\boldsymbol{x}) = [\phi(\boldsymbol{x}), \psi(\boldsymbol{x}), \theta(\boldsymbol{x})]^T \quad \text{[Wb/m]},$$

where each of the three functions is a scalar field. Then $\boldsymbol{A}(\boldsymbol{x}) = [x, xy, xyz]^T$ is a legal vector field that has components analytic in $\boldsymbol{x}$.

Example: From Maxwell's equations, the magnetic flux vector is given by

$$\boldsymbol{B}(\boldsymbol{x}, t) = \nabla \times \boldsymbol{A}(\boldsymbol{x}, t) \quad \text{[Wb/m}^2\text{]}. \tag{5.1.2}$$

We shall see that this is always true because the magnetic charge $\nabla \cdot \boldsymbol{B}(\boldsymbol{x}, t)$ must be 0, which is always true in-vacuo. Feynman (1970b, pp. 14–1 to 14–3) provides an extended and helpful tutorial on the vector potential, with many examples.

To verify that a field is a potential, we may check the units. However, a proper mathematical definition is that the potential must be an analytic function of $\boldsymbol{x}$ and t, so that we can operate on it with $\nabla()$ and $\nabla \times ()$. Note that the divergence of a scalar field is not a legal vector operation.

5.1.1.3 Scalar Potentials

The above discussion describes the utility of potentials for defining vector fields (e.g., Eqs. 5.1.1 and 5.1.2). The key distinction between a potential and a scalar field is that potentials have units and thus have a physical meaning. Scalar potentials (i.e., voltage $\phi(\boldsymbol{x}, t)$ [V], temperature $T(\boldsymbol{x}, t)$ [°C], and pressure $\varrho(\boldsymbol{x}, t)$ [pascals]) are physical scalar fields. All potentials are composed of scalar fields, but not all scalar fields are potentials.

For example, the $\hat{\mathbf{y}}$ component of $\boldsymbol{E}$, $E_y(\boldsymbol{x}, t) = \hat{\mathbf{y}} \cdot \boldsymbol{E}(\boldsymbol{x}, t)$ [V/m], is not a potential. While $\nabla \boldsymbol{E}_y$ is mathematically defined as the gradient of one component of a vector field, it has no physical meaning (as best I know).

5.1.1.4 Vector Potentials

Vector potentials, like scalar potentials, are vector fields with physically meaningful units. They are more complicated than scalar potentials because they are composed of three scalar fields. Vector fields are composed of laminar and rotational flow, which are mathematically described by the fundamental theorem of vector calculus (also called Helmholtz's decomposition theorem). One superficial but helpful comparison is the momentum of a mass, which may be decomposed into its forward (linear) and rotational momentum.

Since we find it useful to analyze problems using potentials (e.g., voltage) and then take the gradient (i.e., voltage difference) to find the flow (i.e., current density $\boldsymbol{J} = \sigma \boldsymbol{E}(\boldsymbol{x}, t)$), the same logic and utility apply when we use the vector potential to describe the magnetic flux (flow) $\boldsymbol{B}(\boldsymbol{x}, t)$ (Feynman 1970c). When operating on a scalar potential, we use a gradient, whereas for the vector potential, we operate with the curl (Eq. 5.1.2).

In Eq. 5.1.1 we assumed that the magnetic flux vector $\boldsymbol{B}(\boldsymbol{x})$ was static, and thus $\boldsymbol{E}(\boldsymbol{x}, t)$ is the gradient of the time-dependent voltage $\phi(\boldsymbol{x}, t)$. However, when the magnetic field is dynamic (*not* static), Eq. 5.1.1 is not valid, due to magnetic induction: A voltage induced into a loop of wire is proportional to the time-varying flux cutting across that loop of wire. This is known as the *Ampere-Maxwell law*. In the static case the induced voltage is zero.

Thus the electric field strength includes both scalar potential $\phi(\boldsymbol{x}, t)$ and magnetic flux vector potential $\boldsymbol{A}(\boldsymbol{x}, t)$ components, while the magnetic field strength depends only on the magnetic potential.

5.1.2 *Gradient* ∇, *Divergence* $\nabla\cdot$, *Curl* $\nabla\times$, *and Laplacian* ∇^2

Three key vector differential operators are used in linear partial differential equations, such as the wave and diffusion equations. All of these begin with the ∇ operator:

Table 5.1 The three vector operators manipulate scalar and vector fields. The gradient converts scalar fields into vector fields. The divergence maps vector fields to scalar fields. The curl maps vector fields to vector fields. Four second-order operators (for example, DoG and **gOd**) are defined in Sect. 5.6.6. Bold mnemonics are reserved for vector-in, vector-out operators, with the curl being an exception.

Name	Input	Output	Operator	Mnemonic
Gradient	Scalar	**Vector**	$\nabla()$	grad
Divergence	**Vector**	Scalar	$\nabla\cdot()$	div
Laplacian	Scalar	Scalar	$\nabla\cdot\nabla = \nabla^2()$	DoG
Wedgie	**Vector**	Scalar	$\nabla\wedge()$	wedge
Curl	**Vector**	**Vector**	$\nabla\times()$	curl
God	**Vector**	**Vector**	$\nabla^2() = \nabla(\nabla\cdot())$	**gOd**
Bull-DoG	**Vector**	**Vector**	$\boldsymbol{\nabla}^2() = \nabla\cdot\nabla()$	**DoG**
Curl of Curl	**Vector**	**Vector**	$\nabla\times\nabla\times = \nabla^2() - \boldsymbol{\nabla}^2()$	**CoC**
Div of Curl	**Vector**	0	$\nabla\cdot\nabla\times()$	**DoC**
Curl of Grad	Scalar	0	$\nabla\times\nabla()$	**CoG**

$$\nabla = \hat{\mathbf{x}}\frac{\partial}{\partial x} + \hat{\mathbf{y}}\frac{\partial}{\partial y} + \hat{\mathbf{z}}\frac{\partial}{\partial z}.$$

As outlined in Table 5.1, the official name of this operator is *nabla*. It has three basic uses: (1) the gradient of a scalar field, (2) the divergence of a vector field, and (3) the curl of a vector field. The shorthand notation $\nabla\phi(\boldsymbol{x}, t) = (\hat{\mathbf{x}}\partial_x + \hat{\mathbf{y}}\partial_x + \hat{\mathbf{z}}\partial_x)\phi(\boldsymbol{x}, t)$ is convenient.

5.1.2.1 Gradient $\nabla()$

As shown in Fig. 5.1, the gradient transforms a complex scalar field $\Phi(\boldsymbol{x}, s) \in \mathbb{C}$ into a vector field ($\mathbb{C}^3$)

$$\begin{aligned}\nabla\Phi(\boldsymbol{x}, s) &= \left(\hat{\mathbf{x}}\frac{\partial}{\partial x} + \hat{\mathbf{y}}\frac{\partial}{\partial y} + \hat{\mathbf{z}}\frac{\partial}{\partial z}\right)\Phi(\boldsymbol{x}, s) \\ &= \hat{\mathbf{x}}\frac{\partial\Phi}{\partial x} + \hat{\mathbf{y}}\frac{\partial\Phi}{\partial y} + \hat{\mathbf{z}}\frac{\partial\Phi}{\partial z}.\end{aligned}$$

The gradient may also be factored into a unit vector $\hat{\mathbf{n}}$, as defined in Fig. 5.1, that gives the direction of the gradient, and the gradient's length $||\nabla()||$, defined in terms of the norm of the gradient. Thus the gradient of $\Phi(\boldsymbol{x})$ may be written in "polar coordinates" as $\nabla\Phi(\boldsymbol{x}) = ||\nabla\Phi||\,\hat{\mathbf{n}}$, which leads to the unit vector

$$\hat{\mathbf{n}} = \frac{\nabla(\Phi(\boldsymbol{x}))}{||\nabla\Phi||}.$$

Consider the paraboloid $z = 1 - (x^2 + y^2)$ as the potential, with isopotential circles of constant z that have a radius of zero at $z = 1$ and unit radius at $z = 0$. The negative gradient

$$\boldsymbol{E}(\boldsymbol{x}) = -\nabla z(x, y) = 2(x\hat{\mathbf{x}} + y\hat{\mathbf{y}} + 0\hat{\mathbf{z}})$$

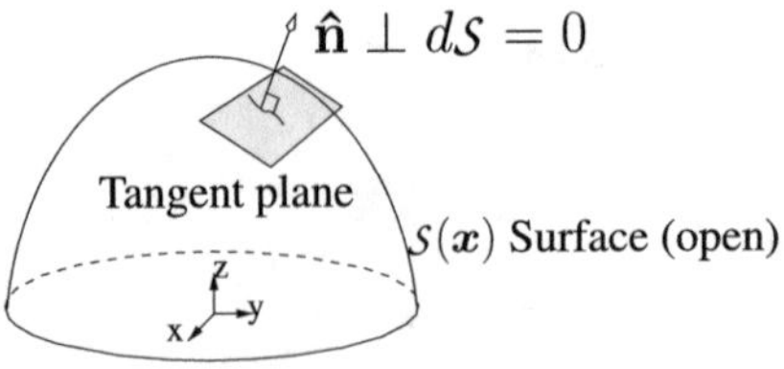

Fig. 5.1 Definition of the unit vector $\hat{\mathbf{n}}$ defined by the gradient $\nabla\mathcal{S} \perp$ to the tangent plane. A bifurcated volume defines surface $\mathcal{S}(\boldsymbol{x})$. At one point a tangent plane (shaded) touches the surface. At that point the gradient $\nabla\mathcal{S}(\boldsymbol{x})$ is normalized to unit length, defining $\hat{\mathbf{n}}$, which is perpendicular ($\perp$) to the shaded tangent plane

is $\perp$ to the circles of constant radius (constant z) and thus points in the direction of the radius.

A skier in free fall on this surface would be the first one down the hill. Normally skiers try to stay close to the isoclines (not in the direction of the gradient) so they can stay in control. If you ski on an isocline, you must walk, since there is no pull due to gravity. The gravitational potential at the surface of the Earth is

$$\phi = G\frac{mM}{r}.$$

5.1.2.2 Divergence $\nabla \cdot ()$

The divergence of a vector field results in a scalar field. For example, the divergence of the electric field flux vector $\boldsymbol{D}(x)$ [C/m^2] equals the scalar field charge density $\rho(\boldsymbol{x})$ [C/m^3]:

$$\nabla \cdot \boldsymbol{D}(\boldsymbol{x}) \equiv \left(\hat{\mathbf{x}}\frac{\partial}{\partial x} + \hat{\mathbf{y}}\frac{\partial}{\partial y} + \hat{\mathbf{z}}\frac{\partial}{\partial z}\right) \cdot \boldsymbol{D}(\boldsymbol{x}) = \frac{\partial D_x}{\partial x} + \frac{\partial D_y}{\partial y} + \frac{\partial D_z}{\partial z} = \rho(\boldsymbol{x}). \tag{5.1.3}$$

Thus the divergence is analogous to the scalar (dot) product (e.g., $\boldsymbol{A} \cdot \boldsymbol{B}$) between two vectors.

Recall that the voltage is the line integral of the electric field,

$$V(a) - V(b) = \int_a^b \boldsymbol{E}(\boldsymbol{x}) \cdot d\boldsymbol{x} = -\int_a^b \nabla V(\boldsymbol{x}) \cdot d\boldsymbol{x} = -\int_a^b \frac{dV}{dx} dx, \tag{5.1.4}$$

which is simply the fundamental theorem of calculus (see Sect. 4.2). In a charge-free region, this integral is independent of the path from a to b, which is a property of a conservative system.

When we work with guided waves (narrow tubes of flux) having rigid walls that block the flow, such that the diameter is small compared with the wavelength (Postulate P10, Sect. 3.10.2), the divergence simplifies to

$$\nabla \cdot \boldsymbol{D}(\boldsymbol{x}) = \nabla_r D_r = \frac{1}{A(r)}\frac{\partial}{\partial r} A(r) D_r(r), \tag{5.1.5}$$

where r is the distance down the horn (range variable), $A(r)$ is the area of the isoresponse surface as a function of the range r, and $D_r(r)$ is the radial component of vector $\boldsymbol{D}$ as a function of the range r. In spherical, cylindrical, and rectangular coordinates, Eq. 5.1.5 provides the correct expression (Table 5.2).

5.1.2.3 Properties of the Divergence

The divergence is a direct measure of the flux density of the vector field. A vector field is said to be *incompressible* if the divergence of that field is zero. It is therefore *compressible* when the divergence is nonzero [e.g., $\nabla \cdot \boldsymbol{D}(x, s) = \rho(\boldsymbol{x}, s)$] (see Table 5.3).

For example, compared to air, water is considered to be incompressible. The stiffness of a fluid (i.e., the bulk modulus) is a measure of its compressibility. At very low frequencies, air may be treated as incompressible (like water), since as $s \to 0$,

$$-\nabla \cdot \boldsymbol{u}(\boldsymbol{x}, s) = \frac{s}{\eta_o P_o} \mathcal{P}(\boldsymbol{x}, s) \to 0.$$

The definition of *compressible* depends on the wavelength in the medium, so the term must be used with some awareness of the frequencies being in the analysis. As a rule of thumb, if the wavelength $\lambda = c_o/f$ is much larger than the size of the system, the medium may be modeled as an incompressible fluid.

5.1.2.4 Curl ∇ × ()

The curl ($\nabla \times ()$) takes a vector in $\mathbb{C}^3$ into a second vector in $\mathbb{C}^3$. For example, in the case of fluids, the vorticity is defined as $\omega = \nabla \times \nu$ and *rotation* as $\boldsymbol{\Omega} = \boldsymbol{\omega/2}$. The curl is a measure of the rotation of a vector field in a plane about the axis perpendicular to that plane. In the case of a liquid, it corresponds to the angular momentum, such as in a whirlpool in water, or a tornado in air. A massive top falls over when not spinning. But once spinning, it can stably stand on its pointed tip. These systems are stable due to conservation of angular momentum.

The curl and the divergence are both key operations when we work with Maxwell's four equations. For example, the curl transforms the vector field $\boldsymbol{H}(\boldsymbol{x}, s) \in \mathbb{C}^3$ [A/m] into a complex vector current density $\boldsymbol{C}(\boldsymbol{x}, s) \in \mathbb{C}^2$ [A/m^2]:

$$\nabla \times \boldsymbol{H}(\boldsymbol{x}, s) \equiv \begin{vmatrix} \hat{\mathbf{x}} & \hat{\mathbf{y}} & \hat{\mathbf{z}} \\ \partial_x & \partial_y & \partial_z \\ H_x & H_y & H_z \end{vmatrix} = \boldsymbol{C}(\boldsymbol{x}, s) \qquad [\text{A/m}^2]. \tag{5.1.6}$$

The notation $|\cdot|$ indicates the determinant (Appendix A.3.1, Sect. A.3.3), ∂_x is shorthand for $\partial/\partial x$, and $\boldsymbol{H} = [H_x, H_y, H_z]^T$.

Exercise #1 If we let $\boldsymbol{H} = -y\hat{\mathbf{x}} + x\hat{\mathbf{y}} + 0\hat{\mathbf{z}}$, $\nabla \times \boldsymbol{H} = 2\hat{\mathbf{z}}$, thus $\boldsymbol{H}$ has a constant rotation; when $\boldsymbol{H} = 0\hat{\mathbf{x}} + 0\hat{\mathbf{y}} + z^2\hat{\mathbf{z}}$, $\nabla \times \boldsymbol{H} = \boldsymbol{0}$ has a curl of zero and thus is irrotational.

There are simple rules that precisely govern when a vector field is rotational versus irrotational, and compressible versus incompressible. These classifications are dictated by Helmholtz's theorem, the fundamental theorem of vector calculus (Eq. 5.6.7).

5.1.2.5 Wedgie $\nabla \wedge ()$

A special case of the curl is the two-dimensional differential wedge products[1]

$$\nabla_x \wedge \boldsymbol{H}(\boldsymbol{x}, t) = \begin{vmatrix} \partial_y & \partial_z \\ H_y & H_z \end{vmatrix} = C_x(\boldsymbol{x}, s) \qquad [\mathrm{A/m^2}].$$

The curl is made up of three such differential wedge-products.[2]

5.1.2.6 Laplacian $\nabla^2()$

The Laplacian operator $\nabla^2 \equiv \nabla\cdot\nabla$ (Table 5.1) is defined as the divergence of the gradient

$$\nabla^2 \equiv \frac{\partial^2}{\partial x^2} + \frac{\partial^2}{\partial y^2} + \frac{\partial^2}{\partial z^2}. \tag{5.1.7}$$

Since the Laplacian does so much common work, we nickname it DoG (Div of Grad).

Starting from a scalar field, the gradient produces a vector, which is then operated on by the divergence to take the output of the gradient back to a scalar field. Thus the Laplacian transforms a scalar field back to a scalar field. We have seen the Laplacian before when we defined complex analytic functions (Eq. 4.2.8).

A classic example of the Laplacian is a voltage scalar field $\Phi(\boldsymbol{x})$ [V], which results in the electric field vector

$$\boldsymbol{E}(\boldsymbol{x}) = [E_x(\boldsymbol{x}), E_y(\boldsymbol{x}), E_z(\boldsymbol{x})]^T = -\nabla\Phi(\boldsymbol{x}) \quad [\mathrm{V/m}].$$

When this is scaled by the permittivity, we obtain the electric flux $\boldsymbol{D} = \epsilon_o \boldsymbol{E}$ [C/m^2], the charge density per unit area. Here ϵ_o [F/m] is the vacuum permittivity $\epsilon_o = 1/c_o r_o \approx 8.85 \times 10^{-12}$ [F/m].

Taking the divergence of $\boldsymbol{D}$ results in the charge density $\rho(\boldsymbol{x})$ [C/m^3] at $\boldsymbol{x}$:

$$\nabla\cdot\boldsymbol{D} = \nabla^2\Phi(\boldsymbol{x}) = \rho(\boldsymbol{x}).$$

Thus the Laplacian of the voltage, scaled by ϵ_o, results in the local charge density.

Another classic example of the Laplacian is an acoustic pressure field $\varrho(\boldsymbol{x}, t)$ [Pa], which defines a vector force density $\mathbf{f}(\boldsymbol{x}, t) = -\nabla\varrho(\boldsymbol{x}, t)$ [N/m^2] (Eq. 5.2.5). When

[1] https://en.wikipedia.org/wiki/Triple_product#As_an_exterior_product.

[2] This notation suggests that $||\nabla\cdot\boldsymbol{E} + \jmath \wedge \boldsymbol{E}||^2 = ||\nabla\cdot\boldsymbol{E}||^2 + || \wedge \boldsymbol{E}||^2$, which is related to Helmholtz's theorem.

this force density [N/m^2] is integrated over an area, the net radial force [N] is

$$F_r = -\int_S \nabla \varrho(\boldsymbol{x}) d\boldsymbol{x} \quad \text{[N]}. \tag{5.1.8}$$

An inflated balloon with a static internal pressure of 3 [atm], in an ambient pressure of 1 [atm] (sea level), forms a sphere due to the elastic nature of the rubber, which acts as a stretched spring under its surface tension. The net normal force on the surface of the balloon is its area times the pressure drop of 2 [atm] across the surface. Thus the static pressure is

$$\varrho(\boldsymbol{x}) = 3u(r_o - r) + 1 \quad \text{[Pa]},$$

where $u(r)$ is a step function of the radius $r = ||\boldsymbol{x}|| > 0$, centered at the center of the balloon, having radius r_o.

Taking the gradient gives the negative[3] of the radial force density (i.e., perpendicular to the surface of the balloon):

$$-f_r(r) = \nabla \varrho(\boldsymbol{x}) = \frac{\partial}{\partial r} 3u(r_o - r) + 1 = -2\delta(r_o - r) \quad \text{[Pa]}.$$

This equation describes a static pressure that is 1 [atm] (10^5 [Pa]) outside the balloon and 3 [atm] inside. The net positive force density is the negative of the gradient of the static pressure.

Finally, taking the divergence of the force produces a double delta function at the balloon's surface. Specifically, $\nabla^2 \varrho(\boldsymbol{x}) = -2\delta^{(1)}(r_o - r)$, where 2 is the pressure drop across the balloon. If we take the thickness of the rubber (l [m]) into account, then $\nabla^2 \varrho = -2(\delta(r_o) - \delta(r_o - l))$.

5.1.2.7 Vector Laplacian $\nabla^2()$

A second form of the Laplacian is the *vector Laplacian* $\nabla^2()$, defined as the divergence of the gradient $\nabla^2() \equiv \nabla \cdot \nabla()$, thus nicknamed **Bull-Dog**, operates on a vector to produce a vector (Table 5.1). We shall need this when working with Maxwell's equations.

5.1.3 Scalar Laplacian Operator in N Dimensions

In general, it may be shown that in $N = 1, 2, 3$ dimensions (Sommerfeld 1949, p. 227),

$$\nabla_r^2 \mathcal{P} \equiv \frac{1}{r^{N-1}} \frac{\partial}{\partial r}\left(r^{N-1} \frac{\partial \mathcal{P}}{\partial r}\right). \tag{5.1.9}$$

[3]The force is pointing out, stretching the balloon.

For each value of N, the area $A(r) = A_o r^{N-1}$. This result will turn out to be useful when we work with the Laplacian in one, two, and three dimensions. This naturally follows from Eq. 5.2.10:

$$\frac{1}{A(r)}\frac{\partial}{\partial r}\left[A(r)\frac{\partial}{\partial r}\right]\varrho(r,t) = \frac{1}{c_o^2}\frac{\partial^2}{\partial t^2}\varrho(r,t) \leftrightarrow \frac{s^2}{c_o^2}\mathcal{P}(r,s).$$

Example: When $N = 3$ (i.e., spherical geometry), $A(r) = A_o r^2$, thus

$$\nabla_r^2\,\mathcal{P} \equiv \frac{1}{r^2}\partial_r r^2 \partial_r\,\mathcal{P} \tag{5.1.10}$$

$$= \frac{1}{r}\frac{\partial^2}{\partial r^2} r\,\mathcal{P}, \tag{5.1.11}$$

resulting in the general d'Alembert solutions (Eq. 4.4.1) for the spherical wave equation,

$$\mathcal{P}^{\pm}(r,s) = \frac{1}{r}e^{\mp\kappa(s)r},$$

where $\kappa(s) = s/c_o$.

Exercise #2 Prove the result of the previous example by expanding Eqs. 5.1.10 and 5.1.11 using the chain rule.

Solution: Expanding Eq. 5.1.10:

$$\begin{aligned}\frac{1}{r^2}\partial_r r^2 \partial_r\,\mathcal{P} =& \frac{1}{r^2}\left(2r + r^2\partial_r\right)\partial_r\,\mathcal{P}\\ =& \frac{2}{r}\,\mathcal{P}_r + \mathcal{P}_{rr}.\end{aligned}$$

Expanding Eq. 5.1.11, we obtain

$$\begin{aligned}\frac{1}{r}\partial_{rr} r\,\mathcal{P} =& \frac{1}{r}\partial_r\left(\mathcal{P} + r\,\mathcal{P}_r\right)\\ =& \frac{1}{r}\left(\mathcal{P}_r + \mathcal{P}_r + r\,\mathcal{P}_{rr}\right)\\ =& \frac{2}{r}\,\mathcal{P}_r + \mathcal{P}_{rr}.\end{aligned}$$

Thus the two are equivalent. ■

5.1.3.1 Summary

The radial component of the Laplacian in spherical coordinates (Eq. 5.1.10) simplifies to

$$\nabla_r^2 \varrho(\boldsymbol{x}) = \frac{1}{r^2}\frac{\partial}{\partial r}r^2\frac{\partial}{\partial r}\varrho(\boldsymbol{x}) = \frac{1}{r}\frac{\partial^2}{\partial r^2}r\,\varrho(\boldsymbol{x}).$$

Since DoG is $\nabla^2 = \nabla\cdot\nabla$, it follows that the net force $\boldsymbol{f}(\boldsymbol{x}) = [F_r, 0, 0]^T$, Eq. 5.1.8 in spherical coordinates has a radial component F_r and angular component of zero. Thus the force across a balloon may be approximated by a delta function across the thin sheet of stretched rubber.

We can extend the preceding example in an interesting way to the case of a rigid hose, a rigid tube, that terminates on the right in an elastic medium (the above example of a balloon), such as an automobile tire. On the far left let's assume there is a pump injecting the fluid into the rigid hose. Consider two different fluids: air and water. Air is treated as a compressible fluid, whereas water is incompressible. However, such a classification is relative, determined by the relative compliance of the balloon (i.e., tire) at the relatively rigid pump and hose.

This is a special case of a more general situation: When a fluid is treated as incompressible (rigid), the speed of sound becomes infinite, and the wave equation is an invalid description. In this case the motion is best approximated by Laplace's equation. This represents the transition from short to long wavelengths, from wave propagation having delay, to quasistatics, having no delay.

This example may be modeled as either an electrical or a mechanical system. While the two systems are very different in their physical realization, they are mathematically equivalent, forming a perfect analog. If we take the electrical analog, the pump is a current source, injecting charge Q_o into the hose, which being rigid cannot expand (has a fixed volume). The hose may be modeled as a resistor and the tire as a capacitor C_o, which fills with charge as it is delivered via the resistor, from the pump. The capacitor obeys the same equation as Hooke's law for a spring, $F = K_o\Delta$, where K_o is the stiffness of the spring, $C_o = 1/K_o$ is the spring's compliance, and Δ is the displacement. In electrical terms, $Q_o = C_o\Phi$, where Φ is the voltage, which acts like a force F; Q_o is the charge, which plays the role of the mass of the fluid. The charge Q is conserved, just as the mass of the fluid is conserved (they cannot be created or destroyed).

The flow of the fluid is called the *flux*, which is the general term for the mass flow, heat or electrical current flow (Table 3.2). The two equations may be rewritten directly in terms of the force (F, Φ) and flow (momentum or current flux). A third alternative is in terms of the electrical

$$I = C_o\frac{d}{dt}\Phi \quad [\text{A}] \tag{5.1.12}$$

and mechanical

$$J = C_o\frac{d}{dt}F \quad [\text{kgm-s/m}] \tag{5.1.13}$$

impedance/admittance relationship.

It is common to treat the stiffness of the balloon, which acts as a spring, as a compliance $C_o = 1/K_o$, in which case the impedance Z is defined in the frequency domain as the ratio of the generalized force over the generalized flow

$$Z(s) = \frac{1}{sC_o} \quad \text{[ohms]}.$$

In the case of mechanical systems $Z_m(s) \equiv F/J$, while for the electrical system, $Z_e(s) \equiv \Phi/I$. In thermodynamics the thermal compliance is $C_o = \mathcal{S}/T$, thus $Z_{themo} = T/s\,\mathcal{S} = 1/sC_o$. It is helpful to use the unit [ohms] when working with any impedance, allowing for a uniform terminology for the different physical situations and many forms of impedance. This greatly simplifies the notation.

In the time domain, Ohm's law becomes Eq. 5.1.13 for the case of a mechanical compliance $C_o = 1/K_o$ and Eq. 5.1.12 for the electrical capacitor C. As shown in Table 3.2, the formula for the generalized impedance is typically expressed in terms of the Laplace frequency s, which of course is the $\mathcal{LT}$of the time variables.

The final solution of this system is solved in the frequency domain. The impedance seen by the source is the sum of the resistance R and the impedance of the load, giving

$$Z = R + \frac{1}{sC}.$$

This results in a simple relationship between the force and the flow, as determined by the action of the source on the load $Z(s)$. The results may be given in terms of the voltage across the compliance in terms of the voltage Φ_s (or current I_s) due to the source. Given some algebra, the voltage across the compliance Φ_c, divided by the voltage of the source, is

$$\frac{\Phi_c}{\Phi_{\text{source}}} = \frac{R}{R + 1/sC}.$$

Thus the calculus reduces to some algebra in the frequency domain, which in this case has a simple pole at $s_p = -1/RC$. The time-domain response is then found by taking the inverse $\mathcal{LT}$.

Cauchy's residue theorem (Sect. 4.5) gives the final answer, which describes how the voltage across the compliance builds exponentially with time, from zero to the final value. Given the voltage, we can also compute the current, as a function of time. This then represents the entire process of either blowing up a balloon, charging a capacitor, or heating water on a stove, the difference being only the physical notation, as the math is identical.

Note that this differential equation is first order in time, which in frequency means the impedance has a single pole. Thus the equation for charging a capacitor or pumping up a balloon describes a diffusion process. If we had taken the impedance of the mass of the fluid in the hose into account, we would have a lumped-parameter model of the wave equation with a second-order system. This is mathematically related to the homework assignment about train cars (masses) connected by springs (Fig. 5.4, Homework DE-3, Problem #2).

Example: The voltage

$$\phi(\boldsymbol{x}, t) = e^{-\boldsymbol{\kappa}\cdot\boldsymbol{x}} u(t - x/c) \leftrightarrow \frac{1}{s} e^{-\boldsymbol{\kappa}\cdot\boldsymbol{x}} \quad [\mathrm{V}] \tag{5.1.14}$$

represents one of d'Alembert's solution (Eq. 4.4.1) of the wave equation (Eq. 3.1.5) as well as an eigenfunction of the gradient operator ∇. From the definition of the scalar (dot) product of two vectors (Fig. 3.4),

$$\boldsymbol{\kappa}\cdot\boldsymbol{x} = \kappa_x x + \kappa_y y + \kappa_z z = ||\boldsymbol{\kappa}||\, ||\boldsymbol{x}|| \cos\theta_{\kappa x},$$

where $||\boldsymbol{\kappa}|| = \sqrt{\kappa_x^2 + \kappa_y^2 + \kappa_z^2}$ and $||\boldsymbol{x}|| = \sqrt{x^2 + y^2 + z^2}$ are the lengths of vectors $\boldsymbol{\kappa}$ and $\boldsymbol{x}$, and $\theta_{\kappa x}$ is the angle between them. As before, $s = \sigma + \omega \jmath$ is the Laplace frequency.

To keep things simple, we let $\boldsymbol{\kappa} = [\kappa_x, 0, 0]^T$ so that $\boldsymbol{\kappa}\cdot\boldsymbol{x} = \kappa_x x\hat{\mathbf{x}}$. We shall soon see that $||\boldsymbol{\kappa}|| = 2\pi/\lambda$ follows from the basic relationship between a wave's radian frequency $\omega = 2\pi f$ and its wavelength λ:

$$\omega\lambda = c_o. \tag{5.1.15}$$

As the frequency increases, the wavelength becomes shorter. This key relationship may have been first researched by Galileo in about 1564, followed by Mersenne[4] in about 1627 (Fig. 1.5).

Exercise #3 Show that Eq. 5.1.14 is an eigenfunction of the gradient operator ∇.

Solution: Taking the gradient of $\phi(\boldsymbol{x}, t)$ gives

$$\begin{aligned}\nabla e^{-\boldsymbol{\kappa}\cdot\boldsymbol{x}} u(t) &= -\nabla\boldsymbol{\kappa}\cdot\boldsymbol{x}\, e^{-\boldsymbol{\kappa}\cdot\boldsymbol{x}} u(t)\\ &= -\boldsymbol{\kappa}\, e^{-\boldsymbol{\kappa}\cdot\boldsymbol{x}} u(t),\end{aligned}$$

or in terms of $\phi(\boldsymbol{x}, t)$,

$$\nabla\phi(\boldsymbol{x}, t) = -\boldsymbol{\kappa}\,\phi(\boldsymbol{x}, t) \leftrightarrow -\frac{s}{c} e^{-\boldsymbol{\kappa}\cdot\boldsymbol{x}}.$$

---

[4] See https://www-history.mcs.st-and.ac.uk/Biographies/Mersenne.html;

> *In the early 1620s, Mersenne listed Galileo among the innovators in natural philosophy whose views should be rejected. However, by the early 1630s, less than a decade later, Mersenne had become one of Galileo's most ardent supporters.*(Garber 2004)

.

Thus $\phi(\boldsymbol{x}, t)$ is an eigenfunction of ∇, having the vector eigenvalue $\boldsymbol{\kappa}$. As before, $\nabla\phi$ is proportional to the current, since ϕ is a voltage, and the ratio (i.e., the eigenvalue) may be thought of as a mass, analogous to the impedance of a mass (or inductor). In general, the units provide the physical interpretation of the eigenvalues and their spectra. ■

Exercise #4 Compute $\hat{\mathbf{n}}$ for $\phi(\boldsymbol{x}, s)$ as given by Eq. 5.1.14.

Solution: $\hat{\mathbf{n}} = \boldsymbol{\kappa}/||\boldsymbol{\kappa}||$ represents a unit vector in the direction of the wave propagation. ■

Exercise #5 If the sign of $\boldsymbol{\kappa}$ is negative, what are the eigenvectors and eigenvalues of $\nabla\phi(\boldsymbol{x}, t)$?

Solution:

$$\begin{aligned}\nabla e^{-\boldsymbol{\kappa}\cdot\boldsymbol{x}}u(t) &= -\boldsymbol{\kappa}\cdot\nabla(\boldsymbol{x})e^{-\boldsymbol{\kappa}\cdot\boldsymbol{x}}u(t)\\ &= -\boldsymbol{\kappa}\; e^{-\boldsymbol{\kappa}\cdot\boldsymbol{x}}u(t).\end{aligned}$$

Nothing changes other than the sign of $\boldsymbol{\kappa}$. Physically this means the wave is traveling in the opposite direction, corresponding to the forward and retrograde d'Alembert waves. ■

Prior to this section, we had considered the Taylor series in only one variable, such as for polynomials $P_N(x), x \in \mathbb{R}$ (Eq. 3.1.7), and $P_N(s), s \in \mathbb{C}$ (Eq. 3.2.20). The generalization from real to complex analytic functions led to the $\mathcal{LT}$ and the host of integration theorems (FTCC, Cauchy CT-1, CT-2, CT-3). What is in store when we generalize from one spatial variable ($\mathbb{R}$) to three ($\mathbb{R}^3$)?

Exercise #6 Find the velocity $v(t)$ of an electron in a field $\boldsymbol{E}$.

Solution: From Newton's second law, $-qE = m_e\dot{v}(t)$ [N], where m_e is the mass of the electron. Thus we must solve this first-order differential equation to find $v(t)$. As before, this is best done in the frequency domain $v(t) \leftrightarrow V(s)$. ■

5.1.3.2 Role of Potentials

Note that the scalar fields (e.g., temperature, pressure, voltage) are all scalar potentials, summarized in Table 3.2. In each case the gradient of the potential results in a vector force field, just as in the electric case above (Eq. 5.1.1).

Table 3.2 is helpful in understanding the physical meaning of the gradient of a potential, which is typically a generalized force (electric field, acoustic force density, temperature flux), that in turn generates a flow (current, velocity, heat flux (entropy)). The ratio of the potential over flow determines the impedance. Four examples are

1. The voltage drop across a resistor causes a current to flow, as described by Ohm's law. The difference in voltage between two points is a crude form of gradient when

the frequency f [Hz] is low, such that the wavelength is much larger than the distance between the two points. This is the essence of the quasistatic approximation (Postulate P10, Sect. 3.10.2).
2. The gradient of the pressure gives rise to a force density in the fluid medium (air, water, oil, etc.), that causes a flow (velocity vector) in the medium.
3. The gradient of the temperature also causes a flow of heat that is proportional to the thermal resistance, given Ohm's law for heat (Feynman 1970b, pp. 3–7).
4. When a solution contains ions, it defines an *electrochemical Nernst potential* $N(\boldsymbol{x}, t)$ (Fermi 1936; Scott 2002). This electrochemical potential is similar to a voltage or temperature field, the gradient of which defines a virtual force on the ions, resulting in an ionic current.

Thus in the above examples there is a potential, the gradient of which is a force, that when applied to the medium (an impedance) causes a flow (flux or current) proportional to that impedance due to the medium, which is the ratio of the gradient of the potential to the current. The product of the force and flow is a power.

Exercise #7 Show that the integral of Eq. 5.1.1 is an antiderivative.
Solution: We use the definition of the antiderivative given by the FTC (Eq. 4.2.2):

$$\begin{aligned}
\phi(\boldsymbol{x}, t) - \phi(\boldsymbol{x}_o, t) &= \int_{\boldsymbol{x}_o}^{\boldsymbol{x}} \boldsymbol{E}(\boldsymbol{x}, t) \cdot d\boldsymbol{x} \\
&= -\int_{\boldsymbol{x}_o}^{\boldsymbol{x}} \nabla\phi(\boldsymbol{x}, t) \cdot d\boldsymbol{x} \\
&= -\int_{\boldsymbol{x}_o}^{\boldsymbol{x}} \left(\hat{\mathbf{x}}\frac{\partial}{\partial x} + \hat{\mathbf{y}}\frac{\partial}{\partial y} + \hat{\mathbf{z}}\frac{\partial}{\partial z}\right) \phi(\boldsymbol{x}, t) \cdot d\boldsymbol{x} \\
&= -\int_{\boldsymbol{x}_o}^{\boldsymbol{x}} \left(\hat{\mathbf{x}}\frac{\partial\phi}{\partial x} + \hat{\mathbf{y}}\frac{\partial\phi}{\partial y} + \hat{\mathbf{z}}\frac{\partial\phi}{\partial z}\right) \cdot (\hat{\mathbf{x}}dx + \hat{\mathbf{y}}dy + \hat{\mathbf{z}}dz) \\
&= -\int_{x_o}^{x} \frac{\partial\phi}{\partial x} dx - \int_{y_o}^{y} \frac{\partial\phi}{\partial y} dy - \int_{z_o}^{z} \frac{\partial\phi}{\partial z} dz \\
&= -\int_{\boldsymbol{x}_o}^{\boldsymbol{x}} d\phi(\boldsymbol{x}, t) \\
&= -\Big(\phi(\boldsymbol{x}, t) - \phi(\boldsymbol{x}_o, t)\Big).
\end{aligned}$$

This may be verified by taking the gradient of both sides:

$$\nabla\phi(\boldsymbol{x}, t) - \overset{0}{\cancel{\nabla\phi(\boldsymbol{x}_o, t)}} = -\nabla \int_{\boldsymbol{x}_o}^{\boldsymbol{x}} \boldsymbol{E}(\boldsymbol{x}, t) \cdot d\boldsymbol{x} = \boldsymbol{E}(\boldsymbol{x}, t).$$

If we apply the FTC, the antiderivative must be $\phi(\boldsymbol{x}, t) = E_x x\hat{\mathbf{x}} + 0\hat{\mathbf{y}} + 0\hat{\mathbf{z}}$. This same point is made by Feynman (1970b, p. 4–1, Eq. 4.28). ■

Given that the force on a charge is proportional to the gradient of the potential, this exercise shows that the integral of the gradient depends on only the end points, the work done in moving a charge depends on only the limits of the integral, which is the definition of a *conservative field* but holds only in the ideal case where $\boldsymbol{E}$ is determined by Eq. 5.1.1—that is, the medium has no friction (there are no other forces on the charge).

5.1.3.3 The Conservative Field

An important question is: When is a field conservative? A field is conservative when the work done by the motion is independent of the path of the motion. Thus the conservative field is related to the FTC, which states that the integral of the work depends on only the end points.

A more complete answer must await the introduction of the fundamental theorem of vector calculus (Eq. 5.6.7). A few examples provide insight:

Example: The gradient of a scalar potential, such as the voltage (Eq. 5.1.1), defines the electric field, which drives a current (flow) across a resistor (impedance). When the impedance is infinite, the flow is zero, leading to zero power dissipation. When the impedance is lossless, the system is conservative.

Example: At audible frequencies the viscosity of air is quite small and thus, for simplicity, it may be taken as zero. However, when the wavelength is small (e.g., at 100 [kHz] $\lambda = c_o/f = 345/10^5 = 3.45$ [mm]) the lossless assumption breaks down, resulting in a significant propagation loss. When the viscosity is taken into account, the field is lossy and thus the field is no longer conservative. In narrow tubes, for example a flute, thermal loss plays a much larger role due to the walls (Appendix D).

Example: If a temperature field is a time-varying constant [i.e., $T(\boldsymbol{x}, t) = T_o(t)$], there is no "heat flux," since $\nabla T_o(t) = 0$. When there is no heat flux [i.e., flux, or flow], there is no heat power, since the power is the product of the force and the flow.

Example: The force of gravity is given by the gradient of Newton's gravitational potential (Eq. 3.1.1):

$$F = -\nabla_r \phi_{\mathrm{N}}(r) = -\frac{\partial}{\partial r}\frac{1}{r} = \frac{1}{r^2}.$$

Historically speaking, $\phi_N(r)$ was the first conservative field, used to explain the elliptic orbits of the planets around the Sun. Galileo's law says that bodies fall with constant acceleration, giving rise to a parabolic path and a time of fall proportional to t^2. This behavior of falling objects directly follows from the Galilean potential:

$$\phi_{\mathrm{G}}(r) = \frac{1}{(r - r_o)} = \frac{-r_o}{1 - r/r_o} \underset{r<r_o}{=} -r_o(1 - r/r_o + (r/r_o)^2 + \cdots) \underset{r \ll r_o}{\approx} r_o - r,$$

which, given the large radius r_o of the Earth and the small distance of the object from the surface of the Earth $r - r_o$, is equal to the distance above the ground. Thus Galileo's law says that the force a falling body sees is constant:

$$F_G = -G_o \nabla_r \phi_G(r) = G_o.$$

This can be scaled by G_o to account for the magnitude of the gravitational force.

Exercise #8 Galileo discovered that the height of a falling object is proportional to the square of the time it falls. Based on Newton's follow-up analysis, today we would say this height $h(t)$ is

$$h(t) = \frac{1}{2} m G_o (t - t_o)^2 \quad [\text{m}],$$

where m is the object's mass and G_o is the gravitational constant for the Earth at its surface r_o. Show that $h(t)$ directly follows from the potential $\phi_G = r_o - r$. This formula applies if you toss a ball into the air or if you drop it from a high place.

Solution: Given Galileo's potential $\phi_G(r) \underset{r \ll r_o}{\approx} m G_o(r_o - r)$, thus $\ddot{h}(t) = m G_o$. Given Galileo's formula for the height $h(t)$, the velocity is $v(t) = \dot{r}(t) = m G_o t$, and the acceleration is $\ddot{r}(t) = m G_o$. ■

Exercise #9 Find the time that it takes to fall from a distance $r = L$. That is solve $h(t) = L$ for the time the object takes to fall from the distance L.

Solution: Setting $t_o = 0$ gives $t^2 = 2L/mG_o$. Thus the time to fall is $T(L) = \sqrt{2L/mG_o}$. ■

5.2 Partial Differential Equations and Field Evolution

The three main classes of partial differential equations (PDEs) are: elliptic, parabolic, and hyperbolic, distinguished by the order of the time derivative. These categories seem to have little mathematical utility (the categories are labels).

5.2.1 *The Laplacian* ∇^2

In the most important case the space operator is the Laplacian ∇^2, the definition of which depends on the dimensionality of the waves—that is, the coordinate system being used. We first discussed the Laplacian as a 2D operator on Sect. 4.2.3 where we studied complex analytic functions, and again on Sect. 5.1. An expression for ∇^2 for one, two, and three dimensions was provided as Eq. 5.1.9. In three-dimensional rectangular coordinates, it is defined as (see Sect. 5.1.2)

$$\nabla^2 T(\boldsymbol{x}) = \left(\frac{\partial^2}{\partial x^2} + \frac{\partial^2}{\partial y^2} + \frac{\partial^2}{\partial z^2}\right) T(\boldsymbol{x}). \tag{5.2.1}$$

The Laplacian operator is ubiquitous in mathematical physics, starting with simple complex analytic functions (Laplace's equation) and progressing to Poisson's equation, the diffusion equation, and finally the wave equation. Only the wave equation results in a delay. The diffusion equation "wave" has an instantaneous spread (the effective "wavefront" velocity is infinite, yet the wavelength is long; it's not a traveling wave).

Examples of elliptic, parabolic, and hyperbolic equations are

1. *Laplace's equation:* The equation

$$\nabla^2 \Phi(\boldsymbol{x}) = 0 \tag{5.2.2}$$

   describes, for example, the voltage inside a closed chamber that has a given voltage on the walls or the steady-state temperature within a closed container having a specified temperature distribution on the walls. There are no dynamics to the potential, even when it is changing, since the potential instantaneously follows the potential on the walls.
2. *Poisson's equation:* In the steady state, the diffusion equation degenerates to either Poisson's or Laplace's equation; both are classified as *elliptic* equations (second order in space, zero-order in time). As in the diffusion equation, the evolution has a wave velocity that is functionally infinite. For example,

$$\nabla^2 \Phi(\boldsymbol{x}, t) = \rho(x, t)$$

   holds for gravitational fields or the voltage around a charge. It does not describe gravity waves, which travel at the speed of light.
3. *Fourier diffusion equation:* Equation 5.2.3 describes the evolution of the scalar temperature $T(\boldsymbol{x}, t)$ (a scalar potential), gradients of solution concentrations (i.e., ink in water), and Brownian motion. Diffusion is first-order in time, which is categorized as *parabolic* (first-order in time, second-order in space). When these equations are Laplace transformed, diffusion has a single real root, resulting in a real solution (e.g., $T \in \mathbb{R}$). There is no wavefront in the case of the diffusion equation. As soon as the source is turned on, the field is nonzero at every point in the bounded container. As an example,

$$\nabla^2 T(\boldsymbol{x}, t) = \kappa_o \frac{\partial T(\boldsymbol{x}, t)}{\partial t} \leftrightarrow s\kappa_o T(\boldsymbol{x}, s) \tag{5.2.3}$$

   describes the temperature $T(\boldsymbol{x}, t) \leftrightarrow T(\boldsymbol{x}, \omega)$, as proposed by Fourier in 1822, or the diffusion of two miscible liquids (Fick 1855) or Brownian motion (Einstein 1905). The diffusion equation is not a wave equation, since the temperature wavefront propagates instantaneously. The diffusion equation does a poor job of representing the velocity of molecules banging into each other, since such collisions have a mean free path, and thus the velocity cannot be infinite.

4. *Wave equations:* There are scalar and vector forms of wave equations.

    1. Scalar wave equations: Equation 3.1.3 describes the evolution of a scalar potential field, such as pressure $\varrho(\boldsymbol{x}, t)$ (sound) or the displacement of a string or membrane under tension. The wave equation is second-order in time. When transformed into the frequency domain, the solution has pairs of complex conjugate roots, leading to two real solutions. The wave equation is classified as *hyperbolic* (second-order in time and space).
    2. Vector wave equations: Maxwell's equations describe the propagation of the electric $\boldsymbol{E}(\boldsymbol{x}, t)$ and magnetic $\boldsymbol{H}(\boldsymbol{x}, t)$ field strength vectors, as well as the electric $\boldsymbol{D}(\boldsymbol{x}, t) = \epsilon_o \boldsymbol{E}(\boldsymbol{x}, t)$ and magnetic $\boldsymbol{B}(\boldsymbol{x}, t) = \mu_o \boldsymbol{H}(\boldsymbol{x}, t)$ flux vectors. ME are antireciprocal (P6).

5.2.1.1 Solution Evolution

The partial differential equation defines the evolution of the scalar field [pressure $\varrho(\boldsymbol{x}, t)$ and temperature $T(\boldsymbol{x}, t)$], or vector field $(\boldsymbol{E}, \boldsymbol{D}, \boldsymbol{B}, \boldsymbol{H})$, as functions of space $\boldsymbol{x}$ and time t. There are two basic categories of field evolution: diffusion and propagation.

1. *Diffusion*: The simplest and easiest PDE example, easily visualized, is a static[5] (time-invariant) scalar temperature field $T(\boldsymbol{x})$ [°C]. Just like an impedance or admittance, a field has regions where it is analytic, and for the same reasons, $T(\boldsymbol{x}, t)$ satisfies Laplace's equation

$$\nabla^2 T(\boldsymbol{x}, t) = 0.$$

Since there is no current when the field is static, such systems are lossless and thus are conservative.

When $T(\boldsymbol{x}, t)$ depends on time (is not static), it is described by the *diffusion equation* (Eq. 5.2.3), a rule for how $T(\boldsymbol{x}, t)$ evolves with time from its initial state $T(\boldsymbol{x}, 0)$. The constant κ_o is called the *thermal conductivity*, which depends on the properties of the fluid in the container, with $s\kappa_o$ being the thermal admittance per unit area. The conductivity is a measure of how the heat gradients induce heat currents $\boldsymbol{J} = -\kappa_o \nabla T$, analogous to Ohm's law for electricity.

Note that when $T(\boldsymbol{x}, t \to \infty)$ the temperature reaches a steady state, $\boldsymbol{J} = 0$ and $\nabla^2 T = 0$. This all depends on what is happening at the boundaries. When the wall temperature of a container is a function of time, the internal temperature $T(\boldsymbol{x}, t)$ will continue to change, but with a frequency-dependent delay that depends on the thermal conductivity κ_o.

[5] Postulate P3, Sect. 3.10.2.

Such a system is analogous to an electrical resistor–capacitor series circuit connected to a battery. For example, the wall temperature (voltage across the battery) represents the potential driving the system. The thermal conductivity κ_o (the electrical resistor) is likewise analogous. The fluid (the electrical capacitor) is being heated (charged) by the heat (charge) flux. In all cases Ohm's law defines the ratio of the potential (voltage) to the flux (current). How this happens can be understood only once the solution to the equations has been established. The fluid has a heat capacity analogous to that of an electrical capacitor (Kirchhoff 1868, 1974). Sometimes the diffusion equation is called the telegraph equation.

2. *Propagation:* Pressure and electromagnetic waves are described by a scalar potential (pressure) (Eq. 3.1.3) and a vector potential (Eq. 5.7.4), leading to scalar and vector wave equations. Sometimes the wave equation is called the telephone equation.

5.2.1.2 The Taylor Series of $f(x)$

Next we extend the concept of the Taylor series of one variable to $\boldsymbol{x} \in \mathbb{R}^3$. Just as we generalized the derivative with respect to a real frequency variable $\omega \in \mathbb{R}$ to complex frequency $s = \sigma + \omega \jmath \in \mathbb{C}$, here we generalize the derivative with respect to $x \in \mathbb{R}$ to the vector $\boldsymbol{x} \in \mathbb{R}^3$.

Since the scalar field is analytic in $\boldsymbol{x}$, it is a good place to start. Assuming we have carefully defined the Taylor series (Eq. 3.2.9) in one and two (Eq. 4.2.7) variables, the Taylor series of $f(\boldsymbol{x})$ in $\boldsymbol{x} \in \mathbb{R}^3$ about $\boldsymbol{x} = 0$ may be defined as

$$f(\boldsymbol{x} + \delta\boldsymbol{x}) = f(\boldsymbol{x}) + \nabla f(\boldsymbol{x}) \cdot \delta\boldsymbol{x} + \frac{1}{2!}\sum_{k=1}^{3}\sum_{l=1}^{3} \frac{\partial^2 f(\boldsymbol{x})}{\partial x_k \partial x_l}\delta x_k \delta x_l + \text{HOT}, \qquad (5.2.4)$$

where HOT stands for Higher Order Terms (Greenberg 1988, p. 639). From this definition, it is clear that the gradient is the generalization of the second term in the one-dimensional Taylor series expansion.

5.2.1.3 Summary

For every potential $\phi(\boldsymbol{x}, t)$ there exists a force density $\boldsymbol{f}(\boldsymbol{x}, t) = -\nabla\phi(\boldsymbol{x}, t)$, proportional to the potentials, that drives a generalized flow $\boldsymbol{u}(\boldsymbol{x}, t)$. If the normal components of the force and flow are averaged over a surface, the mean force and volume flow (i.e., volume velocity for the acoustic case) are defined. In such cases the impedance $Z(s)$ is the net force through the surface force over the net flow, and Gauss's law and quasistatics (Postulate P10, Sect. 3.10.2) come into play (Feynman 1970a). We call this the *generalized impedance*. An example is $Z(s) = \sqrt{s}$.

Assuming linearity (Postulate P2, Sect. 3.10.2), the product of the force and flow is the power, and the ratio (force/flow) is an impedance (Table 3.2). This impedance statement is called Ohm's law, Kirchhoff's laws, Laplace's law, or Newton's laws. In the simplest cases, they are all linearized (proportional) complex relationships between a force and a flow. Very few impedance relationships are inherently linear over a large range of force or current, but for physically useful levels, they are treated as linear. Nonlinear interactions require a more sophisticated approach, typically involving numerical methods, working in the time domain.

In electrical circuits, it is traditional to define a zero potential *ground* point that all voltages use as the reference potential. The ground is a useful convention as a simplifying rule, but it obscures the physics and obscures the fact that the voltage is *not* the force. Rather, the force is the voltage difference, referenced to the ground, which is defined as zero volts. This results in abstracting away (i.e., hiding) the difference in voltage. It seems misleading (more precisely, it is wrong) to state Ohm's law as the voltage over the current, since Ohm's law actually says that the *voltage difference* (i.e., voltage gradient) over the current defines an impedance (Kennelly 1893).

When we measure the voltage between two points, it is a crude approximation to the gradient based on the quasistatic approximation (Postulate P10). The pressure is also a potential, the gradient of which is a force density, which drives the volume velocity (flow).

In Sect. 5.6.5 we introduce the fundamental theorem of vector calculus (otherwise known as Helmholtz's decomposition theorem), which generalizes Ohm's law to include circulation (e.g., angular momentum, vorticity, and the related magnetic effects). To understand these generalizations in flow, we need to understand compressible and rotational fields (Table 5.3), complex analytic functions, and more mathematical physics history.

It is the *difference* in the potential (i.e., voltage, temperature, pressure) that is proportional to the flux. This can be viewed as a major simplification of the gradient relationship, justified by the quasistatic assumption (Postulate P10, Sect. 3.10.2).

The roots of the impedance are related to the eigenmodes of the system equations. The solutions to the equations are the eigenfunctions, evaluated at the eigenvalues (Sect. 4.2.3)

5.2.2 *Scalar Wave Equation (Acoustics)*

In this section we discuss the general solution to the wave equation, which has two forms: scalar waves (acoustics) and vector waves (electromagnetics). These have an important mathematical distinction but a similar solution space, one scalar and the other vector. We start with the scalar wave equation. The vector case will be discussed Sect. 5.7.4.

A good starting point for understanding PDEs is to explore the scalar wave equation (Eq. 3.1.3). Acoustic wave propagation was first analyzed mathematically by Isaac Newton in his famous book *Principia* (1687), in which he first calculated the speed of sound based on the conservation of mass and momentum.

5.2.2.1 Early History

The study of wave propagation begins at least as early as Huygens (ca. 1678) (Pierce 1981, p. 15). The acoustic variables are the *pressure*,

$$\varrho(\boldsymbol{x}, t) \leftrightarrow \mathcal{P}(\boldsymbol{x}, s),$$

and the particle velocity,

$$\nu(\boldsymbol{x}, t) \leftrightarrow \mathcal{U}(\boldsymbol{x}, s).$$

To obtain a wave, we must include two basic components: the stiffness of air and its mass. The two equations are called (1) Newton's second law ($F = ma$) and (2) Hooke's law ($F = kx$), respectively. In vector form these equations are (1) Euler's equation (i.e., conservation of momentum density),

$$-\nabla \varrho(\boldsymbol{x}, t) = \rho_o \frac{\partial}{\partial t} \nu(\boldsymbol{x}, t) \leftrightarrow \rho_o s\, \mathcal{U}(\boldsymbol{x}, s), \tag{5.2.5}$$

which assumes the time-average density ρ_o is independent of time and position $\boldsymbol{x}$, and (2) the continuity equation (i.e., conservation of mass density),

$$-\nabla \cdot \nu(\boldsymbol{x}, t) = \frac{1}{\eta_o P_o} \frac{\partial}{\partial t} \varrho(\boldsymbol{x}, t) \leftrightarrow \frac{s}{\eta_o P_o} \mathcal{P}(\boldsymbol{x}, s) \tag{5.2.6}$$

(Pierce 1981; Morse 1948, p. 295). Here $P_o = 10^5$ [Pa] is the barometric pressure and $\eta_o P_o$ is the adiabatic stiffness, with $\eta_o = 1.4$ (See Sect. D.1.1.1). Combining Eqs. 5.2.5 and 5.2.6 (removing $\nu(\boldsymbol{x}, t)$) results in the three-dimensional scalar pressure wave equation

$$\nabla^2 \varrho(\boldsymbol{x}, t) = \frac{1}{c_o^2} \frac{\partial^2}{\partial t^2} \varrho(\boldsymbol{x}, t) \leftrightarrow \frac{s^2}{c_o^2} \mathcal{P}(\boldsymbol{x}, s) \tag{5.2.7}$$

with $c_o = \sqrt{\eta_o P_o / \rho_o}$ being the air sound velocity, and $\nu = \sqrt{\eta_o P_o \rho_o}$ being the characteristic resistance of air, assuming no visco-thermal losses.

Exercise #10 Show that Eqs. 5.2.5 and 5.2.6 can be reduced to Eq. 5.2.7.

Solution: Taking the divergence of Eq. 5.2.5 gives

$$-\nabla \cdot \nabla \varrho(\boldsymbol{x}, t) = \rho_o \frac{\partial}{\partial t} \nabla \cdot \nu(\boldsymbol{x}, t). \tag{5.2.8}$$

Note that $\nabla \cdot \nabla = \nabla^2$ (Table 5.1). Next, substituting Eq. 5.2.6 into the above equation results in the scalar wave equation, Eq. 5.2.7, since $c_o = \sqrt{\eta_o P_o / \rho_o}$. ■

5.2.3 The Webster Horn Equation (WHEN)

An important generalization of the problem of lossless plane-wave propagation in one-dimensional uniform tubes is known as *transmission line theory*. As depicted in Fig. 5.2, by allowing the area $A(r)$ [e.g., for the conical horn $A(r) = A_o(r/L)^2$ with $L = 1$ [m] and $A_0 \leq 4\pi$] of an acoustical waveguide (horn) to vary along the range axis r (the direction of wave propagation), we can explore general solutions to the wave equation. Classic applications of horns include vocal tract acoustics, loudspeaker design, cochlear mechanics, quantum mechanics (e.g., the hydrogen atom), and wave propagation in periodic media (Brillouin 1953).

We must be more precise when defining the area $A(x)$: The area is *not* the cross-sectional area of the horn; rather it is the wavefront (isopressure) area, which is related to Gauss' law, since the gradient of the pressure defines the force that drives the mass flow (also called volume velocity).

For the scalar wave equation (Eq. 5.1.9), the Webster Laplacian is

$$\nabla_r^2\, \varrho(r,t) = \frac{1}{A(r)}\frac{\partial}{\partial r}\left[A(r)\frac{\partial}{\partial r}\right]\varrho(r,t). \tag{5.2.9}$$

The Webster Laplacian is based on the quasistatic approximation (Postulate P10, Sect. 3.10.2), which requires that the frequency lie below the critical value $f_c = c_o/2d$—namely, that a half wavelength be greater than the horn diameter d (i.e.,

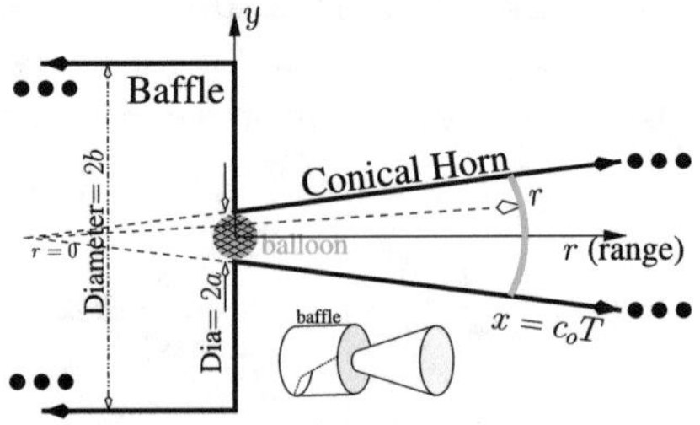

Fig. 5.2 Experimental setup showing a large pipe on the left terminating at the wall containing a small hole with a balloon, shown in green. At time $t = 0$ the balloon is pricked and a pressure pulse is released. The baffle on the left represents a semi-∞ long tube having a large radius compared to the horn input diameter $2a$. At time T the outbound pressure pulse $\varrho(r, T) = \delta(t - x/c_o)/r$ has reached a radius $x = r - r_o = c_o T$, where $r = x$ is the location of the source at the throat of the horn and r is measured from the vertex. At the throat of the horn $\mathcal{V}_+/A_+ = \mathcal{V}_-/A_-$. The term "horn" is used for the case of scalar waves, while the term "wave-guide" is used when speaking of EM waves. When the propagation is constrained by a horn, or wave-guide, the waves are "guided."

$d < \lambda/2$).[6] For the adult human ear canal, $d = 7.5$ [mm] and $f_c = (343/2 \cdot 7.5) \times 10^{-3} \approx 22.87$ [kHz], which is above the upper range of human hearing.

The term on the right of Eq. 5.2.9, which is identical to Eq. 5.1.9, is also the Laplacian for thin tubes (e.g., rectangular, spherical, and cylindrical coordinates). Thus the Webster horn "wave" equation is

$$\frac{1}{A(r)}\frac{\partial}{\partial r}\left[A(r)\frac{\partial}{\partial r}\right]\varrho(r,t) = \frac{1}{c_o^2}\frac{\partial^2}{\partial t^2}\varrho(r,t) \leftrightarrow \frac{s^2}{c_o^2}\mathcal{P}(r,s), \tag{5.2.10}$$

where $\varrho(r,t) \leftrightarrow \mathcal{P}(r,s)$ is the acoustic pressure in Pascals [Pa] (Hanna and Slepian 1924; Mawardi 1949; Eisner 1967; Morse 1948); Olson (1947, p. 101); Pierce (1981, p. 360). Extensive experimental analyses for various types of horns (conical, exponential, parabolic) along with a review of horn theory may be found in Goldsmith and Minton (1924). Of special interest is Eisner (1967) due to his history section and long list of relevant articles.

5.2.3.1 The Limits of the Webster Horn Equation

It is commonly stated that the Webster horn equation (WHEN) is fundamentally limited and thus is an approximation that applies only to frequencies much lower than f_c (Morse 1948; Shaw 1970; Pierce 1981). However, in all these discussions it is assumed that the area function $A(r)$ is the horn's cross-sectional area, not the area of the isopressure wavefront.

In the next section we show that this "limitation" may be avoided (subject to the $f < f_c$ quasistatic limit, Postulate P10, Sect. 3.10.2), making the Webster horn theory an "exact" solution for the lowest order "plane-wave" eigenfunctions of Eq. 5.2.10. The limitation of the quasistatic approximation is that it "ignores" higher order evanescent modes, which are naturally small since, being evanescent modes below their cutoff frequency, the wave number is real and thus they do not propagate (Hahn 1941; Karal 1953). This method is frequently called a *modal analysis* or *eigenanalysis*. This is the same approximation that is required to define an impedance, since every eigenmode has an impedance (Miles 1948). These modes define a Hilbert "vector" space (also called an *eigenspace*).

As derived in Appendix H, the acoustic variables (eigenfunctions) are redefined on the isopressure wavefront boundary for the pressure and the corresponding volume velocity (Hanna and Slepian 1924; Morse 1948; Pierce 1981). The resulting acoustic impedance is then the ratio of the pressure to the volume velocity. This approximation is valid up to the frequency where the first cross-mode begins to propagate ($f > f_c$), which may be estimated given the roots of the Bessel eigenfunctions (Morse

[6]This condition may be written in several ways, the most common being $ka < 1$, where $k = 2\pi/\lambda$ and a is the horn radius. This may be expressed in terms of the diameter as $\frac{2\pi}{\lambda}\frac{d}{2} < 1$, or $d < \lambda/\pi < \lambda/2$. Thus $d < \lambda/2$ may be a more precise metric by the factor $\pi/2 \approx 1.6$. This is called the *half-wavelength assumption*, a synonym for the quasistatic approximation, and the Nyquist theorem (See DE-3, #2).

1948). Perhaps it should be noted that these ideas, which may come from acoustics, apply equally well to electromagnetics and quantum mechanics, and other wave phenomena.

5.2.3.2 Visco-Thermal Losses

When losses are to be included, the wave number $\kappa(s) = s/c_o$ must be replaced with Eq. D.1.5. This introduces dispersion in the wavefront due to the very small dispersive term $2\beta_0\sqrt{s}$, which contains a branch cut. When calculating the losses, we must be careful that they are always on the correct Riemann sheet. In cases where precise estimates of the wave properties and input impedance are required, this term is critical.

The best known examples of wave propagation are electrical and acoustic transmission lines. Such systems are loosely referred to as the *telegraph* or *telephone equations*, harking back to the early days of their discovery (Heaviside 1892; Campbell 1903; Brillouin 1953; Feynman 1970a). The telegraph equation characterizes the large resistance of the wire over long distances along with the stray capacitance of the wire to the ground (which at the time was taken as the second conductor, to save wire). Thus the telegraph equation is best modeled by a diffusion line. The telephone equation included loading coils, consisting of inductors placed periodically in the wire, to increase the circuits inductance. This converted the circuit into a true transmission line. The loading coils were introduced by the AT&T engineer and mathematician George Ashley Campbell (Campbell 1903, 1937), however they were first proposed and promoted by Heaviside.

In acoustics, waveguides are known as horns, such as the horn connected to the first phonographs from around the turn of the century (Webster 1919). Thus the names reflect the historical development, back to a time when mathematics and its applications were related.

5.2.4 Matrix Formulation of the WHEN

Newton's laws of conservation of momentum (Eq. 5.2.5) and mass (Eq. 5.2.6) are modern versions of Newton's starting point for calculating the horn lowest order plane-wave eigenmode wave speed.

The acoustic equations for the average pressure $\mathcal{P}(r,\omega)$ and the volume velocity are derived in Appendix H, where the pressure and particle velocity equations (Eqs. H.1.4 and H.1.6) are transformed into a 2×2 matrix of acoustical variables, average pressure $\mathcal{P}(r,\omega)$ and volume velocity $\mathcal{V}(r,\omega)$:

$$-\frac{d}{dr}\begin{bmatrix}\mathcal{P}(r,\omega)\\ \mathcal{V}(r,\omega)\end{bmatrix} = \begin{bmatrix}0 & \frac{s\rho_0}{A(r)}\\ \frac{sA(r)}{\eta_o P_o} & 0\end{bmatrix}\begin{bmatrix}\mathcal{P}(r,\omega)\\ \mathcal{V}(r,\omega)\end{bmatrix}. \tag{5.2.11}$$

The equations

$$M(r) = \rho_o/A(r) \quad \text{and} \quad C(r) = A(r)/\eta_o P_o \tag{5.2.12}$$

define the per-unit-length mass and compliance of the horn (Ramo et al. 1965, p. 213). The product of $\mathcal{P}(r,s)$ and $\mathcal{V}(r,s)$ defines the acoustic power [W/m^2], while their ratio defines the horn's admittance $Y_{in}^{\pm}(r,s) = \mathcal{V}^{\pm}/\mathcal{P}$, looking in the two directions (Pierce 1981, pp. 37–41).

To obtain the Webster horn pressure equation Eq. 5.2.10 from Eq. 5.2.11, we take the partial derivative of the top equation

$$-\frac{\partial^2 \mathcal{P}}{\partial r^2} = s\frac{\partial M(r)}{\partial r}\mathcal{V} + sM(r)\frac{\partial \mathcal{V}}{\partial r}$$

and then use the lower equation to remove $\partial \mathcal{V}/\partial r$,

$$\frac{\partial^2 \mathcal{P}}{\partial r^2} - s\frac{\partial M(r)}{\partial r}\mathcal{V} = s^2 M(r)C(r)\,\mathcal{P} = \frac{s^2}{c_o^2}\mathcal{P}.$$

Note that $c_o^2 = MC = \left(\frac{\rho_o}{\cancel{A(r)}}\right)\cdot\left(\frac{\cancel{A(r)}}{\eta_o P_o}\right)$. In air $c_o = \sqrt{\eta_0 P_o/\rho_o}$.

We must then use the upper equation a second time to remove $\mathcal{V}$:

$$\frac{\partial^2}{\partial r^2}\mathcal{P} + \frac{1}{A(r)}\frac{\partial A(r)}{\partial r}\frac{\partial}{\partial r}\mathcal{P} = \frac{s^2}{c_o^2}\mathcal{P}(r,s). \tag{5.2.13}$$

By use of the chain rule, equations of this form may be directly integrated, since

$$\begin{aligned}\nabla_r \mathcal{P} &= \frac{1}{A(r)}\frac{\partial}{\partial r}\left[A(r)\frac{\partial}{\partial r}\right]\mathcal{P}(r,s)\\ &= \frac{\partial^2}{\partial r^2}\mathcal{P}(r,s) + \frac{1}{A(r)}\frac{\partial A(r)}{\partial r}\mathcal{P}(r,s).\end{aligned} \tag{5.2.14}$$

This is equivalent to integration by parts, with integration factor $A(r)$. Finally we set $\kappa(s) \equiv s/c_o$, which later may be generalized to include visco-thermal losses (Eq. D.1.5).

Merging Eqs. 5.2.13 and 5.2.14 results in the Webster horn equation (WHEN) (Eq. 5.2.10):

$$\frac{1}{A(r)}\frac{\partial}{\partial r}A(r)\frac{\partial}{\partial r}\mathcal{P}(r,s) = \kappa^2(s)\,\mathcal{P}(r,s) \leftrightarrow \frac{1}{c_o^2}\frac{\partial^2}{\partial t^2}\varrho(r,t). \tag{5.2.15}$$

Equations having this form are known as *Sturm-Liouville equations*. This important class of ordinary differential equations follows from the use of separation of variables of the Laplacian in any (i.e., every) separable coordinate systems (Morse and Feshbach 1953, pp. 494–523). The frequency domain eigensolutions are denoted $\mathcal{P}^{\pm}(r, s)$.

We transform the three-dimensional acoustic wave equation into acoustic variables (Eq. 5.2.7) in Appendix H by the application of Gauss's law, resulting in the one-dimensional WHEN (Eq. 5.2.10), which is a nonsingular Sturm–Liouville equation.[7] It seems significant that the integration factor corresponds to the horn's area function. Thus we have demonstrated that Eqs. 5.2.7 and 5.2.11 reduce to Eq. 5.2.15 in a horn.

5.3 Problems VC-1

5.3.1 *Topics of This Homework*

Vector algebra and fields in $\mathbb{R}^3$, gradient and scalar Laplacian operators, definitions of divergence and curl, Gauss's (divergence) and Stokes's (curl) laws, system classification (postulates).

5.3.2 *Scalar Fields and the* ∇ *Operator*

Problem #1 *Let* $T(x, y) = x^2 + y$ *be an analytic scalar temperature field in two dimensions (single-valued* $\in \mathbb{R}^2$*).*

–1.1: Find the gradient of $T(\boldsymbol{x})$ *and make a sketch of* T *and the gradient.*

–1.2: Compute $\nabla^2 T(\boldsymbol{x})$ *to determine whether* $T(\boldsymbol{x})$ *satisfies Laplace's equation.*

–1.3: Sketch the iso-temperature contours at $T = -10, 0, 10$ *degrees.*

–1.4: The heat flux[8] *is defined as* $\boldsymbol{J}(x, y) = -\kappa(x, y)\nabla T$, *where* $\kappa(x, y)$ *is a constant that denotes thermal conductivity at the point* (x, y). *Given that* $\kappa = 1$ *everywhere (the medium is homogeneous), plot the vector* $\boldsymbol{J}(x, y) = -\nabla T$ *at* $x = 2$, $y = 1$. *Be clear about the origin, direction, and length of your result.*

---

[7] The Webster horn equation is closely related to Schrödinger's equation (Salmon 1946).

[8] The heat flux is proportional to the change in temperature times the thermal conductivity κ of the medium.

–1.5: Find the vector $\perp$ to $\nabla T(x, y)$—that is, tangent to the iso-temperature contours. Hint: Sketch it for one (x, y) point (e.g., 2, 1) and then generalize.

–1.6: The thermal resistance R_T is defined as the potential drop ΔT over the magnitude of the heat flux $|\boldsymbol{J}|$. At a single point the thermal resistance is

$$R_T(x, y) = -\nabla T/|\boldsymbol{J}|.$$

How is $R_T(x, y)$ related to the thermal conductivity $\kappa(x, y)$?

Problem #2 *Acoustic wave equation*
Note: In this problem, we will work in the frequency domain.
–2.1: The basic equations of acoustics in one dimension are

$$-\frac{\partial}{\partial x}\mathcal{P} = \rho_o s \vec{\mathcal{V}} \quad \textit{and} \quad -\frac{\partial}{\partial x}\vec{\mathcal{V}} = \frac{s}{\eta_o P_o}\mathcal{P}.$$

Here $\mathcal{P}(x, \omega)$ is the pressure (in the frequency domain), $\mathcal{V}(x, \omega)$ is the volume velocity (the integral of the velocity over the wavefront with area A), $s = \sigma + \omega \jmath$, $\rho_o = 1.2$ is the specific density of air, $\eta_o = 1.4$, and P_o is the atmospheric pressure (i.e., 10^5 Pa). Note that the pressure field $\mathcal{P}$ is a scalar (pressure does not have direction), while the volume velocity field $\vec{\mathcal{V}}$ is a vector (velocity has direction).

We can generalize these equations to three dimensions using the ∇ operator

$$-\nabla\mathcal{P} = \rho_o s \vec{\mathcal{V}} \quad \textit{and} \quad -\nabla\cdot\vec{\mathcal{V}} = \frac{s}{\eta_o P_o}\mathcal{P}.$$

–2.2: Starting from these two basic equations, derive the scalar wave equation in terms of the pressure $\mathcal{P}$,

$$\nabla^2\mathcal{P} = \frac{s^2}{c_0^2}\mathcal{P},$$

where c_0 is a constant representing the speed of sound.

–2.3: What is c_0 in terms of η_0, ρ_0, and P_0?

–2.4: Rewrite the pressure wave equation in the time domain using the time derivative property of the Laplace transform [e.g., $dx/dt \leftrightarrow sX(s)$]. For your notation, define the time-domain signal using a lowercase letter, $p(x, y, z, t) \leftrightarrow \mathcal{P}$.

5.3.3 Vector Fields and the ∇ Operator

Problem #3 *Let $\boldsymbol{R}(x, y, z) \equiv x(t)\hat{\mathbf{x}} + y(t)\hat{\mathbf{y}} + z(t)\hat{\mathbf{z}}$.*

–3.1: If a, b, and c are constants, what is $\boldsymbol{R}(x, y, z)\cdot\boldsymbol{R}(a, b, c)$?

–3.2: If a, b, and c are constants, what is $\frac{d}{dt}(\boldsymbol{R}(x, y, z)\cdot\boldsymbol{R}(a, b, c))$?

Problem #4 *Find the divergence and curl of the following vector fields:*
–4.1: $\boldsymbol{v} = \hat{\mathbf{x}} + \hat{\mathbf{y}} + 2\hat{\mathbf{z}}$
–4.2: $\boldsymbol{v}(x, y, z) = x\hat{\mathbf{x}} + xy\hat{\mathbf{y}} + z^2\hat{\mathbf{z}}$
–4.3: $\boldsymbol{v}(x, y, z) = x\hat{\mathbf{x}} + xy\hat{\mathbf{y}} + \log(z)\hat{\mathbf{z}}$
–4.4: $\boldsymbol{v}(x, y, z) = \nabla(1/x + 1/y + 1/z)$

5.3.4 Vector and Scalar Field Identities

Problem #5 *Find the divergence and curl of the following vector fields:*
–5.1: $\boldsymbol{v} = \nabla\phi$*, where* $\phi(x, y) = xe^y$
–5.2: $\boldsymbol{v} = \nabla \times \boldsymbol{A}$*, where* $\boldsymbol{A} = x\hat{\mathbf{x}} + y\hat{\mathbf{y}} + z\hat{\mathbf{z}}$
–5.3: $\boldsymbol{v} = \nabla \times \boldsymbol{A}$*, where* $\boldsymbol{A} = y\hat{\mathbf{x}} + x^2\hat{\mathbf{y}} + z\hat{\mathbf{z}}$
–5.4: For any differentiable vector field $\boldsymbol{V}$*, write two vector calculus identities that are equal to zero.*
–5.5: What is the most general form a vector field may be expressed in, in terms of scalar Φ *and vector* $\boldsymbol{A}$ *potentials?*

Problem #6 *Perform the following calculations. If you can state the answer without doing the calculation, explain why.*
–6.1: Let $\boldsymbol{v} = \sin(x)\hat{\mathbf{x}} + y\hat{\mathbf{y}} + z\hat{\mathbf{z}}$*. Find* $\nabla \cdot (\nabla \times \boldsymbol{v})$*.*
–6.2: Let $\boldsymbol{v} = \sin(x)\hat{\mathbf{x}} + y\hat{\mathbf{y}} + z\hat{\mathbf{z}}$*. Find* $\nabla \times (\nabla\sqrt{\boldsymbol{v} \cdot \boldsymbol{v}})$
–6.3: Let $\boldsymbol{v}(x, y, z) = \nabla(x + y^2 + \sin(\log(z))$*. Find* $\nabla \times \boldsymbol{v}(x, y, z)$*.*

5.3.5 Integral Theorems

Problem #7 *For each of the following problems, in a few words, identify either Gauss's or Stokes's law, define what it means, and explain the formula that follows the question.*
–7.1: What is the name of this formula?

$$\int_S \hat{\mathbf{n}} \cdot \boldsymbol{v}\, dA = \int_{\mathcal{V}} \nabla \cdot \boldsymbol{v}\, dV.$$

–7.2: What is the name of this formula?

$$\int_S (\nabla \times \boldsymbol{V}) \cdot d\boldsymbol{S} = \oint_C \boldsymbol{V} \cdot d\boldsymbol{R}$$

Give one important application.
–7.3: Describe a key application of the vector identity

$$\nabla \times (\nabla \times \vec{V}) = \nabla(\nabla \cdot \vec{V}) - \nabla^2\vec{V}.$$

5.3.6 System Classification

Problem #8 *Complete this system classification problem about physical systems using the system postulates.*

–8.1: Provide a brief definition of these classificatioms

L/NL: *linear (L)/nonlinear (NL)*
TI/TV: *time-invariant (TI)/time-varying (TV)*
P/A : *passive (P)/active (A)*
C/NC : *causal (C)/noncausal (NC)*
Re/Clx : *real (Re)/complex (Clx)*

–8.2: Along the rows of the table, classify each system using the abbreviations L/NL, TI/TV, P/A, C/NC, and Re/Clx:

					Category		
#	Case	Definition	L/NL	TI/TV	P/A	C/NC	Re/Clx
1	Resistor	$v(t) = r_0\, i(t)$					
2	Inductor	$v(t) = L\frac{di}{dt}$					
3	Switch	$v(t) \equiv \begin{cases} 0 & t \le 0 \\ V_0 & t > 0. \end{cases}$					
5	Transistor	$I_{out} = g_m(V_{in})$					
7	Resistor	$v(t) = r_0\, i(t+3)$					
8	Modulator	$f(t) = e^{i2\pi t} g(t)$					

–8.3: Classify each equation:

#	Case:	L/NL	TI/TV	P/A	C/NC	Re/Clx
1	$A(x)\frac{d^2y(t)}{dt^2} + D(t)y(x,t) = 0$					
2	$\frac{dy(t)}{dt} + \sqrt{t}\; y(t) = \sin(t)$					
3	$y^2(t) + y(t) = \sin(t)$					
4	$\frac{\partial^2 y}{\partial t^2} + xy(t+1) + x^2y = 0$					
5	$\frac{dy(t)}{dt} + (t-1)\, y^2(t) = ie^t$					

5.4 Three Examples of Finite-Length Horns

Figure 5.3 is taken from the classic book by Olson (1947, p. 101), showing the theoretical radiation impedance $Z_{rad}(r, \omega)$ for five horns. Table 5.2 summarizes the properties of four of these horns: uniform (cylindrical) ($A = A_o$), parabolic ($A(r) = A_o r$), conical (spherical) ($A(r) = A_o r^2$), and exponential ($A(r) = A_o e^{2mr}$).

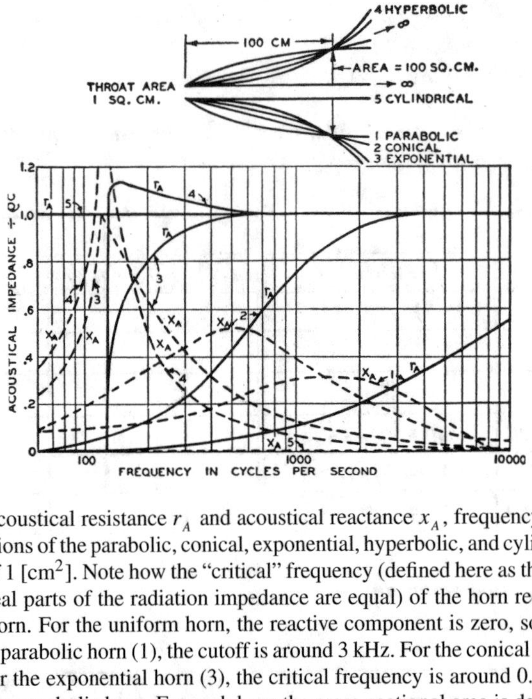

Fig. 5.3 Throat acoustical resistance r_A and acoustical reactance x_A, frequency characteristics of infinite eigenfunctions of the parabolic, conical, exponential, hyperbolic, and cylindrical horns, having a throat area of 1 [cm^2]. Note how the "critical" frequency (defined here as the frequency where the reactive and real parts of the radiation impedance are equal) of the horn reduces dramatically with the type of horn. For the uniform horn, the reactive component is zero, so there is no cutoff frequency. For the parabolic horn (1), the cutoff is around 3 kHz. For the conical horn (2), the cutoff is at 0.6 [kHz]. For the exponential horn (3), the critical frequency is around 0.18 [kHz], which is one-16th that of the parabolic horn. For each horn the cross-sectional area is defined as 100 [cm^2] at a distance of $L = 1$ [m] from the throat (Olson 1947, p. 101); (Morse 1948, p. 283)

Table 5.2 Horns and their properties for $N = 1, 2$, and 3 dimensions, along with the exponential horn (*EXP*). The range variable goes from $r_o \le r \le L$ [m] with area $1 \le A(r) \le 100$ [cm^2]. $F(r)$ is the coefficient on $\mathcal{P}_x$, $\kappa(s) \equiv s/c_o$, where c_o is the speed of sound and $s = \sigma + \omega \jmath$ is the Laplace frequency. The horn's eigenfunctions are $\mathcal{P}^{\pm}(\xi, s) \leftrightarrow \varrho^{\pm}(\xi, t)$. When $\pm$ is indicated, the outbound solution corresponds to the negative sign. Eigenfunctions $H_0^{\pm}(\xi, s)$ are outbound and inbound Hankel functions. The rightmost column is the input radiation admittance normalized by the *characteristic admittance* $\mathcal{Y}_r(r) = A(r)/\rho_o c_o$.

N	Name	radius	Area/A_o	$F(r)$	$\mathcal{P}^{\mp}(r, s)$	$\varrho^{\mp}(r_o, t)$	$Y_{rad}^{\mp}/\mathcal{Y}_r$
1D	uniform	1	1	0	$e^{\mp\kappa(s)r}$	$\delta(t)$	1
2D	parabolic	$\sqrt{r}$	r	$1/r$	$H_0^{\mp}(-\jmath\kappa(s)r)$	—	$\frac{-\jmath r_o H_1^{\mp}}{H_0^{\mp}}$
3D	conical	r	r^2	$2/r$	$e^{\mp\kappa(s)r}/r$	$\delta(t) \pm \frac{c_o}{r_o} u(t)$	$1 \pm c_o/s r_o$
EXP	exponential	e^{mr}	e^{2mr}	$2m$	$e^{-\left(m \mp \sqrt{m^2+\kappa^2}\right)r}$	$e^{-mr} E(t)$	Eq. 5.4.11

5.4.1 *Uniform Horn*

The one-dimensional wave equation $[A(r) = A_o = 1\ [\text{cm}^2]]$ is

$$\frac{d^2}{dr^2}\,\mathcal{P} = \kappa^2(s)\,\mathcal{P},$$

where we set $\kappa^2(s) \equiv s^2/c_o^2$.

5.4.1.1 Solution

The two eigenfunctions of this equation are the two d'Alembert waves (Eq. 4.4.1):

$$\varrho(x,t) = \mathcal{P}_0^+\varrho^+(t - x/c) + \mathcal{P}_0^-\varrho^-(t + (x-L)/c) \;\leftrightarrow\; \mathcal{P}_0^+e^{-\kappa(s)x} + \mathcal{P}_0^-e^{\kappa(s)(x-L)},$$

where $\mathcal{P}_0^\pm \in \mathbb{C}$ are wave amplitudes and $\kappa(s) = s/c_o = \jmath\omega/c$ is called the *propagation function* (also known as the wave-evolution function, propagation constant, and wave number).

Note that for the uniform lossless horn, $\omega/c_o = 2\pi/\lambda$. It is convenient to normalize $\mathcal{P}_0^+ = 1$ and $\mathcal{P}_L^- = 1$.

The characteristic admittance $\mathcal{Y}_r(x)$ (Table 5.2) is independent of direction. The signs must be physically chosen, with the velocity $\mathcal{V}^\pm$ into the port, to ensure that $\mathcal{Y}_r > 0$.

5.4.1.2 Applying the Boundary Conditions

The general solution in terms of the eigenvector matrix, evaluated at $x = L$, is

$$\begin{bmatrix}\mathcal{P}(x)\\ \mathcal{V}(x)\end{bmatrix}_L = \begin{bmatrix}e^{-\kappa x} & e^{\kappa(x-L)}\\ \mathcal{Y}_r e^{-\kappa x} & -\mathcal{Y}_r e^{\kappa(x-L)}\end{bmatrix}_L \begin{bmatrix}\mathcal{P}_0^+\\ \mathcal{P}_0^-\end{bmatrix}_L = \begin{bmatrix}e^{-\kappa L} & 1\\ \mathcal{Y}_r e^{-\kappa L} & -\mathcal{Y}_r\end{bmatrix}\begin{bmatrix}\mathcal{P}_0^+\\ \mathcal{P}_0^-\end{bmatrix}_L, \tag{5.4.1}$$

where $\mathcal{P}_0^+$ and $\mathcal{P}_0^-$ are the relative amplitudes of the two unknown eigenfunctions to be determined by the boundary conditions at $x = 0, L$, $\kappa = s/c$, and $\mathcal{Y}_r = 1/\mathcal{Z}_r = A_o/\rho_o c$.

Solving Eq. 5.4.1 for $\mathcal{P}_0^+$ and $\mathcal{P}_0^-$ with determinant $\Delta = -2\mathcal{Y}_r e^{-\kappa L}$, we get

$$\begin{bmatrix}\mathcal{P}_L^+\\ \mathcal{P}_L^-\end{bmatrix} = \frac{-1}{2\mathcal{Y}_r e^{-\kappa L}}\begin{bmatrix}-\mathcal{Y}_r & -1\\ -\mathcal{Y}_r e^{-\kappa L} & e^{-\kappa L}\end{bmatrix}\begin{bmatrix}\mathcal{P}\\ \mathcal{V}\end{bmatrix}_L = \frac{1}{2}\begin{bmatrix}e^{\kappa L} & -\mathcal{Z}e^{\kappa L}\\ 1 & \mathcal{Z}\end{bmatrix}\begin{bmatrix}\mathcal{P}\\ -\mathcal{V}\end{bmatrix}_L. \tag{5.4.2}$$

In the final step we swapped all the signs, including those on $\mathcal{V}$, and moved $\mathcal{Z}_r = 1/\mathcal{Y}_r$ inside the matrix.

We can uniquely determine these two weights $\mathcal{P}_L^+$, $\mathcal{P}_L^-$ given the pressure and velocity at the boundary $x = L$, which is typically determined by the load impedance ($Z_L(s) = \mathcal{P}_L/\mathcal{V}_L$).

The weights may now be substituted back into Eq. 5.4.1 to determine the pressure and velocity amplitudes at any point $0 \le x \le L$:

$$\begin{bmatrix} \mathcal{P} \\ \mathcal{V} \end{bmatrix}_x = \frac{1}{2}\begin{bmatrix} e^{-\kappa x} & e^{\kappa(x-L)} \\ \mathcal{Y}_r e^{-\kappa x} & -\mathcal{Y}_r e^{\kappa(x-L)} \end{bmatrix}_x \begin{bmatrix} e^{\kappa L} & -\mathcal{Z} e^{\kappa L} \\ 1 & \mathcal{Z} \end{bmatrix} \begin{bmatrix} \mathcal{P} \\ -\mathcal{V} \end{bmatrix}_L . \tag{5.4.3}$$

Setting $x = 0$ and multiplying these out give the final transmission matrix:

$$\begin{bmatrix} \mathcal{P} \\ \mathcal{V} \end{bmatrix}_0 = \frac{1}{2}\begin{bmatrix} e^{\kappa L} + e^{-\kappa L} & \mathcal{Z}_r(e^{\kappa L} - e^{-\kappa L}) \\ \mathcal{Y}_r(e^{\kappa L} - e^{-\kappa L}) & e^{\kappa L} + e^{-\kappa L} \end{bmatrix}_x \begin{bmatrix} \mathcal{P} \\ -\mathcal{V} \end{bmatrix}_L . \tag{5.4.4}$$

Note that the diagonal terms are $\cosh \kappa L$ and the off-diagonal terms are $\sinh \kappa L$.

Applying the last boundary condition, we evaluate Eq. 5.4.2 to obtain the ABCD matrix at the input ($x = 0$) (Pipes 1958),

$$\begin{bmatrix} \mathcal{P} \\ \mathcal{V} \end{bmatrix}_0 = \begin{bmatrix} \cosh \kappa L & \mathcal{Z}_r \sinh \kappa L \\ \mathcal{Y}_r \sinh \kappa L & \cosh \kappa L \end{bmatrix} \begin{bmatrix} \mathcal{P} \\ -\mathcal{V} \end{bmatrix}_L . \tag{5.4.5}$$

Note that the uniform horn is reversible (P7) and reciprocal (P6).

Exercise #11 Evaluate the expression in terms of the load impedance.

Solution: Since $Z_{\text{load}} = -\mathcal{P}_L/\mathcal{V}_L$, we have

$$\left.\frac{\mathcal{P}}{\mathcal{V}}\right|_0 = \frac{Z_{\text{load}} \cosh \kappa L - \mathcal{Z}_r \sinh \kappa L}{Z_{\text{load}} \mathcal{Y}_r \sinh \kappa L - \cosh \kappa L} \tag{5.4.6}$$

■

5.4.1.3 Impedance Matrix

Equation 5.4.5 is an impedance matrix (algebra required)

$$\begin{bmatrix} \mathcal{P}_0 \\ \mathcal{P}_L \end{bmatrix} = \frac{\mathcal{Z}_r}{\sinh(\kappa L)} \begin{bmatrix} \cosh(\kappa L) & 1 \\ 1 & \cosh(\kappa L) \end{bmatrix} \begin{bmatrix} \mathcal{V}_0 \\ \mathcal{V}_L \end{bmatrix} .$$

Exercise #12 Write out the short-circuit ($\mathcal{V}_L = 0$) input impedance $Z_{in}(s)$ for the uniform horn.

Solution:

$$Z_{in}(s) = \left.\frac{\mathcal{P}}{\mathcal{V}}\right|_{\mathcal{V}_L=0} = \mathcal{Z}_r \frac{\cosh \kappa L}{\sinh \kappa L} = \mathcal{Z}_r \tanh \kappa L|_{\mathcal{V}_L=0} .$$

5.4.1.4 Input Admittance Y_{in}

■

Given the input admittance of the horn, it is possible to determine whether it is uniform without further analysis. That is, if the horn is uniform and infinite in length, the input admittance at $x = 0$ is

$$Y_{in}(x = 0, s) \equiv \frac{\mathcal{V}(0, \omega)}{\mathcal{P}(0, \omega)} = \mathcal{Y}_r,$$

since $\mathcal{P}_0^+ = 1$ and $\mathcal{P}_L^- = 0$. For an infinite uniform horn, there are no reflections.

When the uniform horn is terminated with a fixed impedance $\mathcal{Z}_r$ at $x = L$, we can substitute pressure and velocity measurements into Eq. 5.4.2 to find $\mathcal{P}_0$ and $\mathcal{P}_0$, and given these, we can calculate the pressure reflectance at $x = L$ (Eq. 3.4.6),

$$\Gamma_L(s) = \frac{\mathcal{P}_L}{\mathcal{P}_0} = \frac{\mathcal{P}(L, \omega) - \mathcal{Z}_r\, \mathcal{V}(L, \omega)}{\mathcal{P}(L, \omega) + \mathcal{Z}_r\, \mathcal{V}(L, \omega)} = \frac{Z_L - \mathcal{Z}_r}{Z_L + \mathcal{Z}_r}.$$

Given sufficiently accurate measurements of the throat pressure and assuming $\Gamma_L = 0$, the input impedance Z_{in} at the input $\boldsymbol{x} = 0$ is $\mathcal{Z}_r = \rho_o c/A_o$.

5.4.2 *Conical Horn*

Using the conical horn area $A(r) \propto r^2$ in Eq. 5.2.10 (or Eq. 5.2.11) results in the spherical wave equation

$$\mathcal{P}_{rr}(r, \omega) + \frac{2}{r}\, \mathcal{P}_r(r, \omega) = \kappa^2\, \mathcal{P}(r, \omega), \tag{5.4.7}$$

where $\kappa(s) = \pm s/c_o$. The eigensolutions of Eq. 5.4.7 are

$$\mathcal{P}^{\pm}(r, s) = \frac{e^{\mp\kappa r}}{r} \leftrightarrow \frac{1}{r}\delta(t \mp r/c_o).$$

5.4.2.1 Radiation Admittance for the Conical Horn

The conical horn's acoustic input admittance $Y_{in}(r, s)$ at any location r is found by dividing $\mathcal{V}(r, s)$ by $\mathcal{P}(r, s)$:

$$Y_{in}^{\pm}(r, s) = \frac{\mathcal{V}^{\pm}}{\mathcal{P}^{\pm}} = -\frac{A(r)}{s\rho_o}\frac{d}{dr} \ln \mathcal{P}^{\pm}(r, s) \tag{5.4.8}$$

$$= \mathcal{Y}_r(r)\left[1 \pm \frac{c_o}{sr}\right] \leftrightarrow \frac{A(r)}{\rho_o c_o}\left(\delta(t - r/c_o) \pm \frac{c_o}{r} u(t - r/c_o)\right). \tag{5.4.9}$$

The pressure pulse is delayed by r/c_o due to $e^{-\kappa(s)r}$. As the area of the horn increases, the pressure decreases as $1/r = 1/\sqrt{A(r)}$. This results in the uniform backflow $c_o u(t)/r \leftrightarrow c_o/sr$ due to conservation of mass, and the characteristic admittance $\mathcal{Y}_r(r)$ variation with r.

5.4.3 Exponential Horn

If we define the area as $A(r) = A_o e^{2mr}$, the eigenfunctions of the horn are

$$\mathcal{P}^{\pm}(r, \omega \jmath) = e^{-mr} e^{\mp j\sqrt{\omega^2-\omega_c^2}\, r/c}, \tag{5.4.10}$$

which may be shown by the substitution of $\mathcal{P}_c^{\pm}(r, \omega \jmath)$ into Eq. 5.2.10.

This case is of special interest because the radiation impedance is purely reactive below the horn's cutoff frequency ($\omega < \omega_c = mc_o$), as may be seen from curves 3 and 4 of Fig. 5.3. As a result, no energy can radiate from an open horn for $\omega < \omega_c$ because

$$\kappa(s) = -m \pm \frac{\jmath}{c_o}\sqrt{\omega^2 - \omega_c^2} = -m \mp \frac{1}{c_o}\sqrt{\omega_c^2 - \omega^2}$$

is purely real (this is the case of nonpropagating evanescent waves).

If we use Eq. 4.4.7, the input admittance is

$$Y_{in}^{\pm}(x, s) = -\frac{A(x)}{s\rho_o}\left(m \pm \sqrt{m^2 + \kappa^2}\right)x. \tag{5.4.11}$$

Kleiner (2013) gives an equivalent expression for $Y_{in}(x, \omega)$ given area $S(x) = e^{mx}$,

$$Y_{in}(x, \omega) = \frac{S(x)}{\jmath\omega\rho}\left[\frac{m}{2} + \jmath\frac{\sqrt{4\omega^2 - (mc)^2}}{2c}\right],$$

and impedance

$$Z_{in}(r, \omega) = \frac{\rho c}{S_T}\left[\jmath\frac{\omega_c}{\omega} + \sqrt{1 - \left(\frac{\omega_c}{\omega}\right)^2}\right],$$

where $\omega_c(r)$ is the cutoff frequency. Given this exact solution to the exponential horn, we could use a series expansion of the form

$$A(r) = \sum_k a_k e^{b_k x} \quad \in \mathbb{R}$$

to obtain the general solution, for an arbitrary analytic $A(r)$.

5.5 Solution Methods

To model the wave equation, two distinct mathematical techniques are described. The first of these is called separation of variables. This method is limited to a small and restrictive number of separable coordinate systems (SCS). Once the SCS is chosen, the eigenfunctions are known as solutions to that specific Sturm-Liouville equation, which are always scalar (ordinary) differential equation (ODEs).

The second method uses the transmission matrix (a lumped-parameter method). This solution method assumes a limit on the upper frequency response. That is, quasistatics assumes the wavelength is larger than the size of the object being modeled. Thus other than the upper frequency limit, there are no limitations in the analytical and numerical solutions using the transmission matrix method. Furthermore, it may be computed in both the frequency and time domains (Fettweis 1986).

1a. **Separable coordinate systems:** Classically PDEs are solved by *separation of variables*. This method is limited to a few ortho-normal coordinate systems, such as rectangular, cylindrical, and spherical coordinates (Morse 1948, pp. 296–97). Even a slight deviation from separable specific coordinate systems represents a major barrier toward further analysis and understanding, blocking insight into more general cases. Separable coordinate systems have a high degree of symmetry. Note that the solution of the wave equation is not tied to a specific coordinate system.

1b. **Sturm–Liouville methods and eigenvectors:** When the coordinate system is separable, the resulting PDEs are always reduced to a system of Sturm–Liouville equations. The solutions of this important class of Sturm–Liouville eigenfunctions are all tabulated.

    Webster horn theory (Webster 1919; Morse 1948; Pierce 1981) is a generalized Sturm–Liouville equation that adds its physics in the form of the horn's area function. The Webster equation sidesteps the seriously limiting problem of separation of variables by using the alternative quasistatic solution, which ignores the nonpropagating high-frequency evanescent modes. This is essentially a one-dimensional low-pass approximation to the wave equation.

    While mathematics provides rigor, physics provides understanding. While both are important, it is the physical applications than make the theory useful.

2. **Lumped-element method:** As described (see Sect. 3.8) a system may be represented in terms of lumped elements, as either electrical inductors, capacitors, and resistors or their mechanical counterparts, masses, springs, and dashpots. Such systems are represented by 2×2 transmission matrices in the s (i.e., Laplace) domain (Ramo et al. 1965, Appendix IV).

   When a system of lumped-element networks contains only resistors and capacitors or resistors and inductors, the solution is a diffusion equation, which does not support propagated waves. Depending on the elements in the system of equations, there can be an overlap between a diffusion process and scalar waves, represented as transmission lines, both modeled as lumped-element networks of 2×2 matrices (Eq. 3.8.1) (Campbell 1922; Brillouin 1953; Ramo et al. 1965).

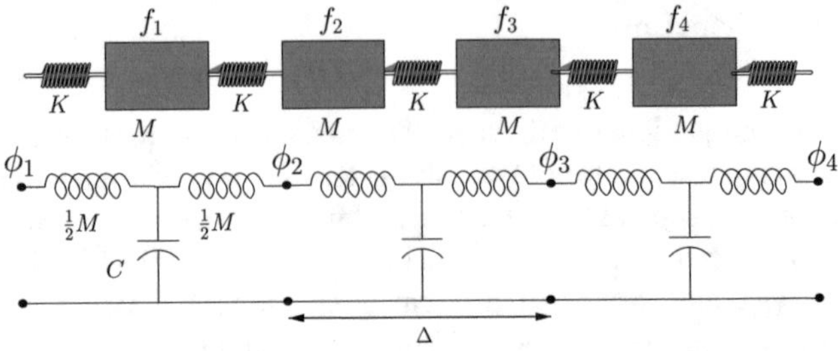

Fig. 5.4 Depiction of a train consisting of cars, treated as a mass M and linkages, treated as springs of stiffness K or compliance $C = 1/K$. The equivalent electrical circuit is shown below the mass-spring system, with the masses modeled by inductors (M) and springs modeled as capacitors (C). For this model to accurately represent a transmission line the frequency must be less than the equivalent Nyquist frequency f_c. The delay of one cell is $T_o = \Delta/c_o$. One can measure either T_o or Δ and $c_o = 1/\sqrt{MC}$. Since $\lambda_c = 2\Delta$, it follows that the cutoff frequency is $f_c = c_o/2\Delta = 1/2T_o$. For $f < f_c$, the frequency response is independent of frequency, thus acting as a pure delay. As the frequency is increased above f_c, the wavelength becomes shorter that the critical wavelength $\lambda_c = 2\Delta$, and the delay line becomes a low-pass filter, strongly departing from transmission line properties

Nyquist sampling and quasistatics: Quasistatic methods provide band-limited solutions below a critical frequency f_c for a much wider class of geometries, by avoiding high-frequency cross-modes. The model of a train is depicted in Fig. 5.4.

Example: Train-mission-line problem. The mechanical mass-spring system of Fig. 5.4 is the electrical equivalent circuit. The mass is modeled as an inductor and the springs as capacitors to ground. The velocity is analogous to a current and the force $f_n(t)$ to the voltage $\phi_n(t)$. The length of each cell is Δ [m].

When the wavelength $\lambda = c_o/f_c$ is greater than twice the physical distance Δ between the elements

$$\lambda > \lambda_c = 2\Delta \quad [\text{m}],$$

the approximation is mathematically equivalent to a transmission line. As described in DE-3, Problem #2, the velocity is $c_o = 1/\sqrt{MC}$ [m/s]. As the frequency increases, the wavelength becomes shorter. When the frequency is equal to the critical frequency f_c the critical wavelength $\lambda_c = c_o/f_c = 2\Delta$. Above the critical frequency the quasistatic (lumped-element) model breaks down and transitions from a delay line to a low-pass filter, as discussed in DE-3, Problem #2.

The frequency is under the control of the modeling process, since more elements may be added to allow for higher frequencies (shorter wavelengths). If the nature of the solution at high frequencies ($f > f_c$) is desired, we may add more sections. For many (perhaps most) problems, lumped elements are easy to use and accurate, as long as we don't violate the Nyquist condition (Brillouin 1953; Ramo et al. 1965).

5.5.1 *Eigenfunctions* $\varrho^{\pm}(r, t)$ *of the WHEN*

Because the wave equation (Eq. 5.2.7) is second order in time, there are two causal independent eigenfunction solutions: an outbound (right–traveling) $\varrho^{+}(r, t)$ wave, and an inbound (left-traveling) $\varrho^{-}(r, t)$ wave.

Every eigenfunction depends on an area function $A(r)$ (Eq. 5.2.10). In theory then, given an eigenfunction, it should be possible to find the area $A(r)$. This is known as the *inverse problem*, which is generally believed to be a difficult problem. Specifically, given the eigenvalues λ_k, how does one determine the corresponding area function $A(r)$?

Because the characteristic impedance $\mathcal{Y}_r(r)$ of the wave in the horn changes with location, there are local reflections due to these area variations. Thus there are fundamental relationships between the area change $dA(r)/dr$, the horn's eigenfunctions $\mathcal{P}^{\pm}(r, s)$, the eigenmodes, and the input impedance.

5.5.1.1 Complex Versus Real Frequency

We shall continue to maintain the distinction that functions of ω are Fourier transforms and causal functions of Laplace frequency s correspond to Laplace transforms, which are necessarily complex analytic in s in the right half-plane (RHP) region of convergence (RoC). This distinction is critical, since we typically describe impedance $Z(s)$ and admittance $Y(s)$ as complex analytic functions in s in terms of their poles and zeros. The eigenfunctions $\mathcal{P}^{\pm}(r, s)$ of Eq. 5.2.10 are also causal complex analytic functions of s.

5.5.1.2 Plane-Wave Eigenfunction Solutions

In 1690, nine years before Newton's publication of *Principia*, Christiaan Huygens was the first to gain insight into wave propagation, today known as *Huygens's principle*. While his concept showed a deep insight, we now know it was flawed, as it ignored the backward-traveling wave (Miller 1991). In 1747 d'Alembert published the first correct solution for theplane-wave scalar wave equation,

$$\varrho(x, t) = f(t - x/c_o) + g(t + x/c_o), \tag{5.5.1}$$

where $f(\cdot)$ and $g(\cdot)$ are general functions of their argument. That this is the solution may be shown by use of the chain rule, by taking partials with respect to x and t.

In terms of physics, d'Alembert's general solution describes two arbitrary waveforms $f(\cdot)$ and $g(\cdot)$ traveling at a speed c_o, one forward and one reversed. This solution is quite easily visualized.

Exercise #13 By the use of the chain rule, prove that d'Alembert's formula satisfies the one-dimensional wave equation.

Solution: Taking a derivative with respect to t and r gives

$$\partial_t \varrho(r,t) = -c_o f'(r - c_o t) + c_o g'(r + c_o t)$$
$$\partial_r \varrho(r,t) = f'(r - c_o t) + g'(r + c_o t),$$

and a second derivative gives

$$\partial_{tt} \varrho(r,t) = c_o^2 f''(r - c_o t) + c_o^2 g''(r + c_o t)$$
$$\partial_{rr} \varrho(r,t) = f''(r - c_o t) + g''(r + c_o t).$$

From these last two equations we have the one-dimensional wave equation

$$\partial_{rr} \varrho(r,t) = \frac{1}{c_o^2} \partial_{tt} \varrho(r,t),$$

which has solutions in Eq. 5.5.1. ■

Exercise #14 Assuming $f(\cdot)$ and $g(\cdot)$ are $\delta(\cdot)$, find the Laplace transform of the solution corresponding to the uniform horn $A(x) = 1$.

Solution: Using Table 3.9 of Laplace transforms on Eq. 5.5.1 gives

$$\varrho(x,t) = \delta(t - x/c_o) + \delta(t + x/c_o) \leftrightarrow e^{-sx/c_o} + e^{sx/c_o}. \tag{5.5.2}$$

Note that the delay $T_o = \pm x/c_o$ depends on the range x. ■

5.5.1.3 Three-Dimensional D'Alembert Spherical Eigenfunctions

We can generalize the d'Alembert solution to spherical waves by changing the area function of Eq. 5.2.10 to $A(r) = A_o r^2$ (see Eq. 5.1.9 and Table 5.2). The wave equation then becomes

$$\nabla_r^2 \varrho(r,t) = \frac{1}{r} \frac{\partial^2}{\partial r^2} r\, \varrho(r,t) = \frac{1}{c_o^2} \frac{\partial^2}{\partial t^2} \varrho(r,t).$$

Multiplying by r results in the general spherical (3D) d'Alembert wave equation solution

$$\varrho(r,t) = \frac{f(t - r/c_o)}{r} + \frac{g(t + r/c_o)}{r}$$

for arbitrary waveforms $f(\cdot)$ and $g(\cdot)$. These are the eigenfunctions for the spherical scalar wave equation.

5.6 Integral Definitions of ∇(), ∇·(), ∇ × (), and ∇ ∧ ()

In Sect. 5.2, we described two forms of wave equations, scalar and vector. Up to now we have only discussed the scalar case. The vector wave equation describes the evolution of a vector field, such as Maxwell's electric field vector $\boldsymbol{E}(\boldsymbol{x}, t)$.

There are two equivalent definitions for each of the four operators: differential and integral. The integral form provides a more intuitive view of the operator, which in the limit converges to the differential form. Following a discussion of the gradient, divergence, and curl integral operators, we discuss these two forms.

In addition there are three fundamental vector theorems: Gauss's law (divergence theorem), Stokes's law (curl theorem), and Helmholtz's decomposition theorem. Without the use of these fundamental vector calculus theorems, we could not understand Maxwell's equations.

5.6.1 *Gradient:* $E = -\nabla\phi(x)$

As shown in Fig. 5.1, the gradient maps $\mathbb{R}^1 \underset{\nabla}{\mapsto} \mathbb{R}^3$. The gradient is defined as the unit-normal $\hat{\mathbf{n}}$ weighted by the potential $\phi(\boldsymbol{x})$ averaged over a closed surface $\mathcal{S}$,

$$\nabla\phi(\boldsymbol{x}) \equiv \lim_{\mathcal{S},\mathcal{V}\to 0} \left\{ \frac{\iint_{\mathcal{S}} \phi(\boldsymbol{x})\,\hat{\mathbf{n}}\,d\mathcal{S}}{\mathcal{V}} \right\} \quad [\mathrm{V/m}], \tag{5.6.1}$$

having area $\mathcal{S}$ and volume $\mathcal{V}$ and centered at $\boldsymbol{x}$ (Greenberg 1988, p. 773).[9] Here $\hat{\mathbf{n}}$ is a dimensionless unit vector perpendicular to the surface $\mathcal{S}$:

$$\hat{\mathbf{n}} = \frac{\nabla\phi}{||\nabla\phi||}. \tag{5.6.2}$$

The dimensions of Eq. 5.6.1 are in the units of the potential times the area, divided by the volume, as needed for a gradient (e.g., [V/m]). The units depend on the potential. If ϕ were temperature, the units would be [deg/m].

Exercise #15 Justify the units of Eq. 5.6.1.

Solution: The units depend on ϕ per unit length. If ϕ is voltage, then the gradient has units of [V/m]. Under the limit, $d|\mathcal{S}|/||\mathcal{S}||$ must have units of m^{-1}. ■

The natural way to define the surface and volume is to place the surface on the isopotential surfaces, forming either a cube or a pill-box-shaped volume. As the volume $||\mathcal{S}||$ goes to zero, so must the area $|\mathcal{S}|$. One must avoid irregular volumes where the area is finite as the volume goes to zero (Greenberg 1988, footnote p. 762).

A well-known example is the potential

---

[9]For further discussions, see Greenberg (1988, pp. 778, 791, 809).

$$\phi(x, y, z) = \frac{Q}{\epsilon_o \sqrt{x^2 + y^2 + z^2}} = \frac{Q}{\epsilon_o R} \quad \text{[V]}$$

around a point charge Q [SI units of coulombs]. The constant ϵ_o is the permittivity [F/m^2]. A second well-known example is the acoustic pressure potential around an oscillating sphere, which has the same form (see Table 5.2).

5.6.1.1 How Does This Work?

To better understand what Eq. 5.6.1 means, consider a three-dimensional Taylor series expansion (See Eq. 5.2.4) of the potential in $\boldsymbol{x}$ about the limit point $\boldsymbol{x}_o$:

$$\phi(\boldsymbol{x}) \approx \phi(\boldsymbol{x}_o) + \nabla\phi(\boldsymbol{x}) \cdot (\boldsymbol{x} - \boldsymbol{x}_o) + \text{HOT}.$$

We could define the gradient using this relationship as

$$\nabla\phi(\boldsymbol{x}_o) = \lim_{\boldsymbol{x}\to\boldsymbol{x}_o} \frac{\phi(\boldsymbol{x}) - \phi(\boldsymbol{x}_o)}{\boldsymbol{x} - \boldsymbol{x}_o}.$$

For this definition to apply, $\boldsymbol{x}$ must approach $\boldsymbol{x}_o$ along $\hat{\mathbf{n}}$. To compute the higher order terms (HOT), we need the Hessian matrix.[10]

The natural way to define a surface $|\mathcal{S}|$ is to find the isopotential contours. The gradient is in the direction of maximum change in the potential, thus perpendicular to the isopotential surface. The key to the integral definition is in taking the limit. As the volume $||\mathcal{S}||$ shrinks to zero, the HOT are small and the integral reduces to the first-order term in the Taylor expansion, since the constant term integrates to zero. Such a construction was used in the proof of the Webster horn equation (Appendix H; Fig. H.1).

One major problem with Eq. 5.6.1 is that this definition is self-referencing, since $\hat{\mathbf{n}}$ is based on the gradient. Thus the integral definition of the gradient is based on the gradient. Equation 5.6.1 is similar to the mean value theorem for the gradient.

5.6.2 *Divergence: $\nabla \cdot \boldsymbol{D} = \rho$ [C/m^3]*

The definition of the divergence at $\boldsymbol{x} = [x, y, z]^T$ is (see Eq. 5.1.3)

$$\nabla\cdot\boldsymbol{D}(\boldsymbol{x}, t) \equiv \left[\partial_x, \partial_y, \partial_z\right] \cdot \boldsymbol{D}(\boldsymbol{x}, t) = \left[\frac{\partial D_x}{\partial x} + \frac{\partial D_y}{\partial y} + \frac{\partial D_z}{\partial z}\right](\boldsymbol{x}, t) = \rho(\boldsymbol{x}, t),$$

which maps $\mathbb{R}^3 \underset{\nabla\cdot}{\mapsto} \mathbb{R}^1$.

[10] $H_{i,j} = \partial^2(\boldsymbol{x})\phi/\partial x_i \partial x_j$, which exists if the potential is analytic in $\boldsymbol{x}$ at $\boldsymbol{x}_o$.

5.6.2.1 Divergence and Gauss's Law

Like the gradient, the divergence of a vector field may be defined as the surface integral of a compressible vector field in the limit as the volume enclosed by the surface goes to zero. As in the case of the gradient, if this definition is to make sense, the surface $\mathcal{S}$ must be closed, defining the volume $\mathcal{V}$. The difference here is that the surface integral is over the normal component of the vector field being operated on (Greenberg 1988, pp. 762–763)

$$\nabla\cdot\boldsymbol{D} = \lim_{\mathcal{V},\mathcal{S}\to 0}\left\{\frac{\iint_{\mathcal{S}} \boldsymbol{D}\cdot\hat{\mathbf{n}}\, d\mathcal{S}}{\mathcal{V}}\right\} = \rho(x,y,z) \quad [\mathrm{C/m^3}]. \tag{5.6.3}$$

As with the gradient, we have defined the surface with area $\mathcal{S}$, and volume $\mathcal{V}$. It is a necessary condition that as the area $\mathcal{S}$ goes to zero, so does the volume $\mathcal{V}$.

As defined previously (Eq. 5.6.2) and shown here in Fig. 5.5, $\hat{\mathbf{n}}$ is a unit vector normal to the closed isopotential surface $\mathcal{S}$. The limit, as the volume and surface simultaneously go to zero, defines the total flux across the surface. Thus the surface integral is a measure of the total flux perpendicular to the surface. It is helpful to compare this formula with that for the gradient, Eq. 5.6.1.

5.6.2.2 Gauss's Law

The definitions in Fig. 5.5 resulted in Gauss's law, a major breakthrough in vector calculus. For the electrical case, this is equivalent to the observation that the total flux across the surface is equal to the net charge enclosed by the surface. Since the volume integral over the charge density $\rho(x,y,z)$ is the total charge enclosed Q_{enc},

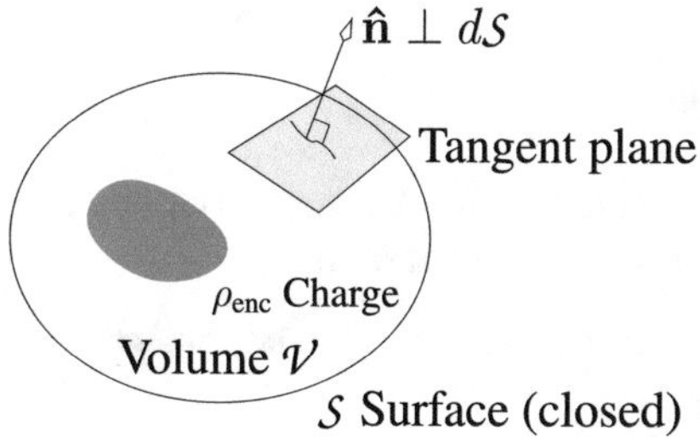

Fig. 5.5 Left: On the left is the physical layout of the integral. Right: The top equation is the integral definition of the divergence of $\boldsymbol{D}$ as an integral over the closed surface $\mathcal{S}$ of the normal component of vector $\boldsymbol{D}$, given the limit as the surface and volume shrink to 0. The middle equation states the enclosed charge Q_{enc} in terms of the surface integral of the normal component of $\boldsymbol{D}$. The bottom equation gives the charge enclosed in terms of a volume integral over the enclosed charge density ρ_{enc}

$$Q_{\text{enc}} = \iiint_{\mathcal{V}} \nabla \cdot \boldsymbol{D} \, d\mathcal{V} = \iint_{\mathcal{S}} \boldsymbol{D} \cdot \hat{\boldsymbol{n}} \, d\mathcal{S} \quad [\text{C}]. \tag{5.6.4}$$

When the surface integral over the normal component of $\boldsymbol{D}(\boldsymbol{x})$ is zero, the total charge enclosed is zero. If there is only positive (or negative) charge inside the surface, $\nabla \cdot \boldsymbol{D} = \rho(\boldsymbol{x}) = 0$. It is clear that this result only holds in the quasistatic limit, which is always satisfied because $\mathcal{S} \to 0$.

Taking the derivative with respect to time gives the total current normal to the surface:

$$I_{\text{enc}} = \iint_{\mathcal{S}} \boldsymbol{D} \cdot \hat{\boldsymbol{n}} \, d\mathcal{S} = \dot{Q}_{\text{enc}} = \iiint_{\mathcal{V}} \dot{\rho}_{\text{enc}} \, d\mathcal{V} \quad [\text{A}]. \tag{5.6.5}$$

As summarized by Feynman (1970b, p. 13-2):

> *The current leaving the closed surface $\mathcal{S}$ equals the rate of the charge leaving that volume $\mathcal{V}$, defined by that surface.*

The integral definition reduces to a common-sense summary that can be grasped intuitively.

5.6.3 Integral Definition of the Curl: $\nabla \times \boldsymbol{H} = \boldsymbol{C}$

As briefly summarized on Sect. 5.1.2.3, the differential definition of the curl maps $\mathbb{R}^3 \underset{\nabla\times}{\mapsto} \mathbb{R}^3$. The curl of the magnetic field strength $\boldsymbol{H}(\boldsymbol{x})$ is the current density $\boldsymbol{C} = \sigma \boldsymbol{E} + \dot{\boldsymbol{D}}$:

$$\nabla \times \boldsymbol{H} \equiv \begin{vmatrix} \hat{\mathbf{x}} & \hat{\mathbf{y}} & \hat{\mathbf{z}} \\ \partial_x & \partial_y & \partial_z \\ H_x & H_y & H_z \end{vmatrix} = \boldsymbol{C} \quad [\text{A/m}^2].$$

5.6.3.1 Curl and Stokes's Law

Like the gradient and divergence, the curl may be written in integral form, allowing for the physical interpretation of its meaning:

> *The surface integral definition of $\nabla \times \boldsymbol{H} = \boldsymbol{C}$ [A/m^2], where the current density $\boldsymbol{C}$ is perpendicular to the rotation plane of $\boldsymbol{H}$.*

Stokes's law states that the open surface integral over the normal component of the curl of the magnetic field strength ($\hat{\mathbf{n}} \cdot \nabla \times \boldsymbol{H}$ [A/m^2]) is equal to the line integral $\oint_{\mathcal{B}} \boldsymbol{H} \cdot d\boldsymbol{l}$ along the boundary $\mathcal{B}$. As summarized in Fig. 5.6, Stokes's law is

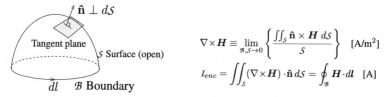

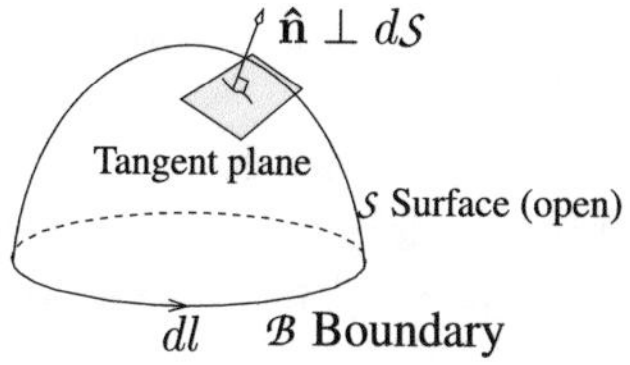

$$\nabla\times\boldsymbol{H} \equiv \lim_{\mathcal{B},\mathcal{S}\to 0}\left\{\frac{\iint_{\mathcal{S}} \hat{\mathbf{n}}\times\boldsymbol{H}\, d\mathcal{S}}{\mathcal{S}}\right\} \quad [\mathrm{A/m^2}]$$

$$I_{enc} = \iint_{\mathcal{S}} (\nabla\times\boldsymbol{H})\cdot\hat{\mathbf{n}}\, d\mathcal{S} = \oint_{\mathcal{B}} \boldsymbol{H}\cdot d\boldsymbol{l} \quad [\mathrm{A}]$$

Fig. 5.6 Left: The integral definition of the curl is related to that of the divergence (Eq. 5.6.3), as an integration over the tangent to the surface, except: (1) the curl is defined as the cross-product $\hat{\mathbf{n}}\times\boldsymbol{H}$ [A/m^2] of unit vector $\hat{\mathbf{n}}$ with the current density $\boldsymbol{H}$ (Greenberg 1988, p. 823), and (2) the surface is open, leaving a boundary $\mathcal{B}$ along the open edge. As with the divergence, which leads to Gauss's law, this definition leads to a second fundamental theorem of vector calculus: *Stokes's law* (also called the *curl theorem*). Right: Equations that summarize Stokes theorem (law)

$$\begin{aligned} I_{\mathrm{enc}} &= \iint_{\mathcal{S}} (\nabla\times\boldsymbol{H})\cdot\hat{\mathbf{n}}\, d\mathcal{S} \\ &= \iint_{\mathcal{S}} \boldsymbol{C}\cdot\hat{\mathbf{n}}\, d\mathcal{S} \\ &= \oint_{\mathcal{B}} \boldsymbol{H}\cdot d\boldsymbol{l} \quad [\mathrm{A}]. \end{aligned} \tag{5.6.6}$$

That is: *he line integral of* $\boldsymbol{H}$ *along the open surface's boundary* $\mathcal{B}$ *is equal to the total current enclosed* I_{enc} .

In many texts the normalization (denominator under the integral) is a volume $\mathcal{V}$ (Greenberg 1988, pp. 778, 823–4). However, because the surface is open, this volume does not exist (when we define a volume, the surface must be closed). The definition must hold even in the limit when the curved surface $\mathcal{S}$ degenerates to a plane, with the boundary $\mathcal{B}$ enclosing $\mathcal{S}$. In this limit there is no volume.

To resolve this problem, we take the normalization to be the surface $\mathcal{S}$ of Fig. 5.6. Note that in the limit $\mathcal{B}\to 0$, the limiting definition is independent of any curvature, since the integral is over the normal component of $\boldsymbol{H}$ (i.e., $\hat{\mathbf{n}}\perp\boldsymbol{H}(\boldsymbol{x},t)$) at the limit point $\boldsymbol{x}_o$. The net flux is independent of the curvature of $\mathcal{S}$ as $\mathcal{B}\to 0$.

Finally, the curl of a vector is composed of three wedge-products. However, the curl is defined only in $\mathbb{R}^3$. Since the wedge-product is defined in $\mathbb{R}^2$, for vectors of any length, it can be used to generalize the curl to every vector space, independent of its dimension.

The wedge-product (Fig. 3.4) is also a natural tool for dealing with the Hall effect and superconductivity, since these are two-dimensional phenomena.

5.6.4 Summary

Since integration is a linear process (sums of smaller elements), we can tile (tessellate) the surface, breaking it up into smaller surfaces and their boundaries, the sum of which is equal to the integral over the original boundary. This is an important concept that leads to the proof of Stokes's law.

Table 5.1 provides a description of the three basic integration theorems along with their mapping domains. The integral formulations of Gauss's and Stokes's laws use $\hat{\mathbf{n}} \cdot \boldsymbol{D}$ and $\boldsymbol{H} \times \hat{\mathbf{n}}$ in the integrands. The key distinction between the two laws naturally follows from the properties of the scalar ($\boldsymbol{A} \cdot \boldsymbol{B}$) and vector ($\boldsymbol{A} \times \boldsymbol{B}$) products, as discussed in Fig. 3.4. To fully appreciate the differences between Gauss's and Stokes's laws, we must master these two types of vector products.

Paraphrasing Feynman (1970b, 3–12), we have

1. $\Phi_2 = \Phi_1 + \int_1^2 \nabla\Phi \cdot d\mathcal{S}$
   The line integral of an analytic function depends only on the end points.
2. $\oint \boldsymbol{D} \cdot \hat{\mathbf{n}}\, d\mathcal{S} = \oint \nabla \cdot \boldsymbol{D}\, d\mathcal{V} = \oint \rho\, d\mathcal{V}$

   The normal component over a surface integral equal the divergence over the volume integral.
3. $\oint_{\mathcal{B}} \boldsymbol{E} \cdot d\boldsymbol{l} = \oint_{\mathcal{S}} (\nabla \times \boldsymbol{E}) \cdot \hat{\mathbf{n}}\, d\mathcal{S} = -\oint \dot{\mathcal{B}} \cdot \hat{\mathbf{n}}\, d\mathcal{S}$

   The line integral of the electric field (the induced EMF in any loop) equals the integral of the normal component of time-rate-of-change of the magnetic flux. The induced EMF is the Thévenin voltage in series with the loop. This explains the working of an electrical transformer.

5.6.5 *Helmholtz's Decomposition Theorem*

This is the time to rethink everything we have defined in terms of the two types of vector fields that decompose every analytic vector field as shown in Table 5.3. The *irrotational field* is defined as one that is curl free. The *incompressible field* is one that is divergence free.

Table 5.3 The four possible classifications of scalar and vector potential fields: rotational/irrotational and compressible/incompressible. Rotational fields are generated by the vector potential (e.g., $\boldsymbol{A}(\boldsymbol{x}, t)$), while compressible fields are generated by the scalar potentials (e.g., voltage $\phi(\boldsymbol{x}, t)$, velocity ψ, pressure $\varrho(\boldsymbol{x}, t)$, or temperature $T(\boldsymbol{x}, t)$).

Field: $\nu(\boldsymbol{x}, t)$	Compressible $\nabla \cdot \nu \neq 0$	Incompressible $\nabla \cdot \nu = 0$
Rotational $\nabla \times \nu \neq 0$	$\nu = \nabla\phi + \nabla \times \omega$ Vector wave Eq. $\nabla^2 \nu = \frac{1}{c^2}\ddot{\nu}$	$\nu = \nabla \times \boldsymbol{w}$ Lubrication theory Boundary layers
Irrotational conservative $\nabla \times \nu = 0$	Acoustics $\nu = \nabla\psi$ $\nabla^2 \varrho(\boldsymbol{x}, t) = \frac{1}{c^2}\ddot{\varrho}(\boldsymbol{x}, t)$	Statics $\nabla^2\phi = 0$ Laplace's Eq. ($c \to \infty$)

Fig. 5.7 Left: Hermann von Helmholtz (taken from Helmholtz (1978)), Right: Gustav Kirchhoff. Together they were the first to account for viscous (Helmholtz 1858, 1978, 1863b) and thermal (Kirchhoff 1868, 1974) losses in the acoustic propagation of airborne sound, as first experimentally verified by Mason (1928), p. 241)

According to Helmholtz's decomposition theorem, every analytic vector field may be decomposed into independent rotational and compressible components (Helmholtz 1978). An alternative name for Helmholtz's decomposition theorem is the fundamental theorem of vector calculus (FTVC). Gauss's and Stokes's theorems[11] along with Helmholtz's decomposition theorem are the three fundamental theorems of vector calculus. Portraits of Helmholtz and Kirchhoff are provided in Fig. 5.7.

5.6.5.1 The Four Categories of Linear Fluid Flow

The following is a summary of the four cases for fluid flow, as shown in Table 5.3:

1,1 Compressible, rotational fluid (general case): $\nabla\psi \neq 0$, $\nabla\times\boldsymbol{w} \neq 0$. This is wave propagation in a medium where viscosity cannot be ignored, as in acoustics close to the boundaries, where viscosity contributes to losses (Batchelor 1967).

1,2 Incompressible, rotational fluid (lubrication theory): $\nu = \nabla\times\boldsymbol{w} \neq 0$, $\nabla\cdot\nu = 0$, $\nabla^2\psi = 0$. In this case, the flow is dominated by the walls, while the viscosity and heat transfer introduce shear. This is typical of lubrication theory (solenoidal fields).

2,1 Compressible, irrotational fluid (acoustics): $\nu = \nabla\psi$, $\nabla\times\boldsymbol{w} = 0$. Here losses (viscosity and thermal diffusion) are small (assumed to be zero). We can define a velocity potential ψ, the gradient of which gives the air particle velocity, thus $\nu = -\nabla\psi$. For an irrotational fluid, $\nabla\times\nu = 0$ (Greenberg 1988, p. 826). This is the case for the conservative field, where $\int \nu\cdot\hat{\mathbf{n}}dR$ depends on only the end points and $\oint_{\mathcal{B}} \nu\cdot\hat{\mathbf{n}}dR = 0$. When a fluid may be treated as having no viscosity, it is typically assumed to be irrotational, since it is viscosity and shear that lead to

[11]These theorems are mathematical relationships that follow from physical principles.

rotation (Greenberg 1988, p. 814). A fluid's angular velocity is $\Omega = \frac{1}{2}\nabla \times \nu = 0$; thus irrotational fluids have zero angular velocity ($\Omega = 0$).

2,2 Incompressible, irrotational fluid (statics): $\nabla \cdot \nu = 0$ and $\nabla \times \nu = 0$; thus $\nu = \nabla\psi$ and $\nabla^2\psi = 0$. An example is water in a small space at low frequencies, where the wavelength is long compared to the size of the container; the fluid may be treated as incompressible. When $\nabla \times \nu = 0$, the effects of viscosity may be ignored, as it is the viscosity that creates the shear that leads to rotation. This is the case in modeling the cochlea, where fluid losses are ignored and the quasistatic limit is justified.

In summary, each of these cases is an approximation that best applies in the low-frequency limit. This is why it is called *quasistatic*, meaning low but not zero frequency, where the wavelength is more than twice the diameter.

A magnetic solenoidal field is a uniform-flux field $\boldsymbol{B}_z(\boldsymbol{x})$ that is generated by a solenoidal coil as shown in Fig. 5.8, and to an excellent approximation is uniform inside the coil, making it similar to that of a permanent magnet. As a result, the divergence of a solenoidal field is approximately zero, which makes it incompressible ($\nabla\cdot = 0$) and rotational ($\nabla\times \neq 0$).

You need to know the term *solenoidal* since it is widely used. However the preferred terms are *incompressible* and *rotational*. Strictly speaking, the term *solenoidal field* applies to only a magnetic field produced by a solenoid.

5.6.5.2 Application and Derivation

Helmholtz's decomposition theorem is expressed as the linear sum of a scalar potential $\phi(x, y, z)$ (think voltage) and a vector potential (think magnetic vector potential). Specifically,

$$\boldsymbol{E}(\boldsymbol{x}, s) = -\nabla\phi(\boldsymbol{x}, s) + \nabla \times \boldsymbol{A}(x, s), \tag{5.6.7}$$

where ϕ is the scalar and $\boldsymbol{A}$ is the vector potential as a function of the Laplace frequency s. Of course, this decomposition is general (not limited to the electromagnetic case). It applies to linear fluid vector fields, which include linear fluids (water, air). If rotational and dilation become coupled, this relationship will break down.

This theorem is easily stated (and proved) but less easily appreciated (Heras 2016). A physical description is helpful: Every vector field may be split into two independent parts: translation and rotation.[12] We have seen this same idea in vector algebra, where the scalar and wedge-products of two vectors are perpendicular (Fig. 3.4).

For example, think of linear versus angular momentum, which is independent in that they represent different ways of delivering kinetic energy via different modalities (degrees of freedom, DoF). Linear and rotational motions are a common theme in physics, rooted in geometry. Thus it seems a natural extension to split a vector field into independent dilation and rotation parts.

---

[12] Actually it may be split into three independent parts: translation, dilation, and rotation.

Example: A fluid with mass and momentum can be moving along a path and independently be rotating. These independent modes of motion correspond to different types of kinetic energy (modes), such as translational, compressional, and rotational. Each eigenmode of vibration can be viewed as an independent degree of freedom (DoF).

Helmholtz's decomposition theorem (FTVC) quantifies these degrees of freedom. To prove the construction, second-order vector identities **DoC**: $\nabla\cdot\nabla\times() = 0$ and **CoG**: $\nabla\times\nabla() = 0$ may be used to verify the FTVC. The role of the FTVC is especially useful when applied to Maxwell's equations.

5.6.5.3 Helmholtz's Decomposition Theorem Proof

To prove Eq. 5.6.7 we must understand how it splits the vector field into parts. The identities have a physical meaning: Every vector field may be split into its translational and rotational parts. If $\boldsymbol{E}$ is the electric field [V/m], ϕ is the voltage, and $\boldsymbol{A}$ is the current induced rotational part.

To do this we need the two key vector identities that are always zero for analytic fields: the curl of the gradient ($\nabla\times\nabla()$),

$$\nabla\times\nabla\phi(\boldsymbol{x}) = 0, \tag{5.6.8}$$

and the divergence of the curl[13] ($\nabla\cdot\nabla\times()$)

$$\nabla\cdot(\nabla\times\boldsymbol{A}) = 0. \tag{5.6.9}$$

These identities are easily verified by working out a few specific examples based on the definitions of the three operators, gradient, divergence, and curl, or in terms of the operator's integral definitions (see Sect. 5.6).

By applying these two identities to Helmholtz's decomposition, we can appreciate the theorem's significance. We can work backward via a physical argument that rotational momentum (rotational energy) is independent of translational momentum. Once these forces are made clear, the vector operations take on meaning. One might conclude that the physics is simply related to geometry via the scalar and vector products.

Specifically, if we take the divergence of Eq. 5.6.7 and use the DoG, then

$$-\frac{1}{\epsilon_o}\dot{\rho} = \nabla\cdot\boldsymbol{E} = \nabla\cdot\{-\nabla\phi + \cancel{\nabla\times\boldsymbol{A}}\} \overset{0}{=} -\nabla\cdot\nabla\phi = -\nabla^2\phi,$$

[13] Helmholtz was the first person to apply mathematics in modeling the eye and the ear (Helmholtz 1863a).

If we take the curl, then

$$-\dot{\boldsymbol{B}} = \nabla\times\boldsymbol{E} = \nabla\times\{\overset{0}{\cancel{-\nabla\phi}} + \nabla\times\boldsymbol{A}\} = \nabla\times\nabla\times\boldsymbol{A} = \text{v}^2\boldsymbol{A} - \boldsymbol{\nabla}^2\boldsymbol{A},$$

since the $\nabla\times\nabla$ zeros the scalar field $\phi(x, y, z)$.

5.6.6 *Second-Order Operators*

In addition to the first-order vector derivatives are second-order operators, the most important being the scalar Laplacian $\nabla^2() = \nabla\cdot\nabla()$ and vector Laplacian $\boldsymbol{\nabla}^2() = \nabla\cdot\nabla()$, which operates on vectors.[14]

5.6.6.1 Terminology

There are six second-order combinations of ∇, requiring six mnemonics (Table 5.1):

1. DoG: Divergence of the gradient (*scalar Laplacian* operates on scalar potentials (Greenberg 1988, p. 779)):

$$\begin{aligned}\nabla^2\phi &= (\nabla\cdot\nabla)\phi \\ &= \frac{\partial^2\phi}{\partial x^2} + \frac{\partial^2\phi}{\partial y^2} + \frac{\partial^2\phi}{\partial z^2}.\end{aligned} \tag{5.6.10}$$

2. **DoG**: Divergence of the Gradient (Bull-Dog, the *vector Laplacian* $\boldsymbol{\nabla}^2$ (Sommerfeld 1952, p. 33)):

$$\begin{aligned}\boldsymbol{\nabla}^2\boldsymbol{A} &= (\boldsymbol{\nabla}\cdot\nabla)\boldsymbol{A} \\ &= \frac{\partial^2\boldsymbol{A}}{\partial x^2} + \frac{\partial^2\boldsymbol{A}}{\partial y^2} + \frac{\partial^2\boldsymbol{A}}{\partial z^2} \\ &= \text{v}^2\boldsymbol{A} - \nabla\times\nabla\times\boldsymbol{A}.\end{aligned} \tag{5.6.11}$$

3. **gOd**: Gradient of the Divergence ($\text{v}^2\boldsymbol{A} = \nabla(\nabla\cdot\boldsymbol{A})$)

14 https://en.wikipedia.org/wiki/Del_in_cylindrical_and_spherical_coordinates#Non-trivial_calculation_rules.

$$
\begin{aligned}
\nabla^2 \boldsymbol{A} &= \nabla(\nabla\cdot \boldsymbol{A}) \\
&= \nabla\left(\frac{\partial a_x}{\partial x}+\frac{\partial a_y}{\partial y}+\frac{\partial a_z}{\partial z}\right) \\
&= \left(\hat{\mathbf{x}}\frac{\partial}{\partial x}+\hat{\mathbf{y}}\frac{\partial}{\partial y}+\hat{\mathbf{z}}\frac{\partial}{\partial z}\right)\left(\frac{\partial a_x}{\partial x}+\frac{\partial a_y}{\partial y}+\frac{\partial a_z}{\partial z}\right) \\
&= \hat{\mathbf{x}}\frac{\partial}{\partial x}\nabla\cdot A+\hat{\mathbf{y}}\frac{\partial}{\partial y}\nabla\cdot A+\hat{\mathbf{z}}\frac{\partial}{\partial z}\nabla\cdot A. \qquad (5.6.12)
\end{aligned}
$$

4. **CoC**: Curl of the curl (**CoC** = **gOd**- **DoG** (Sommerfeld 1952, p. 33, Eq. 2b))

$$
\begin{aligned}
\nabla\times\nabla\times \boldsymbol{A} &= \nabla(\nabla\cdot \boldsymbol{A}) - (\boldsymbol{\nabla}\cdot\boldsymbol{\nabla})\boldsymbol{A} \\
&= \nabla^2 \boldsymbol{A} - \boldsymbol{\nabla}^2 \boldsymbol{A}. \qquad (5.6.13)
\end{aligned}
$$

5. DoC: Divergence of the curl ($\nabla\cdot\nabla\times = 0$)
6. CoG: Curl of the gradient ($\nabla\times\nabla = 0$)

DoC(·) and CoG(·) are special because they are always zero:

$$\nabla\cdot\nabla\times \boldsymbol{A} = 0, \qquad \nabla\times\nabla\phi = 0, \qquad (5.6.14)$$

making them useful in proving the FTVC (Eq. 5.6.7).

A third special vector identity $\nabla\times\nabla\times$ is Eq. 5.6.13. which operates on vector fields and is useful for defining the vector Laplacian **DoG** as the difference between **gOd** and **CoC** (i.e., **DoG** = **gOd** − **CoC**):

$$\boldsymbol{\nabla}^2() = \nabla^2() - \nabla\times\nabla\times().$$

The role of **gOd** (∇^2) is commonly ignored because it is zero for the magnetic wave equation, due to there being no magnetic charge [$\nabla\cdot\boldsymbol{B}(x,t) = 0$; thus $\nabla^2 \boldsymbol{B}(x,t) \equiv 0$]. However for the electric vector wave equation it plays a role as a source term:

$$\nabla^2 \phi(\boldsymbol{x},t) = -\nabla \boldsymbol{E}(\boldsymbol{x},t) = -\frac{1}{\epsilon_o}\nabla^2 \boldsymbol{D}(x,t) = -\frac{1}{\epsilon_o}\nabla\rho(\boldsymbol{x},t),$$

or since $\nabla\cdot\boldsymbol{D} = \rho$,

$$\nabla^2 \boldsymbol{D}(\boldsymbol{x},t) = \nabla\nabla\cdot\boldsymbol{D} = -\nabla\rho(\boldsymbol{x},t).$$

When the charge density is inomogeneous, such as the case of a plasma (e.g., the Sun), this term plays an important role as a source term in the electric wave equation. This case needs to be further explored through some physical examples.

Exercise #16 Show that **DoG**: $\boldsymbol{\nabla}^2$ and **gOd**: ∇^2 differ.

Solution: Use $\nabla\times\nabla\times()$ to explore this relationship. If **DoG** and **gOd** were the same, $\nabla\times\nabla\times()$ would be null. ∎

Exercise #17 What is the difference between **DoG**: ∇^2 and Bull-DoG: $\boldsymbol{\nabla}^2$?

Solution: DoG operates on scalar functions while Bull-DoG operates on vector functions. ■

5.6.6.2 Discussion

It is helpful to view these two groups as playing fundamentally different roles:

utility operators $\nabla \times \nabla \times ()$: **DoG**: $\nabla^2()$, and gOd: $\boldsymbol{\nabla}^2()$,

and

identity operators DoC: $DoC() = 0$ and CoG: $\nabla \times \nabla()$=0 (Eq. 5.6.14).

When using second-order differential operators, we must be careful with the order of operations, which can be subtle. Most of this is common sense. For example, don't operate on a scalar field with $\nabla\times$, and don't operate on a vector field with ∇.[15] **DoG** acts on each vector component $\boldsymbol{\nabla}^2 \boldsymbol{A} = \nabla^2 A_x \hat{\mathbf{x}} + \nabla^2 A_y \hat{\mathbf{y}} + \nabla^2 A_z \hat{\mathbf{z}}$ (Eq. 5.6.11), which is very different from the action of the Laplacian (DoG).

5.7 The Unification of Electricity and Magnetism

Once we have mastered the three basic vector operations—gradient, divergence, and curl—we are ready to appreciate Maxwell's equations. Like the vector operations, these equations may be written in integral or differential form. An important difference is that with Maxwell's equations, we are dealing with well-defined physical quantities. The scalar and vector fields take on meaning and units. Thus, to understand these important equations, we must master both the names and units of the four fields $\boldsymbol{E}, \boldsymbol{H}, \boldsymbol{B}, \boldsymbol{D}$, as described in Table 5.4.

Table 5.4 The variables of Maxwell's equations have names (e.g., EF, MI) and units (in square brackets [SI units]). The units are necessary to obtain a full understanding of each of the four variables and their corresponding equations. For example, Eq. *EF* has units [V/m]. By integrating $\boldsymbol{E}$ from $x = a, b$, one obtains the voltage difference between the two points. The *speed of light in-vacuo* is $c = 3 \times 10^8 = 1/\sqrt{\mu_o \epsilon_o}$ [m/s], and the *characteristic resistance* of light $r_o = 377 = \sqrt{\mu_o/\epsilon_o}$ [Ω] (i.e., ohms).

Symbol	Name	Units	Maxwell's Eq.
$\boldsymbol{E}$	EF: Electric field strength	[V/m]	$\nabla \times \boldsymbol{E} = -\partial_t \boldsymbol{B}$
$\boldsymbol{D} = \epsilon_o \boldsymbol{E}$	ED: Electric displacement (flux density)	[C/m^2]	$\nabla \cdot \boldsymbol{D} = \rho$
$\boldsymbol{H}$	MF: Magnetic field strength	[A/m]	$\nabla \times \boldsymbol{H} = \boldsymbol{J_m} + \partial_t \boldsymbol{D}$
$\boldsymbol{B} = \mu_o \boldsymbol{H}$	MI: Magnetic induction (flux density)	[Wb/m^2]	$\nabla \cdot \boldsymbol{B} = 0$

[15] This operation defines a dyadic tensor, a generalization of the vector.

Fig. 5.8 A solenoid is a uniform coil of wire. The term comes from the Greek meaning "shaped like a pipe or channel." When a steady current is passed through the wire, a uniform magnetic field intensity $\boldsymbol{H}$ is created. Such a coil is indistinguishable from a permanent bar magnet with its north and south poles. Depending on the direction of the current, one end of a finite solenoidal coil is the north pole of the magnet, and the other end is the south pole. The uniform field inside the coil is called *solenoidal*, a confusing synonym for *rotational* (Figure from Wikipedia).

5.7.1 Field Strength $\boldsymbol{E}, \boldsymbol{H}$

As summarized by Eq. 5.7.1, there are two field strengths: the electric $\boldsymbol{E}$ with units of [V/m] and the magnetic $\boldsymbol{H}$ with units of [A/m]. The ratio $|\boldsymbol{E}|/|\boldsymbol{H}| = \sqrt{\mu_o/\epsilon_o} = 377$ [ohms] for in-vacuo plane-waves (μ_o, ϵ_o).

If two conducting plates are placed 1 [m] apart, with 1 [V] across them, the electric field is $\boldsymbol{E} = 1$ [V/m]. If a charge (i.e., an electron) is placed in an electric field, it feels a force $\boldsymbol{f} = q\boldsymbol{E}$ [N], where q is the magnitude of the charge [C].

To help us understand the meaning of $\boldsymbol{H}$, consider the solenoid made of wire, as shown in Fig. 5.8, that carries a current of I_θ [A]. The axial (along the long axis) magnetic field H_z inside such a solenoid is uniform, with a direction that depends on the polarity of I_θ, is

$$H_z = \frac{N}{L} I_\theta \quad \text{[Wb]},$$

where L is the length of the coil, N is the number of turns.

5.7.2 Flux $\boldsymbol{D}, \boldsymbol{B}$

Flux is a flow, such as the mass flux of water flowing in a pipe [kg/s] driven by a force (pressure drop) across the ends of the pipe, or the heat flux in a thermal conductor, that has a temperature drop across it (i.e., a window or a wall). The flux is the same as the flow, be it charge, mass, or heat (Table 3.2). In Maxwell's equations there are also two fluxes: the electric flux $\boldsymbol{D}$ and the magnetic flux $\boldsymbol{B}$. The flux density units for $\boldsymbol{D}$ is [C/m^2], and the magnetic flux density $\boldsymbol{B}$ is measured in either weber [Wb/m^2]) or [tesla] [T].

5.7.3 Maxwell's Equations

Maxwell's equations (ME) consist of two curl equations (Eq. 5.7.1) operating on the field strengths EF $\boldsymbol{E}$ and MF $\boldsymbol{H}$, and two divergence equations (Eq. 5.7.2) operating on the field fluxes ED $\boldsymbol{D}$ and MI $\boldsymbol{B}$. In matrix format, the ME are

$$\begin{aligned}\nabla\times\begin{bmatrix}\boldsymbol{E}(\boldsymbol{x},t)\\ \boldsymbol{H}(\boldsymbol{x},t)\end{bmatrix} &= \partial_t\begin{bmatrix}-\boldsymbol{B}(\boldsymbol{x},t)\\ \boldsymbol{D}(\boldsymbol{x},t)\end{bmatrix}\\ &= \begin{bmatrix}0 & -\mu_o\\ \epsilon_o & 0\end{bmatrix}\partial_t\begin{bmatrix}\boldsymbol{E}(\boldsymbol{x},t)\\ \boldsymbol{H}(\boldsymbol{x},t)\end{bmatrix}\\ &\leftrightarrow \begin{bmatrix}0 & -s\mu_o\\ \sigma_o+s\epsilon_o & 0\end{bmatrix}\begin{bmatrix}\mathbf{E}(\boldsymbol{x},\omega)\\ \mathbf{H}(\boldsymbol{x},\omega)\end{bmatrix}.\end{aligned} \tag{5.7.1}$$

When the medium is conducting, $\partial_t \boldsymbol{D}$ must be replaced by $\boldsymbol{C} = \sigma_o E + \partial_t \boldsymbol{D} \leftrightarrow (\sigma_o + s\epsilon_o)\boldsymbol{E}(\boldsymbol{x},\omega)$ where $\sigma_o + s\epsilon_o$ is an admittance density [℧/m^2].

There are also two auxiliary equations:

$$\nabla\cdot\begin{bmatrix}\boldsymbol{D}\\ \boldsymbol{B}\end{bmatrix} = -\partial_t\begin{bmatrix}\rho(\boldsymbol{x})\\ 0\end{bmatrix}. \tag{5.7.2}$$

The top equation states conservation of charge, while the lower states that there is no magnetic charge. When expressed in integral format, Stokes's law follows from the curl equations and Gauss's law from the divergence equations.

Exercise #18 When a static current is flowing in a wire in the $\hat{\mathbf{z}}$ direction, the magnetic flux is determined by Stokes's theorem (Fig. 5.6). Thus, just outside the wire we have

$$I_{\text{enc}} = \iint_S (\nabla\times\boldsymbol{H})\cdot\hat{\mathbf{n}}\, d|S| = \oint_{\mathcal{B}} \boldsymbol{H}\cdot d\boldsymbol{l} \quad [\text{A}].$$

For this simple geometry, the current in a wire is related to $\boldsymbol{H}(\boldsymbol{x},t)$ by

$$I_{\text{enc}} = \oint_{\mathcal{B}} \boldsymbol{H}\cdot d\boldsymbol{l} = H_\phi 2\pi r.$$

Here H_ϕ is perpendicular to both the radius r and the direction of the current $\hat{\mathbf{z}}$. Thus

$$\boldsymbol{H}_\phi = \frac{I_{\text{enc}}}{2\pi r},$$

where $\boldsymbol{H}_\phi$ is attenuated by $1/r$ (Ramo et al. 1965, Eq. 9, p. 244).

Exercise #19 Explain how Stokes's theorem may be applied to $\nabla\times\boldsymbol{E} = -\dot{\boldsymbol{B}}$, and explain what it means. Hint: This is the same argument given above for the current in a wire, but for the electric case.

Solution: Integrating the left side of equation Eq. 5.7.3 over an open surface results in a voltage (emf) induced in the loop closing the boundary $\mathcal{B}$ of the surface

$$\phi_{\text{induced}} = \iint_S (\nabla\times\boldsymbol{E})\cdot\hat{\mathbf{n}}\, d|S| = \oint_{\mathcal{B}} \boldsymbol{E}\cdot d\boldsymbol{l} \quad [\text{V}]. \tag{5.7.3}$$

The emf (electromagnetic force) is the same as the Thévenin source voltage induced by the rate of change of the flux. Integrating Eq. 5.7.3 over the same open surface $\mathcal{S}$ results in the source of the induced voltage ϕ_{induced}, which is proportional to the rate of change of the flux [weber]:

$$\phi_{\text{induced}} = -\frac{\partial}{\partial t} \iint_{\mathcal{S}} \boldsymbol{B} \cdot \hat{\mathbf{n}}\, dA = L\dot{\psi} \quad \text{[Wb/s] or [V]},$$

where L [H] is the inductance of the wire, with impedance $Z_L(s) = sL$. The area integral on the right is the total flux crossing normal to the surface ψ [Wb]. The rate of change of the total flux [Wb/s] is the induced (Thévenin) voltage [V]. ■

If we apply Gauss's theorem to the divergence equations, we find the total flux that crosses the closed surface.

Exercise #20 Apply Gauss's theorem to equation ED and explain what it means in physical terms.

Solution: The area of the normal component of $\boldsymbol{D}$ is equal to the volume integral over the charge density. Thus Gauss's theorem says that the total charge within the volume Q_{enc}, found by integrating the charge density $\rho(\boldsymbol{x})$ over the volume $\mathcal{V}$, is equal to the normal component of the flux $\boldsymbol{D}$ that crosses the surface $\mathcal{S}$:

$$Q_{\text{enc}} = \iiint_{\mathcal{V}} \nabla\cdot\boldsymbol{D}\, dV = \iint_{\mathcal{S}} \boldsymbol{D}\cdot\hat{\boldsymbol{n}}\, dA.$$

When equal amounts of positive and negative charge exist within the volume, regardless of its distribution, the integral is zero. ■

5.7.3.1 Summary

Maxwell's four equations relate the field strengths to the flux densities. There are two types of variables: field strengths ($\boldsymbol{E}, \boldsymbol{H}$) and flux densities ($\boldsymbol{D}, \boldsymbol{B}$). There are two classes: electric ($\boldsymbol{E}, \boldsymbol{D}$) and magnetic ($\boldsymbol{H}, \boldsymbol{B}$). This is a 2×2 matrix, with column being field strength and flux densities and rows being electric and magnetic variables.

	Strength	Flux density
Electric	$\boldsymbol{E}$ [V/m]	$\boldsymbol{D}$ [C/m^2]
Magnetic	$\boldsymbol{H}$ [A/m]	$\boldsymbol{B}$ [Wb/m^2]

Applying Stokes's curl theorem to the forces induces a Thévenin voltage (emf) or Norton current source. Applying Gauss's divergence theorem to the flows gives the total charge enclosed. The magnetic charge is zero ($\nabla\cdot\boldsymbol{B} = 0$) because magnetic monopoles do not exist. However, magnetic dipoles do exist, as in the example of the electron, which contains a magnetic dipole.

5.7.4 Derivation of the Vector Wave Equation

Next we provide the derivation of the vector wave equation starting from Maxwell's equations (Eq. 5.7.1), which is reminiscent of the derivation of the Webster horn equation (Eq. 5.2.2). Working in the frequency domain and taking the curl of both sides give

$$\begin{aligned}\nabla\times\nabla\times\begin{bmatrix}\mathbf{E}\\ \mathbf{H}\end{bmatrix} &= \begin{bmatrix}0 & -s\mu_o\\ s\epsilon_o & 0\end{bmatrix}\nabla\times\begin{bmatrix}\mathbf{E}\\ \mathbf{H}\end{bmatrix}\\ &= \begin{bmatrix}0 & -s\mu_o\\ s\epsilon_o & 0\end{bmatrix}\begin{bmatrix}0 & -s\mu_o\\ s\epsilon_o & 0\end{bmatrix}\begin{bmatrix}\mathbf{E}\\ \mathbf{H}\end{bmatrix}\\ &= -\frac{s^2}{c_o^2}\begin{bmatrix}\mathbf{E}\\ \mathbf{H}\end{bmatrix}.\end{aligned}$$

Using the CoC identity $\nabla\times\nabla\times() = \nabla^2\,() - \boldsymbol{\nabla}^2()$ (Eq. 5.6.13) gives

$$\boldsymbol{\nabla}^2\begin{bmatrix}\mathbf{E}\\ \mathbf{H}\end{bmatrix} - \nabla^2\begin{bmatrix}\mathbf{E}\\ \mathbf{H}\end{bmatrix} = \frac{s^2}{c_o^2}\begin{bmatrix}\mathbf{E}\\ \mathbf{H}\end{bmatrix}$$

or finally Maxwell's vector wave equation

$$\begin{aligned}\boldsymbol{\nabla}^2\begin{bmatrix}\mathbf{E}\\ \mathbf{H}\end{bmatrix} - \frac{s^2}{c_o^2}\begin{bmatrix}\mathbf{E}\\ \mathbf{H}\end{bmatrix} &= \nabla\begin{bmatrix}\frac{1}{\epsilon_o}\nabla\cdot\mathbf{D}\\ \cancel{\frac{1}{\mu_o}\nabla\cdot\mathbf{B}}^{\,0}\end{bmatrix}\\ &= \frac{1}{\epsilon_o}\begin{bmatrix}\nabla\rho(\boldsymbol{x},s)\\ 0\end{bmatrix}\end{aligned} \tag{5.7.4}$$

with the electric excitation term $\nabla\rho(\boldsymbol{x},s)$. Note that if μ and ϵ depended on $\boldsymbol{x}$, the terms on the right would not be zero. In deep outer space with its black holes and plasma everywhere (e.g., inside the Sun), this seems possible, even likely.

Recall the d'Alembert solutions of the scalar wave equation (Eq. 4.4.1)

$$\boldsymbol{E}(\boldsymbol{x},t) = \boldsymbol{f}(\boldsymbol{x}-ct) + g(\boldsymbol{x}+ct),$$

where $\boldsymbol{f}$ and g are arbitrary vector fields. This result applies to the vector case, since it represents three identical, yet independent, scalar wave equations in the three dimensions.

5.7.4.1 Poynting Vector

The EM power flux density $\mathcal{P}$ [W/m^2] is perpendicular to $\boldsymbol{E}$ and $\boldsymbol{B}$, denoted as

$$\mathcal{P} = \frac{1}{\mu_o}\boldsymbol{E}\times\boldsymbol{B} = \boldsymbol{E}\times\boldsymbol{H} \qquad [\text{W/m}^2].$$

The corresponding EM momentum flux density $\mathcal{M}$ (hence ME are related to mass, thus gravity) is

$$\mathcal{M} = \epsilon_o\boldsymbol{E}\times\boldsymbol{B} = \boldsymbol{D}\times\boldsymbol{B} \qquad [\text{C/m}^2 \cdot \text{Wb/m}^2].$$

Since the speed of light is $c_o = 1/\sqrt{\mu_o\epsilon_o}$, dividing by the momentum flux density gives

$$\mathcal{P} = c_o^2\mathcal{M} \qquad [\text{W/m}^2],$$

which is related to the Einstein energy–mass equivalence formula $\mathcal{E} = mc_o^2$ (Sommerfeld 1952).

For example, the power emitted by the Sun is about 1360 [W/m^2], with a radiation pressure of 4 [μN/m^2] (i.e., 4 [μPa]) (Fitzpatrick 2008). By way of comparison, the threshold audible acoustic pressure at the human eardrum at 1 [kHz] is 20 [μPa], which is 14 [dB] (a factor of 5) the solar radiation pressure. Also:

> *The lasers used in Inertial Confinement Fusion (e.g., the NOVA experiment in Lawrence Livermore National Laboratory) typically have energy fluxes of* 10^{18} [W/m^2]. *This translates to a radiation pressure of about* 10^4 *atmospheres!*
>
> – Fitzpatrick (2008, p. 291)

One [atm] is 10^5 [Pa].

5.7.4.2 Electrical Impedance Seen by an Electron

Up to now we have considered only the Brune impedance, which is a special case with no branch points or branch cuts. We can define *impedance* for diffusion, as in the diffusion of heat. There is also the diffusion of electrical and magnetic fields at the surface of a conductor, where the resistance of the conductor dominates the dielectric properties. This is called the electrical *skin effect*, where the conduction currents are dominated by the conductivity of the metal rather than the displacement currents. In such cases, the impedance is proportional to $\sqrt{s}$, which requires that it has a branch cut. Still, the real part of the impedance must be positive in the right s half-plane, the required condition of all impedances, such that Postulate P3 (Sect.3.10.2) is satisfied. The same effect is observed in acoustics (see Appendix D).

When we deal with Maxwell's equations, the force is defined by the Lorentz force,

$$\boldsymbol{f} = q\boldsymbol{E} + q\nu\times\boldsymbol{B} = q\boldsymbol{E} + \boldsymbol{C}\times\boldsymbol{B},$$

which is the force on a charge (e.g., electron) due to the electric $\boldsymbol{E}$ and magnetic $\boldsymbol{B}$ fields. The magnetic field plays a role when the charge has a velocity ν. When a charge is moving with velocity ν, it may be viewed as a current $\boldsymbol{C} = q\nu$ (see the discussion on Sect.3.10.2).

The complex admittance density is

$$Y(s) = \sigma_o + s\epsilon_o \quad [\mho/\mathrm{m}^2]$$

(Feynman 1970b, p. 13–1). Here σ_o is the electrical conductivity and ϵ_o is the electrical permittivity. Since $\omega\epsilon_o \ll \sigma_o$, the conductivity reduces to the resistance of the wire per unit length.[16]

5.8 Potential Solutions of Maxwell's Equations

One primary reason for using potentials is to generate solutions to Maxwell's equations. For example, if we extend Eq. 5.1.1, we can express Maxwell's equations in terms of scalar and vector potentials. These relationships are (Sommerfeld (1952, p. 146); Feynman (1970d, pp. 18–10)):

$$\boldsymbol{E}(\boldsymbol{x}, t) = -\nabla\phi(\boldsymbol{x}, t) - \frac{\partial \boldsymbol{A}(\boldsymbol{x}, t)}{\partial t} \quad [\mathrm{V/m}] \tag{5.8.1}$$

and

$$\boldsymbol{H}(\boldsymbol{x}, t) = \frac{1}{\mu_o}\left[\nabla\times\boldsymbol{A}(\boldsymbol{x}, t) + \frac{\partial \boldsymbol{D}(\boldsymbol{x}, t)}{\partial t}\right] \quad [\mathrm{A/m}]. \tag{5.8.2}$$

We have extended $\boldsymbol{H}(\boldsymbol{x}, t)$ to include the electric potential term

$$\boldsymbol{D}(\boldsymbol{x}, t) = \epsilon(\boldsymbol{x}, t)\boldsymbol{E}(\boldsymbol{x}, t) = -\epsilon(\boldsymbol{x}, t)\nabla\phi(\boldsymbol{x}, t),$$

normally taken to be zero because taking the curl of $\boldsymbol{H}(t)$ naturally removes any electrical potential term due to CoG.[17] When the permittivity ($\epsilon_o(\boldsymbol{x}, t)$ [F/m^2]) is both in-homogeneous and time dependent,

$$\nabla\cdot\boldsymbol{E} = -\nabla^2\Phi - \nabla\cdot\dot{\boldsymbol{A}} = \rho(\boldsymbol{x}, t)/\epsilon_o(\boldsymbol{x}, t)$$

---

[16]For copper $\omega \ll \omega_c = \sigma_o/\epsilon_o \approx 6\times10^7/9\times10^{-12} \approx 6.66\times10^{18}$ [rad/s], or $f_c = 10^{18}$ [Hz]. This corresponds to a wavelength of $\lambda_o \approx c_o/f_c = 0.30$ [nm]. For comparison, the Bohr radius (hydrogen) is ≈0.053 [nm] (5.66 times smaller) and the Lorentz radius (of the electron) is estimated to be 2.8×10^{-15} [m] (2.8 [femtometers])..

[17]In-vacuo $\epsilon_o = 8.85\times10^{-12}$ [F/m^2] is the capacitance,and $s\epsilon_o$ is the electric compliance-density of light. The related magnetic mass-density is the permeability $\mu_o = 4\pi\times10^{-7}$ [H/m^2] having an inductive impedance of $s\mu_o$ [Ω/m]. It is helpful to think of ϵ_o as a capacitance per unit area and μ_o as an inductance per unit area (consistent with their units). The in-vacuo speed of light is $c_o = 1/\sqrt{\epsilon_o\mu_o} = 3\times10^8$ [m/s], but is slower when traveling in matter (Brillouin 1960).

and

$$\nabla\times[\epsilon(\boldsymbol{x},\omega)\nabla\phi(\boldsymbol{x},\omega)] = \cancelto{0}{\epsilon(\boldsymbol{x},t)\nabla\times\phi} + \nabla\epsilon(\boldsymbol{x},t)\times\nabla\phi \neq 0.$$

The extension makes the potential solutions symmetric so that $\boldsymbol{E}$ and $\boldsymbol{H}$ each have electrical and magnetic excitation.

Exercise #21 Explain why some dependence on $\phi(\boldsymbol{x},t)$ do not appear in Eq. 5.8.2 but do in 5.8.1.

Solution: For $\boldsymbol{H}(\boldsymbol{x},t)$ to depend on $\phi(x,t)$ it must appear through the electric strength, as $\boldsymbol{E}(x,t) = -\nabla\Phi(x,t)$. But then $\nabla\times\boldsymbol{H}(x,t)$ would mean applying CoG (i.e., $\nabla\times\nabla\phi = 0$) on the right side of the equation. Since this term would be zero; it is assumed to be zero, thus $\boldsymbol{H}(x,t)$ dependents on only $\boldsymbol{A}(x,t)$. To fill out the symmetry, we have added $\partial_t\boldsymbol{D}(\boldsymbol{x},t)$ to Eq. 5.8.2, to see what might happen in a more general case. ■

5.8.1 *Use of Helmholtz's Theorem on Potential Solutions*

The generalized solutions to Maxwell's equations (Eqs. 5.8.1 and 5.8.2) have been expressed in terms of EM potentials $\phi(\boldsymbol{x})$ and $\boldsymbol{A}(\boldsymbol{x})$ and Helmholtz's theorem. These are solutions to Maxwell's equations expressed in terms of the potentials $\phi(\boldsymbol{x},s)$ and $\boldsymbol{A}(\boldsymbol{x},s)$, as determined at the boundaries (Sommerfeld 1952, p. 146). These relationships are invariant to certain functions added to each potential, as shown below. They are equivalent to Maxwell's equations following the application of $\nabla\cdot$ and $\nabla\times$.

Next we show that the potential equations (Eqs. 5.8.1 and 5.8.2) are consistent with Maxwell's equations (Eq. 5.7.1).

5.8.2 *ME for $\boldsymbol{E}(\boldsymbol{x},t)$*

Taking the curl of Eq. 5.8.1, applying CoG = 0, and using Eq. 5.8.2, we find that

$$\begin{aligned}\nabla\times\boldsymbol{E} &= -\cancelto{0}{\nabla\times\nabla\Phi} - \nabla\times\frac{\partial\boldsymbol{A}}{\partial t}\\ &= -\frac{\partial\boldsymbol{B}}{\partial t}\end{aligned} \tag{5.8.3}$$

recovers Maxwell's equation for $\boldsymbol{E}(\boldsymbol{x})$ (Eq. 5.7.1).

Taking the divergence of Eq. 5.8.2 and applying DoC = 0 give Eq. 5.7.2 for $\boldsymbol{B}(\boldsymbol{x})$:

$$\nabla\cdot\boldsymbol{B}(\boldsymbol{x}) = \cancelto{0}{\nabla\cdot\nabla\times\boldsymbol{A}(\boldsymbol{x})} = 0.$$

5.8.3 ME for $H(x, t)$

To recover Maxwell's equation for $\boldsymbol{H}(\boldsymbol{x})$ (Eq. 5.7.1, $\nabla\times\boldsymbol{H} = \boldsymbol{C}$) from the potential equation (Eq. 5.8.2), we take the curl and use $\boldsymbol{B} = \epsilon_o\boldsymbol{H}$ (Table 5.4):

$$\begin{aligned}
\nabla\times\boldsymbol{B}(\boldsymbol{x}) &= \mu_o\nabla\times\boldsymbol{H}(\boldsymbol{x})\\
&= \nabla\times\nabla\times\boldsymbol{A}(\boldsymbol{x})\\
&= \mathrm{v}^2\,\boldsymbol{A}(\boldsymbol{x},t) - \nabla^2\boldsymbol{A}(\boldsymbol{x},t)\\
&= \nabla\nabla\cdot\boldsymbol{A}(\boldsymbol{x},t) - \frac{1}{c_o^2}\frac{\partial^2}{\partial_t^2}\boldsymbol{A}(\boldsymbol{x},t)\\
&= -\frac{1}{c^2}\left(\ddot{\mathrm{A}} + \nabla\dot{\Phi}\right) + \mu_o\mathbf{J}.
\end{aligned}$$

This last equation may be split into two independent equations by the use of Helmholtz's theorem:

$$\nabla^2\boldsymbol{A} - \frac{1}{c_o^2}\ddot{\boldsymbol{A}} = -\mu_o\mathbf{J} \quad \text{and} \quad \nabla\cdot\boldsymbol{A} + \frac{1}{c_o^2}\ddot{\Phi} = 0.$$

Taking the divergence of Eq. 5.8.2 and applying DoC = 0 gives Eq. 5.7.2 ($\nabla\cdot\boldsymbol{D} = -\dot{\rho}$). Alternatively,

$$\nabla^2\Phi - \frac{1}{c_o^2}\Phi = -\frac{\rho}{\epsilon_o},$$

which is the scalar potential wave equation driven by the charge (Sommerfeld 1952, p. 146).

5.8.4 Summary

In conclusion, Eq. 5.8.1, along with DoC $= 0$ and CoG $= 0$, gives Maxwell's Eqs. 5.7.1 and 5.7.2 for $\boldsymbol{E}$. Likewise, Eq. 5.8.2, along with DoC $= 0$ and CoG $= 0$, gives Maxwell's Eqs. 5.7.1 and 5.7.2 for $\boldsymbol{H}$. The above derivation for $H(x, t)$ from A and Φ derives the magnetic component of the field, expressed in terms of its vector potential, in the same way as Eq. 5.7.1 describes $\boldsymbol{E}(x, t)$ in terms of the potentials.

We may view the potential equations (Eqs. 5.8.1 and 5.8.2) as equivalent to Maxwell's equations; thus they are the solutions to ME in terms of potentials.

Exercise #22 Starting with the values of the speed of light $c_o = 3\times 10^8$ [m/s] and the characteristic resistance of light waves $r_o = 377$ [ohms], use the formulas $c_0 = 1/\sqrt{\mu_o\epsilon_o}$ and $r_o = \sqrt{\epsilon_o/\mu_o}$ to find values for ϵ_o and μ_o.

Solution: Squaring $c_o^2 = 1/\mu_o\epsilon_o$ and $r_o^2 = \mu_o/\epsilon_o$, we may solve for the two unknowns: $c_o^2r_o^2 = \frac{1}{\not{\mu_o}\epsilon_o}\frac{\not{\mu_o}}{\epsilon_o} = 1/\epsilon_o^2$; thus $\epsilon_o = 1/c_or_o = 10^{-8}/2.998\cdot 377 \approx 8.85\times$

10^{-12} [F/m]. Likewise, $\mu_o = r_o/c_o = (377/2.998) \times 10^{-8} \approx 125.75 \times 10^{-8}$ [H/m]. The value of μ_o is defined in the international SI units as $4\pi \times 10^{-7} \approx 12.56610^{-7}$ [H/m].

It is more productive to memorize c_0 and r_0, from which ϵ_o and μ_o may be quickly derived. ■

Exercise #23 Starting from Eq. 5.7.1, with $\boldsymbol{J_m} = \sigma_o \boldsymbol{E}$, Maxwell's equation, including the magnetic intensity, is

$$\nabla \times \boldsymbol{H}(\boldsymbol{x}, t) = \boldsymbol{J_m}(\boldsymbol{x}, t) + \frac{\partial}{\partial t}\boldsymbol{D}(\boldsymbol{x}, t).$$

Find the equation for the magnetically induced current $\boldsymbol{J_m}(\boldsymbol{x}, t)$.

Solution: The divergence of the curl is zero (DoC = 0),

$$\cancelto{0}{\nabla\cdot\nabla\times \boldsymbol{H}}(\boldsymbol{x}, t) = \nabla\cdot \boldsymbol{J_m}(\boldsymbol{x}, t) + \frac{\partial}{\partial t}\rho(\boldsymbol{x}, t) = 0, \qquad (5.8.4)$$

which is conservation of charge (i.e., Gauss's theorem). ■

5.9 Problems VC-2

5.9.1 *Topics of This Homework*

Partial differential equations; fundamental theorem of vector calculus (Helmholtz's theorem); wave equation; Maxwell's equations (ME) and variables ($\boldsymbol{E}$, $\boldsymbol{D}$; $\boldsymbol{B}$, $\boldsymbol{H}$); Second-order vector differentials; Webster horn equation.

Notation: The following notation is used in this homework:

1. $s = \sigma + j\omega$ is the Laplace frequency, as used in the Laplace transform.
2. A Laplace transform pair is indicated by the symbol $\leftrightarrow$: for example, $f(t) \leftrightarrow F(s)$.
3. π_k is the kth prime; for example, $\pi_k \in \mathbb{P}$, $\pi_k = [2, 3, 5, 7, 11, 13, \ldots]$ for $k = 1, \ldots, 6$).

5.9.2 *Partial Differential Equations (PDEs): Wave Equation*

Problem #1 *Solve the wave equation in one dimension by defining* $\xi = t \mp x/c$.

–1.1: Show that d'Alembert's solution, $\varrho(x, t) = f(t - x/c) + g(t + x/c)$, *is a solution to the acoustic pressure wave equation in one dimension:*

$$\frac{\partial^2 \varrho(x,t)}{\partial x^2} = \frac{1}{c^2}\frac{\partial^2 \varrho(x,t)}{\partial t^2},$$

where $f(\xi)$ and $g(\xi)$ are arbitrary functions.

Problem #2 *Solving the wave equation in spherical coordinates (i.e., three dimensions)*

–2.1: Write the wave equation in spherical coordinates $\varrho(r, \theta, \phi, t)$. Consider only the radial term r (i.e., dependence on angles θ and ϕ is assumed to be zero). Hint: The form of the Laplacian as a function of the number of dimensions is given in *Eq. 5.1.9*). Alternatively, look it up on the Internet or in a calculus book.

–2.2: Show that this equation is true:

$$\nabla_r^2 \varrho(r) \equiv \frac{1}{r^2}\frac{\partial}{\partial r} r^2 \frac{\partial}{\partial r}\varrho(r) = \frac{1}{r}\frac{\partial^2}{\partial r^2} r\,\varrho(r). \tag{VC-2.1}$$

Hint: Expand both sides of the equation.

–2.3: Use the results from Eq. VC-2.1 to show that the solution to the spherical wave equation is

$$\nabla_r^2\, \varrho(r,t) = \frac{1}{c^2}\frac{\partial^2}{\partial t^2}\,\varrho(r,t) \tag{VC-2.2}$$

$$\varrho(r,t) = \frac{f(t - r/c)}{r} + \frac{g(t + r/c)}{r}. \tag{VC-2.3}$$

–2.4: Using $f(\xi) = \sin(\xi)u(\xi)$ and $g(\xi) = e^{\xi}u(\xi)$, write the solutions to the spherical wave equation, where $u(\xi)$ is the Heaviside step function.

–2.5: Sketch this $f(\xi)$ and $g(\xi)$ for several times (e.g., 0, 1, and 2 seconds), and describe the behavior of the pressure $\varrho(r,t)$ as a function of time t and radius r.

–2.6: What happens when the inbound wave reaches the center at $r = 0$?

5.9.3 Helmholtz's Formula

Every differentiable vector field may be written as the sum of a scalar potential ϕ and a vector potential $\boldsymbol{w}$. This relationship is best known as the fundamental theorem of vector calculus (also called Helmholtz's formula):

$$\boldsymbol{v} = -\nabla\phi + \nabla\times\boldsymbol{w}. \tag{VC-2.4}$$

This formula seems to be a natural extension of the algebraic products $\boldsymbol{A}\cdot\boldsymbol{B} \perp \boldsymbol{A}\times\boldsymbol{B}$, since $\boldsymbol{A}\cdot\boldsymbol{B} \propto \|\boldsymbol{A}\|\,\|\boldsymbol{B}\|\cos(\theta)$ and $\boldsymbol{A}\times\boldsymbol{B} \propto \|\boldsymbol{A}\|\,\|\boldsymbol{B}\|\sin(\theta)$, as developed in Appendix A.3.1. Thus these orthogonal components have magnitude 1 when we take the norm, due to Euler's identity ($\cos^2(\theta) + \sin^2(\theta) = 1$).

As shown in Table 5.1, Helmholtz's formula separates a vector field (i.e., $\mathbf{v}(\mathbf{x})$) into compressible and rotational parts:

1. The rotational (e.g., angular) part is defined by the vector potential $\boldsymbol{w}$, which requires that $\nabla \times \nabla \times \boldsymbol{w} \neq 0$. A field is irrotational (conservative) when $\nabla \times \boldsymbol{v} = 0$, meaning that the field $\boldsymbol{v}$ can be generated using only a scalar potential, $\boldsymbol{v} = \nabla\phi$ (note that this is how a conservative field is usually defined, by saying there exists some ϕ such that $\boldsymbol{v} = \nabla\phi$).[18]
2. The compressible (e.g., radial) part of a field is defined by the scalar potential ϕ, which requires that $\nabla \cdot \nabla\phi = \nabla^2\phi \neq 0$. A field is incompressible (solenoidal) when $\nabla \cdot \boldsymbol{v} = 0$, meaning that the field $\boldsymbol{v}$ can be generated using only a vector potential, $\boldsymbol{v} = \nabla \times \boldsymbol{w}$.

The definitions and generating potential functions of irrotational (conservative) and incompressible (solenoidal) fields naturally follow from two key vector identities: (1) $\nabla \cdot (\nabla \times \boldsymbol{w}) = 0$ and (2) $\nabla \times (\nabla\phi) = 0$.

Problem #3 *Define the following:*
–*3.1: A conservative vector field*
–*3.2: An irrotational vector field*
–*3.3: An incompressible vector field*
–*3.4: A solenoidal vector field*
–*3.5: When is a conservative field irrotational?*
–*3.6: When is an incompressible field irrotational?*

Problem #4 *For each of the following, (i) compute $\nabla \cdot \boldsymbol{v}$, (ii) compute $\nabla \times \boldsymbol{v}$, and (iii) classify the vector field (e.g., conservative, irrotational, incompressible, etc.).*
–*4.1:* $\boldsymbol{v}(x, y, z) = -\nabla(3yx^3 + y\log(xy))$
–*4.2:* $\boldsymbol{v}(x, y, z) = xy\hat{\mathbf{x}} - z\hat{\mathbf{y}} + f(z)\hat{\mathbf{z}}$
–*4.3:* $\boldsymbol{v}(x, y, z) = \nabla \times (x\hat{\mathbf{x}} - z\hat{\mathbf{y}})$

[18] A note about the relationship between the generating function and the test: You might imagine special cases where $\nabla \times \boldsymbol{w} \neq 0$ but $\nabla \times \nabla \times \boldsymbol{w} = 0$ (or $\nabla\phi \neq 0$ but $\nabla^2\phi = 0$). In these cases, the vector (or scalar) potential can be recast as a scalar (or vector) potential. For example, consider a field $\boldsymbol{v} = \nabla\phi_0 + \boldsymbol{b}$, where $\boldsymbol{b} = x\hat{\mathbf{x}} + y\hat{\mathbf{y}} + z\hat{\mathbf{z}}$. Note that $\boldsymbol{b}$ can actually be generated by either a scalar potential ($\phi_1 = \frac{1}{2}[x^2 + y^2 + z^2]$, such that $\nabla\phi_1 = \boldsymbol{b}$) or a vector potential ($\boldsymbol{w}_0 = \frac{1}{2}[z^2\hat{\mathbf{x}} + x^2\hat{\mathbf{y}} + y^2\hat{\mathbf{z}}]$, such that $\nabla \times \boldsymbol{w}_0 = \boldsymbol{b}$). We find that $\nabla \times \boldsymbol{v} = 0$; therefore $\boldsymbol{v}$ must be irrotational. We say this irrotational field is generated by $\nabla\phi = \nabla(\phi_0 + \phi_1)$.

5.9.4 *Maxwell's Equations*

The variables have the following names and defining equations (see Table 5.4):

Symbol	Equation	Name	Units
$\boldsymbol{E}$	$\nabla \times \boldsymbol{E} = -\dot{\boldsymbol{B}}$	Electric field strength	[volts/m]
$\boldsymbol{D} = \epsilon_o \boldsymbol{E}$	$\nabla \cdot \boldsymbol{D} = \rho$	Electric displacement (flux density)	[coul/m^2]
$\boldsymbol{H}$	$\nabla \times \boldsymbol{H} = \boldsymbol{J} + \dot{\boldsymbol{D}}$	Magnetic field strength	[amps/m]
$\boldsymbol{B} = \mu_o \boldsymbol{H}$	$\nabla \cdot \boldsymbol{B} = 0$	Magnetic induction (flux density)	[webers/m^2]

Note that $\boldsymbol{J} = \sigma \boldsymbol{E}$ is the *current density* (which has units of [amps/m^2]). Furthermore, the *speed of light in-vacuo* is $c_o = 3 \times 10^8 = 1/\sqrt{\mu_o \epsilon_o}$ [m/s], and the *characteristic resistance* of light $r_o = 377 = \sqrt{\mu_o/\epsilon_o}$ [Ω (i.e., ohms)].

5.9.4.1 Speed of Light

Problem #5 *The* speed of light *in-vacuo is* $c_o = 1/\sqrt{\mu_o \epsilon_o} \approx 3 \times 10^8$ *[m/s]. The* characteristic resistance *in-vacuo is* $r_o = \sqrt{\mu_o/\epsilon_o} \approx 377$ [Ω].

–5.1: Find a formula for the in-vacuo permittivity ϵ_o *and* permeability *in terms of* c_o *and* r_o*. Based on your formula, what are the numeric values of* ϵ_o *and* μ_o*?*

–5.2: In a few words, identify the law given by this equation, define what it means, and explain the formula:]

$$\int_S \hat{\mathbf{n}} \cdot \boldsymbol{v}\, dA = \int_{\mathcal{V}} \nabla \cdot \boldsymbol{v}\, dV.$$

5.9.4.2 Application of Maxwell's Equations

Problem #6 *The electric Maxwell equation is* $\nabla \times \mathbf{E} = -\dot{\boldsymbol{B}}$*, where* $\mathbf{E}$ *is the* electric field strength *and* $\dot{\boldsymbol{B}}$ *is the time rate of change of the* magnetic induction field, *or simply the magnetic flux density. Consider this equation integrated over a two-dimensional surface S, where* $\hat{\mathbf{n}}$ *is a unit vector normal to the surface (you may also find it useful to define the closed path C around the surface):*

$$\iint_S [\nabla \times \boldsymbol{E}] \cdot \hat{\mathbf{n}} dS = -\frac{\partial}{\partial t} \iint_S \boldsymbol{B} \cdot \hat{\mathbf{n}} dS.$$

–6.1: Apply Stokes's theorem to the left-hand side of the equation.

–6.2: Consider the right-hand side of the equation. How is it related to the magnetic flux Ψ *through the surface* S*?*

–6.3: Assume the right-hand side of the equation is zero. Can you relate your answer in question 6.1 to one of Kirchhoff's laws?

Problem #7 *The magnetic Maxwell equation is* $\nabla \times \boldsymbol{H} = \boldsymbol{C} \equiv \boldsymbol{J} + \dot{\boldsymbol{D}}$, *where* $\boldsymbol{H}$ *is the* magnetic field strength, $\boldsymbol{J} = \sigma \boldsymbol{E}$ *is the conductive (resistive) current density, and the* displacement current $\dot{\boldsymbol{D}}$ *is the time rate of change of the* electric flux density $\boldsymbol{D}$. *Here we defined a new variable* $\boldsymbol{C}$ *as the total current density.*

–*7.1: First consider the equation over a two-dimensional surface* S*:*

$$\iint_S [\nabla \times \boldsymbol{H}] \cdot \hat{\mathbf{n}} dS = \iint_S [\boldsymbol{J} + \dot{\boldsymbol{D}}] \cdot \hat{\mathbf{n}} dS = \iint_S \boldsymbol{C} \cdot \hat{\mathbf{n}} dS.$$

Then apply Stokes's theorem to the left-hand side of this equation. In a sentence or two, explain the meaning of the resulting equation. Hint: What is the right-hand side of the equation?

Problem #8 *Consider the next equation in three dimensions. Take the divergence of both sides and integrate over a volume* V *(closed surface* S*):*

$$\iiint_V \nabla \cdot [\nabla \times \mathbf{H}] dV = \iiint_V \nabla \cdot \mathbf{C} dV.$$

–*8.1: What happens to the left-hand side of this equation?* Hint: Can you apply a vector identity? *Apply the divergence theorem (sometimes known as Gauss's theorem) to the right-hand side of the equation, and interpret your result.* Hint: Can you relate your result to one of Kirchhoff's laws?

5.9.5 *Second-Order Differentials*

Problem #9 *This problem is about second-order vector differentials.*

–*9.1: If* $\boldsymbol{v}(x, y, z) = \nabla \phi(x, y, z)$, *then what is* $\nabla \cdot \boldsymbol{v}(x, y, z)$*?*

–*9.2: Evaluate* $\nabla^2 \phi$ *and* $\nabla \times \nabla \phi$ *for* $\phi(x, y) = xe^y$.

–*9.3: Evaluate* $\nabla \cdot (\nabla \times \boldsymbol{v})$ *and* $\nabla \times (\nabla \times \boldsymbol{v})$ *for* $\boldsymbol{v} = x\hat{\mathbf{x}} + y\hat{\mathbf{y}} + z\hat{\mathbf{z}}$.

–*9.4: When* $\boldsymbol{V}(x, y, z) = \nabla(1/x + 1/y + 1/z)$, *what is* $\nabla \times \boldsymbol{V}(x, y, z)$*?*

–*9.5: When was Maxwell born and when did he die? How long did he live (within* ± 10 *years)?*

5.9.6 *Capacitor Analysis*

Problem #10 *Find the solution to the Laplace equation between two infinite*[19] *parallel plates separated by a distance* d. *Assume that the left plate at* $x = 0$ *is at voltage* $V(0) = 0$ *and the right plate at* $x = d$ *is at voltage* $V_d \equiv V(d)$.

[19]We study plates that are infinite because this means the electric field lines are perpendicular to the plates, running directly from one plate to the other. However, we solve for per-unit-area characteristics of the capacitor.

–10.1: Write Laplace's equation in one dimension for $V(x)$.

–10.2: Write the general solution to your differential equation for $V(x)$.

–10.3: Apply the boundary conditions $V(0) = 0$ and $V(d) = V_d$ to solve for the constants in your equation from question 10.2.

–10.4: Find the charge density per unit area ($\sigma = Q/A$, where Q is charge and A is area) on the surface of each plate. Hint: $\boldsymbol{E} = -\nabla V$, and Gauss's law states that $\iint_S \boldsymbol{D} \cdot \hat{\mathbf{n}} dS = Q_{enc}$.

–10.5: Determine the per-unit-area capacitance C of the system.

5.9.7 Webster Horn Equation

Problem #11 *Horns illustrate an important generalization of the solution of the one-dimensional wave equation in regions where the properties (i.e., area of the tube) vary along the axis of wave propagation. Classic applications of horns are in vocal tract acoustics, loudspeaker design, cochlear mechanics, and any case that has wave propagation. Write the formula for the Webster horn equation, and explain the variables.*

5.10 Further Readings

The above concepts come straight from mathematical physics, as developed in the seventeenth–nineteenth centuries. Much of this was first developed in acoustics by Helmholtz, Stokes, and Rayleigh, following through Green's footsteps, as described by Rayleigh (1896). When it comes to fully appreciating Green's theorem and reciprocity, I have found Rayleigh (1896) to be a key reference. To repeat my reading experience, start with Brillouin (1953, 1960), followed by Sommerfeld (1952), Pipes (1958). Second-tier reading contains many items: Morse (1948), Sommerfeld (1949), Morse and Feshbach (1953), Ramo et al. (1965), Feynman (1970a); Boas (1987). A third tier might include Helmholtz (1863a), Fry (1928), Lamb (1932), Bode (1945), Montgomery et al. (1948), Beranek (1954), Fagen (1975), Lighthill (1978), Hunt (1952), Olson (1947). Other physics writings include the impressive series of mathematical-physics textbooks by authors J.C. Slater, and the Landau and Lifshitz.[20]

[20] https://www.amazon.com/Mechanics-Third-Course-Theoretical-Physics/dp/0750628960.

Appendix A
Notation

A.1 Number Systems

The notation used in this book is defined in this appendix so that it may be quickly accessed.[1] Where the definition is sketchy, page numbers are provided where these concepts are fully explained, along with many other important and useful definitions. For example, a discussion of $\mathbb{N}$ may be found on Sect. 2.1. Math symbols such as $\mathbb{N}$ may be found at the top of the index, since they are difficult to alphabetize.

A.1.1 Units

Strangely, or not, classical mathematics, as taught today in schools, does not seem to acknowledge the concept of physical units. Units seem to have been abstracted away. This makes mathematics distinct from physics, where almost everything has units. Presumably this makes mathematics more general (i.e., abstract). But for the engineering mind, this is not ideal, or worse, as it necessarily means that important physical meaning, by design, has been surgically removed. We shall use SI units whenever possible, which means this book is not a typical book on mathematics. Spatial coordinates are quoted in meters [m], and time in seconds [s]. Angles in degrees have no units, whereas radians have units of inverse-seconds [s^{-1}]. A complete list of SI units may be found at https://physics.nist.gov/cuu/pdf/sp811.pdf and Graham et al. (1994) for a discussion of basic math notation.

When writing a complex number we shall adopt $1\jmath$ to indicate $\sqrt{-1}$. MATLAB/Octave allows either $1\imath$ or $1\jmath$.

Units are SI; angles are in degrees [deg] unless otherwise noted. The units for π are always radians [rad]. For example $\sin(\pi)$, $e^{\jmath 90^\circ} = e^{\jmath\pi/2}$.

[1] https://en.wikipedia.org/wiki/List_of_mathematical_symbols_by_subject#Definition_symbols.

J. Allen, *An Invitation to Mathematical Physics and Its History*,
https://doi.org/10.1007/978-3-030-53759-3

A.1.2 Symbols and Functions

We use ln as the log function base e, log as base 2, and π_k to indicate the kth prime (e.g., $\pi_1 = 2, \pi_2 = 3$).

It is helpful to know where the letters of the alphabet are. Everyone knows the first letter is /a/ and the last /z/, but what is the 10th or 20th letter? The table below shows that /j/ is the 10th letter, /t/ the 20th, /o/ the 15th, and /z/ the 26th.

index:	1	10	20	+1
+0	a	j	t	—
+5	e	o	y	z

This is helpful for quickly moving around the alphabet, when looking up words in the dictionary or an alphabetized list. If you forget how many letters there are in the English alphabet, this will help you recall it is 26.

When working with Fourier $\mathcal{FT}$ and Laplace $\mathcal{LT}$ transforms, lowercase symbols are in the time domain while uppercase indicates the frequency domain, as $f(t) \leftrightarrow F(\omega)$. An important exception is Maxwell's equations because they are so widely used as uppercase bold letters (e.g., $\boldsymbol{E}(\boldsymbol{x}, \omega)$). It would seem logical to change this to $\boldsymbol{e}(\boldsymbol{x}, t) \leftrightarrow \boldsymbol{E}(\boldsymbol{x}, \omega)$, to conform.

A.1.3 Common Mathematical Symbols

There are many pre-defined symbols in mathematics, too many to summarize here. We shall only use a small subset, defined here.

- $\forall$ *set* is a collection of objects that have a common property, defined by braces. For example, if set $P = \{a, b, c\}$ such that $a^2 + b^2 = c^2$, then members of P obey the Pythagorean theorem. Thus we could say that $\{1, 1, \sqrt{2}\} \in P$.
- Number sets: $\mathbb{N}, \mathbb{P}, \mathbb{Z}, \mathbb{Q}, \mathbb{F}, \mathbb{I}, \mathbb{R}, \mathbb{C}$ are briefly discussed below, and in greater detail in Sect. 2.1.
- One can define sets of sets and subsets of sets, and this is prone (in my experience) to error. For example, what is the difference between the number 0 and the null set $\varnothing = \{0\}$? Is $0 \in \varnothing$? Ask a mathematician. This seems a lackluster construction in the world of engineering.
- A vector is a column n-tuple. For example $[3, 5]^T = \begin{bmatrix} 3 \\ 5 \end{bmatrix}$.
- The symbol $\perp$ is used in different ways to indicate two things are perpendicular, orthogonal, or in disjoint sets. In set theory $A \perp B$ is equivalent to $A \cap B = \varnothing$. If two vectors $\boldsymbol{E}, \boldsymbol{H}$ are perpendicular $\boldsymbol{E} \perp \boldsymbol{H}$, then their inner product $\boldsymbol{E} \cdot \boldsymbol{H} = 0$ is zero. One must infer the meaning of $\perp$ from its context.

Table A.1 List of all upper- and lowercase Greek letters used in the text

A	A	α	alpha	N	N	ν	nu
B	B	β	beta	Ξ	Ξ	ξ	xi
Γ	Γ	γ	gamma	O	O	o	omicron
Δ	Δ	δ	delta	Π	Π	π	pi
E	E	ε	epsilon	P	P	ρ	rho
Z	Z	ζ	zeta	Σ	Σ	σ	sigma
H	H	η	eta	T	T	τ	tau
Θ	Θ	θ	theta	Υ	Υ	υ	upsilon
I	I	ι	iota	Φ	Φ	ϕ	phi
K	K	κ	kappa	X	X	χ	chi
Λ	Λ	λ	lambda	Ψ	Ψ	ψ	psi
M	M	μ	mu	Ω	Ω	ω	omega

A.1.3.1 Greek Letters

Frequently Greek letters, as provided in Table A.1, are associated in engineering and physics with a specific physical meaning. For example, ω [rad] is the radian frequency $2\pi f$, ρ [kgm/m^3] is commonly the density. ϕ, ψ are commonly used to indicate angles of a triangle, and $\zeta(s)$ is the Riemann zeta function. Many of these are so well established it makes no sense to define new terms, so we will adopt these common terms (and define them).

Likely you do not know all of these Greek letters, commonly used in mathematics. Some of them are pronounced in strange ways. The symbol ξ is pronounced "see," ζ is "zeta," β is "beta," and χ is "kie" (rhymes with pie and sky). I will assume you know how to pronounce the others, which are more phonetic in English. One advantage of learning LaTeX the powerful open-source math-oriented word-processing system used to write this book is that math symbols are included, making then easily learned.

A.1.3.2 Double-Bold Notation

Table A.2 indicates the symbol followed by a page number indication where it is discussed, and the Genus (class) of the number type. For example, $\mathbb{N} > 0$ indicates the infinite set of *counting numbers* $\{1, 2, 3, \cdots\}$, not including zero. Starting from any counting number, you get the next one by adding 1. Counting numbers are sometimes called the natural or cardinal numbers.

We say that a number is in the set with the notation $3 \in \mathbb{N} \subset \mathbb{R}$, which is read as "3 is in the set of counting numbers, which in turn is in the set of real numbers," or in vernacular language "3 is a real counting number."

Prime numbers ($\mathbb{P} \subset \mathbb{N}$) are taken from the counting numbers, but do not include 1.

Table A.2 Double-bold notation for the types of numbers. (#) is a page number. Symbol with an exponent denote the dimensionality. Thus $\mathbb{R}^2$ represents the real plane. An exponent of 0 denotes point, e.g., $\jmath \in \mathbb{C}^0$

Symbol (Section)	Genus	Examples	Counter examples
$\mathbb{N}$ (Sect. 2.1)	Counting	1,2,17,3, 10^{20}	–5, 0, π, –10.3, 5j
$\mathbb{P}$ (Sect. 2.1)	Prime	2, 3, 17, 199, 23993	0, 1, 4, 3^2, 12
$\mathbb{Z}$ (Sect. 2.1)	Integer	–1, 0, 17, 5j, -10^{20}	1/2,π, $\sqrt{5}$
$\mathbb{Q}$ (Sect. 2.1)	Rational	2/1, 3/2, 1.5, 1.14	$\sqrt{2}, 3^{-1/3}, \pi$
$\mathbb{F}$ (Sect. 2.1)	Fractional	1/2, 7/22	$1/\sqrt{2}$
$\mathbb{I}$ (Sect. 2.1)	Irrational	$\sqrt{2}, 3^{-1/3}, \pi, e$	Vectors
$\mathbb{R}$ (Sect. 2.1)	Reals	$\sqrt{2}, 3^{-1/3}, \pi$	$2\pi j$
$\mathbb{C}$ (Sect. 2.1)	Complex	$1, \sqrt{2}j, 3^{-j/3}, \pi^j$	Vectors
$\mathbb{G}$	Gaussian integers	$3 - 2\jmath \in \mathbb{Z} \cup \mathbb{C}$	Complex integers

The signed integers $\mathbb{Z}$ include 0 and negative integers. Rational numbers $\mathbb{Q}$ are historically defined to include $\mathbb{Z}$, a somewhat inconvenient definition, since the more interesting class are the *fractionals* $\mathbb{F}$, a subset of rationals $\mathbb{F} \in \mathbb{Q}$ that exclude the integers (i.e., $\mathbb{F} \perp \mathbb{Z}$). This is a useful definition because the rationals $\mathbb{Q} = \mathbb{Z} \cup \mathbb{F}$ are formed from the union of integers and fractionals.

The rationals may be defined, using set notation (a very sloppy notation with an incomprehensible syntax), as

$$\mathbb{Q} = \{p/q : q \neq 0 \;\&\; p, q \in \mathbb{Z}\},$$

which may be read as "the set '$\{\cdots\}$' of all p/q such that ':' $q \neq 0$, 'and' $p, q \subset \mathbb{Z}$." The translation of the symbols is in single ('$\cdots$') quotes.

Irrational numbers $\mathbb{I}$ are very special: They are formed by taking a limit of fractionals, as the numerator and denominator $\to \infty$, and approach a limit point. It follows that irrational numbers must be approximated by fractionals.

The reals ($\mathbb{R}$) include complex numbers ($\mathbb{C}$) having a zero imaginary part (i.e., $\mathbb{R} \subset \mathbb{C}$).

The *size* of a set is denoted by taking the absolute value (e.g., $|\mathbb{N}|$). Normally in mathematics this symbol indicates the cardinality, so we are defining it differently from the standard notation.

A.1.4 Classification of Numbers

From the above definitions there exists a natural hierarchical structure of numbers:

$$\mathbb{P} \subset \mathbb{N}, \quad \mathbb{Z} : \{\mathbb{N}, 0, -\mathbb{N}\}, \quad \mathbb{F} \perp \mathbb{Z}, \quad \mathbb{Q} : \mathbb{Z} \cup \mathbb{F}, \quad \mathbb{R} : \mathbb{Q} \cup \mathbb{I} \subset \mathbb{C}$$

1. The primes are a subset of the counting numbers: $\mathbb{P} \subset \mathbb{N}$.
2. The signed integers $\mathbb{Z}$ are composed of $\pm\mathbb{N}$ and 0, thus $\mathbb{N} \subset \mathbb{Z}$.
3. The fractionals $\mathbb{F}$ do not include of the signed integers $\mathbb{Z}$.
4. The rationals $\mathbb{Q} = \mathbb{Z} \cup \mathbb{F}$ are the union of the signed integers and fractionals.
5. Irrational numbers $\mathbb{I}$ have the special properties $\mathbb{I} \perp \mathbb{Q}$.
6. The reals $\mathbb{R} : \mathbb{Q}, \mathbb{I}$ are the union of rationals and *irrationals* $\mathbb{I}$.
7. Reals $\mathbb{R}$ may be defined as a subset of those complex numbers $\mathbb{C}$ having zero imaginary part.

A.1.5 Rounding Schemes

In MATLAB/Octave there are five different rounding schemes (i.e., mappings): `round(x)`, `fix(x)`, `floor(x)`, `ceil(x)`, `roundb(x)`, with input $x \in \mathbb{R}$ and output $k \in \mathbb{N}$. For example $3 = \lfloor \pi \rceil$, $3 = \lfloor e^1 \rceil = \lfloor 2.7183 \rceil$ rounds to the nearest integer, whereas 3=`floor(`π`)` rounds down while 3=`ceil(`e^1`)` $= \lceil e^1 \rceil$ rounds up. Rounding schemes are used for quantizing a number and generating a remainder. For example: `y=rem(x)` is equivalent to $y = x - \lfloor x \rceil$. Note `round(`π`)` $\equiv \lfloor \pi \rceil$ introduces negative remainders whenever a number rounds up ($\pi = \lceil \pi \rceil - 0.8541$).

The *continued fraction algorithm* (CFA), Sect. 2.4.4 is a recursive rounding scheme, operating on the reciprocal of the remainder. For example:

$$\begin{aligned}\exp(1) &= 3 + 1/(-4 + 1/(2 + 1/(5 + 1/(-2 + 1/(-7))))) - o\left(1.75 \times 10^{-6}\right), \\ &= [3; -4, 2, 5, -2, -7] - o(1.75 \times 10^{-6}).\end{aligned}$$

The expressions in square brackets is a notation for the CFA integer coefficients. The Octave/MATLAB function having $x \in \mathbb{R}$ is either `rat(x)` with output $\in \mathbb{N}$, or `rats(x)`, with output $\in \mathbb{F}$.

A.1.6 Periodic Functions

Fourier series tells us that periodic functions are discrete in frequency, with frequencies given by nT_s, where T_s is the sample period. The discrete $\mathcal{FT}$ (DFT) is a good example. When using the DFT, the sample period is $T_s = 1/2F_{max}$ and the minimum and maximum frequencies are given by $F_{min} = F_{max}/\text{NFT}$, where NFT is the size of the DFT.

This concept is captured by the *Fourier series*, which is a frequency expansion of a periodic function. This concept is quite general. Periodic in frequency implies discrete in time. Periodic and discrete in time requires periodic and discrete in frequency (the case of the DFT). The modulo function $x = \mod(x, y)$ is periodic with period y $(x, y \in \mathbb{R})$.

A periodic function may be conveniently indicated using double-parentheses notation. This is sometimes known as modular arithmetic. For example,

$$f((t))_T = f(t) = f(t \pm kT)$$

is periodic on $t, T \in \mathbb{R}$ with a period of T and $k \in \mathbb{Z}$. This notation is useful when dealing with Fourier series of periodic functions such as $\sin(\theta)$, where $\sin(\theta) = \sin((\theta))_{2\pi} = \mod(\sin(\theta), 2\pi)$.

When a discrete valued (e.g., time $t \in \mathbb{N}$) sequence is periodic with period $N \in \mathbb{Z}$, we use square brackets

$$f[[n]]_N = f[n] = f[n \pm kN],$$

with $k \in \mathbb{Z}$. This notation will be used with discrete-time signals that are periodic, such as the case of the DFT.

It is common for fractions to repeat. For example $1/7 = 0.((142857))$ where the double brackets indicates this number repeats. That is $1/7 = 0.142857$, 142857, 142857, 142857, ...

A.2 Differential Equations Versus Polynomials

A polynomial has *degree* N defined by the largest power. A quadratic equation is degree 2, and a cubic has degree 3. We shall indicate a polynomial by the notation

$$P_N(z) = z^N + a_{N-1}z^{N-1} \cdots a_0.$$

It is a good practice to normalize the polynomial so that $a_N = 1$. This will not change the roots, defined by Eq. 3.1.7 (Sect. 3.1.1). The coefficient on z^{N-1} is always the sum of the roots z_n ($a_{N-1} = \sum_n^N z_n$), and a_0 is always their product ($a_0 = \prod_n^N z_n$). Polynomials with $a_N = 1$ are denoted *monic polynomials*. Because that is a mouthful, it is helpful to call $M_N(s) = P_N(s)/a_N$ a monic.

Differential equations have *order* (polynomials have degree). If a second-order differential equation is Laplace transformed (Sect. 3.10), one is left with a degree 2 polynomial. For example,

$$\frac{d^2}{dt^2}y(t) + b\frac{d}{dt}y(t) + cy(t) = \alpha\left(\frac{d}{dt}x(t) + \beta x(t)\right) \leftrightarrow \tag{A.2.1}$$

$$(s^2 + bs + c)Y(s) = \alpha(s + \beta)X(s). \tag{A.2.2}$$

$$\frac{Y(s)}{X(s)} = \alpha\frac{s + \beta}{s^2 + bs + c} \equiv H(s) \leftrightarrow h(t). \tag{A.2.3}$$

As with monics, the lead coefficient must always be 1. The complex variable $s \in \mathbb{C}$ is the *Laplace frequency*.

The ratio of the output $Y(s)$ over the input $X(s)$ is called the system *transfer function* $H(s)$. The coefficient $\alpha \in \mathbb{R}$ is called the *gain*. The roots of the numerator are called the *zeros* and those of the denominator, the *poles*. When $H(s)$ is the ratio of two degree 1 monics, the transfer function is said to be *bilinear*, since it is linear in both the input and output. In such cases $H(s)$ has only one pole and one zero. The inverse Laplace transform of the transfer function is called the system *impulse response*, which describes the system's output signal $y(t)$ for any given input signal $x(t)$, via convolution (i.e., $y(t) = h(t) \star x(t)$).

A.3 Matrix Algebra: Systems

A.3.1 Vectors

Vectors as columns of ordered sets of scalars $\in \mathbb{C}$. When we write them out in text, we typically use row notation, with the *transpose* symbol:

$$[a, b, c]^T = \begin{bmatrix} a \\ b \\ c \end{bmatrix}.$$

This is strictly to save space on the page. The notation for *conjugate transpose is* $\dagger$, for example

$$\begin{bmatrix} a \\ b \\ c \end{bmatrix}^\dagger = \begin{bmatrix} a^* \; b^* \; c^* \end{bmatrix}.$$

The above example is said to be a *three-dimensional* vector because it has three components.

A.3.2 Vector Products

A *scalar product* (aka dot product) is defined to "weight" vector elements before summing them, resulting in a scalar. The transpose of a vector (a *row-vector*) is typically used as a *scale factor* (i.e., weights) on the elements of a vector. For example,

$$\begin{bmatrix} 1 \\ 2 \\ -1 \end{bmatrix} \cdot \begin{bmatrix} 1 \\ 2 \\ 3 \end{bmatrix} = \begin{bmatrix} 1 \\ 2 \\ -1 \end{bmatrix}^T \begin{bmatrix} 1 \\ 2 \\ 3 \end{bmatrix} = \begin{bmatrix} 1 \; 2 \; -1 \end{bmatrix} \begin{bmatrix} 1 \\ 2 \\ 3 \end{bmatrix} = 1 + 2 \cdot 2 - 3 = 2.$$

A more interesting example is

$$\begin{bmatrix}1\\2\\4\end{bmatrix} \cdot \begin{bmatrix}1\\s\\s^2\end{bmatrix} = \begin{bmatrix}1\\2\\4\end{bmatrix}^T \begin{bmatrix}1\\s\\s^2\end{bmatrix} = \begin{bmatrix}1 & 2 & 4\end{bmatrix} \begin{bmatrix}1\\s\\s^2\end{bmatrix} = 1 + 2s + 4s^2.$$

Polar scalar product: The vector-scalar product in polar coordinates is (Fig. 3.4, Sect. 3.5.1)

$$\boldsymbol{B} \cdot \boldsymbol{C} = \|\boldsymbol{B}\| \, \|\boldsymbol{C}\| \cos\theta \in \mathbb{R},$$

where $\cos\theta \in \mathbb{R}$ is called the *direction-cosine* between $\boldsymbol{B}$ and $\boldsymbol{C}$.

Polar wedge-product: The vector wedge-product in polar coordinates is (Fig. 3.4, Sect. 3.5.1)

$$\boldsymbol{B} \wedge \boldsymbol{C} = \jmath \|\boldsymbol{B}\| \, \|\boldsymbol{C}\| \sin\theta \in \mathbb{R},$$

where $\sin\theta \in \mathbb{R}$ is therefore the *direction-sine* between $\boldsymbol{B}$ and $\boldsymbol{C}$.

Complex polar vector product: From these two polar definitions and $e^{\jmath\theta} = \cos\theta + \jmath \sin\theta$,

$$\boldsymbol{B} \cdot \boldsymbol{C} + \jmath \boldsymbol{B} \wedge \boldsymbol{C} = ||\boldsymbol{B}||||\boldsymbol{C}|| \, e^s.$$

Hence

$$|\boldsymbol{B} \cdot \boldsymbol{C}|^2 + |\boldsymbol{B} \wedge \boldsymbol{C}|^2 = |\|\boldsymbol{B}\|^2 \, \|\boldsymbol{C}\|^2 \cos^2\theta| + |\|\boldsymbol{B}\|^2 \, \|\boldsymbol{C}\|^2 \sin^2\theta| = ||\boldsymbol{B}||^2||\boldsymbol{C}||^2.$$

This relationship holds true in any vector space, of any number of dimensions, containing vectors $\boldsymbol{B}$ and $\boldsymbol{C}$. In this case $s = \sigma + \omega\jmath \in \mathbb{C}$ can be the Laplace frequency. Jaynes (1991) has an relevant discussion about this type of vector product.

A.3.3 *Norms of Vectors*

The norm of a vector is the scalar product of the vector with itself

$$\|\boldsymbol{A}\| = \sqrt{\boldsymbol{A} \cdot \boldsymbol{A}} \geq 0,$$

forming the Euclidean length of the vector.

Euclidean distance between two points in $\mathbb{R}^3$: The scalar product of the difference between two vectors $(\boldsymbol{A} - \boldsymbol{B}) \cdot (\boldsymbol{A} - \boldsymbol{B})$ is the Euclidean distance between the points they define

$$\|\boldsymbol{A} - \boldsymbol{B}\| = \sqrt{(a_1 - b_1)^2 + (a_2 - b_2)^2 + (a_3 - b_3)^2}.$$

Triangle inequality

$$\|\boldsymbol{A}+\boldsymbol{B}\| = \sqrt{(a_1+b_1)^2+(a_2+b_2)^2+(a_3+b_3)^2} \le \|\boldsymbol{A}\| + \|\boldsymbol{B}\|.$$

In terms of a right triangle this says the sum of the lengths of the two sides is greater to the length of the hypotenuse, and equal when the triangle degenerates into a line.

Vector cross product: The *vector product* (aka cross product) $\boldsymbol{A}\times\boldsymbol{B} = \|\boldsymbol{A}\|\,\|\boldsymbol{B}\|\sin\theta$ is defined between the two vectors $\boldsymbol{A}$ and $\boldsymbol{B}$. In Cartesian coordinates

$$\boldsymbol{A}\times\boldsymbol{B} = \det\begin{vmatrix} \hat{\mathbf{x}} & \hat{\mathbf{y}} & \hat{\mathbf{z}} \\ a_1 & a_2 & a_3 \\ b_1 & b_2 & b_3 \end{vmatrix}.$$

The triple product: This is defined between three vectors as

$$\boldsymbol{A}\cdot(\boldsymbol{B}\times\boldsymbol{C}) = \det\begin{vmatrix} a_1 & a_2 & a_3 \\ b_1 & b_2 & b_3 \\ c_1 & c_2 & c_3 \end{vmatrix}.$$

This may be indicated without the use of parentheses, since there can be no other meaningful interpretation. However for clarity, parentheses should be used. The triple product is the volume of the parallelepiped (3D-crystal shape) outlined by the three vectors, as shown in Fig. 3.4 Sect. 3.5.1.

Dialects of vector notation: Physical fields are, by definition, functions of space $\boldsymbol{x}$ [m], and in the most general case, time t[s]. When Laplace transformed, the fields become functions of space and complex frequency (e.g., $\boldsymbol{E}(\boldsymbol{x},t) \leftrightarrow \boldsymbol{E}(\boldsymbol{x},s)$). As before, there are several equivalent vector notations. For example, $\boldsymbol{E}(\boldsymbol{x},t) = \begin{bmatrix} E_x, & E_y, & E_z \end{bmatrix}^T = E_x(\boldsymbol{x},t)\hat{\mathbf{x}} + E_y(\boldsymbol{x},t)\hat{\mathbf{y}} + E_z(\boldsymbol{x},t)\hat{\mathbf{z}}$ is "in-line," to save space. The same equation may written in "displayed" notation as:

$$\boldsymbol{E}(\boldsymbol{x},t) = \begin{bmatrix} E_x(\boldsymbol{x},t) \\ E_y(\boldsymbol{x},t) \\ E_z(\boldsymbol{x},t) \end{bmatrix} = \begin{bmatrix} E_x \\ E_y \\ E_z \end{bmatrix}(\boldsymbol{x},t) = \begin{bmatrix} E_x, & E_y, & E_z \end{bmatrix}^T \equiv E_x\hat{\mathbf{x}} + E_y\hat{\mathbf{y}} + E_z\hat{\mathbf{z}}.$$

Note the three notations for vectors, bold font, element-wise columns, element-wise transposed rows and dyadic format. These are all shorthand notations for expressing the vector. Such usage is similar to a dialect in a language.

Complex elements: When the elements are complex ($\in \mathbb{C}$), the transpose is defined as the complex conjugate of the elements. In such complex cases the transpose conjugate may be denoted with a $\dagger$ rather than T

$$\begin{bmatrix} -2\jmath \\ 3\jmath \\ 1 \end{bmatrix}^\dagger = \begin{bmatrix} 2\jmath & -3\jmath & 1 \end{bmatrix} \in \mathbb{C}.$$

For this case when the elements are complex, the dot product is a real number

$$\boldsymbol{a} \cdot \boldsymbol{b} = \boldsymbol{a}^\dagger \boldsymbol{b} = \begin{bmatrix} a_1^* \ a_2^* \ a_3^* \end{bmatrix} \begin{bmatrix} b_1 \\ b_2 \\ b_3 \end{bmatrix} = a_1^* b_1 + a_2^* b_2 + a_3^* b_3 \in \mathbb{R}.$$

Norm of a complex vector: The dot product of a vector with itself is called the *norm* of $\boldsymbol{a}$

$$\|\boldsymbol{a}\| = \sqrt{\boldsymbol{a}^\dagger \boldsymbol{a}} \geq 0.$$

which is always nonnegative.

Such a construction is useful when $\boldsymbol{a}$ and $\boldsymbol{b}$ are related by an impedance matrix

$$\boldsymbol{V}(s) = \boldsymbol{Z}(s)\boldsymbol{I}(s)$$

and we wish to compute the power. For example, the impedance of a mass is ms and a capacitor is $1/sC$. When given a system of equations (a mechanical or electrical circuit) one may define an impedance matrix.

Complex power: In this special case, the *complex power* $\mathcal{P}(s) \in \mathbb{R}(s)$ is defined, in the complex frequency domain (s), as

$$\mathcal{P}(s) = \boldsymbol{I}^\dagger(s)\boldsymbol{V}(s) = \boldsymbol{I}^\dagger(s)\boldsymbol{Z}(s)\boldsymbol{I}(s) \leftrightarrow p(t) \quad [\text{W}].$$

The real part of the complex power must be positive. The imaginary part corresponds to available stored energy.

The case of three dimensions is special, allowing definitions that are not easily defined in more than three dimensions. A vector in $\mathbb{R}^3$ labels the point having the coordinates of that vector.

A.3.4 Matrices

When working with matrices, the role of the weights and vectors can change, depending on the context. A useful way to view a matrix is as a set of column-vectors, weighted by the elements of the column-vector of weights multiplied from the right. For example,

$$\begin{bmatrix} a_{11} & a_{12} & a_{13} & \cdots & a_{1M} \\ a_{21} & a_{22} & a_{23} & \cdots & a_{2M} \\ & & \ddots & & \\ a_{N1} & a_{N2} & a_{N3} & \cdots & a_{NM} \end{bmatrix} \begin{bmatrix} w_1 \\ w_2 \\ w_3 \\ \vdots \\ w_M \end{bmatrix} = w_1 \begin{bmatrix} a_{11} \\ a_{21} \\ a_{31} \\ \vdots \\ a_{N1} \end{bmatrix} + w_2 \begin{bmatrix} a_{12} \\ a_{22} \\ a_{32} \\ \vdots \\ a_{N2} \end{bmatrix} + \cdots + w_M \begin{bmatrix} a_{1M} \\ a_{2M} \\ a_{3M} \\ \vdots \\ a_{NM} \end{bmatrix},$$

where the weights are $\left[w_1, w_2, \ldots, w_M\right]^T$. Alternatively, the matrix is a set of row vectors of weights, each of which is applied to the column-vector on the right ($[w_1, w_2, \ldots, W_M]^T$).

The determinant of a matrix is denoted as either $\det \boldsymbol{A}$ or simply $|\boldsymbol{A}|$ (as in the absolute value). The inverse of a square matrix is $\boldsymbol{A}^{-1}$ or $\mathrm{inv}\boldsymbol{A}$. If $|A| = 0$, the inverse does not exist. $\boldsymbol{A}\boldsymbol{A}^{-1} = \boldsymbol{A}^{-1}\boldsymbol{A}$.

MATLAB/Octave's notional convention for a row-vector is $[a, b, c]$ and a column-vector is $[a; b; c]$. A prime on a vector takes the complex conjugate transpose. To suppress the conjugation, place a period before the prime. The argument converts the array into a column-vector, without conjugation. A tacit notation in MATLAB is that *vectors* are columns and the index to a vector is a row vector. MATLAB defines the notation 1:4 as the "row-vector" (Ambaum, 2010; Apte, 2009; Arnold and Rogness, 2019; Batchelor, 1967), which is unfortunate as it leads users to assume that the default vector is a row. This can lead to serious confusion later, as MATLAB's default vector is a column. I have not found the above convention explicitly stated, and it took me years to figure this out for myself.

When writing a complex number we shall adopt $1\jmath$ to indicate $\sqrt{-1}$. MATLAB/Octave allows either $1\imath$ or $1\jmath$.

Units are SI; angles are in degrees [deg] unless otherwise noted. The units for π are always radians [rad]. For example $\sin(\pi)$, $e^{\jmath 90^\circ} = e^{\jmath\pi/2}$.

A.3.5 2 × 2 *Complex Matrices*

Here are some definitions to learn:

1. *Scalar*: A number— for example, $\{a, b, c, \alpha, \beta, \ldots\} \in \{\mathbb{Z}, \mathbb{Q}, \mathbb{I}, \mathbb{R}, \mathbb{C}\}$
2. *Vector*: A quantity having direction as well as magnitude, often denoted by a bold letter $\mathbf{x}$, or with an arrow over the top $\vec{x}$. In matrix notation, this is typically represented as a single row $[x_1, x_2, x_3, \ldots]$ or single column $[x_1, x_2, x_3 \ldots]^T$ (where T indicates the transpose). In this text we will typically use column-vectors. The vector may also be written out using unit vector notation to indicate direction. For example: $\mathbf{x}_{3\times 1} = x_1\hat{\mathbf{x}} + x_2\hat{\mathbf{y}} + x_3\hat{\mathbf{z}} = [x_1, x_2, x_3]^T$, where $\hat{\mathbf{x}}, \hat{\mathbf{y}}, \hat{\mathbf{z}}$ are unit vectors in the x, y, z Cartesian directions (here the vector's subscript 3×1 indicates its dimensions). The type of notation used may depend on the engineering problem you are solving.
3. *Matrix:* $A = \left[\mathbf{a}_1, \mathbf{a}_2, \mathbf{a}_3, \ldots, \mathbf{a}_M\right]_{N\times M} = \{a_{n,m}\}_{N\times M}$ can be a non-square matrix if the number of elements in each of the vectors (N) is not equal to the number of vectors (M). When $M = N$, the matrix is square. It may be inverted if its determinant $|A| = \prod \lambda_k \neq 0$ (where λ_k are the eigenvalues). In this text we work only with 2×2 and 3×3 square matrices.
4. *Linear system of equations*: $A\mathbf{x} = \mathbf{b}$, where $\mathbf{x}$ and $\mathbf{b}$ are vectors and matrix A is a square.

(a) *Inverse:* The solution of this system of equations may be found by finding the inverse $\mathbf{x} = A^{-1}\mathbf{b}$.
(b) *Equivalence:* If two systems of equations $A_0\mathbf{x} = \mathbf{b}_0$ and $A_1\mathbf{x} = \mathbf{b}_1$ have the same solution (i.e., $\mathbf{x} = A_0^{-1}\mathbf{b}_0 = A_1^{-1}\mathbf{b}_1$), they are said to be equivalent.
(c) *Augmented matrix:* The first type of augmented matrix is defined by combining the matrix with the right-hand side. For example, given the linear system of equations of the form $A\mathbf{x} = \mathbf{y}$

$$\begin{bmatrix} a & b \\ c & d \end{bmatrix} \begin{bmatrix} x_1 \\ x_2 \end{bmatrix} = \begin{bmatrix} y_1 \\ y_2 \end{bmatrix},$$

the augmented matrix is

$$[A|y] = \left[\begin{array}{cc|c} a & b & y_1 \\ c & d & y_2 \end{array}\right].$$

A second type of augmented matrix may be used for finding the inverse of a matrix (rather than solving a specific instance of linear equations $Ax = b$). In this case the augmented matrix is

$$[A|I] = \left[\begin{array}{cc|cc} a & b & 1 & 0 \\ c & d & 0 & 1 \end{array}\right].$$

Performing Gaussian elimination on this matrix, until the left side becomes the identity matrix, yields A^{-1}. This is because multiplying both sides by A^{-1} gives $A^{-1}A|A^{-1}I = I|A^{-1}$.

5. *Permutation matrix* (P): A matrix that is equivalent to the identity matrix, but with scrambled rows (or columns). Such a matrix has the properties $\det(P) = \pm 1$ and $P^2 = I$. For the 2×2 case, there is only one permutation matrix:

$$P = \begin{bmatrix} 0 & 1 \\ 1 & 0 \end{bmatrix} \qquad P^2 = \begin{bmatrix} 0 & 1 \\ 1 & 0 \end{bmatrix} \begin{bmatrix} 0 & 1 \\ 1 & 0 \end{bmatrix} = \begin{bmatrix} 1 & 0 \\ 0 & 1 \end{bmatrix}.$$

A permutation matrix P swaps rows or columns of the matrix it operates on. For example, in the 2×2 case, pre-multiplication swaps the rows,

$$PA = \begin{bmatrix} 0 & 1 \\ 1 & 0 \end{bmatrix} \begin{bmatrix} a & b \\ \alpha & \beta \end{bmatrix} = \begin{bmatrix} \alpha & \beta \\ a & b \end{bmatrix},$$

whereas post-multiplication swaps the columns,

$$AP = \begin{bmatrix} a & b \\ \alpha & \beta \end{bmatrix} \begin{bmatrix} 0 & 1 \\ 1 & 0 \end{bmatrix} = \begin{bmatrix} b & a \\ \beta & \alpha \end{bmatrix}.$$

For the 3×2 case there are $3 \cdot 2/2 = 3$ such matrices (swap a row with the other 2, then swap the remaining two rows).

6. *Gaussian elimination (GE) operations* G_k: There are three types of elementary row operations, which may be performed without fundamentally altering a system of equations (e.g., the resulting system of equations is *equivalent*). These operations are (1) swap rows (e.g., using a permutation matrix), (2) scale rows, or (3) perform addition/subtraction of two scaled rows. All such operations can be performed using matrices.
   For lack of a better term, we'll describe these as "Gaussian elimination" or "GE" matrices.[2] We will categorize any matrix that performs only elementary row operations (but any number of them) as a "GE" matrix. Therefore, a cascade of GE matrices is also a GE matrix.
   Consider the GE matrix
   $$G = \begin{bmatrix} 1 & 0 \\ 1 & -1 \end{bmatrix}.$$
   (a) This pre-multiplication scales and subtracts row (1) from (2) and returns it to row (2).
   $$GA = \begin{bmatrix} 1 & 0 \\ 1 & -1 \end{bmatrix} \begin{bmatrix} a & b \\ \alpha & \beta \end{bmatrix} = \begin{bmatrix} a & b \\ a - \alpha & b - \beta \end{bmatrix}.$$
   The shorthand for this Gaussian elimination operation is $(1) \leftarrow (1)$ and $(2) \leftarrow (1) - (2)$.
   (b) Post-multiplication adds and scales *columns*.
   $$AG = \begin{bmatrix} a & b \\ \alpha & \beta \end{bmatrix} \begin{bmatrix} 1 & 0 \\ -1 & 1 \end{bmatrix} = \begin{bmatrix} a - b & b \\ \alpha - \beta & \beta \end{bmatrix}.$$
   Here the second column is subtracted from the first, and placed in the first. The second column is untouched. This operation is *not* a Gaussian elimination. Therefore, to put Gaussian elimination operations in matrix form, we form a cascade of pre-multiply matrices.
   Here $\det(G) = 1$, $G^2 = I$, which won't always be true if we scale by a number greater than 1. For instance, if $G = \begin{bmatrix} 1 & 0 \\ m & 1 \end{bmatrix}$ (scale and add), then we have $\det(G) = 1$, $G^n = \begin{bmatrix} 1 & 0 \\ n \cdot m & 1 \end{bmatrix}$.

A.3.6 Derivation of the Inverse of a 2 × 2 Matrix

1. Step 1: To derive (*ii*) starting from (*i*), normalize the first column to 1.

[2]The term "elementary matrix" may also be used to refer to a matrix that performs an elementary row operation. Typically, each elementary matrix differs from the identity matrix by a single row operation. A cascade of elementary matrices could be used to perform Gaussian elimination.

$$\begin{bmatrix} 1 & \frac{b}{a} \\ 1 & \frac{d}{c} \end{bmatrix} \begin{bmatrix} x_1 \\ x_2 \end{bmatrix} = \begin{bmatrix} \frac{1}{a} & 0 \\ 0 & \frac{1}{c} \end{bmatrix} \begin{bmatrix} y_1 \\ y_2 \end{bmatrix}.$$

2. Step 2: Subtract row (1) from row (2): (2) ← (2)–(1)

$$\begin{bmatrix} 1 & \frac{b}{a} \\ 0 & \frac{d}{c} - \frac{b}{a} \end{bmatrix} \begin{bmatrix} x_1 \\ x_2 \end{bmatrix} = \begin{bmatrix} \frac{1}{a} & 0 \\ -\frac{1}{a} & \frac{1}{c} \end{bmatrix} \begin{bmatrix} y_1 \\ y_2 \end{bmatrix}.$$

3. Step 3: Multiply row (2) by ca and express result in terms of the determinate $\Delta = ad - bc$.

$$\begin{bmatrix} 1 & \frac{b}{a} \\ 0 & \Delta \end{bmatrix} \begin{bmatrix} x_1 \\ x_2 \end{bmatrix} = \begin{bmatrix} \frac{1}{a} & 0 \\ -c & a \end{bmatrix} \begin{bmatrix} y_1 \\ y_2 \end{bmatrix}.$$

4. Step 4: Solve row (2) for x_2: $x_2 = -\frac{c}{\Delta} y_1 + \frac{a}{\Delta} y_2$.
5. Step 5: Solve row (1) for x_1:

$$x_1 = \frac{1}{a} y_1 - \frac{b}{a} x_2 = \left[\frac{1}{a} + \frac{b}{a} \frac{c}{\Delta} \right] y_1 - \frac{b}{a} \frac{a}{\Delta} y_2.$$

Rewriting in matrix format, in terms of $\Delta = ad - bc$, gives

$$\begin{bmatrix} x_1 \\ x_2 \end{bmatrix} = \begin{bmatrix} \frac{1}{a} + \frac{bc}{a\Delta} & -\frac{b}{\Delta} \\ -\frac{c}{\Delta} & \frac{a}{\Delta} \end{bmatrix} \begin{bmatrix} y_1 \\ y_2 \end{bmatrix} = \frac{1}{\Delta} \begin{bmatrix} \frac{\Delta + bc}{a} & -b \\ -c & a \end{bmatrix} \begin{bmatrix} y_1 \\ y_2 \end{bmatrix} = \frac{1}{\Delta} \begin{bmatrix} d & -b \\ -c & a \end{bmatrix} \begin{bmatrix} y_1 \\ y_2 \end{bmatrix},$$

since $d = (\Delta + bc)/a$.

Summary: This is a lot of messy algebra, so it is essential that you memorize the final result:
(1) swap the diagonal, (2) change the off-diagonal signs, and (3) normalize by the determinant Δ.

Appendix B
Eigenanalysis

Eigenanalysis is ubiquitous in engineering applications. It is useful in solving differential and difference equations, data-science applications, numerical approximation and computing, and linear algebra applications. Typically one must take a course in linear algebra to become knowledgeable in the inner workings of this method. In this appendix we intend to provide sufficient basics to allow one to read the text.

B.1 The Eigenvalue Matrix (Λ)

Given 2×2 matrix $\boldsymbol{A}$, the related matrix eigenequation is

$$\boldsymbol{A}\boldsymbol{E} = \boldsymbol{E}\Lambda. \tag{B.1.1}$$

Pre-multiplying by $\boldsymbol{E}^{-1}$ diagonalizes $\boldsymbol{A}$, resulting in the *eigenvalue matrix*

$$\Lambda = \boldsymbol{E}^{-1}\boldsymbol{A}\boldsymbol{E} \tag{B.1.2}$$

$$= \begin{bmatrix} \lambda_1 & 0 \\ 0 & \lambda_2 \end{bmatrix}. \tag{B.1.3}$$

Post-multiplying by $\boldsymbol{E}^{-1}$ recovers $\boldsymbol{A}$

$$\boldsymbol{A} = \boldsymbol{E}\Lambda\boldsymbol{E}^{-1} = \begin{bmatrix} a_{11} & a_{12} \\ a_{21} & a_{22} \end{bmatrix}. \tag{B.1.4}$$

Matrix product formula:
This last relation is the entire point of the eigenvector analysis, since it shows that any power of $\boldsymbol{A}$ may be computed from powers of the eigenvalues. Specifically,

$$\boldsymbol{A}^n = \boldsymbol{E}\Lambda^n\boldsymbol{E}^{-1}. \tag{B.1.5}$$

J. Allen, *An Invitation to Mathematical Physics and Its History*,
https://doi.org/10.1007/978-3-030-53759-3

For example, $A^2 = AA = E\Lambda \underbrace{(E^{-1}E)}_{1} \Lambda E^{-1} = E\Lambda^2 E^{-1}$.

Equations B.1.1, B.1.3, and B.1.4 are the key to eigenvector analysis, and you need to memorize them. You will use them repeatedly throughout this text.

$A - \lambda_\pm I_2$ is singular:
If we restrict Eq. B.1.1 to a single eigenvector (one of $e_\pm$), along with the corresponding eigenvalue $\lambda_\pm$, we obtain the two matrix equations

$$Ae_\pm = e_\pm \lambda_\pm = \lambda_\pm e_\pm.$$

Note the swap in the order of $E_\pm$ and $\lambda_\pm$. Since $\lambda_\pm$ is a scalar, this is legal (and critically important), since this allows us to factor out $e_\pm$

$$(A - \lambda_\pm I_2)e_\pm = 0. \tag{B.1.6}$$

The matrix $A - \lambda_\pm I_2$ must be singular because when it operates on $e_\pm$, having nonzero norm, it must be zero.

It follows that its determinant (i.e., $|(A - \lambda_\pm I_2)| = 0$) must be zero. This equation uniquely determines the eigenvalues $\lambda_\pm$.

B.1.1 *Calculating the Eigenvalues* $\lambda_\pm$

The eigenvalues $\lambda_\pm$ of A may be determined from $|(A - \lambda_\pm I_2)| = 0$. As an example we let A be Pell's equation (Eq. 2.6.9, Sect. 2.5.2). In this case the eigenvalues may be found from

$$\begin{vmatrix} 1-\lambda_\pm & N \\ 1 & 1-\lambda_\pm \end{vmatrix} = (1-\lambda_\pm)^2 - N = 0,$$

thus $\lambda_\pm = (1 \mp \sqrt{N})$.[3]

B.1.2 *Calculating the Eigenvectors* $e_\pm$

Once the eigenvalues have been determined, they are substituted into Eq. B.1.6, which determines the eigenvectors $E = \left[e_+, e_-\right]$, by solving

$$(A - \lambda_\pm)\mathbf{e}_\pm = \begin{bmatrix} 1-\lambda_\pm & 2 \\ 1 & 1-\lambda_\pm \end{bmatrix} e_\pm = 0, \tag{B.1.7}$$

[3] It is a convention to order the eigenvalues from largest to smallest.

where $1 - \lambda_\pm = 1 - (1 \mp \sqrt{N}) = \pm\sqrt{N}$, thus the Pell equation eigenvalues are

$$\lambda_\pm = 1 \mp \sqrt{N}.$$

Recall that Eq. B.1.6 is singular because we are using an eigenvalue, and each eigenvector is pointing in a unique direction (this is why it is singular). You might expect that this equation has no solution. In some sense you would be correct. When we solve for $\boldsymbol{e}_\pm$, the two equations defined by Eq. B.1.6 are *co-linear* (the two equations describe parallel lines so their scalar-product is one). This follows from the fact that there is only one eigenvector for each eigenvalue.

Since there is only one eigenvalue we are expecting trouble, yet we may proceed to solve for $\boldsymbol{e}_+ = [e_1^+, e_2^+]^T$ with eigenvalue $+\sqrt{N}$

$$\begin{bmatrix} \sqrt{N} & N \\ 1 & \sqrt{N} \end{bmatrix} \begin{bmatrix} e_1^+ \\ e_2^+ \end{bmatrix} = 0.$$

If we divide the top row by $\sqrt{N}$ the two rows are identical, since the matrix must be singular. Thus this matrix equation gives two identical equations. This is the price of an over-specified equation (the singular matrix is degenerate).

We can determine each eigenvectors direction, but not their magnitudes.

Following the same procedure for $\lambda_- = -\sqrt{N}$, the equation for $\boldsymbol{e}_-$ is

$$\begin{bmatrix} -\sqrt{N} & N \\ 1 & -\sqrt{N} \end{bmatrix} \begin{bmatrix} e_1^- \\ e_2^- \end{bmatrix} = 0.$$

As before, this matrix is singular. Here $e_1^- - \sqrt{N} e_2^- = 0$, thus the eigenvector is $e^- = c\,[\sqrt{N}, 1]^T$, where c is a normalization constant.

Thus the *unnormalized* eigenmatrix is

$$\boldsymbol{E} = \begin{bmatrix} e_1^+ & e_2^- \\ e_2^+ & e_2^- \end{bmatrix} = \begin{bmatrix} \sqrt{N} & -\sqrt{N} \\ 1 & 1 \end{bmatrix}.$$

B.1.3 Normalization of the Eigenvectors

The constant c may be determined by normalizing the eigenvectors to have unit length. Since we cannot determine the length, we set it to 1. Thus the degeneracy may be resolved by the one degree of freedom normalization

$$\left(\pm\sqrt{N}\right)^2 + 1^2 = N + 1 = 1/c^2.$$

Thus the normalization factor to force each eigen vector to have length 1 is $c = 1/\sqrt{N+1}$.

B.2 Pell Equation Solution Example

Section 2.6.2 showed that the solution $[x_n, y_n]^T$ to Pell's equation is given by powers of the Pell matrix $\boldsymbol{A}$. For $N = 2$, in Sect. 2.6.2 we found the explicit formula for $[x_n, y_n]^T$, based on powers of the Pell matrix

$$\boldsymbol{A} = 1_J \begin{bmatrix} 1 & 2 \\ 1 & 1 \end{bmatrix}. \tag{B.2.1}$$

This recursive solution to Pell's equation (Eq. 2.6.7) is Eq. 2.6.9 (Sect. 2.5.2). Thus we need powers of A, that is A^n, which gives an explicit expression for $[x_n, y_n]^T$. By the diagonalization of A, its powers are simply the powers of its eigenvalues.

From MATLAB/Octave with $N = 2$ the eigenvalues of Eq. B.2.1 are $\lambda_\pm \approx [2.4142_J, -0.4142_J]$ (i.e., $\lambda_\pm = 1_J(1 \pm \sqrt{2})$). The solution for $N = 3$ is shown on Sect. B.2.1.

Once the matrix has been diagonalized, one may compute powers of that matrix as powers of the eigenvalues. This results in the general solution given by

$$\begin{bmatrix} x_n \\ y_n \end{bmatrix} = 1_J{}^n \boldsymbol{A}^n \begin{bmatrix} 1 \\ 0 \end{bmatrix} = 1_J{}^n \boldsymbol{E} \Lambda^n \boldsymbol{E}^{-1} \begin{bmatrix} 1 \\ 0 \end{bmatrix}.$$

The eigenvalue matrix D is diagonal with the eigenvalues sorted, largest first. The MATLAB/Octave command `[E,D]=eig(A)` is helpful to find D and E given any A. As we saw above,

$$\Lambda = 1_J \begin{bmatrix} 1+\sqrt{2} & 0 \\ 0 & 1-\sqrt{2} \end{bmatrix} \approx \begin{bmatrix} 2.414_J & 0 \\ 0 & -0.414_J \end{bmatrix}.$$

B.2.1 *Pell Equation Eigenvalue–Eigenvector Analysis*

Here we show how to compute the eigenvalues and eigenvectors for the 2×2 Pell matrix for $N = 2$

$$\boldsymbol{A} = \begin{bmatrix} 1 & 2 \\ 1 & 1 \end{bmatrix}.$$

The MATLAB/Octave command `[E,D]=eig(A)` returns the eigenvector matrix $\boldsymbol{E}$

$$\boldsymbol{E} = [\boldsymbol{e}_+, \boldsymbol{e}_-] = \frac{1}{\sqrt{3}} \begin{bmatrix} \sqrt{2} & -\sqrt{2} \\ 1 & 1 \end{bmatrix} = \begin{bmatrix} 0.8165 & -0.8165 \\ 0.5774 & 0.5774. \end{bmatrix}$$

and the eigenvalue matrix Λ (Matlab/Octave's D)

Table B.1 Summary of the solution of Pell's equation due to the Pythagoreans using matrix recursion, for the case of N = 3. The integer solutions are shown on the right. Note that $x_n/y_n \to \sqrt{3}$, in agreement with the Euclidean algorithm. The MATLAB/Octave program for generating this data is `PellSol3.m`. It seems likely that the powers of β_0 could be absorbed in the starting solution, and then be removed from the recursion

Pell's Equation for $N = 3$

Case of $N = 3$ & $[x_0, y_0]^T = [1, 0]^T$, $\beta_0 = \jmath/\sqrt{2}$

Note: $x_n^2 - 3y_n^2 = 1$, $\quad x_n/y_n \underset{\infty}{\longrightarrow} \sqrt{3}$

$$\begin{bmatrix} x_1 \\ y_1 \end{bmatrix} = \beta_0 \begin{bmatrix} 1 \\ 1 \end{bmatrix} = \beta_0 \begin{bmatrix} 1 & 3 \\ 1 & 1 \end{bmatrix} \begin{bmatrix} 1 \\ 0 \end{bmatrix} \qquad (1\beta_0)^2 - 3(1\beta_0)^2 = 1$$

$$\begin{bmatrix} x_2 \\ y_2 \end{bmatrix} = \beta_0^2 \begin{bmatrix} 4 \\ 2 \end{bmatrix} = \beta_0^2 \begin{bmatrix} 1 & 3 \\ 1 & 1 \end{bmatrix} \begin{bmatrix} 1 \\ 1 \end{bmatrix} \qquad \left(4\beta_0^2\right)^2 - 3\left(2\beta_0^2\right)^2 = 1$$

$$\begin{bmatrix} x_3 \\ y_3 \end{bmatrix} = \beta_0^3 \begin{bmatrix} 10 \\ 6 \end{bmatrix} = \beta_0^3 \begin{bmatrix} 1 & 3 \\ 1 & 1 \end{bmatrix} \begin{bmatrix} 4 \\ 2 \end{bmatrix} \qquad \left(10\beta_0^3\right)^2 - 3\left(6\beta_0^3\right)^2 = 1$$

$$\begin{bmatrix} x_4 \\ y_4 \end{bmatrix} = \beta_0^4 \begin{bmatrix} 28 \\ 16 \end{bmatrix} = \beta_0^4 \begin{bmatrix} 1 & 3 \\ 1 & 1 \end{bmatrix} \begin{bmatrix} 10 \\ 6 \end{bmatrix} \qquad \left(28\beta_0^4\right)^2 - 3\left(16\beta_0^4\right)^2 = 1$$

$$\begin{bmatrix} x_5 \\ y_5 \end{bmatrix} = \beta_0^5 \begin{bmatrix} 76 \\ 44 \end{bmatrix} = \beta_0^5 \begin{bmatrix} 1 & 3 \\ 1 & 1 \end{bmatrix} \begin{bmatrix} 28 \\ 16 \end{bmatrix} \qquad \left(76\beta_0^5\right)^2 - 3\left(44\beta_0^5\right)^2 = 1$$

$$\Lambda \equiv \begin{bmatrix} \lambda_+ & 0 \\ 0 & \lambda_- \end{bmatrix} = \begin{bmatrix} 1+\sqrt{2} & 0 \\ 0 & 1-\sqrt{2} \end{bmatrix} = \begin{bmatrix} 2.4142 & 0 \\ 0 & -0.4142 \end{bmatrix}.$$

The factor $\sqrt{3}$ on $\boldsymbol{E}$ normalizes each eigenvector to 1 (i.e., MATLAB/Octave's command `norm([`$\sqrt{2}$`, 1])` gives $\sqrt{3}$).

In the following discussion we show how to determine $\boldsymbol{E}$ and D (i.e., Λ), given $\boldsymbol{A}$.

B.2.2 Pell's Equation for N = 3

In Table B.1, Pell's equation for $N = 3$ is given, with $\beta_0 = \jmath/\sqrt{2}$. Perhaps try other trivial solutions such as $[-1, 0]^T$ and $[\pm\jmath, 0]^T$, to provide clues to the proper value of β_0 for cases where $N > 3$.[4]

Example: I suggest that you verify $\boldsymbol{E}\Lambda \neq \Lambda\boldsymbol{E}$ and $\boldsymbol{AE} = \boldsymbol{E}\Lambda$ with MATLAB/ Octave. Here is the MATLAB/Octave program which does this:

---

[4]My student Kehan found the general formula for β_o.

```
A = [1 2; 1 1]; %define the matrix
[E,D] = eig(A); %compute the eigenvector and eigenvalue matrices
A*E-E*D      %This should be $\approx 0$, within numerical error.
E*D-D*E      %This is not zero
```

Summary:
Thus far we have shown that for the case of Pell matrix with $N = 2$, the normalized eigenmatrix and its inverse is

$$\boldsymbol{E} = [\boldsymbol{e}_+, \boldsymbol{e}_-] = \frac{1}{\sqrt{3}}\begin{bmatrix} \sqrt{2} & -\sqrt{2} \\ 1 & 1 \end{bmatrix} \qquad \boldsymbol{E}^{-1} = \frac{\sqrt{6}}{4}\begin{bmatrix} 1 & \sqrt{2} \\ -1 & \sqrt{2} \end{bmatrix}$$

and the eigenmatrix is

$$\Lambda = \begin{bmatrix} \lambda_+ & 0 \\ 0 & \lambda_- \end{bmatrix} = \begin{bmatrix} 1+\sqrt{2} & 0 \\ 0 & 1-\sqrt{2} \end{bmatrix}.$$

Note that when working with numeric data it is not necessary to normalize $\boldsymbol{E}$. For example, the form of $\mathbf{e}_1^{\pm} = [1 \pm \lambda^+, 1]^T$ is very simple, and easy to work with. Once normalize it becomes ($N = 2$) $[\sqrt{2}/\sqrt{3}, 1/\sqrt{3}]^T = [0.8165, 0.57735]^T$, obscuring its natural simplicity. The normalization buys little in terms of function.

B.3 Symbolic Analysis of $\mathcal{T}E = E\Lambda$

B.3.1 *The 2 x 2 Transmission Matrix*

Here we assume

$$\mathcal{T} = \begin{bmatrix} \mathcal{A} & \mathcal{B} \\ \mathcal{C} & \mathcal{D} \end{bmatrix}$$

with $\Delta_{\mathcal{T}} = 1$.

The eigenvectors $\boldsymbol{e}_\pm$ of $\mathcal{T}$ are

$$\boldsymbol{e}_\pm = \begin{bmatrix} \frac{1}{2\mathcal{C}}\left[(\mathcal{A}-\mathcal{D}) \mp \sqrt{(\mathcal{A}-\mathcal{D})^2 + 4\mathcal{BC}}\,\right] \\ 1 \end{bmatrix} \tag{B.3.1}$$

and eigenvalues are

$$\lambda_\pm = \frac{1}{2}\left[(\mathcal{A}+\mathcal{D}) \mp \sqrt{(\mathcal{A}-\mathcal{D})^2 + 4\mathcal{BC}}\right]. \tag{B.3.2}$$

Thus the expression under the radical may be rewritten in terms of the determinant of $\mathcal{T}$ (i.e., $\Delta_{\mathcal{T}} = \mathcal{AD} - \mathcal{BC}$) since

$$(\mathcal{A} - \mathcal{D})^2 - (\mathcal{A} + \mathcal{D})^2 = -4\mathcal{A}\mathcal{D}.$$

Thus for the ABCD matrix the expression under the radical becomes

$$\begin{aligned}(\mathcal{A} - \mathcal{D})^2 + 4\mathcal{B}\mathcal{C} &= \mathcal{A}^2 + \mathcal{D}^2 - 4\mathcal{A}\mathcal{D} + 4\mathcal{B}\mathcal{C} \\ &= \mathcal{A}^2 + \mathcal{D}^2 - 4\Delta_{\mathcal{T}}.\end{aligned}$$

Rewriting the eigenvectors and eigenvalues in terms of $\Delta_{\mathcal{T}} = \pm 1$, we find

$$\boldsymbol{e}_{\pm} = \begin{bmatrix} \frac{1}{\mathcal{C}}\left[\frac{\mathcal{A}-\mathcal{D}}{2} \mp \sqrt{\left(\frac{\mathcal{A}+\mathcal{D}}{2}\right)^2 \mp \Delta_{\mathcal{T}}}\right] \\ 1 \end{bmatrix} \tag{B.3.3}$$

and

$$\lambda_{\pm} = \left[\frac{\mathcal{A}+\mathcal{D}}{2} \mp \sqrt{\left(\frac{\mathcal{A}+\mathcal{D}}{2}\right)^2 \mp \Delta_{\mathcal{T}}}\right]. \tag{B.3.4}$$

Note this may be further simplified since the radical is the same.

B.3.2 *Matrices with Symmetry*

For the case of the ABCD matrix the eigenvalues depend on reciprocity, since $\Delta_{\mathcal{T}} = 1$ if $\mathcal{T}(s)$ is reciprocal, and $\Delta_{\mathcal{T}} = -1$ if it is antireciprocal. Thus it is helpful to display the eigenfunctions and values in terms of $\Delta_{\mathcal{T}}$ so this distinction is explicit.

B.3.2.1 Reversible Systems

When $\mathcal{A} = \mathcal{D}$

$$\boldsymbol{E} = \begin{bmatrix} -\sqrt{\frac{\mathcal{B}}{\mathcal{C}}} & +\sqrt{\frac{\mathcal{B}}{\mathcal{C}}} \\ 1 & 1 \end{bmatrix} \qquad \Lambda = \begin{bmatrix} \mathcal{A} - \sqrt{\mathcal{B}\mathcal{C}} & 0 \\ 0 & \mathcal{A} + \sqrt{\mathcal{B}\mathcal{C}} \end{bmatrix} \tag{B.3.5}$$

the transmission matrix is said to be reversible, and the properties greatly simplify.

Note that $r_o = \sqrt{\mathcal{B}/\mathcal{C}}$ is the characteristic impedance and $\kappa(s) = 1/\sqrt{\mathcal{B}\mathcal{C}}$ is the propagation function.

B.3.2.2 Reciprocal Systems

When the matrix is symmetric ($\mathcal{B} = \mathcal{C}$), the corresponding system is said to be *reciprocal*. Most physical systems are reciprocal. The determinant of the transmission matrix of a reciprocal network $\Delta_\mathcal{T} = \mathcal{AD} - \mathcal{BC} = 1$. For example, electrical networks composed of inductors, capacitors, and resistors are always reciprocal. It follows that the complex impedance matrix is symmetric (Van Valkenburg 1964a).

Magnetic systems such as dynamic loudspeakers are antireciprocal, and correspondingly $\Delta_\mathcal{T} = -1$. The impedance matrix of these loudspeakers is skew symmetric (Kim and Allen 2013). All impedance matrices are either symmetric or antisymmetric, depending on whether they are reciprocal (LRC networks) or antireciprocal (magnetic networks). These systems have complex eigenvalues with negative real parts, corresponding to damped (lossy) systems. This follows from conservation of energy. The impedance matrix cannot be Hermitian, or the losses would be zero, because Hermitian matrices have real eigenvalues. Thus a physical system having loss cannot be Hermitian because the eigenvalues must have a negative real parts.

In summary, given a reciprocal system, the $\mathcal{T}$ matrix has $\Delta_\mathcal{T} = 1$, and the corresponding impedance matrix is symmetric (*not* Hermitian).

B.3.3 Impedance Matrix

As previously discussed in Sect. 3.8, the $\mathcal{T}$ matrix corresponding to an impedance matrix is

$$\begin{bmatrix} V_1 \\ V_2 \end{bmatrix} = \mathbf{Z}(s) \begin{bmatrix} I_1 \\ I_2 \end{bmatrix} = \frac{1}{\mathcal{C}} \begin{bmatrix} \mathcal{A} & \Delta_\mathcal{T} \\ 1 & \mathcal{D} \end{bmatrix} \begin{bmatrix} I_1 \\ I_2 \end{bmatrix}.$$

Reciprocal systems have skew-symmetric impedance matrices, namely, $z_{12} = z_{21}$ (i.e., $\Delta_\mathcal{T} = 1$). This condition is best understood using the T form of the impedance matrix, as shown in Fig. 3.9 (Sect. 3.8.2). When the system is both reversible $\mathcal{A} = \mathcal{D}$ and reciprocal, the impedance matrix simplifies to

$$\mathbf{Z}(s) = \frac{1}{\mathcal{C}} \begin{bmatrix} \mathcal{A} & 1 \\ 1 & \mathcal{A} \end{bmatrix}.$$

For such systems there are only two degrees of freedom, $\mathcal{A}$ and $\mathcal{C}$. As discussed previously in Sect. 3.8, each of these has a physical meaning: $1/\mathcal{A}$ is the Thévenin source voltage given a voltage drive and $\mathcal{B}/\mathcal{A}$ is the Thévenin impedance (Sect. 3.8.3).

Impedance is not Hermitian: By definition, when a system is Hermitian its matrix is conjugate symmetric

$$\mathbf{Z}(s) = \mathbf{Z}^\dagger(s),$$

a stronger condition than reciprocal, but not the symmetry of the Brune impedance matrix. A reciprocal Brune impedance is symmetric (not Hermitian).

Appendix C
Laplace Transforms $\mathcal{LT}$

Laplace transforms are discussed in Sect. 3.10, with the definition of the $\mathcal{LT}$ in Eq. 3.10.1 (Sect. 3.10). Level-I (basic) $\mathcal{LT}$s are listed in Table 3.6 (Sect. 3.9.1).

C.1 Tables of Laplace Transforms

The following tables of $\mathcal{LT}$ and $\mathcal{LT}^{-1}$ are a convenient summary of their properties and evaluations for many different functions. Table 3.8 gives basic function properties such as convolution and the properties of step functions and frequency scaling. Table C.1 provides the commands for doing symbolic (computer algebra and calculus) transformations, which includes some unusual $\mathcal{LT}$s and Taylor series of the $\Gamma(s)$ function (Graham et al. 1994), a complex analytic extension of the factorial. Table 3.9 gives the basic transforms typically used for more common calculations. Table C.2 provides extended less common transforms, such as the half-derivative and integration and Bessel functions.

These tables are available in most books on differential equations and remain a core technology for analytic methods for solving differential equations.

C.1.1 $\mathcal{LT}^{-1}$ of the Riemann Zeta Function

The analytic properties of the zeta function $\zeta(s)$ have been a holy grail for mathematicians, starting with Euler, all of whom have made their reputation on that function. For the neophyte, $\zeta(s)$ is important because it is an analytic extension of the sieve, which is the prime identification method. Analytic continuation of the $\zeta(s)$ function was first stated by Riemann, as described in his 1851 paper.[5] But is his definition correct? To resolve this one needs to improve our understanding of the core definition

[5]https://www.youtube.com/watch?v=v9nyNBLCPks.

J. Allen, *An Invitation to Mathematical Physics and Its History*,
https://doi.org/10.1007/978-3-030-53759-3

Table C.1 Symbolic relationships among Laplace transforms. K_3 is a constant

Syms	Command	Result
Syms t s p	Laplace($t^{(p-1)}$)	$\Gamma(p)s^{-p}$
Syms s	Laplace(gamma(s))	$e^{e^{-t}}$
Syms s t a	Laplace(exp(-a*s)/s,s,t)	Heaviside($t - a$)
Syms Gamma s t	Taylor(Gamma,s,t)	$\frac{1}{s} - \gamma + s\left(\frac{\gamma^2}{2} + \frac{\pi^2}{12}\right) +$ $s^2\left(+\frac{1}{6}\text{ polygamma }(2,1) - \frac{\gamma\pi^2}{12} - \frac{\gamma^3}{6}\right) +$ $s^3 K_3 + \cdots$

Table C.2 An extended table of Laplace transforms. J_0, K_1 are Bessel functions of the first and second kind

$f(t) \leftrightarrow F(s)$	Name
$\frac{d^{1/2}}{dt^{1/2}} f(t)u(t) \leftrightarrow \sqrt{s}F(s)$	Half derivative
$\frac{d^{1/2}}{dt^{1/2}} u(t) \leftrightarrow \sqrt{s}$	Half derivative
$\frac{d}{dt}\frac{1}{\sqrt{\pi t}}u(t) \leftrightarrow \frac{s}{\sqrt{s}} = \sqrt{s}$	Semi-inductor
$\frac{1}{\sqrt{\pi t}}u(t) \leftrightarrow \frac{1}{\sqrt{s}}$	Half integration
$\text{erfc}(\alpha\sqrt{t}) \leftrightarrow \frac{1}{s}e^{-2\alpha\sqrt{s}}$ (Morse and Feshbach 1953, p. 1582)	$\alpha > 0$; erfc
$J_0(at)u(t) \leftrightarrow \frac{1}{\sqrt{s^2+a^2}}$	J-Bessel
$J_n(\omega_o t)\, u(t) \leftrightarrow \frac{\left(\sqrt{s^2+\omega_o^2}-s\right)^n}{\omega_o^n\sqrt{s^2+\omega_o^2}}$	
$J_1(t)u(t)/t \leftrightarrow \sqrt{s^2+1} - s$	
$J_1(t)u(t)/t + 2u(t) \leftrightarrow \sqrt{s^2+1} + s = e^{\sinh^{-1}(s)}$	
$\delta(t) + J_1(t)u(t)/t \leftrightarrow \sqrt{s^2+1}$	
$I_0(t)u(t) \leftrightarrow 1/\sqrt{s^2-1}$	I-Bessel
$u(t)/\sqrt{t+1} \leftrightarrow e^s\sqrt{\frac{\pi}{s}}\text{erfc}(\sqrt{s})$	Error function
$\sqrt{t}u(t) * \sqrt{1+t}u(t) \leftrightarrow e^{s/2}K_1(s/2)/2s$	K-Bessel
$\text{Zeta}(t) \leftrightarrow \zeta(s)$	Riemann zeta function

of analytic continuation, as it relates to the $\mathcal{LT}^{-1}$. In the case of the geometric series the analytic continuation is the closed-form expression $f(s) = 1/(1-s)$ which is valid for all $s \neq 1$. This is not Riemann's definition of analytic continuation.

This section is a beginners review of $\zeta(s)$, building on the developments of analytic functions from Chap. 3, especially in Sects. 3.2.6 and 3.3.6. Well understood are the locations of the poles of zeta, which depend on the prime numbers. Not so well understood are the remaining analytic properties over the entire plane, such as the zeros of $\zeta(x)$, namely, the poles of $1/\zeta(s)$. A key function is $\ln \zeta(s)$.

Consider $z \equiv e^{sT}$, where T is the sample period at which data are taken (every T seconds). For example, if $T = 22.676 = 10^6/44,100$ [μs], then the data is sampled at 44.10 [kHz]. This is how modern digital audio works, for CD-quality music. The unit-time delay time operator z^{-1} is

$$\delta(t - T) \leftrightarrow e^{-sT}.$$

When we deal with the Euler and Riemann zeta functions, the only sampling period that makes sense is $T = 1$ [s] or 1 [Hz] (i.e., $n \in \mathbb{Z}$). In this case, the samples of interest are $\text{mod}(n, \pi_k)$. Starting from the sieve of Eratosthenes, Euler showed that the counting numbers $n \in \mathbb{Z}$, presented at a rate of one per second [1-Hz], may be uniquely reduced to multiples of the primes. This is the basis for the fundamental theorem of arithmetic, the theorem of the concept of the prime number, which states that every integer may be uniquely factored into a product of prime numbers.

The zeta function The poles of the zeta function depend explicitly on the primes, which makes it a very special function. In 1737 Euler proposed the real-valued function $\zeta(x) \in \mathbb{R}$, and $x \in \mathbb{R}$ to prove that the number of primes is infinite (Goldstein 1973). Euler's definition of $\zeta(x) \in \mathbb{R}$ is given by the analytic power series,

$$\zeta(x) = \sum_{n=1}^{\infty} \frac{1}{n^x} \qquad \text{for } x > 1 \in \mathbb{R}. \tag{C.1.1}$$

This series converges for $x > 0$, since $R = n^{-x} < 1, n > 1 \in \mathbb{N}$.[6]

In 1860 Riemann extended the zeta function into the complex plane, resulting in $\zeta(s)$, defined by the complex analytic power series, identical to the Euler formula except $x \in \mathbb{R}$ has been replaced by $s \in \mathbb{C}$;

$$\text{Zeta}(t) \leftrightarrow \zeta(s) \equiv \frac{1}{1^s} + \frac{1}{2^s} + \frac{1}{3^s} + \frac{1}{4^s} + \cdots = \sum_{n=1}^{\infty} \frac{1}{n^s} = \sum_{n=1}^{\infty} n^{-s} \qquad \text{for } \Re\{s\} = \sigma > 1. \tag{C.1.2}$$

This formula converges for $\Re\{s\} > 1$ (Goldstein 1973). To determine the formula in other regions of the s plane, we need to extend the series via analytic continuation. As it turns out, Euler's formulation provided detailed information about the structure of primes, going far beyond his original goal.

C.1.1.1 Euler Product Formula

As first published by Euler in 1737, we can recursively factor out the leading prime term, which results in Euler's product formula. Euler's procedure is an algebraic implementation of the sieve of Eratosthenes (Fig. 2.3, Sect. 2.4.1).

Multiplying $\zeta(s)$ by the factor $1/2^s$ and subtracting from $\zeta(s)$ remove all the powers of 2: $1/2^0 + 1/2^s + 1/2^{2s} + 1/2^{3s} + \cdots$

[6]Sanity check: For example, let $n = 2$ and $x > 0$. Then $R = 2^{-\epsilon} < 1$, where $\epsilon \equiv \lim x \to 0^+$. Taking the log gives $\log_2 R = -\epsilon \log_2 2 = -\epsilon < 0$. Since $\log R < 0$, $R < 1$.

$$\left(1-\frac{1}{2^s}\right)\zeta(s)=1+\frac{1}{2^s}+\frac{1}{3^s}+\frac{1}{4^s}+\frac{1}{5^s}+\cdots-\left(\frac{1}{2^s}+\frac{1}{4^s}+\frac{1}{8^s}+\frac{1}{16^s}+\cdots\right), \quad \text{(C.1.3)}$$

which results in

$$\zeta_1(s)=\left(1-2^{-s}\right)\zeta(s)=1+\frac{1}{3^s}+\frac{1}{5^s}+\frac{1}{7^s}+\frac{1}{9^s}+\frac{1}{11^s}+\frac{1}{13^s}+\cdots. \quad \text{(C.1.4)}$$

Repeating this with a lead factor $1/3^s$ applied to Eq. C.1.4 gives[7]

$$\frac{1}{3^s}\left(1-2^{-s}\right)\zeta(s)=\frac{1}{3^s}+\frac{1}{9^s}+\frac{1}{15^s}+\frac{1}{21^s}+\frac{1}{27^s}+\frac{1}{33^s}+\cdots. \quad \text{(C.1.5)}$$

Subtracting Eq. C.1.5 from Eq. C.1.4 cancels the terms on the right-hand side of Eq. C.1.4, giving

$$\zeta_2(s)=\left(1-3^{-s}\right)\left(1-2^{-s}\right)\zeta(s)=1+\frac{1}{5^s}+\frac{1}{7^s}+\frac{1}{11^s}+\frac{1}{13^s}+\frac{1}{17^s}+\frac{1}{19^s}+\cdots.$$

If we express this in terms of the primes π_k, we can better visualize the structure:

$$\zeta_2(s)=\left(1-\pi_2^{-s}\right)\left(1-\pi_1^{-s}\right)\zeta(s)=1+\frac{1}{\pi_3^s}+\frac{1}{\pi_4^s}+\frac{1}{\pi_5^s}+\frac{1}{\pi_6^s}+\frac{1}{\pi_7^s}+\frac{1}{\pi_8^s}+\cdots.$$

Thus ζ_2 has removed primes π_1, π_2, leaving π_3 as the lead term in the series on the right-hand side.

This leads to a recursion in ζ_k,

$$\zeta_k(s)=\zeta(s)\prod_{l=1}^{k}\zeta_l(s)=1+\sum_{l=k+1}^{\infty}\pi_l^{-s}.$$

The series on the right-hand side converges rapidly to 1 as each prime is removed, because the RoC is becoming much larger with each recursion. Each recursive step in this construction ensures that the lead term, along with all of its multiplicative factors, is subtracted out, just like the cancellations with the sieve of Eratosthenes. It is instructive to compare each iteration with that of the sieve (see Fig. 2.3, Sect. 2.4.1).

Repeating this process with the remaining primes removes all the terms on the right-hand side but the first (leaving 1), which results in *Euler's analytic product formula* ($s = x \in \mathbb{R}$), or *Riemann's complex analytic generating function* ($s \in \mathbb{C}$):

[7]This is known as *Euler's sieve*, as distinguished from the *Eratosthenes sieve*.

$$1 = \zeta(s)(1-2^{-s})\cdot(1-3^{-s})\cdot(1-5^{-s})\cdot(1-7^{-s})\cdot\,\cdots\,\cdot(1-\pi_n^{-s})\cdot\,\cdots.$$

$$= \zeta(s)\prod_{k=1}^{\infty}(1-\pi_k^{-s}) \tag{C.1.6}$$

$$\zeta(s) = \frac{1}{\prod_k \mathcal{P}_k(s)}, \quad \Re\{s\} = \sigma > 0, \tag{C.1.7}$$

where the zeros of $\mathcal{P}_k(s) = 1 - \pi_k^{-s}$ define the poles of $\zeta(s)$ for prime π_k.

Finding the RoC of the product formula: It would be interesting to find the RoC for $\mathcal{P}_k(s)$, and for rigor, this question demands further investigation. To find the RoC, we need to evaluate

$$|\pi_k^{-s}| = |e^{-sT_k}| = |e^{-\sigma T_k}| = \left(\frac{1}{\pi_k}\right)^{\sigma} < 1 \quad \text{for} \quad \sigma > 0,$$

where $T_k = \ln \pi_k$. For example,

$$\frac{1}{\mathcal{P}_5(s)} = \frac{1}{1-\left(\frac{1}{5}\right)^s} = 1 + \frac{1}{5^s} + \frac{1}{5^{2s}} + \frac{1}{5^{3s}}\cdots, \quad \Re\{s\} = \sigma > 0.$$

The RoC for each root is $\sigma \geq 0$ since when $\sigma < 0$.

Since $1/\pi_k < 1$ for all $k \in \mathbb{N}$, the Taylor series of $\zeta_k(x)$ is entire except at its poles. Note that the RoC of a Taylor series in powers of π_k^{-s} increases with k.

Exercise #1 Work out the RoC for $k = 2$.

Solution: The formula for the RoC is given above, which for $\pi_2 = 3$ is

$$|\pi_k^{-s_r}| = \begin{cases} \left(\frac{1}{3}\right)^{\sigma_r} < 1 & \text{for} \quad \sigma_r > 0, \\ \left(\frac{1}{3}\right)^{-\sigma_r} < 1 & \text{for} \quad \sigma_r < 0, \end{cases}$$

where σ_r is the boundary of the RoC. ■

Exercise #2 Show how to construct $\text{Zeta}_2(t) \leftrightarrow \zeta_2(s)$ by working in the time domain.

Solution: The basic procedure for building a sieve is to sum the integers

$$S_1 = \sum_{n=1}^{\infty} n2^{n-1} = 1\cdot 2^0 + 2\cdot 2^1 + 3\cdot 2^2 + \cdots,$$

while the sieve for the kth prime π_k is

$$S_k = \sum_{n=1}^{\infty} n\pi_k^{n-1} = 1\cdot \pi_k^0 + \pi_k\cdot 2^1 + \pi_k\cdot 2^2 + \cdots.$$

This sum may be written in terms of the convolution with the Heaviside step function u_k, since

$$u_k \star u_k = nu_k = 0 \cdot u_0 + 1 \cdot u_1 + 2u_2 + \cdots + ku_k + \cdots .$$

■

C.1.1.2 Poles of $\zeta_k(s)$

Riemann proposed that Euler's zeta function $\zeta(s) \in \mathbb{C}$ has a complex argument [first explored by Chebyshev in 1850 (Bombieri 2000)] that extends $\zeta(s)$ into the complex plane ($s \in \mathbb{C}$), thus making it a complex analytic function. Thus we might presume that $\zeta(s)$ has an inverse Laplace transform. There seems to be very little written on this topic (Hill 2007). We explore this question further here.

We can now identify the poles of $\zeta_k(s)$ ($p \in \mathbb{N}$), which are required to determine the RoC. For example, the kth factor of Eq. C.1.7 expressed as an exponential, is

$$\zeta_k(s) \equiv \frac{1}{1 - \pi_k^{-s}} = \frac{1}{1 - e^{-sT_k}} = \sum_{k=0}^{\infty} e^{-skT_k}, \tag{C.1.8}$$

where $T_k \equiv \ln \pi_k$. Thus $\zeta_p(s)$ has poles at $-s_n T_p = 2\pi n \jmath$ (when $e^{-sT_p} = 1$), and

$$\omega_n = \frac{2\pi n}{T_k},$$

with $-\infty < n \in \mathbb{Z} < \infty$. These poles are the eigenmodes of the zeta function. Figure C.1 is a domain-colorized plot of this function. It is clear that the RoC of ζ_k is > 0. It would be helpful to determine why $\zeta(s)$ has a more restrictive RoC than each of its factors.

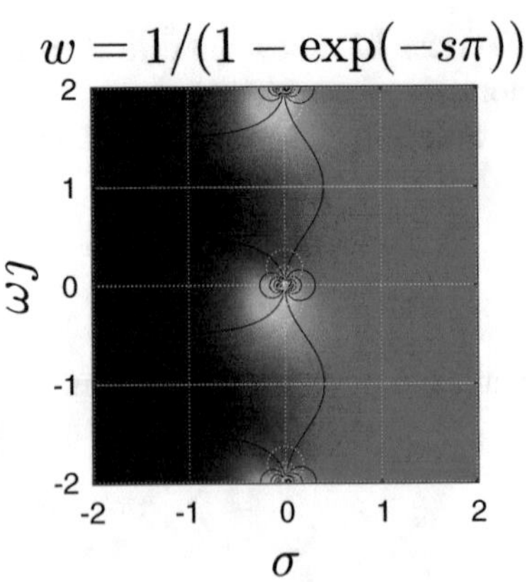

Fig. C.1 Plot of $w(s) = \frac{1}{1-e^{-s\pi}}$. Here $w(s)$ has poles where $e^{s_n \ln 2} = 1$—namely, where $\omega_n \ln 2 = n2\pi$, as seen in the colorized map ($s = \sigma + \omega\jmath$ is the Laplace frequency [rad])

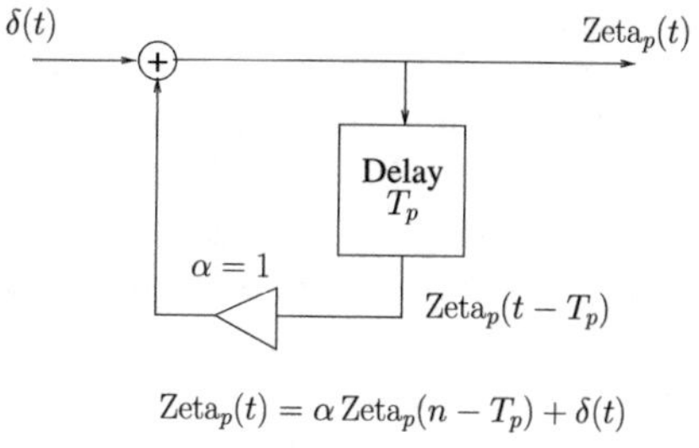

Fig. C.2 This feedback network is described by a time-domain difference equation with delay $T_p = \ln \pi_k$. It has an all-pole transfer function given by Eq. C.1.11. Physically this delay corresponds to T_p [s]

C.1.1.3 Inverse Laplace Transform

The inverse Laplace transform of Eq. C.1.8 is an infinite series of delays T_p (Table 3.9)[8]

$$\mathrm{Zeta}_p(t) = \delta(t))_{T_p} \equiv \sum_{k=0}^{\infty} \delta(t - kT_p) \leftrightarrow \frac{1}{1 - e^{-sT_p}}. \tag{C.1.9}$$

C.1.1.4 Inverse Transform of the Product of Factors

The time-domain version of Eq. C.1.7 may be written as the convolution of all the $\mathrm{Zeta}_k(t)$ factors

$$\mathrm{Zeta}(t) \equiv \mathrm{Zeta}_2 \star \mathrm{Zeta}_3(t) \star \mathrm{Zeta}_5(t) \star \mathrm{Zeta}_7(t) \star \cdots \star \mathrm{Zeta}_p(t) \star \cdots, \tag{C.1.10}$$

where $\star$ represents time convolution (Table 3.8).

Such functions may be generated in the time domain as shown in Fig. C.2, using a feedback delay of T_p [s] as described by the equation in the figure, with a unity feedback gain $\alpha = 1$,

$$\mathrm{Zeta}(t) = \mathrm{Zeta}(t - T_p) + \delta(t).$$

Taking the Laplace transform of the system equation, we see that the transfer function between the state variable $q(t)$ and the input $x(t)$ is given by $\underset{p}{\mathrm{Zeta}}(t)$. Taking the $\mathcal{LT}$, we see that $\zeta(s)$ is an all-pole function,

$$\zeta_p(s) = e^{-sT_p}\zeta_p(s) + 1(t) \text{ or } \zeta_p(s) = \frac{1}{1 - e^{-sT_p}}. \tag{C.1.11}$$

In terms of the physics, these transmission line equations are telling us that $\zeta(s)$ may be decomposed into an infinite cascade of transmission lines (Eq. C.1.10), each having a unique delay given by $T_k = \ln \pi_k$, $\pi_k \in \mathbb{P}$, the log of the primes. The input

[8]Here we use a shorthand double-parentheses notation $f(t))_T \equiv \sum_{k=0}^{\infty} f(t - kT)$ to define the one-sided infinite sum.

admittance of this cascade may be interpreted as an analytic continuation of $\zeta(s)$ that defines the eigenmodes of that cascaded impedance function.

Working in the time domain provides a key insight, as it allows us to determine the analytic continuation of the infinity of possible continuations, which may not be obvious in the frequency domain. Transforming to the time domain is a variant of analytic continuation of a function $Z(s) \leftrightarrow \text{Zeta}(t)$ that depends on the assumption that $\text{Zeta}(t)$ is one-sided in time (causal). It may be helpful to compare this variant to Euler's continuation of $\zeta(s)$, and later Riemann's classic 1851 definition of complex analytic continuation $\text{Zeta}(s)$.

Additional relationships: We need to know some important relationships provided by both Euler and Riemann (1859) when we study $\zeta(s)$.

With the goal of generalizing his result, Euler extended the definition with the functional equation

$$\zeta(s) = 2^s \pi^{s-1} \sin\left(\frac{\pi s}{2}\right) \Gamma(1-s)\, \zeta(1-s). \tag{C.1.12}$$

This seems closely related to Riemann's time reversal symmetry properties (Bombieri 2000),

$$\pi^{-s/2}\Gamma\left(\frac{s}{2}\right)\zeta(s) = \pi^{-(1-s)/2}\Gamma\left(\frac{1-s}{2}\right)\zeta(1-s).$$

This equation is of the form $F\left(\frac{s}{2}\right)\zeta(s) = F\left(\frac{1-s}{2}\right)\zeta(1-s)$, where $F(s) = \Gamma(s)/\pi^s$.

As shown in Table 3.8, the $\mathcal{LT}^{-1}$ of $f(-t) \leftrightarrow F(-s)$ represents time reversal. This leads to causal and anticausal functions that are symmetric about $\Re\{s\} = 1/2$ (Riemann 1859) leading to an interpretation of Euler's functional equation.

Riemann (1859, page 2) provides an alternative integral definition of $\zeta(s)$, based on the complex contour integration,[9]

$$2\sin(\pi s)\Gamma(s-1)\zeta(s) = \jmath \oint_{x=-\infty}^{\infty} \frac{(-x)^{s-1}}{e^x - 1} dx \overset{-x \to y\jmath}{=} \oint_{y=-\infty\jmath}^{\infty\jmath} \frac{(y\jmath)^{s-1}}{e^{-y\jmath} - 1} dx\jmath.$$

Given $\zeta_k(s)$, it seems important to look at the inverse $\mathcal{LT}$of $\zeta_k(1-s)$ to gain insight into the analytically extended $\zeta(s)$.

What is the RoC of $\zeta(s)$? It is commonly stated that Euler's and thus Riemann's product formulas are valid only for $\Re s > 1$; however, this does not seem to be actually proved (I could be missing this proof). Here I argue that the product formula is entire except at the poles—namely, that the formula is valid everywhere other than at the poles.

The argument goes as follows: Starting from the product formula (Eq. C.1.7, Sect. C.1.1.1), we form the log-derivative and study the poles and residues:

---

[9]We can verify Riemann's use of x, which is taken to be real rather than complex. This could be more natural (i.e., modern Laplace transformation notation) if $-x \to y\jmath \to z$.

$$
\begin{aligned}
D(s) &\equiv \frac{d}{ds} \ln \Pi_k \frac{1}{1-e^{-sT_k}} \\
&= -\sum_{k=1}^{\infty} \frac{1}{1-e^{-sT_k}} \frac{d}{ds} 1 - e^{-sT_k} \\
&= -\sum_k \frac{T_k e^{-sT_k}}{1-e^{-sT_k}} \leftrightarrow \sum_{k=1}^{\infty} \sum_{n=1}^{\infty} \delta(t - nT_k).
\end{aligned}
$$

Here $T_k = \ln \pi_k$, as previously defined, and $\leftrightarrow$ denotes the inverse Laplace transform, transforming $D(s) \leftrightarrow d(t)$ into the time domain. Note that $d(t)$ is a causal function, composed of an infinite number of delta functions (i.e., time delays), as shown in Fig. C.2 (Sect. C.1.1.4).

Zeros of $\zeta(s)$ We are still left with the most important question: Where are the zeros of $\zeta(s)$? Equation C.1.11 has no zeros; it is an all-pole system. The cascade of many such systems is also all-pole. As I see it, the issue is: What is the actual formula for $\zeta(s)$?

To answer this question, we need to study the properties of the reflectance function $\Gamma(s)$. Frequency-domain transfer functions having unity magnitude on the $j\omega$ axis are called *all-pass filters* in the engineering literature. When the reflectance is lossless, it is therefore all-pass since $|\Gamma(j\omega| = 1$. An important property of all-pass filters is that they may be accurately approximated by pole-zero pairs straddling the $j\omega$ axis, with the poles to the left (as required by causality) and the zeros to the right. Given this placement, the phases of the poles and zeros add. The group delay gives the net delay of the all-pass filter, which is twice the delay of the poles alone. It would seem that this careful placement of the zeros exactly across from the poles provides the requirement that the zeros all line up parallel to the $j\omega$ axis, as deemed by the Riemann hypothesis. Could this be the resolution of this long-standing mystery? An alternative possibility is that the convergent product formula has zeros that are obscured by the lack of convergence of Eq. C.1.2.

Filter properties Given the function

$$
F(s) = \frac{(s+1)(s-1)}{(s+2)},
$$

1. Find the minimum phase $M(s)$ and all-pass $A(s)$ parts: The minimum phase part has all of its poles and zeros in the left half-plane (LHP), while the all-pass part has its poles in the LHP and mirrored zeros in the RHP. Thus we place a removable pole zero pair symmetrically across from the RHP zero, and then write the expression as the product, that is $F(s) = M(s) \cdot A(s)$:

$$
F(s) = \frac{(s+1)(s-1)}{(s+2)} \cdot \frac{s+1}{s+1} = \frac{(s+1)^2}{s+2} \cdot \frac{s-1}{s+1}
$$

Thus $M(s) \equiv \frac{(s+1)^2}{s+2}$ and $A(s) \equiv \frac{s-1}{s+1}$.

2. Find the magnitude of $M(s)$: Take the real part of the log of M and then the anti-log. Thus $|M| = e^{\Re \ln M(s)}$.
3. Find the phase of $M(s)$: In this case we use the imaginary part: $\angle M = \Im \ln M(s)$.
4. Find the magnitude of $A(s)$: 1, by definition.
5. Find the phase of $A(s)$: $\angle A = \Im \ln(A)$.

C.1.1.5 More Questions

There are a number of question to be addressed:

1. Can we interpret the zeta function as a frequency-domain quantity, and then inverse transform it into the time domain?
   The answer to this is yes, and the results are quite interesting.
2. Make a histogram of the entropy for the first million integers.
   This is a 5 min job in MATLAB/Octave. It goes something line this:

```
K=1e5; N=1:K; F=zeros(K,10);
 for n=1:K;
  f=factor(n);
  F(n,1:length(f))=f;
 end;
hist(F);
```

Appendix D
Visco-Thermal Losses

D.1 Adiabatic Approximation at Low Frequencies

Newton's early development understandably ignored viscous and thermal losses, by assuming iso-thermal conditions. But starting at very low frequencies, the isothermal assumption breaks down. Modern theory, for audio frequencies, assumes the adiabatic approximation, and is thus described by the scalar wave equation (Pierce 1981). But it turns out that even at audio frequencies, the adiabatic approximation is invalid. This was first shown by Kirchhoff, but was not fully appreciated for more than a century, due to mathematical difficulties, which we now believe can be explained, as discussed next.

Following Helmholtz (1858), as extended by Kirchhoff (1868), visco-thermal loss mechanisms are related. The full theory was first worked out by Kirchhoff (1868, 1974). To understand how the are related is complicated, due to both the history and the mathematics, as briefly discussed by Pierce (1981). Both forms of damping are caused by two different, but coupled, diffusion effects: (1) viscous effects, due to shear at the container walls, and (2) thermal effects, due to deviations from adiabatic expansion (Kirchhoff 1868, 1974). I believe that Einstein was eventually involved, following his studies on Brownian motion (Einstein 1905).[10,11]

These two loss mechanism are restricted to a thin region called the *boundary layer*, which critically depends on the square root of the Laplace frequency. A key transition region is where the acoustic wavelength approaches the complex boundary layer thickness. When the radius of the container (a horn) approaches the viscous boundary layer, the theory breaks down.

[10]See Ch. 3 of https://www.ks.uiuc.edu/Services/Class/PHYS498/.

[11]https://en.wikipedia.org/wiki/Einstein_relation_(kinetic_theory)

J. Allen, *An Invitation to Mathematical Physics and Its History*,
https://doi.org/10.1007/978-3-030-53759-3

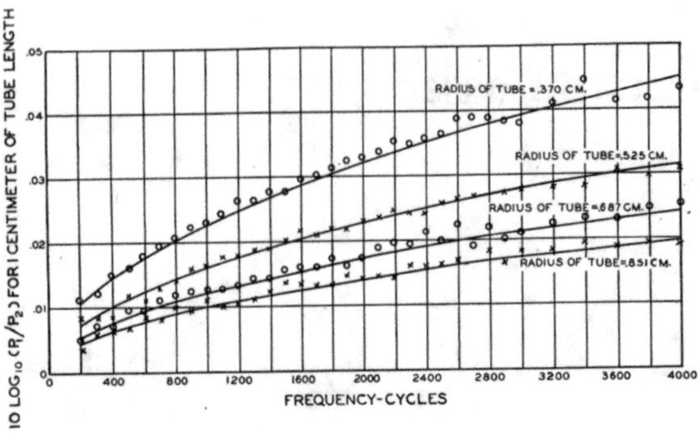

Fig. D.1 This figure, taken from Mason (1928), compares the Helmholtz-Kirchhoff theory for $|\kappa(f)|$ to Mason's 1928 experimental measurements of the loss. The ratio of two powers (P_1 and P_2) is plotted (see Mason's discussion immediately below his Fig. 4), and as indicated in the label: "$10 \log_{10} P_1/P_2$ for 1 [cm] of tube length." This is a plot of the transmission power ratio in [dB/cm] which is $10 \log |\Gamma(\omega)|^2$ where $\Gamma(i\omega)$ the reflection coefficient. For a discussion of the reflection coefficient and its properties, refer to Sect. 4.4.1, p. 189

D.1.1 Lossy Wave-Guide Propagation

The formulation of visco-thermal loss in air transmission was first worked out by Helmholtz (1863a) and then extended by Kirchhoff (1868) to include thermal damping (Rayleigh 1896, Vol. II, p. 319). These losses are accurately represented by the complex analytic propagation function $\kappa(s)$ (Eq. D.1.5). Following his review of these theories, Crandall (1926, Appendix A), the head of the 1926 Acoustic Research Department at that time, noted that the "Helmholtz-Kirchhoff" theory had never been experimentally verified. Acting on Crandall's suggestion, Physicist Warren Mason set out to experimentally verify Kirchhoff's 60 year old theory. Mason's analysis consumed several years Mason 1928.

This was a continued effort. Mason (1927) extended earlier work of Stewart's on acoustic transmission lines, by including viscous and thermal losses. Stewart's acoustic theory Stewart 1922; Stewart and Lindsay 1930 was acoustic version of the work of George Campbell (1904–1923) on electrical wave filters. If today you design earphones and hearing aids, or otoacoustic research, the works of Steward and Mason are relevant.

D.1.1.1 Mason's specification of the propagation function

Mason's results are reproduced in Fig. D.1 for tubes of radii between 3.7 and 8.5 [mm] of lengths L, having power reflectance

$$|\Gamma_L(f)|^2 = \left|e^{-\kappa(f)\,L}\right|^2 \ [\text{cm}^{-1}]. \tag{D.1.1}$$

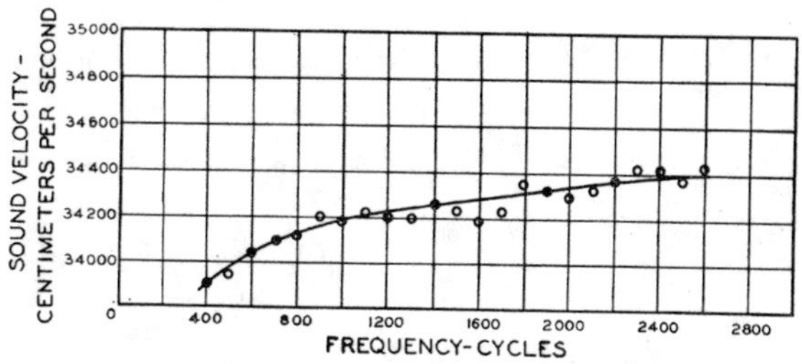

Fig. D.2 Figure 5 from (Mason 1928) showing the velocity of sound as a function of frequency, when visco-thermal losses are included. Note the change in the slope around 0.8 [kHz]

The complex propagation function use by Mason (1928), as taken from (Rayleigh 1896, p. 319), was

$$\kappa(\omega) = \frac{P\eta_o'\sqrt{\omega}}{2c_o S\sqrt{2\rho_o}} + \frac{i\omega}{c_o}\left\{1 + \frac{P\eta_o'}{2S\sqrt{2\omega\rho_o}}\right\}, \tag{D.1.2}$$

and the characteristic impedance was

$$z_o(\omega) = \sqrt{P_o\eta_o\rho_o}\left\{1 + \frac{P\eta_o'}{2S\sqrt{2\omega\rho_o}} - \jmath\frac{P\eta_o'}{2S\sqrt{2\omega\rho_o}}\right\}, \tag{D.1.3}$$

where $S = \pi R^2$ is the tube area and $P = 2\pi R$ is its perimeter.

Following Mason (1928, Fig. 5), the measured speed of sound

$$c_o'(\omega) = c_o\left\{1 - \frac{P\eta_o'}{2S\sqrt{2\omega\rho_o}}\right\}. \tag{D.1.4}$$

depends on frequency ω, as derived from the imaginary part of Eq. D.1.2. Figure D.2 directly compares Mason's measured sound speed with this equation.

Reduction of Kirchhoff's equations to complex analytic form: Using the thermodynamic constant β_1 defined by Eq. (D.1.7), we may rewrite κ and z_o in terms of $\beta_0 = \beta_1/2$ as

$$c_o\kappa(s) = s + \beta_1\sqrt{s} = \left(\beta_o + \sqrt{s}\right)^2 - \beta_o^2. \tag{D.1.5}$$

The two boxed equations below provide the derivation for the reduction of Mason's formula, for $c_o\kappa(\omega)$, to its complex analytic form (Eq. D.1.5). The first step is to define β_1 and $c_o\kappa(\omega) - s$. The second box shows it is equal to $\beta_1\sqrt{s}$. Thus $\kappa(s)$ is complex analytic in the Laplace frequency s, and following a completion of squares in $\sqrt{s}$, is given by Eq. D.1.5. The inverse Laplace transforms of $\sqrt{s}$ and $1/\sqrt{s}$ are provided in Table C.2 of Appendix C and Table 3.9 of Sect. 3.10.1.

STARTING FROM MASON (1928) $c_o\kappa(\omega)$:

$$c_o\kappa(\omega) = \frac{P\eta_o'\sqrt{\omega}}{2S\sqrt{2\rho_o}} + j\omega\left\{1+\frac{P\eta_o'}{2S\sqrt{2\omega\rho_o}}\right\} \tag{D.1.6}$$

DEFINE VARIABLES β_1 AND $s = j\omega$:

$$c_o\kappa(\omega) = s + \underbrace{\frac{P\eta_o'}{2S\sqrt{\rho_o}}}_{\beta_1}\left[\sqrt{\frac{\omega}{2}}+\frac{s}{\sqrt{2\omega}}\right] \tag{D.1.7}$$

THUS

$$c_o\kappa(\omega) - s = \frac{\beta_1}{\sqrt{2}}\left[\sqrt{\omega}+\frac{s}{\sqrt{\omega}}\right] \tag{D.1.8}$$

MULTIPLYING TOP AND BOTTOM ON RIGHT BY $\sqrt{j}$:

$$c_o\kappa(\omega) - s = \frac{\beta_1}{\sqrt{2}}\left[\frac{\sqrt{j\omega}}{\sqrt{j}}+\frac{s\sqrt{j}}{\sqrt{j\omega}}\right] \qquad \text{SET } \sqrt{j\omega}=\sqrt{s}$$

$$= \frac{\beta_1}{\sqrt{2}}\left[\frac{\sqrt{s}}{\sqrt{j}}+\frac{s\sqrt{j}}{\sqrt{s}}\right], \qquad \text{CROSS MULTIPLY}$$

$$= \frac{\beta_1}{\sqrt{2}}\left[\frac{s+js}{\sqrt{j}\cdot\sqrt{s}}\right], \qquad \text{FACTOR OUT } \frac{s}{\sqrt{s}}$$

$$= \beta_1\left[\frac{1+j}{\sqrt{2j}}\right]\frac{s}{\sqrt{s}} \qquad \text{REPLACE } \frac{s}{\sqrt{s}}=\sqrt{s}$$

$$= \beta_1\sqrt{s}$$

TO SHOW $1+j=\sqrt{2j}$, SQUARE BOTH SIDES: $1-1+2j=2j$.

Acoustic constants for air: Assuming $\eta_o = 1.4$ (ratio of specific heats), $\rho_o = 1.2$ [kgm/m^3] (density), a temperature of 23.5 [°C], and $P_o = 10^5$ [Pa] (atmospheric pressure), the lossless sound velocity is $c_o = \sqrt{P_o\eta_o/\rho_o} = 341.57$ [m/s]. By a comparison of this value of c_o to Fig. D.2, it is clear that this value does not apply to Mason's measurements. Thus to agree with his experimental results, either the ratio P_o/ρ_o or η_o need to be corrected. Since β_1 depends on η_o', ρ_o and μ_o this is a good place to look for the discrepancy.

Correction for η_o': The dimensionless constant $\eta_o'/\sqrt{\mu}$ is defined as the *composite thermodynamic constant* (Kirchhoff 1868; Rayleigh 1896)

$$\frac{\eta_o'}{\sqrt{\mu_o}} = \left[1+\sqrt{5/2}\left(\eta_o^{1/2}-\eta_o^{-1/2}\right)\right] = 1.5345.$$

Mason (1928) assumed the viscosity to be $\mu_o = 18.6\times10^{-6}$ [Pa-s] (viscosity), thus the dynamic-to-adiabatic visco-thermal elastic ratio is

$$\frac{\eta_o'}{\eta_o} = \frac{1.5345}{1.4}\sqrt{\mu_o} = 1.0961\sqrt{\mu_o},$$

giving

$$\beta_1 = 1.0961\frac{P}{2S}\sqrt{\frac{\mu_o}{\rho_o}}. \tag{D.1.9}$$

Here μ_o/ρ_o is known as the *kinematic viscosity* and μ_o as the *dynamic viscosity*.

Reduction of the lossy characteristic impedance: Rendering Eq. D.1.3 dimensionless, its complex analytic expression greatly simplifies

$$\frac{z_o(\omega)}{\sqrt{P_o\eta_o\rho_o}} - 1 = \frac{\beta_1}{\sqrt{\omega}}\frac{1-\jmath}{\sqrt{2}} = \frac{\beta_1}{\sqrt{s}}\sqrt{\jmath}\sqrt{-\jmath} = \frac{\beta_1}{\sqrt{s}},$$

thus the lossy normalized characteristic impedance is $z_o(\omega)/r_o = 1 + \frac{\beta_1}{\sqrt{s}}$, where $r_o = \sqrt{P_o\eta_o\rho_o}$ is the lossless characteristic resistance, and β_1 is defined by Eq. D.1.9.

Reduction of the lossy speed of sound: Finally, starting from Eq. D.1.4,

$$c_o'(\omega) = c_o\left\{1 - \frac{P\eta_o'}{2S\sqrt{2\omega\rho_o}}\right\}, \tag{D.1.10}$$

note that for $\omega > 0$, $\left(1 - \frac{c_o'}{c_o}\right) > 0$, thus

$$1 - \frac{c_o'(\omega)}{c_o} = \frac{P\eta_o'}{2S\sqrt{2\omega\rho_o}} = \frac{P}{2S}\frac{\eta_o'\sqrt{\jmath}}{\sqrt{2s\rho_o}} = \frac{P}{2S}1.0961\frac{\eta_o}{\sqrt{s}}\sqrt{\frac{\mu_o}{\rho_o}}\sqrt{\frac{\jmath}{2}} = \beta_1\frac{\eta_o}{\sqrt{s}}\sqrt{\frac{\jmath}{2}}.$$

For $\omega < 0$, this will differ, and for $\omega = 0$ it is singular. This is because of the singular branch cut at $s = 0$, due to $1/\sqrt{s}$. The inverse $\mathcal{LT}$ of $1/\sqrt{s}$ is provided in Appendix C, Table C.2.

Case of the cylindrical guide: For the case of a cylindrical wave guide, $P/2S = 1/R$. Thus

$$\beta_1 R = \eta_o'/\sqrt{\rho_o} = 1.0961\,\eta_o\sqrt{\mu_o/\rho_o} = 1.0961 \times 1.4\sqrt{\frac{18.6\times10^{-6}}{1.2}} = 1.9105\times10^{-3}.$$

It is well documented in the literature that the boundary layer thickness is proportional to the square root of the kinematic viscosity (μ/ρ) over frequency. Our result, Eq. D.1.5, is a complex-analytic extension of this classic boundary layer equation, derived from the classic results of Kirchhoff (1868) and Rayleigh (1896), as verified by Mason (1928).

Roots of the lossy wave equation: We may factor Eq. D.1.5 to reveal the mathematical impact of the damping on $\kappa(s)$ ($\beta_1 = 2\beta_o$)

$$\sqrt{s_o} = -\beta_o \pm \beta_o,$$

namely $\sqrt{s_o} = \{0, -\beta_1\}$. In general there must be four roots, but since $\beta_1 \in \mathbb{R} > 0$, the roots in this case are degenerate, and in the left half plane.

The smaller the radius the greater the damping ($\beta_1 = 1.1 \times 10^{-3}/R$). Also note that the propagation function $\kappa(\omega)$ has a Helmholtz-Kirchhoff correction for both the real and imaginary parts. Thus both the speed of sound and the damping are dependent on frequency, and in a similar way.

Pressure Eigen-solutions: The forwarded P_- and backward P_+ pressure waves propagate as

$$P_{\pm}(s,x) = e^{-\kappa(s)x}, \quad e^{-\overline{\kappa}(s)x}, \tag{D.1.11}$$

where $\overline{\kappa}(s)$ the complex conjugate of $\kappa(s)$, such that $\Re\kappa(s) > 0$. The term $\beta_o\sqrt{s}$ affects both the real and imaginary parts of $\kappa(s)$. The real part is a frequency-dependent loss, and the imaginary part introduces a frequency-dependent speed of sound (Mason 1928).

D.1.2 *Impact of Viscous and Thermal Losses*

Equation D.1.2 and the measured data are compared in Fig. D.1, as reproduced from Mason's Fig. 4, which shows that the wave speed drops from 344 m/s at 2.6 kHz to 339 m/s at 0.4 kHz, a 1.5% reduction. At 1 kHz the loss is 1 dB/m for a 7.5-mm tube. Note that the loss and the speed of sound vary inversely with the radius. As the radius approaches the *boundary layer thickness* (i.e., the radial distance such that the loss is e^{-1}), the effect of the damping dominates the propagation.

Cut-off frequency s_o**:** The frequency where the lossless part equals the lossy part is defined as $\kappa(s_o) = 0$, namely, $\sqrt{s_o} = -\beta_o$, or $s_o = \beta_o^2$.

To get a feeling for the magnitude of s_o, let $R = 0.75/2$ [cm] (i.e., the average radius of the adult ear canal). Then for $R = 3.75 \times 10^{-3}$ [cm]

$$s_o = (1.9 \times 10^{-3}/3.75 \times 10^{-3})^2 = 1/4.$$

We conclude that the losses are insignificant in the audio range, since for the human ear canal, $f_o = \beta_o^2/\pi \approx 0.25/\pi = 0.08$ Hz.[12] This frequency represents the lower bound of the transition from adiabatic to iso-thermal equilibrium. It should be clear that acoustic frequencies do not actually obey the adiabatic approximation, due to the thin boundary layer. Both the real and imaginary part of propagation function $\kappa(s)$, the characteristic impedance $z_o(s)$ and the speed of sound $c_o'(s)$ all depended on frequency in the auditory range of human hearing.

Summary: The Helmholtz-Kirchhoff theory of viscous and thermal losses results in a frequency-dependent speed of sound that has a frequency dependence proportional to $1/\sqrt{s}$ rather than $1/\sqrt{\omega}$ (Mason 1928, Eq. 4). This corresponds to a 2% change in the sound velocity over the decade from 0.2 to 2 kHz (Mason 1928, Fig. 5), in agreement with Mason's experimental results.

---

[12] /home/jba/Mimosa/2C-FindLengths.16/doc.2-c_calib.14/m/MasonKappa.m.

Appendix E
Thermodynamic Systems

Many people find thermodynamics difficult to understand. Here we explore the reasons behind this lack of transparency, and propose a solution. To understand the problem it is helpful to compare the nonlinear energy-equilibrium methods used in thermodynamic, to linear impedance methods, used in acoustics, mechanics, and electrical circuit theory.

Acoustics, mechanics, and electrical circuits are explained in terms of linear systems of equations, such as as Kirchhoff's and Ohm's laws, as discussed in Appendix D. The linear impedance formulation leads to a system of equations that is easily solved, using standard methods of linear algebra. Thermodynamics, on the other hand, is formulated in terms of equilibrium energies, resulting in nonlinear systems of equations. This explains the difficulty in understanding the relationships. A second issue is that these equations are frequently over-specified.

Traditionally thermodynamics has been formulated with two types of variables, those that are proportional to the mass, called *extensive* such as volume and mass, and the those that are independent of mass, called *intensive* variables, such as temperature and density (Ambaum 2010).

A more modern and transparent notation is to work with two *conjugate* variables (CV),[13] *force density* and *flux*. Examples of the CV force density include pressure $\mathcal{P}$, temperature T, the Nernst potential, and electrical voltage. Fluxes CVs include mass flux and its area integral, volume velocity $\mathcal{V}$, heat flux, called entropy S, and electrical current.

Impedance is the ratio of CV, and the power is their product. For example, in electrical network theory, the impedance is the voltage over the current and the power is the product. Working with impedance always results in systems of linear equations. *linear conjugate variables* (LCV) are a generalization of force, the gradient of a potential, and flow. Namely, the formulation in terms of LCV simplifies the thermodynamic system of nonlinear equations.

[13]https://en.wikipedia.org/wiki/Conjugate_variables_(thermodynamics).

J. Allen, *An Invitation to Mathematical Physics and Its History*,
https://doi.org/10.1007/978-3-030-53759-3

E.1 Summary of the Thermodynamic Relations

To reduce the number of variables we remove mass from the system by working with *specific* variables (those that are normalized by the mass). The ratio of the volume $\mathcal{V}$ and mass M defines the specific volume $\nu \equiv \mathcal{V}/M = 1/\rho$, where ρ is the density. Throughout this discussion we shall use SI units, with $\mathcal{V}$ [m^3], M [kgm], $\rho = 1/\nu$ [kg/m^3].

Four classical *thermodynamic energies* are defined:

1. *internal energy* u,
2. *specific enthalpy* $h = u + \mathcal{P}\nu$,
3. *Helmholtz free energy* $f = u - TS$, and
4. *Gibbs function* $g = u - TS + \mathcal{P}\nu$.

The Gibbs function is a linear combination of the other three (Ambaum 2010, Sect. 3.1). As explained by Tolman (1948), Gibbs viewed a phase transition as a membrane separating two states of a gas. These four relations could be combined into a 3×3 nonlinear energy matrix relating $[h, f, g]^T$ to $[u, \mathcal{P}\nu, TS]^T$. The entries of the matrix elements would all be either $0, 1, -1$.

When many variables are used to describe a system, as in the case of thermodynamics, a matrix formulation is more compact representation. Here traditional thermodynamic relations are reformulated using linear 2×2 matrix algebra.[a] Starting from the state equation and the Helmholtz expression for the free energy, we rewrite the system as a linear two-port network. By evaluating the impedance matrix elements, we may obtain the standard thermodynamic relations in terms of the two heat capacities c_p and c_v. Due to this unification of variables, the 2×2 formulation, such a formulation manipulation of thermodynamic relations becomes transparent.

[a] An alternative approach is to use a 3×3 scattering matrix in more than 2 variables (Bilbao 2004).

In acoustics, the gradient of the pressure $-\nabla\mathcal{P}$ is the acoustic force density, which is proportional to the volume velocity $\mathcal{V}$ (the acoustic flux, or mass flow). Thus the potential and its corresponding mass flux are linear conjugate variables, the ratio of which defines the specific acoustic impedance, which for air is ρc_o. When working with impedances it is standard practice to work with the Laplace complex frequency domain s. The $\mathcal{LT}$ transforms differential equations in time into algebraic relations (i.e., polynomials). In thermodynamics the term *potential* has a very different meaning. Here we shall not refer to energies as potentials. When speaking of conjugate variables, pressure, voltage and temperature are potentials, and velocity, current and entropy are flows.

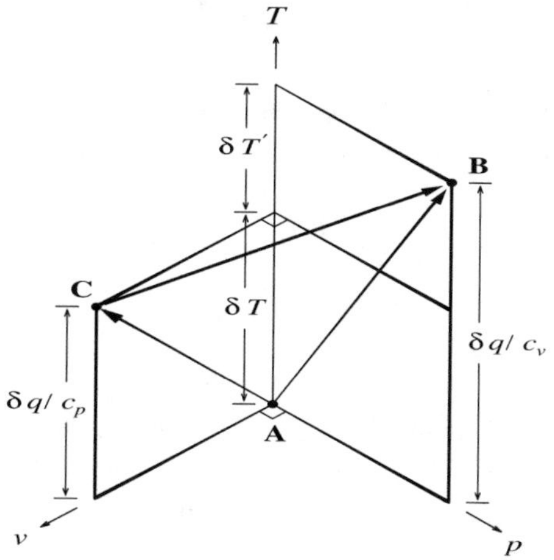

Fig. E.1 Diagram showing the PVT diagram for an ideal gas. The axes are volume (v), pressure (p), and temperature (T). Shown are the definitions for the two latent heat capacities c_p and c_v. From this figure we can more easily identify the meanings of the various matrix elements. Figure 3.1 from Ambaum (2010, p. 45)

Using modern engineering terminology, the gradient of every potential is a generalized force F, which according to Ohm's law, is proportional to a generalized flow J. Working with LCV $\mathcal{P}$ and the volume velocity $s\,\mathcal{V}$ gives

$$-\nabla\,\mathcal{P} = \rho c_o s\,\mathcal{V}.$$

Here ρc_o is the impedance of air, defined as the ratio of the generalized force $F = -\nabla\,\mathcal{P}$ over the volume velocity (flow) $s\,\mathcal{V}$ (Fig. E.1).

Ohm's law defines a linear relation, with the impedance being the complex proportional factor. The power is the product of LCV, which is a nonlinear relation, since it is quadratic in the conjugate variables. It follows from linear algebra that linear systems of equations are more transparent, compared, say, to the thermodynamic energy-based nonlinear of equations.

In terms of the four basic energy definitions, the force-flow relations for the four power relations are defined by four differentials (Ambaum 2010).

$$\begin{aligned} du &= \quad TdS - pd\nu \\ dh &= \quad TdS + \nu d\,\mathcal{P} \\ df &= -SdT - \mathcal{P}d\nu \\ dg &= -SdT + \nu d\,\mathcal{P}. \end{aligned}$$

Linear 2×2 **systems of equations:** The LCV thermodynamic variables are the mechanical (acoustical) pressure $\mathcal{P}$, volume velocity $\mathcal{V}$ and the heat variables, temperature T, and entropy rate $\dot{S}$.

The linear transmission and impedance matrices $\mathcal{T}$ and $\mathcal{Z}$ relate the acoustic variables $\mathcal{P}$, $\mathcal{V}$ to the thermodynamic variables T, $\dot{S}$ in two very different ways:

$$\begin{bmatrix} T \\ \dot{\mathcal{S}} \end{bmatrix} = \mathcal{T}\begin{bmatrix} \mathcal{P} \\ -\mathcal{V} \end{bmatrix} = \begin{bmatrix} \mathcal{A} & \mathcal{B} \\ \mathcal{C} & \mathcal{D} \end{bmatrix}\begin{bmatrix} \mathcal{P} \\ -\mathcal{V} \end{bmatrix} \quad \text{and} \quad \begin{bmatrix} \mathcal{P} \\ T \end{bmatrix} = \mathcal{Z}\begin{bmatrix} \mathcal{V} \\ \dot{\mathcal{S}} \end{bmatrix} = \begin{bmatrix} z_{11} & z_{12} \\ z_{21} & z_{22} \end{bmatrix}\begin{bmatrix} \mathcal{V} \\ \dot{\mathcal{S}} \end{bmatrix}.$$

On the left is the *transmission matrix* $\mathcal{T}$ and on the right is the corresponding *impedance matrix* $\mathcal{Z}$. By inspection, the $\mathcal{T}$ and $\mathcal{Z}$ matrix elements are

$$\mathcal{T} = \begin{bmatrix} \frac{T}{\mathcal{P}}\Big|_{\mathcal{V}=0} & \frac{T}{\mathcal{V}}\Big|_{\mathcal{P}=0} \\ \frac{\dot{\mathcal{S}}}{\mathcal{P}}\Big|_{\mathcal{V}=0} & \frac{\dot{\mathcal{S}}}{\mathcal{V}}\Big|_{\mathcal{P}=0} \end{bmatrix}, \qquad \mathcal{Z} = \begin{bmatrix} \frac{\mathcal{P}}{\mathcal{V}}\Big|_{\dot{\mathcal{S}}=0} & \frac{\mathcal{P}}{\dot{\mathcal{S}}}\Big|_{\mathcal{V}=0} \\ \frac{T}{\mathcal{V}}\Big|_{\dot{\mathcal{S}}=0} & \frac{T}{\dot{\mathcal{S}}}\Big|_{\mathcal{V}=0} \end{bmatrix}.$$

Due to reciprocity, the determinant of $\mathcal{T} = 1$ and $z_{12} = z_{21}$ (Postulate P8). When using this matrix representation, any algebra is explicit. The transmission matrix $\mathcal{T}$ and impedance matrix $\mathcal{Z}$ are related by Eq. 3.8.6, as discussed on Sect. 3.8.2.

Summary of Maxwell relations

There are six variants on the transmission matrix (Van Valkenburg 1964a). The most common are the transmission $\mathcal{T}$ and the impedance $\mathcal{Z}$ matrices. Others, such as the admittance matrix $\mathcal{Y}$ is simply the inverse of $\mathcal{Z}$. But other variants are also used for special applications, such scattering matrices.

All the matrices used in thermodynamics are reciprocal, meaning $z_{12} = z_{21}$ or $\Delta_{\mathcal{T}} = 1$. Each matrix form show reciprocity in a different way. These expressions for reciprocity were first worked out by Maxwell, where they are called Maxwell relations (Ambaum 2010). From the Maxwell relations, it may be shown that the specific heat capacities are equal to $c_v = y_{22} = z_{11}/\Delta_{\mathcal{Z}}$ and $c_p = g_{22} = 1/y_{22} = \Delta_{\mathcal{Z}}/z_{11}$.

E.1.1 Physical Meaning of the Matrix Elements

The utility of the $\mathcal{T}$ and $\mathcal{Z}$ representations are very different. The impedance matrix $\mathcal{Z}$ always has a physically measurable interpretation. The diagonal elements z_{11}, z_{22} represent the input impedance looking into the two ports, while the off-diagonal elements are transfer impedances, related to the transfer function between input and output fluxes and potentials. If the system is reciprocal they are equal ($z_{12} = z_{21}$).

The transmission matrix plays a theoretical and modeling role. Thus $\mathcal{Z}$ is used for experiments and $\mathcal{T}$ for calculations. Other important matricides are the admittance matrix $\mathcal{Y}$ and $\mathcal{G}$. There are simple relationships between all these matrices (Van Valkenburg 1964a, p. 310), a few of which are

$$\mathcal{T} = \frac{1}{z_{21}}\begin{bmatrix} z_{11} & \Delta_{\mathcal{Z}} \\ 1 & z_{22} \end{bmatrix}, \quad \mathcal{Z} = \frac{1}{\mathcal{C}}\begin{bmatrix} \mathcal{A} & \Delta_{\mathcal{T}} \\ 1 & \mathcal{D} \end{bmatrix}, \quad \mathcal{Y} = \frac{1}{\Delta_{\mathcal{T}}}\begin{bmatrix} z_{22} & -z_{12} \\ -z_{21} & z_{11} \end{bmatrix}, \quad \mathcal{G} = \frac{1}{z_{11}}\begin{bmatrix} 1 & -z_{12} \\ z_{21} & \Delta_{\mathcal{Z}} \end{bmatrix}.$$

Physical description of the 2x2 conjugate variables

To determine the physical meanings of the matrix elements we must start with the equation of state for the ideal gas $\mathcal{P}\mathcal{V} = RT$, where R is the specific gas constant. Two key thermodynamic constants are

$$c_v \equiv \left.\frac{\partial u}{\partial T}\right|_{\mathcal{V}} = \left.\frac{\partial \mathcal{S}}{\partial T}\right|_{\mathcal{V}} \qquad c_p \equiv \left.\frac{\partial h}{\partial T}\right|_{\mathcal{P}} = \left.\frac{\partial \mathcal{S}}{\partial T}\right|_{\mathcal{P}}.$$

To identify c_p and c_v in terms of these matrix elements we need to do some algebra. The equations need to be reformulated as

$$\begin{bmatrix} \mathcal{P} \\ \dot{\mathcal{S}} \end{bmatrix} = \begin{bmatrix} z_{11} & \beta \\ \gamma & \delta \end{bmatrix} \begin{bmatrix} \mathcal{V} \\ T \end{bmatrix} \quad \text{and} \quad \begin{bmatrix} \mathcal{V} \\ \dot{\mathcal{S}} \end{bmatrix} = \begin{bmatrix} y_{11} & y_{12} \\ sc_v & sc_p \end{bmatrix} \begin{bmatrix} \mathcal{P} \\ T \end{bmatrix}. \tag{E.1.1}$$

Here $s = \sigma + \omega \jmath$ is the Laplace frequency, common to the Laplace transform. Both c_p and c_v have units of Farads.

For incompressable liquids these are equal, thus we may define $c_l = c_p = c_v$. In this case the internal energy u and enthalpy h are given by

$$u \approx h = u_o + c_l T,$$

which means that any volume change is driven only by temperature rather than by work (Ambaum 2010).

For water in the vapor state they are distinct with $c_p = c_v + R$ and $c_v = 5R/2$. It follows that a mixture of dry air with 5% water vapor is mainly diatomic

$$(c_v/c_p)^2 = 1.4 = (5+2)/5.$$

From Eq. E.1.1, z_{11} is the reciprocal of the *bulk modulus*, defined as ratio of the volume over the pressure compliance for constant entropy (iso-thermal). Likewise, z_{22} is the ratio of $\mathcal{T}/\dot{\mathcal{S}}$ for constant volume (i.e., isobaric).

E.1.2 SI Units

The pressure $\mathcal{P}$ [Pa] is a potential and the volume velocity $\mathcal{V}$ [kgm/m^2] is a flux having power $\mathcal{P}\mathcal{V}$. The corresponding thermal power is $Q = T\dot{\mathcal{S}}$ where $\dot{\mathcal{S}}$ is the entropy-rate (Q is the absorbed heat power) in Watts. Since the temperature T is a potential, analogous to a voltage, then $d\mathcal{S}/dt$ must be a flux, analogous to the current in an electrical system.

The total internal energy of a substance is defined by the enthalpy $H = PV - Q$, where $Q = T\mathcal{S}$ is the energy due to heat and PV is the mechanical energy. If H is constant and we vary the heat by a small amount, then $0 = \delta H \delta Q - \delta(PV) = \delta Q - P\delta V - V\delta P$. The thermodynamic constant c_p is defined as $\partial Q/\partial T|_{\mathcal{P}_o}$ (constant $\mathcal{P}$) and c_v is defined as $\partial Q/\partial T|_{\mathcal{V}_o}$ (constant $\mathcal{V}$).

E.1.3 Summary and Conclusions

Since $\mathcal{P}\,\mathcal{V}$ is the acoustic power and $\mathcal{P}/\,\mathcal{V}$ the acoustic impedance, $\dot{Q} = T\,\dot{\mathcal{S}}$ is the thermal power, thus z_{22} is the associated thermal impedance. What is unique about this set of equations is that they are linear in their variables, and that they must obey reciprocity, defined as either $|\mathcal{T}| = 1$ or $z_{12} = z_{21}$.

Appendix F
Number Theory Applications

F.1 Division with Rounding Method

We want to show that the GCD for $m, n, k \in \mathbb{N}$ (Eq. 2.4.2, Sect. 2.4.3) may be written in matrix form as

$$\begin{bmatrix} m \\ n \end{bmatrix}_{k+1} = \begin{bmatrix} 0 & 1 \\ 1 & -\left\lfloor \frac{m}{n} \right\rfloor \end{bmatrix} \begin{bmatrix} m \\ n \end{bmatrix}_k . \tag{F.1.1}$$

Equation F.1.1 implements the $\texttt{gcd}(m, n)$ for $m > n$.

This starts with $k = 0$, $m_0 = a$, and $n_0 = b$. With this method there is no need to test whether $n_n < m_n$, as it is built into the procedure. The method uses the floor function $\lfloor x \rfloor$, which finds the integer part of x ($\lfloor x \rfloor$ rounds toward $-\infty$). After each step we will see that the value $n_{k+1} < m_{k+1}$. The method terminates when $n_{k+1} = 0$ with $\gcd(a, b) = m_{k+1}$.

The following vectorized code is more efficient than the direct matrix method:

```
function n=gcd2(a,b)

M=[abs(a);abs(b)]; %Save (a,b) in array M(2,1)

% done when M(1) = 0
while M(1) ~= 0
disp(sprintf('M(1)=%g, M(2)=%g ',M(1),M(2)));
M=[M(2)-M(1)*floor(M(2)/M(1)); M(1)]; %automatically sorted
end  %done

n=M(2); %GCD is M(2)
```

With a minor extension in the test for "end," this code can be made to work with irrational inputs (e.g., $(n\pi, m\pi)$).

This method calculates the number of times $n < m$ must subtract from m using the floor function. This operation is the same as the mod function.[14] Specifically,

[14] https://en.wikipedia.org/wiki/Modulo_operation.

J. Allen, *An Invitation to Mathematical Physics and Its History*,
https://doi.org/10.1007/978-3-030-53759-3

$$n_{k+1} = m_k - \left\lfloor \frac{m}{n} \right\rfloor n_k \tag{F.1.2}$$

so that the output is the definition of the remainder of modular arithmetic. This would have been obvious to anyone using an abacus, which explains why it was discovered so early.

Note that the next value of $m = M(1)$ is always less than $n = M(2)$ and must remain greater than or equal to zero. This one-line vector operation is then repeated until the remainder $M(1)$ is 0. The gcd is then $n = M(2)$. When we use irrational numbers, the code still works except the error is never exactly zero due to IEEE 754 rounding. Thus the criterion must be that the error is within some small factor times the smallest number (which in MATLAB/Octave is the number `eps` = $2.220446049250313 \times 10^{-16}$, as defined in the IEEE 754 standard).

Thus, without factoring the two numbers, Eq. F.1.2 recursively finds the gcd. Perhaps this is best seen with some examples.

The GCD is an important and venerable method, useful in engineering and mathematics but, as best I know, not typically taught in the traditional engineering curriculum.

GCD applied to polynomials: An interesting generalization is to work with polynomials rather than numbers and apply the Euclidean algorithm.

The GCD may be generalized in several significant ways. For example, what is the GCD of two polynomials? To answer this question, we must factor the two polynomials to identify common roots.

F.2 Derivation of the CFA Matrix

We can define the continued fraction algorithm (CFA) starting from the basic definitions of the floor and remainder formulas. Starting with a decimal number x, we split it into the decimal and remainder parts.[15] If we start with $n = 0$ and $x_o = x \in \mathbb{I}$, the integer part is

$$m_0 = \lfloor x \rfloor \in \mathbb{N}$$

and the remainder is

$$r_0 = x - m_0.$$

Corresponding to the CFA, the next target x_1 for $n = 1$ is

$$x_1 = r_0^{-1}$$

and the integer part is $m_1 = \lfloor x_1 \rfloor$. As in the case of $n = 0$, the integer part is

[15] The method presented here was developed by Yiming Zhang as a student project in 2019.

$$m_1 = \lfloor x_1 \rfloor$$

and the remainder is

$$r_1 = x_1 - m_1.$$

The recursion for $n = 2$ is similar.

For us to better appreciate what is happening, it is helpful to write these recursions in matrix format. Rewriting the case of $n = 1$ and using the remainder formula for the ratio of two numbers $p \geq q \in \mathbb{N}$ with $q \neq 0$, we have

$$\begin{bmatrix} p \\ q \end{bmatrix} = \begin{bmatrix} u_1 & 1 \\ 1 & 0 \end{bmatrix} \begin{bmatrix} r_0 \\ r_1 \end{bmatrix}.$$

From the remainder formula, $u_1 = \lfloor p/q \rfloor$. Continuing with $n = 2$:

$$\begin{bmatrix} r_0 \\ r_1 \end{bmatrix} = \begin{bmatrix} u_2 & 1 \\ 1 & 0 \end{bmatrix} \begin{bmatrix} r_1 \\ r_2 \end{bmatrix},$$

where $u_1 = \lfloor r_0/r_1 \rfloor$. Continuing with $n = 3$:

$$\begin{bmatrix} r_1 \\ r_2 \end{bmatrix} = \begin{bmatrix} u_3 & 1 \\ 1 & 0 \end{bmatrix} \begin{bmatrix} r_2 \\ r_3 \end{bmatrix},$$

where $u_2 = \lfloor r_1/r_2 \rfloor$.

For arbitrary n we find

$$\begin{bmatrix} r_{n-2} \\ r_{n-1} \end{bmatrix} = \begin{bmatrix} u_n & 1 \\ 1 & 0 \end{bmatrix} \begin{bmatrix} r_{n-1} \\ r_n \end{bmatrix}, \tag{F.2.1}$$

where $u_n = \lfloor r_{n-1}/r_n \rfloor$. This terminates when $r_n = 0$ in the above nth step:

$$\begin{bmatrix} r_{n-2} \\ r_{n-1} \end{bmatrix} = \begin{bmatrix} u_n & 1 \\ 1 & 0 \end{bmatrix} \begin{bmatrix} r_{n-1} \\ r_n = 0 \end{bmatrix}.$$

Example: We let $p = 355$ and $q = 113$, which are coprime, and set $n = 1$. Then Eq. F.2.1 becomes

$$\begin{bmatrix} 355 \\ 113 \end{bmatrix} = \begin{bmatrix} 3 & 1 \\ 1 & 0 \end{bmatrix} \begin{bmatrix} r_0 \\ r_1 \end{bmatrix},$$

since $u_1 = \lfloor \frac{355}{113} \rfloor = 3$. Solving for the RHS gives $[r_0; r_1] = [113; 16]$ ($355 = 113 \cdot 3 + 16$). To find $[r_0; r_1]$, we take the inverse:

$$\begin{bmatrix} r_0 \\ r_1 \end{bmatrix} = \begin{bmatrix} 0 & 1 \\ 1 & -3 \end{bmatrix} \begin{bmatrix} 355 \\ 113 \end{bmatrix}.$$

For $n = 2$, with the RHS from the previous step,

$$\begin{bmatrix} 113 \\ 16 \end{bmatrix} = \begin{bmatrix} u_2 & 1 \\ 1 & 0 \end{bmatrix} \begin{bmatrix} r_1 \\ r_2 \end{bmatrix},$$

since $u_2 = \lfloor \frac{113}{16} \rfloor = 7$. Solving for the RHS gives $[r_1; r_2] = [16; 1]$ $(113 = 16 \cdot 7 + 1)$. It seems we are done, but let's go one step further.

For $n = 3$ we now have

$$\begin{bmatrix} 16 \\ 1 \end{bmatrix} = \begin{bmatrix} u_3 & 1 \\ 1 & 0 \end{bmatrix} \begin{bmatrix} r_1 \\ r_2 \end{bmatrix},$$

since $u_3 = \lfloor \frac{16}{1} \rfloor = 16$. Solving for the RHS gives $[r_1; r_2] = [1; 0]$. This confirms that we are done, since $r_2 = 0$.

Derivation of Eq. F.2.1: Equation Eq. F.2.1 is derived as follows: Starting from the target $x \in \mathbb{R}$, we define

$$p = \lfloor x \rfloor \quad \text{and} \quad q = \frac{1}{x - p} \in \mathbb{R}.$$

These two relationships for truncation and remainder allow us to write the general matrix recursion relation for the CFA (Eq. F.2.1). Given $\{p, q\}$, we continue with the above CFA method.

One slight problem with the above is that the output is on the right and the input on the left. Thus we need to take the inverse of these relationships to turn this into a composition.

F.3 Taking the Inverse to Get the gcd

Variables p and q are the remainders r_{n-1} and r_n, respectively. Using this notation with $n - 1$ gives Eq. F.2.1. Inverting this gives the formula for the GCD:

$$\begin{bmatrix} r_{n-1} \\ r_n \end{bmatrix} = \begin{bmatrix} 0 & 1 \\ 1 & -\lfloor \frac{r_{n-2}}{r_{n-1}} \rfloor \end{bmatrix} \begin{bmatrix} r_{n-2} \\ r_{n-1} \end{bmatrix}.$$

This terminates when $r_n = 0$ and the `gcd(p,q)` is r_{n-1}. Not surprisingly these equations mirror Eq. 2.4.3 (Sect. 2.4.3), but with a different indexing scheme and interpretation of the variables.

This then explains why Gauss called the CFA the *Euclidean algorithm*. He was not confused. But since the equations have an inverse relationship, they are not strictly the same.

Appendix G
Eleven Postulates of Systems of Algebraic Networks

Physical systems obey basic rules that follow from the physics. It is helpful to summarize these restrictions as postulates presented in terms of a taxonomy, or categorization method, of the fundamental properties of physical systems. Eleven of these are listed in this appendix from an article by Kim and Allen (2013).

G.1 Representative System

A taxonomy of physical systems comes from a systematic summary of the laws of physics, which includes at least the eleven basic network postulates, described in Sect. 3.10.

To describe the network postulates, it is helpful to start from a two-port matrix representation as discussed in Sect. 3.8.

As shown in Fig. G.1, the two-port transmission matrix for an acoustic transducer (loudspeaker) is characterized by the equation

$$\begin{bmatrix} \Phi_i \\ I_i \end{bmatrix} = \begin{bmatrix} A(s) & B(s) \\ C(s) & D(s) \end{bmatrix} \begin{bmatrix} F_l \\ -U_l \end{bmatrix} = \frac{1}{T} \begin{bmatrix} z_m(s) & Z_e(s) z_m(s) + T^2 \\ 1 & Z_e(s) \end{bmatrix} \begin{bmatrix} F_l \\ -U_l \end{bmatrix}, \quad \text{(G.1.1)}$$

shown as a product of three 2×2 matrices in the figure, with each factor representing one of the three Hunt parameters of the loudspeaker.

This figure represents the electromechanical motor of the loudspeaker and consists of three elements: the electrical input impedance $Z_e(s)$, a gyrator, which is similar to a transformer that relates current to force, and the output mechanical impedance $z_m(s)$. This circuit describes what is needed to fully characterize its operation, from electrical input to mechanical (acoustical) output.

The input is electrical (voltage and current) $[\Phi_i, I_i]$ and the output (load) is the mechanical (force and velocity) $[F_l, U_l]$. The first matrix is the general case, expressed in terms of four unspecified functions $A(s)$, $B(s)$, $C(s)$, and $D(s)$, while the second matrix is for the specific example of Fig. G.1. The three entries are the

J. Allen, *An Invitation to Mathematical Physics and Its History*,
https://doi.org/10.1007/978-3-030-53759-3

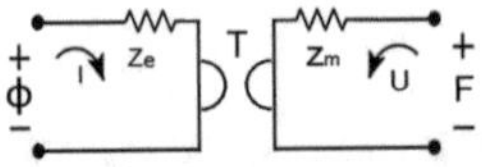

$$\begin{bmatrix}\Phi\\ I\end{bmatrix} = \begin{bmatrix}1 & Z_e\\ 0 & 1\end{bmatrix}\begin{bmatrix}0 & T\\ T^{-1} & 0\end{bmatrix}\begin{bmatrix}1 & z_m\\ 0 & 1\end{bmatrix}\begin{bmatrix}F\\ -U\end{bmatrix}$$

Fig. G.1 The schematic representation of an algebraic network, defined by its two-port ABCD transmission, that has three elements called the Hunt parameters (Hunt 1952): $Z_e(s)$, the electrical impedance, $z_m(s)$, the mechanical impedance, and $T(s)$, the transduction coefficient matrix of an electromechanical transducer network. The port variables are $\Phi(f)$ and $I(f)$: the frequency-domain voltage and current, and $F(f)$ and $U(f)$: the force and velocity (Hunt 1952; Kim and Allen 2013). This matrix factors the two-port model into three 2×2 matrices, separating the three physical elements as matrix algebra. It is a standard impedance convention that the flows $I(f) and U(f)$ are defined into each port. Thus it is necessary to apply a negative sign on the velocity $-U(f)$ so that it has an outward flow, as required to match the next cell with its inward flow

electrical driving-point impedance $Z_e(s)$, the mechanical impedance $z_m(s)$, and the transduction $T = B_o l$, where B_o is the magnetic flux strength and l is the length of the wire crossing the flux. Since the transmission matrix is antireciprocal, its determinant $\Delta_{\mathcal{T}} = -1$, as is easily verified.

Other common examples of cross-modality transduction and current–thermal (thermoelectric effect) and force–voltage (piezoelectric effect). These systems are all reciprocal: thus the transduction has the same sign.

G.2 Impedance Matrix

These eleven network postulates describe the properties of a system that has an input and an output. For an electromagnetic transducer (loudspeaker) the system is described by the two-port transmission matrix, as shown in Fig. G.1. The electrical input impedance of a loudspeaker is $Z_e(s)$, defined by

$$Z_e(s) = \left.\frac{V(\omega)}{I(\omega)}\right|_{U=0}.$$

Note that this driving-point impedance must be causal since it is a function of s; thus it has a Laplace transform. The corresponding two-port impedance matrix for Fig. G.1 is

$$\begin{bmatrix}\Phi_i\\ F_l\end{bmatrix} = \begin{bmatrix}z_{11}(s) & z_{12}(s)\\ z_{21}(s) & z_{22}(s)\end{bmatrix}\begin{bmatrix}I_i\\ U_l\end{bmatrix} = \begin{bmatrix}Z_e(s) & -T(s)\\ T(s) & z_m(s)\end{bmatrix}\begin{bmatrix}I_i\\ U_l\end{bmatrix}. \tag{G.2.1}$$

Such a description allows us to define Thévenin parameters, a concept used widely in circuit analysis and network models from other modalities.

The impedance matrix is an alternative description of the system but with generalized forces $[\Phi_i, F_l]$ on the left and generalized flows $[I_i, U_l]$ on the right. A rearrangement of terms allows us to go from the ABCD to the impedance parameters

(Van Valkenburg 1964b). The electromagnetic transducer is antireciprocal (Postulate P6), $z_{12} = -z_{21} = T = B_o l$.

G.3 Taxonomy of Algebraic Networks

The postulates are extended beyond those defined by Carlin and Giordano (Sect. 3.10) when there is an interaction of waves and a structured medium, along with other properties not covered by classic network theory. Assuming quasistatics (QS), the wavelength must be large relative to the medium's lattice constants. Thus the QS property must be extended to three dimensions and possibly to the cases of anisotropic and random media.

Causality: P1 As we stated, due to causality the negative properties (e.g., negative refractive index) must be limited in bandwidth as a result of the Cauchy–Riemann conditions. However, even causality needs to be extended to include the delay, as quantified by the d'Alembert solution to the wave equation, which means that the delay is proportional to the distance. Thus we generalize Postulate P1 to include the space-dependent delay. When we wish to discuss this property, we call it Einstein causality, which says that the delay must be proportional to the distance x, with impulse response $\delta(t - x/c)$.

Linearity: P2 The wave properties of a system may be nonlinear. This is not restrictive, as most physical systems are naturally nonlinear. For example, a capacitor is inherently nonlinear: As the charge builds up on the plates of the capacitor, a stress is applied to the intermediate dielectric due to the electrostatic force $F = qE$. In a similar manner, an inductor is nonlinear. Two wires carrying a current are attracted or repelled due to the force created by the flux. The net force is the product of the two fluxes due to each current.

In summary, most physical systems are naturally nonlinear; it's simply a matter of degree. An important counterexample is an amplifier with negative feedback and a very large open-loop gain. There are, therefore, many types of nonlinearity, both instantaneous types and those with memory (e.g., hysteresis). Given the nature of Postulate P1, even an instantaneous nonlinearity may be ruled out. The linear model is so critical for our analysis, providing fundamental understanding, that we frequently take Postulates P1 and P2 for granted.

Passive/Active impedances: P3 This postulate is about conservation of energy and Otto Brune's positive-real (PR, also called physically realizable) condition that every passive impedance must obey. Following on the work of his primary Ph.D. thesis advisor Wilhelm Cauer (1900–1945) and Ernst Guillemin, along with Norbert Wiener and Vannevar Bush at MIT, Brune mathematically characterized the properties of every PR one-port driving-point impedance (Brune 1931b).

When the input resistance of the impedance is real, the system is said to be passive, which means the system obeys conservation of energy. The real part of $Z(s)$

is positive if and only if the corresponding reflectance is less than 1 in magnitude. The reflectance of $Z(s)$ is defined as a bilinear transformation of the impedance, normalized by its surge resistance r_o (Campbell 1903):

$$\Gamma(s) = \frac{Z(s) - r_o}{Z(s) + r_o} = \frac{\hat{Z} - 1}{\hat{Z} + 1},$$

where $\hat{Z} = Z/r_o$. The surge resistance is defined in terms of the inverse Laplace transform of $Z(s) \leftrightarrow z(t)$, which must have the form

$$z(t) = r_o\delta(t) + \rho(t),$$

where $\rho(t) = 0$ for $t < 0$. It naturally follows that $\gamma(t) \leftrightarrow \Gamma(s)$ is zero for negative and zero time—namely, $\gamma(0) = 0, t \leq 0$.

Given any linear PR impedance $Z(s) = R(\sigma, \omega) + jX(\sigma, \omega)$ that has real part $R(\sigma, \omega)$ and imaginary part $X(\sigma, \omega)$, the impedance is defined as being PR (Brune 1931b) if and only if

$$\Re Z(s) = R(\sigma \geq 0, \omega) \geq 0. \tag{G.3.1}$$

That is, the real part of any PR impedance is nonnegative everywhere in the right half-plane ($\sigma \geq 0$). This is a very strong condition on the complex analytic function $Z(s)$ of a complex variable s. This condition is equivalent to any of the following statements (Van Valkenburg 1964a):

1. There are no poles or zeros in the right half-plane ($Z(s)$ may have poles and zeros on the $\sigma = 0$ axis).
2. If $Z(s)$ is PR, then its reciprocal $Y(s) = 1/Z(s)$ is PR (the poles and zeros swap).
3. If the impedance can be written as the ratio of two polynomials (a limited case related to the quasistatics approximation, Postulate P9) that have degrees N and L, then $|N - L| \leq 1$.
4. The angle of the impedance $\theta \equiv \angle Z$ lies within $[-\pi \leq \theta \leq \pi]$.
5. The impedance and its reciprocal are complex analytic in the right half-plane; thus each obeys the Cauchy–Riemann conditions there.

Energy and power: Since Postulate P3 requires the impedance PR condition, it ensures that every impedance is positive-definite (PD), thus guaranteeing that conservation of energy is obeyed (Schwinger and Saxon 1968, p.17). This means that the total energy absorbed by any PR impedance must remain positive for all time. Mathematically we can state this as

$$\mathcal{E}(t) = \int_{-\infty}^{t} v(t)i(t)\, dt = \int_{-\infty}^{t} i(t){\star}z(t)\, i(t)\, dt > 0,$$

where $i(t)$ is any current, $v(t) = z(t) \star i(t)$ is the corresponding voltage, and $z(t)$ is the real causal impulse response of the impedance [e.g., $z(t) \leftrightarrow Z(s)$ are a Laplace transform pair]. In summary, if $Z(s)$ is PR, then $\mathcal{E}(t)$ is PD.

As discussed in detail by Van Valkenburg (1964b), any rational PR impedance can be represented as a partial fraction expansion, which can be expanded into first-order poles as

$$Z(s) = K\frac{\Pi_{i=1}^{L}(s-n_i)}{\Pi_{k=1}^{N}(s-d_k)} = \sum_n \frac{\rho_n}{s-s_n}e^{j(\theta_n-\theta_d)}, \tag{G.3.2}$$

where ρ_n is a complex scale factor (residue). Every pole in a PR function has only simple poles and zeros, which requires that $|L-N| \le 1$ (Van Valkenburg 1964b).

Whereas the PD property clearly follows from Postulate P3 (conservation of energy), the physics is not so clear. Specifically, what is the physical meaning of the constraints on $Z(s)$? In many ways, the impedance concept is highly artificial, as expressed by Postulates P1–P7.

When the impedance is not rational, special care must be taken. An example of this is the semi-inductor $M\sqrt{s}$ and the semicapacitor $K/\sqrt{s}$ due, for example, to the skin effect in EM theory and viscous and thermal losses in acoustics, both of which are frequency-dependent boundary-layer diffusion losses (Vanderkooy 1989). They remain positive-real but have a branch cut and thus are double-valued in frequency.

Real-time response: P4 The impulse response of every physical system is real, not complex. This requires that the Laplace transform have conjugate symmetry $H(s) = H^*(s^*)$, where the $*$ indicates conjugation [e.g., $R(\sigma,\omega) + X(\sigma,\omega) = R(\sigma,\omega) - X(\sigma,-\omega)$].

Time invariance: P5 The meaning of time-invariant requires that the impulse response of a system does not change over time. This requires that the system coefficients of the differential equation describing the system are constant (independent of time).

Rayleigh reciprocity: P6 Reciprocity is defined in terms of the unloaded output voltage that results from an input current. Specifically (same as Eq. 3.8.6, Sect. 3.8.2),

$$\begin{bmatrix} z_{11}(s) & z_{12}(s) \\ z_{21}(s) & z_{22}(s) \end{bmatrix} = \frac{1}{C(s)}\begin{bmatrix} A(s) & \Delta_{\mathcal{T}} \\ 1 & D(s) \end{bmatrix}, \tag{G.3.3}$$

where $\Delta_{\mathcal{T}} = A(s)D(s) - B(s)C(s) = \pm 1$ for the reciprocal and antireciprocal systems, respectively. This is best understood in terms of Eq. G.2.1. The off-diagonal coefficients $z_{12}(s)$ and $z_{21}(s)$ are defined as

$$z_{12}(s) = \left.\frac{\Phi_i}{U_l}\right|_{I_i=0} \qquad z_{21}(s) = \left.\frac{F_l}{I_i}\right|_{U_l=0}.$$

When these off-diagonal elements are equal [$z_{12}(s) = z_{21}(s)$], the system is said to obey Rayleigh reciprocity. If they are opposite in sign [$z_{12}(s) = -z_{21}(s)$], the system is said to be antireciprocal. If a network has neither reciprocal nor antireciprocal characteristics, then we denote it nonreciprocal (McMillan 1946). The most comprehensive discussion of reciprocity, even to this day, is that of Rayleigh (1896,

Vol. I). The reciprocal case may be modeled as an ideal transformer (Van Valkenburg 1964a), while for the antireciprocal case the generalized force and flow are swapped across the two-port. An electromagnetic transducer (e.g., a moving-coil loudspeaker or electrical motor) is antireciprocal (Kim and Allen 2013; Beranek and Mellow 2012); it requires a gyrator rather than a transformer, as shown in Fig. G.1.

Reversibility: P7 A second two-port property is the reversible/nonreversible postulate. A reversible system is invariant to the input and output impedances being swapped. This property is defined by the input and output impedances being equal.

Referring to Eq. G.3.3, when the system is reversible, $z_{11}(s) = z_{22}(s)$ or, in terms of the transmission matrix variables, $\frac{A(s)}{C(s)} = \frac{D(s)}{C(s)}$ or simply $A(s) = D(s)$, assuming $C(s) \neq 0$.

An example of a nonreversible system is a transformer with a turns ratio that is not 1. Also, an ideal operational amplifier (when the power is turned on) is nonreversible due to the large impedance difference between the input and output. Furthermore, it is active; it has a power gain due to the current gain at constant voltage (Van Valkenburg 1964b).

Generalizations of this lead to group theory and Noether's theorem. These generalizations apply to systems that have many modes, whereas quasistatics holds when they operate below a cutoff frequency (Table G.1), meaning that, as in the case of the transmission line, there are no propagating transverse modes. While this assumption is never exact, it leads to highly accurate results because the nonpropagating evanescent transverse modes are attenuated over a short distance and thus, in practice, may be ignored (Montgomery et al. 1948; Schwinger and Saxon 1968, Chaps. 9–11).

We extend the Carlin and Giordano postulate set to include Postulate P7, reversibility, which was refined by Van Valkenburg (1964a). To satisfy the reversibility condition, the diagonal components in a system's impedance matrix must be equal. In other words, the input force and flow are proportional to the output force and flow, respectively (i.e., $Z_e = z_m$).

Spatial invariance: P8 The characteristic impedance and wave number $\kappa(s, x)$ may be strongly frequency- and/or spatially dependent or even be negative over some limited frequency ranges. Due to causality, the concept of a negative group velocity must be restricted to a limited bandwidth (Brillouin 1960). As Einstein's theory of relativity makes clear, all materials must be strictly causal (Postulate P1), a view that must therefore apply to acoustics but at a very different time scale. We first discuss generalized postulates, expanding on those of Carlin and Giordano.

Deterministicity (randomness): P9 When the media are uniform and time-invariant, the impedance and transfer functions will be deterministic. When the media are turbulent, the response will be random. When light propagates through the universe, it is strongly time varying. Thus, astrophysics will be seen as random, whereas experiments in the shelter of the relatively uniform environment of the Earth will be deterministic. Model calculation will also be deterministic unless one is trying to create a random, time varying, and turbulent medium.

Table G.1 Several ways of indicating the quasistatic (QS) approximation. For network theory there is only one lattice constant a, which must be much less than the wavelength (wavelength constraint). These three constraints are not equivalent when the object may be a larger structured medium, spanning many wavelengths but with a cell structure size much smaller than the wavelength. For example, each cell could be a Helmholtz resonator or an electromagnetic transducer (e.g., an earphone)

Measure	Domain
$ka < 1$	Wave number constraint
$\lambda > 2\pi a$	Wavelength constraint
$f_c < c/2\pi a$	Bandwidth constraint

The quasistatic constraint: P10 When a system is described by the wave equation, delay is introduced between two points in space and depends on the wave speed. When the wavelength is large compared to the delay, we can successfully apply the quasistatic (QS) approximation. This method has widespread application and is frequently used without mention of the assumption. This can lead to confusion, since the limitations of the approximation may not be appreciated. An example is the use of quasistatics in quantum mechanics. The QS approximation has widespread use when the signals may be accurately approximated by a band-limited signal. Examples include KCL, KVL, wave-guides, transmission lines, and most important, impedance. The QS property is not mentioned in the six postulates of Carlin and Giordano (1964); thus they need to be extended in some fundamental ways.

When the dimensions of a cellular structure in the material are much smaller than the wavelength, can the QS approximation be valid? This effect can be viewed as a mode filter that suppresses unwanted (or conversely enhances the desired) modes (Ramo et al. 1965). QS may be applied to a three-dimensional specification, as in a semiconductor lattice. But such applications fall outside the scope of this text (Schwinger and Saxon 1968).

Although I have never seen the idea discussed in the literature, the QS approximation is applied when Green's theorem is defined. For example, Gauss's law is not true when the volume of the container violates QS, since changes in the distribution of the charge have not reached the boundary, when doing the integral. Thus such integral relationships assume that the system is in quasi-steady-state (i.e., that QS holds).

Formally, QS is defined as $ka < 1$, where $k = 2\pi/\lambda = \omega/c$ and a is the cellular dimension or the size of the object. Other ways of expressing this include $\lambda/4 > a$, $\lambda/2\pi > a$, $\lambda > 4a$, or $\lambda > 2\pi a$. It is not clear whether it is better to normalize λ by 4 (quarter wavelength constraint) or $2\pi \approx 6.28 > 4$, which is more conservative by a factor of $\pi/2 \approx 1.6$. Also k and a can be vectors (e.g., Eq. 3.1.5, Sect. 3.1).

Sergei Schelkunoff may have been the first to formalize this concept (Schelkunoff 1943), but he was not the first to use it, as exemplified by the Helmholtz resonator. George Ashley Campbell was the first to use the concept in the important application of a wave filter, some 30 years before Schelkunoff (Campbell 1903). These two men were 40 years apart and both worked for the telephone company (after 1929, called AT&T Bell Labs) (Fagen 1975).

There are alternative definitions of the QS approximation, depending on the geometric cell structure. The alternatives are listed in Table G.1.

The quasistatic approximation: Since the velocity perpendicular to the walls of a horn must be zero, any radial wave propagation is exponentially attenuated ($\kappa(s)$ is real and negative, i.e., the propagation function $\kappa(s)$ (Sect. 4.4) will not describe radial wave propagation), with a space constant of about 1 diameter. The assumption that these radial waves can be ignored (i.e., more than 1 diameter from their source) is called the quasistatic approximation. As the frequency is increased and once $f \geq f_c = 2c_o/\lambda$, the radial wave can satisfy the zero normal velocity wall boundary condition and therefore will not be attenuated. Thus above this critical frequency, radial waves (also known as higher order modes) are supported (κ becomes imaginary). Thus for Eq. 5.2.10 (Sect. 5.2.2) to describe guided wave propagation, $f < f_c$. But even under this condition, the solution is not precise within a diameter (or so) of any discontinuities (i.e., rapid variations) in the area.

Each horn, as determined by the area function $A(r)$, has a distinct wave equation and thus a distinct solution. Note that the area function determines the upper cutoff frequency via the quasistatic approximation, since $f_c = c_o/\lambda_c$, $\lambda_c/2 > d$, and $A(r) = \pi(d/2)^2$. Thus to satisfy the quasistatic approximation, the frequency f must be less than the cutoff frequency:

$$f < f_c(r) = \frac{c_o}{4}\sqrt{\frac{\pi}{A(r)}}. \tag{G.3.4}$$

We have discussed two alternative matrix formulations of these equations: the ABCD transmission matrix, used for computation, and the impedance matrix, used when working with experimental measurements (Pierce 1981, Chap. 7). For each formulation, reciprocity and reversibility show up as different matrix symmetries, as addressed in Sect. 3.10 (Pierce 1981, pp. 195–203).

Periodic $\leftrightarrow$ discrete: P11 As has been shown in the discussion on the Fourier transform, when the time (or frequency) domain response is periodic, the frequency (or time) domain is discrete. This is a fundamental symmetry property that must always be obeyed. This is closely related to the causal $\leftrightarrow$ complex analytic property of the Laplace and z transforms.

Summary

A transducer converts between modalities. We propose the general definitions of the eleven system postulates that include all transduction modalities, such as electrical, mechanical, and acoustical. It is necessary to generalize the concept of the QS approximation (Postulate P9) to allow for guided waves.

Given the combination of the important QS approximation and these space-time, linearity, and reciprocity properties, a rigorous definition and characterization of a system can thus be established. It is based on a taxonomy of such materials formulated in terms of material and physical properties and extended network postulates.

Appendix H
Webster Horn Equation Derivation

H.1 Overview

In this appendix we transform the acoustic equations, Eqs. 5.2.5 and 5.2.6 (Sect. 5.2.1), into their equivalent integral form, Eq. 5.2.10 (Sect. 5.2.2). This derivation is similar (but not identical) to that of Hanna and Slepian (1924) and Pierce (1981, p. 360).

H.1.1 Conservation of Momentum

The first step is to integrate the normal component of Eq. 5.2.5 (Sect. 5.2.1) over the isopressure surface $\mathcal{S}$, defined by $\nabla p = 0$

$$-\int_{\mathcal{S}} \nabla p(\mathbf{x}, t) \cdot d\mathbf{A} = \rho_o \frac{\partial}{\partial t} \int_{\mathcal{S}} \mathbf{u}(\mathbf{x}, t) \cdot d\mathbf{A}.$$

The average pressure $\varrho(x, t)$ is defined by dividing by the total area:

$$\varrho(x, t) \equiv \frac{1}{A(x)} \int_{\mathcal{S}} p(x, t)\, \hat{\boldsymbol{n}} \cdot d\boldsymbol{A}. \tag{H.1.1}$$

From the definition of the gradient operator, we have

$$\nabla p = \frac{\partial p}{\partial x} \hat{\boldsymbol{n}}, \tag{H.1.2}$$

where $\hat{\boldsymbol{n}}$ is a unit vector perpendicular to the isopressure surface $\mathcal{S}$. Thus the left side of Eq. 5.2.5 reduces to $\partial \varrho(x, t)/\partial x$.

The integral on the right side defines the volume velocity,

J. Allen, *An Invitation to Mathematical Physics and Its History*,
https://doi.org/10.1007/978-3-030-53759-3

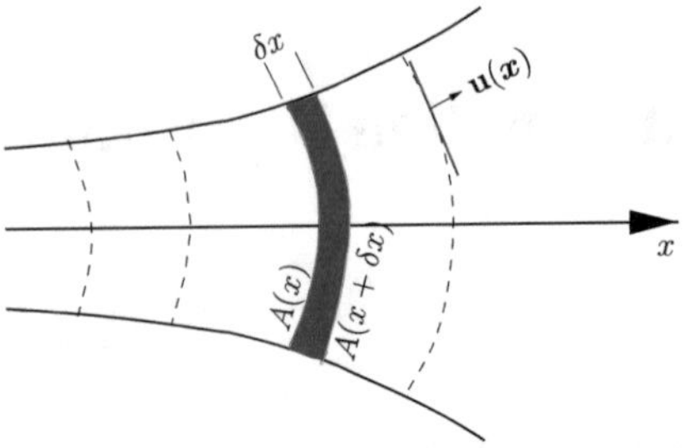

Fig. H.1 Derivation of the horn equation using Gauss's law: The divergence of the velocity $\nabla \cdot \mathbf{u}$ within δx, shown as the shaded region, is integrated over the enclosed volume. Next the divergence theorem is applied, transforming the integral to a surface integral normal to the surface of propagation. This results in the difference of the two volume velocities $\delta\nu = \nu(x+\delta x) - \nu(x) = [\mathbf{u}(x+\delta x)\cdot\mathbf{A}(x+\delta x) - \mathbf{u}(x)\cdot\mathbf{A}(x)]$. The flow is always perpendicular to the isopressure contours

$$\nu(x,t) \equiv \int_S \mathbf{u}(x,t)\cdot d\mathbf{A}. \tag{H.1.3}$$

Thus the integral form of Eq. 5.2.5 becomes

$$-\frac{\partial}{\partial x}\varrho(x,t) = \frac{\rho_o}{A(x)}\frac{\partial}{\partial t}\nu(x,t) \leftrightarrow \mathcal{Z}(x,s)\,\mathcal{V}, \tag{H.1.4}$$

where

$$\mathcal{Z}(s,x) = s\rho_o/A(x) = sM(x) \tag{H.1.5}$$

and $M(x) = \rho_o/A(x)$ [kgm/m^5] is the per-unit-length mass density of air.

H.1.2 Conservation of Mass

Integrating Eq. 5.2.6 over the volume V gives

$$-\int_V \nabla\cdot\boldsymbol{u}\,dV = \frac{1}{\eta_o P_o}\frac{\partial}{\partial t}\int_V p(\boldsymbol{x},t)dV.$$

The volume V is defined by two isopressure surfaces between x and $x+\delta x$ (the shaded region of Fig. H.1). On the right-hand side we use the definition of the average pressure (i.e., Eq. H.1.1) integrated over the volume dV.

Applying Gauss's law to the left-hand side[16] and using the definition of ϱ (on the right) in the limit $\delta x \to 0$ give

[16]As shown in Fig. H.1, taking the limit of the difference between the two volume velocities $\nu(x+\delta x) - \nu(x)$ divided by δx results in $\partial\nu/\partial x$.

$$-\frac{\partial}{\partial x}\nu(x,t) = \frac{A(x)}{\eta_o P_o}\frac{\partial}{\partial t}\varrho(x,t) \leftrightarrow \mathcal{Y}(x,s)\,\mathcal{P}(x,s), \tag{H.1.6}$$

where

$$\mathcal{Y}(x,s) = sA(x)/\eta_o P_o = sC(x). \tag{H.1.7}$$

$C(x) = A(x)/\eta_o P_o$ [m^4/N] is the per-unit-length compliance of the air. Equations H.1.4 and H.1.6 accurately characterize the Webster plane-wave mode up to the frequency where the higher order eigenmodes begin to propagate (i.e., $f > f_c$).

H.1.3 Horn Properties

H.1.3.1 Speed of Sound c_o

In terms of $M(x)$ and $C(x)$, the speed of sound and the acoustic admittance are

$$c_o = \sqrt{\frac{\text{stiffness}}{\text{mass}}} = \frac{1}{\sqrt{C(x)M(x)}} = \sqrt{\frac{\eta_o P_o}{\rho_o}}. \tag{H.1.8}$$

This assumes the medium is lossless. For a discussion of lossy propagation, see Appendix D.

H.1.4 Characteristic Admittance $\mathcal{Y}_r(x)$

Since the horn equation (Eq. 5.2.10) is second order, it has two eigenfunction solutions $\mathcal{P}^{\pm}$. The ratios of Eq. H.1.7 to Eq. H.1.5 are determined by the local stiffness $1/C(x)$ and mass $M(x)$. The ratio C/M determines the area-dependent characteristic admittance $\mathcal{Y}_r(x)$ ($\in \mathbb{R}$):

$$\mathcal{Y}_r(x) = \frac{1}{\sqrt{\text{stiffness}\cdot\text{mass}}} = \sqrt{\frac{\mathcal{Y}(x,s)}{\mathcal{Z}(x,s)}} = \sqrt{\frac{C(x)}{M(x)}} = \sqrt{\frac{A(x)}{\not s\rho_o}\frac{\not s A(x)}{\eta_o \mathcal{P}_o}} = \frac{A(x)}{\rho_o c_o} > 0 \tag{H.1.9}$$

Campbell (1903, 1910, 1922). The characteristic impedance is $\mathcal{Z}_r(x) = 1/\mathcal{Y}_r(x)$. Based on a physical argument, $\mathcal{Y}_r(x)$ must be positive and real; thus only the positive square root is allowed. As long as $A(x)$ has no jumps (is continuous), $\mathcal{Y}_r(x)$ must be the same in both directions. It is locally determined by the isopressure surface and its volume velocity.

H.1.4.1 Radiation Admittance

The radiation admittance is defined looking into a horn with no termination (infinitely long) from the input at $x = 0$:

$$Y^{\pm}_{rad}(s) = \frac{\mathcal{V}^{\pm}}{\mathcal{P}^{\pm}} \in \mathbb{C}. \tag{H.1.10}$$

The impedance depends on the direction, with $+$ looking to the right and $-$ to the left.

The input admittance $Y^{\pm}_{in}(x, s)$ is computed using the upper equation of Eq. 5.2.11 (Sect. 5.2.3) for $\mathcal{V}(x, s)$ and then dividing by the pressure eigenfunction $\mathcal{P}^{\pm}$. This results in the logarithmic derivative of $\mathcal{P}^{\pm}(x, s)$:

$$Y^{\pm}_{in}(x, s) \equiv \frac{\mathcal{V}^{\pm}}{\mathcal{P}^{\pm}} = \frac{-1}{sM(x)} \frac{\partial}{\partial r} \ln \mathcal{P}^{\pm}(x, s).$$

For example, for the conical horn (last column of Table 5.2, Sect. 5.4)

$$Y^{\pm}_{in} = \mathcal{Y}_r(1 \pm c_o/sr_o). \tag{H.1.11}$$

Note that $Y^{+}_{in}(x, s) + Y^{-}_{in}(x, s) = 2\mathcal{Y}_r = 2A_0r^2/\rho_o c_o \in \mathbb{R}$, which shows that the frequency-dependent parts of the two admittances, being equal and opposite in sign, exactly cancel.

As the wavefront travels down the variable-area horn, there is a mismatch in the characteristic admittance due to the change in area. This mismatch creates a reflected wave, which in the case of the conical horn is $= -c_o/sr_o$. Due to conservation of volume, there is a corresponding identical forward component that travels forward, equal to $+c_o/sr_o$. The sum of these two responses to the change in area must be zero in order to conserve volume velocity.

The resulting equation for the velocity eigenfunctions is therefore

$$\mathcal{V}^{\pm}(x, s) = Y^{\pm}_{in}(x, s)\, \mathcal{P}^{\pm}(x, s).$$

Propagation function $\kappa(s)$ The eigenfunctions of the lossless wave equation propagate as

$$\mathcal{P}^{\pm}(x, s) = \frac{e^{\mp\kappa(s)x}}{\sqrt{A(x)}},$$

where $\kappa(s) = \sqrt{\mathcal{Z}(x, s)\mathcal{Y}(x, s)} = \pm s\sqrt{MC}$. The velocity eigenfunctions $\mathcal{V}^{\pm}(x, s)$ may be computed from Eq. H.1.4.

From the above definitions,

$$\kappa(s) = \sqrt{\frac{s\rho_o}{\cancel{A(x)}} \frac{s\cancel{A(x)}}{\eta_o P_o}} = \frac{s}{c_o}.$$

Thus $\kappa(s)$ and s are the eigenvalues of the differential operations $\partial/\partial x$ and $\partial/\partial t$ on the pressure $\mathcal{P}(x, s)$. See Appendix D for the inclusion of visco-thermal losses.

Appendix I
Quantum Mechanics and the WHEN

While it is clear that both Schrödinger's equation and Dirac's equations are highly accurate, after about 100 years, it is not clear why. Both of these theories seem to violate classical electromagnetics (EM), such as Ohm's law, since they are built on energy principles rather than electric and magnetic fields. Here we delve into this question, by providing a classical (i.e., EM-based) derivation for the hydrogen atom, one of the most important and obvious successes of quantum mechanics (QM). The problem with QM is not that it fails—rather, it succeeds, without obvious basis. The problem is that we cannot understand the basic principles, and it seems to be in contradiction with any principles of a physical theory.

Based on the Rydberg series, we determine the reflection coefficient, and thus the radiation impedance seen by the electron, in a radial coordinate system centered on the proton. Since the electron and proton both have spin $\frac{1}{2}$, their magnetic fields must attractively align, accounting for the near-field vector potential, and complementing the far-field attraction due to their opposite signs. As the electron and proton approach each other, due to their far-field potential attraction, the magnetic near field becomes more attractive at close range, due to the magnetic dipoles of the two "particles," causing them to merge with neutral net magnetic moment and neutral charge, giving a highly stable hydrogen atom. However, given a sufficiently strong distorting field, this highly symmetric state could be disturbed, leading to photon radiation, constrained by the radial eigenstates. It seems more clear than ever that photons and electrons are in a state of equilibrium at the outskirts of very large Rydberg atoms.[17]

I.1 Equation for Rydberg Eigenmodes

Like every tuned resonant circuit, atoms have well-defined resonant frequencies, or eigenmodes. Figure I.1 shows the observed radiation spectra for hydrogen. From the very beginning, it has been clear that there is a pattern to these spectral lines. In 1880

[17] https://physics.aps.org/synopsis-for/10.1103/PhysRevLett.121.193401.

J. Allen, *An Invitation to Mathematical Physics and Its History*,
https://doi.org/10.1007/978-3-030-53759-3

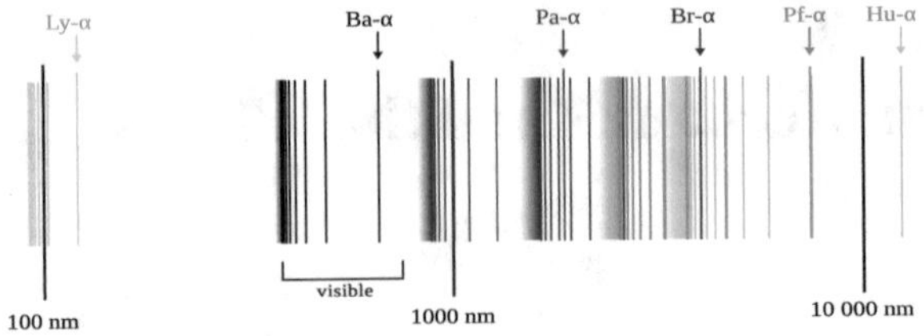

Fig. I.1 Diagram of the wavelength spectrum of hydrogen for the Lyman, Balmer, and Paschen series, as a function of each line's wavelength. The notation "Ly-α" indicates the longest wavelength $\lambda_{11} = 122$ [nm] (i.e., lowest frequency of 2.46 [GHz]) for the Lyman series. Figure citation: https://en.wikipedia.org/wiki/Hydrogen_spectral_series

Rydberg easily fitted a formula that quantifies the observed eigen spectral lines of hydrogen in terms of the reciprocals of the radiated wavelengths:

$$\frac{1}{\lambda_{nm}} = R_\infty \left(\frac{1}{n^2} - \frac{1}{m^2} \right), \qquad \frac{f_{nm}}{c_o R_\infty} = \frac{1}{n^2} - \frac{1}{m^2}, \tag{I.1.1}$$

all based on these simple observations. Here $R_\infty = 1.097 \times 10^7$ [m^{-1}]; $c_o = 3 \times 10^8$ [m/s] is the speed of light; f_{nm} are the dimensionless *Rydberg integer frequencies*; and $n, m \in \mathbb{N}$ are positive integers $\in \mathbb{N}$, where n labels the series and $m > n$ ($\lambda > 0$) describes the transition from orbit m to orbit n, as described in the caption of Fig. I.3.

I.1.1 The Rydberg Atom Model

In 1909 Rutherford demonstrated that the atom consisted of a dense core (the proton) surrounded by electrons. This view was supported by the spectrum of the atom, which allows for a radiation spectrum caused by electrons jumping from one energy level to another. It was then noted by Bohr in 1913 (Bohr 1954) and others that the wavelengths of hydrogen, as described by Eq. I.1.1, are consistent with Fig. I.2, where the reciprocal wavelength [m^{-1}] is given by Eq. I.1.1, having frequencies $f_{nm} = c/\lambda_{nm}$ [Hz]. The challenge of the 1920s was to explain these intuitive and simple models of hydrogen. This gave rise to the birth of quantum mechanics, the history of which is nicely summarized in Condon and Morse (1929).

It was clear from the days of Bohr that the Rydberg formula did not follow the typical rules of eigenspectra, so much so that Arnold Sommerfeld wrote (Sommerfeld 1949, p. 201):

> The lines of this spectrum cumulate at the limit given by the Rydberg constant R. The adjoining *continuum* lies in the near ultraviolet range. Both the discrete and the continuous spectrum are given by the Schrödinger equation. This equation reduces to a simple mathematical formula the enigma of the spectral lines, with their finite cumulation point, the behavior of which differs so fundamentally from that of all mechanical systems.

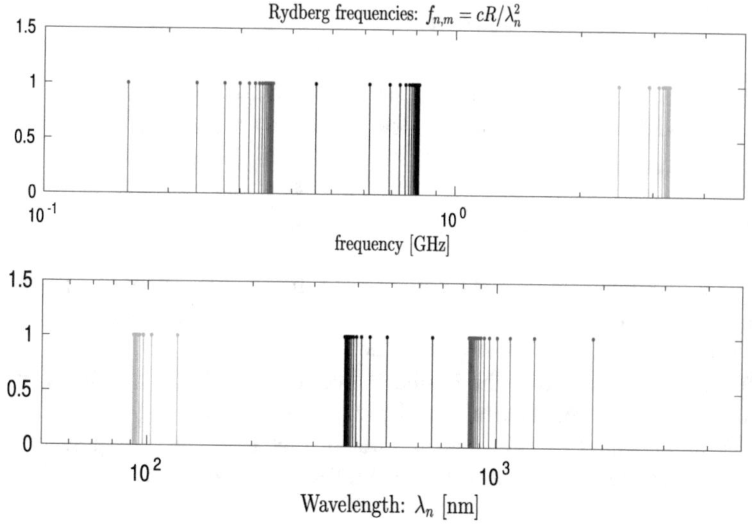

Fig. I.2 Rydberg frequencies in [GHz] and the corresponding wavelengths, computed from the Rydberg formula $\lambda_{nm}^{-1} = R\left(\frac{1}{n^2} - \frac{1}{m^2}\right)$, where integer n defines the series (Lyman: $n = 1$, Balmer: $n = 2$, Paschen: $n = 3$, etc.) and integer $m > n$ defines the outer transition line (see Fig. I.3). For example, according to the lower panel (green series), the Lyman series line $\lambda_{1,2} = 122$ [nm] ($n = 1$ and $m = 2$), in agreement with the lower panel of this figure, Figs. I.3 and I.1. The frequency of the Paschen series line (3,6) is at 1.094 [μm] (0.3 [GHz]) (upper panel) (http://www.physics.drexel.edu/%7Etim/open/hydrofin/)

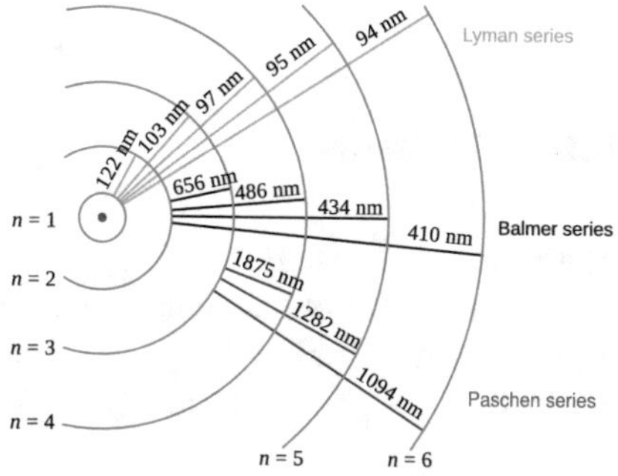

Fig. I.3 This diagram defines hydrogen's allowed electron transitions, defining the Lyman ($n = 1$), Balmer ($n = 2$), and Paschen ($n = 3$) series. The numbers represent the wavelengths λ [nm] of the photons having frequencies $f_{nm} = c_o/\lambda_{nm}$, following an electron transition from level n to m (taken from: https://en.wikipedia.org/wiki/Hydrogen_spectral_series)

I.1.2 Rydberg Wave Equation

The objective of this analysis is to demonstrate that one can define a classical Sturm–Liouville model of the *enigmatic* Rydberg atom, by the use of the Webster horn equation

$$\frac{1}{A(r)}\frac{\partial}{\partial r}A(r)\frac{\partial}{\partial r}\psi(r,t) = \frac{1}{c_o^2}\frac{\partial^2}{\partial t^2}\psi(r,t), \tag{I.1.2}$$

which is a one-dimensional wave equation for the electric potential $\psi(r, t)$ propagating in a wave-guide having area $A(r)$ as a function of the range, where r is the *range* variable (the axis of wave propagation).

We shall show that given the Rydberg spectrum (Eq. I.1.1), we may accurately estimate the electric reflectance $\Gamma(s)$ looking out from the origin (i.e., the proton location, as indicated by the small red dot in Fig. I.3). The radiation impedance $Z_{\text{rad}}(s)$ seen by the proton is related to the reflectance $\Gamma(s)$ by the relation

$$Z_{\text{rad}}(s) = r_o\frac{1+\Gamma(s)}{1-\Gamma(s)}. \tag{I.1.3}$$

This formula is the basis of the *Smith chart* used in both physics and engineering studies. It follows that once $\Gamma(s)$ is known (i.e., evaluated given Eq. I.1.1), the radiation impedance may be computed. It has been shown that the area function $A(r)$ may be found given the radiation impedance (Sondhi and Gopinath 1971; Youla 1964).

I.2 Rydberg Solution Methods

The basic idea behind the method is to use Eq. I.1.3, by noting that the poles of the impedance are determined by the roots of the denominator of Z_{rad}. Specifically, if s_p is an impedance pole, then it must satisfy $\Gamma(s_p) \approx 1$. Except for losses due to radiation, the atom is lossless; thus $|\Gamma(s)| = 1$. Namely, it must be of the form

$$\Gamma(s) = e^{-\jmath\phi(f)}, \tag{I.2.1}$$

where the *phase* $\phi(f) \in \mathbb{R}$ and $s = \sigma + \omega\jmath$ is the complex Laplace radian frequency, with $\omega = 2\pi f$ [Hz]. Since we know the eigenmode frequencies, which obey $\phi(f_{n_o,m}) = 2\pi m$, we may find $\phi(f)$, as follows: For a given series index n_o, and given the eigenfrequencies f_m, we seek the phase mode function $\phi_{n_o}(f)$ that maps the eigenfrequencies to their mode index m, i.e.,

$$\phi_{n_o}(f_m) = 2\pi m.$$

I.2.1 Group Delay $\tau(s)$

The phase $\phi(\omega)$ is related to the group delay $\tau(\omega)$ by the relation

$$\tau(\omega) = -\frac{\partial}{\partial \omega}\phi(\omega).$$

Here one may assume that the phase is complex analytic,[18] thus allowing a causal damping term into the reflectance phase Eq. I.2.1. This follows naturally because the reflectance must be causal (Postulate 3.10.2, Sect. 3.10.1). In the time domain the delay may be written in terms of the inverse $\mathcal{LT}$of the group delay,

$$\Gamma(s) = e^{-\jmath \int_o^s \tau(s)ds}.$$

Typically one uses the reflectance phase $2\pi\phi(f)$; thus the group delay is $\tau(f) = -\partial\phi(f)/\partial f$, which is physically interpreted here as the frequency-dependent delay from the proton to the radius of the electron's orbit. Thus this delay is given by

$$\tau(f) = n\frac{\partial}{\partial f}\left(1 - \frac{n^2}{c_o R}f\right)^{-1/2} = \frac{n^3}{2c_o R}\left(1 - \frac{n^2}{c_o R}f\right)^{-3/2},$$

which is constant for low frequencies and then rises to ∞ as frequency approaches the Rydberg frequency ($f \to c_o R/n^2$).

One may solve Eq. I.1.1 for m, for the case of the Lyman series ($n_o = 1$), by the use of the following identity for the Rydberg eigenfrequencies f_{nm}, which follow directly from Eq. I.1.1, with $m = n_o + l$ (with $n_o, m, l \in \mathbb{N}$)

$$\begin{aligned} f_{nm} &= \frac{c_o}{\lambda_{nm}} = c_o R\left(\frac{1}{n_o^2} - \frac{1}{(n_o + l)^2}\right) \\ &= \frac{c_o R}{n_o^2}\left(1 - \frac{1}{(1 + l/n_o)^2}\right). \end{aligned} \tag{I.2.2}$$

Note that as $l \to \infty$, $f_{n_o,l} \to= c_o R/n_o^2$, which is Sommerfeld's "finite cumulation point" [Hz] $f_{n_o,\infty}$ for the Lyman series ($n_o = 1$).

We can solve Eq. I.2.2 for the mode number $l/n < 1$ as a function of mode frequency:

[18] It follows that these relationships are related by a Hilbert transform.

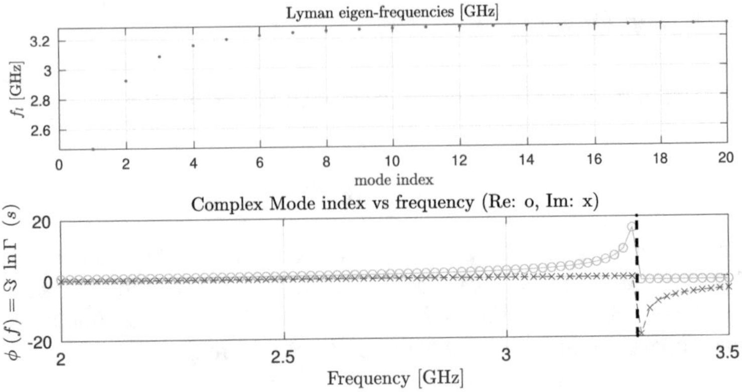

Fig. I.4 The top panel is a plot of Eq. I.1.1, showing how the eigenmode frequencies f_l depend on the eigen-number index l. As the mode number increases, the frequency reaches an asymptote at $c_o R \approx 3.29$ [GHz], with a wavelength limit near $1/R \approx 91.2$ [nm]. The lower panel shows the inverse mapping from frequency to the mode index number $\phi(f)$. This figure is for the Lyman series ($n = 1$ and $m = 1, \ldots, 20$). The inverse of this relationship is $l = \phi^{-1}(f_l)$ may be derived from Eq. I.1.1, which provides the pole frequencies required to satisfy Eq. I.1.3. Note that for frequencies greater than c_o/R the phase switches from purely real to imaginary, accounting for the free electrons that must exist above the upper accumulation frequency (i.e., 3.29 [GHz] for this example)

$$
\begin{aligned}
n^2 \frac{f_{nl}}{c_o R} &= 1 - \frac{1}{(1+l/n)^2} && \text{Starting from Eq. I.2.2} \\
\frac{1}{(1+l/n)^2} &= 1 - n^2 \frac{f_{nl}}{c_o R} && \text{Solving for } l/n \\
(1+l/n)^2 &= \frac{1}{1 - n^2 \frac{f_{nl}}{c_o R}} \\
\frac{l}{n} &= \pm \frac{1}{\sqrt{1 - n^2 \frac{f_{nl}}{c_o R}}} - 1 && \phi(f_{nl})/2\pi = l = m - n_o \in \mathbb{N}, \qquad \text{(I.2.3)}
\end{aligned}
$$

as summarized in the lower panel of Fig. I.4.

I.2.2 Finding the Area Function

Once the phase has been determined, we can compute the impedance using Eq. I.1.3. We may then decompose the impedance by using the analytic continued fraction algorithm (or Cauer synthesis), discussed in Sect. 3.8.

I.3 Euclid's Formula and the Rydberg Atom Model

Fundamental to quantum mechanics is the Rydberg series, which describes the quantized energy levels of atoms[19]:

$$\nu_{n,m} = c_o R_\infty Z_n^2 \left(\frac{1}{n^2} - \frac{1}{m^2} \right), \tag{I.3.1}$$

where $\nu_{n,m}$ are the possible eigenfrequencies, c_o is the speed of light, $R_\infty \approx 1.097 \times 10^7$ is the Rydberg constant, Z_n is the atomic number, along with positive integers $m > n \in \mathbb{N}$, which represent the *principal quantum numbers* that label all possible allowed atomic eigenstates. Integer n indicates the lowest (rest) atomic eigenstate while m labels the higher (excited) state.[20] When $n = 1$, the series is the Lyman series corresponding to hydrogen ($Z_1 = 1$). When $n = 1$, $m = 2$, and $Z_1 = 1$, the frequency is

$$\nu_{1,1} = c_o R_\infty \overset{\frac{3}{4}}{\left(\frac{1}{1^2} - \frac{1}{2^2} \right)} = 2.5 \times 10^{15} \text{ [Hz]}, \tag{I.3.2}$$

and the wavelength is $\lambda = c_o/\nu = 4 \times 10^8/R_\infty = 36.36$ [m].

An open question in this model is: *Why are states either empty or filled?* The amplitudes of the modes of a string or organ pipe are not quantized. What is it about the atom that forces the energy to be quantized?

Given observed frequencies $\nu_{n,m}$ it is possible to find the area function that traps the photons into the Rydberg eigenstates. Equation I.3.1 may be rewritten as

$$\nu_{n,m} = c_o R Z_n^2 4 \left(\frac{m^2 - n^2}{(2nm)^2} \right).$$

It is interesting to compare Eq. I.3.1 to Euclid's formula Eq. 2.6.6 (Sect. 2.5.1):

$$a = m^2 - n^2, \qquad b = 2mn, \qquad c = m^2 + n^2, \tag{I.3.3}$$

where $m > n \in \mathbb{N}$. Euclid's formula is equivalent to the Pythagorean theorem for integers, since

$$c^2 = a^2 + b^2, \tag{I.3.4}$$

with $\{a, b, c\} \in \mathbb{N}$. Here $a < b < c$.

[19] https://www.youtube.com/watch?v=e0IWPEhmMho.

[20] http://en.wikipedia.org/w/index.php?title=Rydberg_formula.

If we interpret the quantum numbers as multiples of a quarter wavelength, then the Rydberg formula is congruent to the Pythagorean theorem. Given the symmetry, this cannot be an accident.

In terms of the lengths of the right triangle $\{a, b, c\}$, Rydberg's formula becomes

$$\nu_{n,m} = c_o R Z_n^2 4 \left(\frac{a}{b^2}\right).$$

But since $b^2 = c^2 - a^2$,

$$\begin{aligned}\nu_{n,m} &= c_o \frac{R Z_n^2}{a} 4 \left(\frac{a^2}{c^2 - a^2}\right)\\ &= c_o \frac{R Z_n^2}{a} 4 \frac{a^2}{c^2} \left(\frac{1}{1 - (a/c)^2}\right).\end{aligned}$$

In terms of quantized (discrete) angles, $\sin(\theta_{n,m}) = a/c$,

$$\begin{aligned}\nu_{n,m} &= c_o \frac{R Z_n^2}{a} 4 \left(\frac{\sin^2\theta}{1 - \sin^2\theta}\right)\\ &= c_o \frac{R Z_n^2}{a} 4 \left(\frac{\sin^2\theta}{\cos^2\theta}\right)\\ &= c_o \frac{R Z_n^2}{a} 4 \tan^2\theta_{n,m}.\end{aligned}$$

I.3.1 Eigenmodes of the Rydberg Atom

One way to think of eigenmodes is to make an analogy to a piano string or an organ pipe. In these much simpler systems, there is an almost constant delay, say τ, due to a characteristic length, say $L = \tau c_o$, such that the eigenmodes of a string are given by integer multiples of a half wavelength $\nu_n = n c_o/2L$, while the eigenmodes of the organ pipe are multiples of a quarter wavelength. The distinction is the boundary conditions. For the string the end point boundary conditions are pinned displacement (i.e., zero velocity). The organ pipe is closed at one end and open at the other, resulting in multiples of a quarter wavelength $\nu_n = n c_o/4L$. In each case $\nu = n/\tau$, where $\tau = 2L/c_o$ is the round-trip delay; thus $\nu = n c_o/2L$. We suggest looking at the Rydberg series in the same way, but with very different eigenfrequencies (Eq. I.3.1). Sommerfeld (1949, p. 201) makes a very interesting comment regarding Eq. I.3.1:

> This equation reduces to a simple mathematical formula the enigma of the spectral lines, with their finite cumulation point, the behavior of which differs so fundamentally from that of all mechanical systems.

I.3.2 Discussion

The Rydberg frequencies f_{nl} ($n = 1, l = 1, \ldots, \infty$) has poles in the radiation impedance (Eq. I.1.3) when $\phi_l(f_{nl}) \in \mathbb{N}$. Working backwards from the Rydberg formula (Eq. I.2.1), we have solved for $\phi(f_{nl})$ indicating where this condition is valid (Eq. I.2.3). Since the reflectance and the impedance must be causal complex analytic functions of Laplace frequency s, we must replace the discrete frequency f_{nl} with s:

$$\jmath 2\pi f_{nl} \rightarrow s = \sigma + \omega\jmath,$$

thereby forcing $l(s)$ to be a complex analytic function of s. Then the poles of the radiation impedance must satisfy

$$\Gamma(s_{nl}) = e^{\jmath 2\pi l(f_{nl})} = 1,$$

resulting in eigenfrequencies at f_{nl}.

The next step in this analysis is to determine the area function $A(r)$ given Z_{rad} (Eq. I.1.3). To do this we must solve an integral equation for $A(r)$, as discussed by Sondhi and Gopinath (1971) and by Youla (1964).

Perhaps this could be done using an analytic representation for the area function,

$$A(r) = \sum_k a_k r^k.$$

I.4 Relations Between Sturm–Liouville and Quantum Mechanics

If we compare the Schrödinger equation (SE) for hydrogen with the corresponding Sturm–Liouville equation we can begin to appreciate their differences. The QM equation for hydrogen is

$$\begin{aligned}
\imath\hbar\frac{\partial}{\partial t}\psi(x,t) &= -\frac{\hbar^2}{2m_o}\nabla_r^2\psi(x,t) + V(r)\psi(x,t) \\
&= -\frac{\hbar^2}{2m_o}\frac{1}{r^2}\frac{\partial}{\partial r}r^2\psi(x,t) + V(r)\,\psi(x,t) && \text{(I.4.1)} \\
&= -\frac{\hbar^2}{2m_o}\left[\frac{2}{r}\frac{\partial}{\partial r}\psi(x,t) + \frac{\partial^2}{\partial r^2}\psi(x,t)\right] + V(r)\,\psi(x,t), && \text{(I.4.2)}
\end{aligned}$$

whereas the horn equation is given by Eq. I.1.2.

There are several obvious and disturbing differences between these two equations. First, the SE is, of course, first order in time. Diffusion equations have no delay and thus cannot have traditional eigenmodes, which result from standing waves in a wave equation, due to boundary conditions. Second, the EM horn equation is of Sturm–Liouville (SL) form, which is a true wave equation (versus the SE, which is a diffusion equation). The obvious question arises: Is there a potential V that would allow these two formulations to be equivalent? If so, then this would provide an explanation as to why the SE is successful in explaining the properties of Rydberg atoms.

To explore this possibility we may expand the two differential equations and directly compare them. Expanding Eq. I.1.2 gives

$$\frac{1}{c_o^2}\frac{\partial^2}{\partial t^2}\psi(r,t) = \frac{1}{A(r)}\frac{\partial}{\partial r}A(r)\frac{\partial}{\partial r}\psi(r,t) \tag{I.4.3}$$

$$= \frac{\partial^2}{\partial r^2}\psi(r,t) + \frac{1}{A(r)}\frac{\partial A(r)}{\partial r}\psi(r,t). \tag{I.4.4}$$

Between these two equations we may remove ψ'':

$$\imath\hbar\frac{\partial}{\partial t}\psi(x,t) = -\frac{\hbar^2}{2m_o}\left[\frac{2}{r}\frac{\partial}{\partial r}\psi(x,t) + \frac{1}{c_o^2}\frac{\partial^2}{\partial t^2}\psi(r,t) - \frac{1}{A(r)}\frac{\partial A(r)}{\partial r}\psi(r,t)\right] + V(r)\,\psi(x,t). \tag{I.4.5}$$

It seems that this may isolate the time and spatial parts (as in separation of variables).

I.4.1 The Exponential Horn

A relevant and motivational example is the solution of the exponential horn, having area function $A(r) = A_o e^{2mr}$. This case is interesting because it has a closed-form solution, which seems relevant and perhaps even related to the hydrogen atom.

Assuming that $\varrho(r,t) \leftrightarrow \mathcal{P}(r,\omega)$ are a Fourier transform pair, with $A(r) = A_o e^{2mr}$, Eq. I.1.2 reduces to

$$\frac{\partial^2 \mathcal{P}(r,\omega)}{\partial r^2} + 2m\frac{\partial \mathcal{P}(r,\omega)}{\partial r} = \kappa^2\,\mathcal{P}(r,\omega) \leftrightarrow \frac{1}{c_o^2}\frac{\partial^2 \varrho}{\partial t^2}, \tag{I.4.6}$$

with $\kappa(s) = s/c_o$.

Exercise #1 Show that Eq. I.4.6 follows from Eq. I.1.2.
Solution: Starting from Eq. I.1.2 with area $A(r) = A_o e^{2mr}$

$$\frac{1}{\cancel{A_o}e^{2mr}}\frac{\partial}{\partial r}\left(\cancel{A_o}e^{2mr}\frac{\partial \varrho}{\partial r}\right) = \frac{1}{c_o^2}\frac{\partial^2 \varrho}{\partial t^2}$$

$$\varrho_{rr}(r,t) + 2m\varrho_r(r,t) = \frac{1}{c_o^2}\frac{\partial^2 \varrho}{\partial t^2} \leftrightarrow \kappa^2\,\mathcal{P}(r,\omega),$$

which is the time-domain version of Eq. I.4.6. ■

Since this equation is second order in time with constant coefficients, it has two closed-form solutions:

$$\begin{aligned} \mathcal{P}_c^{\pm}(r) &= \mathcal{P}_o^{\pm}(\omega)\, e^{-mr}\, e^{\mp\sqrt{m^2+\kappa^2}\, r} \\ &= \mathcal{P}_o^{\pm}(\omega)\, e^{-mr}\, e^{\mp j \frac{r}{c_o} \sqrt{\omega^2-\omega_c^2}}, \end{aligned}$$

with $\omega_c = mc_o$. The two wave amplitudes $\mathcal{P}_0^{\pm}(\omega)$ must be determined from the boundary conditions.

Exercise #2 Shown that $\mathcal{P}^{\pm}(r,\omega)$ satisfy Eq. I.4.6.
Solution: Taking partials with respect to r,

$$\begin{aligned} \partial_r\, \mathcal{P}^{\pm}(r,\omega) &= \left(-m \mp \sqrt{m^2+\kappa^2}\right) \mathcal{P}^{\pm}(r,\omega) \\ \partial_{rr}\, \mathcal{P}^{\pm}(r,\omega) &= \left(-m \mp \sqrt{m^2+\kappa^2}\right)^2 \mathcal{P}^{\pm}(r,\omega) \\ &= \left(2m^2 + \kappa^2 \pm 2m\sqrt{m^2+\kappa^2}\right) \mathcal{P}^{\pm}(r,\omega). \end{aligned}$$

Thus Eq. I.4.6 reduces to

$$\left(2m^2 + \kappa^2 \pm 2m\sqrt{m^2+\kappa^2}\right) + 2m\left(-m \mp \sqrt{m^2+\kappa^2}\right) = \kappa^2,$$

which is an identity. ■

Next consider the Fourier series (or Fourier transform) of the area function,

$$A(r) = \sum_k a_k e^{2m_k r}.$$

It follows from the linearity of the wave equation that the general solution of Eq. I.4.6 is

$$\mathcal{P}^{\pm}(r,\omega) = \sum_k a_k^{\pm}(\omega)\, e^{-m_k r}\, e^{\mp\sqrt{m_k^2+\kappa^2}\, r}.$$

Here we have combined $\mathcal{P}^{\pm}(\omega)$ and a_k as coefficients $a_k^{\pm}(\omega)$.

References

Ambaum, M. H. (2010). *Thermal physics of the atmosphere*. Wiley Online Library.

Apte, S. (2009). The *genius germs* hypothesis: Were epidemics of leprosy and tuberculosis responsible in part for the great divergence? *Hypothesis*, *7*(1):e3. https://www.hypothesisjournal.com/wp-content/uploads/2011/09/vol7no1-hj003.pdf.

Arnold, D., & Rogness, J. (2019). Möbius transformations revealed. *Youtube.com*. video explaining the Riemann sphere: https://www.youtube.com/watch?v=0z1fIsUNhO4.

Batchelor, G. (1967). *An introduction to fluid dynamics*. Cambridge, UK: Cambridge University Press.

Beranek, L. L. (1954). *Acoustics* (451 pages). McGraw-Hill, New York.

Beranek, L. L., & Mellow, T. J. (2012). *Acoustics: Sound fields and transducers*. Waltham, MA: Elsevier; Academic.

Bilbao, S. (2004). *Wave and scattering methods for numerical simulation* (398 pages). London: Wiley.

Boas, R. (1987). *Invitation to complex analysis*. New York: The Random House.

Bode, H. (1945). *Network analysis and feedback amplifier design*. New York: Van Nostrand.

Bohr, N. (1954). *Rydberg's discovery of the spectral laws*. Gleerup, Sweden: C.W.K.

Bombieri, E. (2000). Problems of the millennium: The Riemann hypothesis. In *Publications of the Clay institute* (pp. 1–11). Clay institute. "RiemannZeta-ClayDesc.pdf".

Boyer, C., & Merzbach, U. (2011). *A history of mathematics*. Hoboken, NJ: Wiley. https://books.google.com/books?id=BokVHiuIk9UC.

Brillouin, L. (1953). *Wave propagation in periodic structures*. London: Dover. Updated 1946 edition with corrections and added appendix.

Brillouin, L. (1960). *Wave propagation and group velocity*. Pure and applied physics (Vol. 8, 154 pages). Cambridge, MA: Academic.

Britannica, (2004). *Encyclopædia Britannica Online*. Britannica Online. https://en.wikipedia.org/wiki/Pierre-Simon_Laplace.

Brune, O. (1931a). Synthesis of a finite two-terminal network whose driving point impedance is a prescribed function of frequency. *Journal of Mathematics and Physics*, *10*, 191–236.

Brune, O. (1931b). *Synthesis of a finite two-terminal network whose driving point impedance is a prescribed function of frequency*. PhD thesis. Cambridge, MA: MIT. https://dspace.mit.edu/handle/1721.1/10661.

Burton, D. M. (1985). The history of mathematics: An introduction. *Group*, *3*(3):35.

Calinger, R. S. (2015). *Leonhard Euler: Mathematical genius in the enlightenment*. Princeton, NJ: Princeton University Press.

J. Allen, *An Invitation to Mathematical Physics and Its History*,
https://doi.org/10.1007/978-3-030-53759-3

Campbell, G. (1937). *The collected papers of George Ashley Campbell*. AT&T, Archives, 5 Reiman Rd, Warren, NJ. Forward by Vannevar Bush; Introduction by E.H. Colpitts.

Campbell, G. A. (1903). On loaded lines in telephonic transmission. *Philosophical Magazine*, *5*(27), 313–30.

Campbell, G. A. (1910). Telephonic intelligibility. *Philosophical Magazine*, *19*(6), 152–59.

Campbell, G. A. (1922). Physical theory of the electric wave-filter. *Bell System Technical Journal*, *1*(2), 1–32.

Carlin, H. J., & Giordano, A. B. (1964). *Network theory, an introduction to reciprocal and nonreciprocal circuits*. Englewood Cliffs, NJ: Prentice Hall.

Cauer, W. (1932). The Poisson integral for functions with positive real part. *Bulletin of the American Mathematical Society*, *38*(1919), 713–717. https://www.ams.org/bull/1932-38-10/S0002-9904-1932-05510-0/S0002-9904-1932-05510-0.pdf.

Cauer, W. (1958). *Synthesis of linear communication networks* (Vols. 1 and 2). New York: McGraw-Hill.

Cauer, W., Klein, W., Pelz, F., Knausenberger, G., & Warfield, J. (1958). *Synthesis of linear communication networks*. New York: McGraw-Hill.

Condon, E., & Morse, P. (1929). *Quantum mechanics*. New York: McGraw-Hill.

Crandall, I. B. (1926). *Theory of vibrating systems and sound (272 pages)*. New York: Van Nostrand.

D'Angelo, J. P. (2017). *Linear and complex analysis for applications*. Boca Raton, FL: CRC Press, Taylor & Francis Group.

Einstein, A. (1905). Über die von der molekularkinetischen theorie der wärme geforderte bewegung von in ruhenden flüssigkeiten suspendierten teilchen. *Annalen der Physik*, *322*(8), 549–60.

Eisner, E. (1967). Complete solutions of the Webster horn equation. *Journal of the Acoustical Society of America*, *41*(4B), 1126–46.

Fagen, M. (Ed.). (1975). *A history of engineering and science in the bell system - the early years (1875–1925)*. Murray Hill, NJ, Cambridge, MA: Bell Telephone Laboratories.

Fermi, E. (1936). *Thermodynamics*. New York, NY: Dover.

Fettweis, A. (1986). Wave digital filters: Theory and practice. *Proceedings of the IEEE*, *74*, 270–327.

Feynman, R. (1968). *The character of physical law*. Cambridge, MA: MIT press.

Feynman, R. (1970a). *Feynman lectures on physics*. Boston, MA: Addison-Wesley.

Feynman, R. (1970b). *Feynman lectures on physics* (Vol. II). Boston, MA: Addison-Wesley.

Feynman, R. (1970c). *Feynman lectures on physics* (Vol. III). Boston, MA: Addison-Wesley.

Feynman, R. (1970d). *Feynman lectures on physics* (Vol. I). Boston, MA: Addison-Wesley.

Feynman, S. O. L. (2019). Youtube citation. *Youtube*. How the speed of light was first measured: https://en.wikipedia.org/wiki/Speed_of_light On the speed of light: https://www.youtube.com/watch?v=b9F8Wn4vf5Y How to measure the speed of light: https://www.youtube.com/watch?v=rZ0wx3uD2wo "Greek" versus "Babylonian" math; Mathematical deduction (8:41): https://www.youtube.com/watch?v=YaUlqXRPMmY Take another point of view (2/4): https://www.youtube.com/watch?v=xnzB_IHGyjg.

Fick, A. (1855). Ueber diffusion. *Annalen der Physik*, *170*(1), 59–86.

Fitzpatrick, R. (2008). *Maxwell's equations and the principles of electromagnetism*. Jones & Bartlett.

Flanders, H. (1973). Differentiation under the integral sign. *American Mathematical Monthly*, *80*(6), 615–27.

Fletcher, N., & Rossing, T. (2008). *The physics of musical instruments*. New York: Springer. https://books.google.com/books?id=9CRSRYQlRLkC.

Fry, T. (1928). *Probability and its engineering uses*. Princeton NJ: Van Nostrand.

Galileo, (1638). *Two new sciences* 1914. New York: Macmillan & Co. https://files.libertyfund.org/files/753/0416_Bk.pdf.

Gallagher, T. F. (2005). *Rydberg atoms* (Vol. 3). Cambridge: Cambridge University Press.

Garber, D. (2004). On the frontlines of the scientific revolution: How mersenne learned to love galileo. *Perspectives on Science*, *12*(2), 135–63.

Goldsmith, A. N., & Minton, J. (1924). The performance and theory of loud speaker horns. *Proceedings of the Institute of Radio Engineers, 12*(4), 423–78.

Goldstein, L. (1973). A history of the prime number theorem. *American Mathematical Monthly, 80*(6), 599–615.

Graham, R. L., Knuth, D. E., & Patashnik, O. (1994). *Concrete mathematics: A foundation for computer science* (2n ed.). Boston, MA: Addison-Wesley Longman.

Greenberg, M. D. (1988). *Advanced engineering mathematics*. Upper Saddle River, NJ: Prentice Hall.

Grosswiler, P. (2004). Dispelling the alphabet effect. *Canadian Journal of Communication, 29*(2),

Hahn, W. (1941). A new method for the calculation of cavity resonators. *Journal of Applied Physics, 12*, 62–68.

Hamming, R. W. (2004). *Methods of mathematics applied to calculus, probability, and statistics*. North Chelmsford, MA: Courier Corporation.

Hanna, C. R., & Slepian, J. (1924). The function and design of horns for loud speakers. *Transactions of the American Institute of Electrical Engineers, XLIII*, 393–411.

Heaviside, O. (1892). On resistance and conductance operators, and their derivatives inductance and permittance, especially in connection with electric and magnetic energy. *Electrical Papers, 2*, 355–74.

Helmholtz, H. (1858). Ueber integrale der hydrodynamichen gleichungen welche den wirelbewegungen entsprechen (on integrals of the hydrodynamic equations that correspond to vortex motions). *Journal fur die reine und angewandte Mathematik, 55*, 25–55.

Helmholtz, H., & (1863a). *On the sensations of tone*. 1954 (300 pages), New York: Dover.

Helmholtz, H. (1863b). Ueber den einfluss der reibung in der luft auf die schallbewegung (on the influence of friction in the air on the sound movement). *Verhandlungen des naturhistorisch-medicinischen Vereins zu Heildelberg, III*(17), 16–20. Sitzung vom 27.

Helmholtz, H. (1978). On integrals of the hydrodynamic equations that correspond to vortex motions. *International Journal of Fusion Energy, 1*(3–4), 41–68. English translation of Helmholtz's 1858 paper, translated by Uwe Parpart (1978). The image of Helmholtz is on page 41. https://www.wlym.com/archive/fusion/fusionarchive_ijfe.html.

Heras, R. (2016). The Helmholtz theorem and retarded fields. *European Journal of Physics, 37*(6), 065204. https://stacks.iop.org/0143-0807/37/i=6/a=065204.

Hestenes, D. (2003). Spacetime physics with geometric algebra. *American Journal of Physics, 71*(7), 691–714.

Hill, J. M. (2007). Laplace transforms and the riemann zeta function. *Integral Transforms and Special Functions, 18*(3), 193–205. jhill@uow.edu.au.

Horn, R., & Johnson, C. (1988). *Matrix analysis*. Cambridge, England: Cambridge University Press.

Hunt, F. V. (1952). *Electroacoustics, the analysis of transduction, and its historical background* (260 pages). Woodbury, NY: The Acoustical Society of America. 11797.

Jaynes, E. (1991). Scattering of light by free electrons as a test of quantum theory. In *The electron* (pp. 1–20). Berlin: Springer.

Johnson, D. H. (2003). Origins of the equivalent circuit concept: The voltage-source equivalent. *Proceedings of the IEEE, 91*(4), 636–640. https://www.ece.rice.edu/~dhj/paper1.pdf.

Johnson, F. S., Cragin, B. L., & Hodges, R. R. (1994). Electromagnetic momentum density and the poynting vector in static fields. *American journal of physics, 62*(1), 33–41.

Karal, F. C. (1953). The analogous acoustical impedance for discontinuities and constrictions of circular cross section. *Journal of the Acoustical Society of America, 25*(2), 327–334.

Kelley, C. T. (2003). *Solving nonlinear equations with Newton's method* (Vol. 1). Siam.

Kennelly, A. E. (1893). Impedance. *Transactions of the American Institute of Electrical Engineers, 10*, 172–232.

Kim, N., & Allen, J. B. (2013). Two-port network analysis and modeling of a balanced armature receiver. *Hearing Research, 301*, 156–167.

Kim, N., Yoon, Y.-J., & Allen, J. B. (2016). Generalized metamaterials: Definitions and taxonomy. *Journal of the Acoustical Society of America, 139*, 3412–3418.

Kirchhoff, G. (1868). On the influence of heat conduction in a gas on sound propagation. *Annual Review of Physical Chemistry, 134*, 177–193.

Kirchhoff, G. (1974). On the influence of heat conduction in a gas on sound propagation (*English*). In Lindsay, R. B., (Ed.), *Physical acoustics, benchmark papers in acoustics* (Vo. 4, pp. 7–19). Stroudsburg, Pennsylvania: Dowden, Hutchinson & Ross, Inc.

Kleiner, M. (2013). *Electroacoustics*. Boca Raton, FL, USA: CRC Press, Taylor & Francis Group. https://www.crcpress.com/authors/i546-mendel-kleiner.

Kuhn, T. (1978). *Black-body theory and the quantum discontinuity, 1894–1912; (371 pages)*. Oxford University Press (1978) & University of Chicago Press (1987), Oxford, Oxfordshire; Chicago, IL.

Kusse, B. R., & Westwig, E. A. (2010). *Mathematical physics: Applied mathematics for scientists and engineers*. New Jersey: Wiley.

Lamb, H. (1932). *Hydrodynamics*. New York: Dover Publications.

Lighthill, S. M. J. (1978). *Waves in fluids*. England: Cambridge University Press.

Lin, J. Y. (1995). The Needham Puzzle: Why the Industrial Revolution did not originate in China. *Economic Development and Cultural Change, 43*(2), 269–292.

Lundberg, K. H., Miller, H. R., & Trumper, R. L. (2007). Initial conditions, generalized functions, and the laplace transform: Troubles at the origin. *IEEE Control Systems Magazine, 27*(1), 22–35.

Mason, W. (1927). A study of the regular combination of acoustic elements with applications to recurrent acoustic filters, tapered acoustic filters, and horns. *Bell System Technical Journal, 6*(2), 258–294.

Mason, W. (1928). The propagation characteristics of sound tubes and acoustic filters. *Physical Review, 31*, 283–295.

Mawardi, O. K. (1949). Generalized Solutions of Webster's Horn Theory. *Journal of the Acoustical Society of America, 21*(4), 323–30.

Maxwell, J. C. (1865). Viii. a dynamical theory of the electromagnetic field. *Philosophical transactions of the Royal Society of London, 32*(155), 459–512.

Mazur, J. (2014). *Enlightening symbols: A short history of mathematical notation and its hidden powers*. Princeton, NJ: Princeton University Press.

McMillan, E. M. (1946). Violation of the reciprocity theorem in linear passive electromechanical systems. *Journal of the Acoustical Society of America, 18*, 344–347.

Menzies, G. (2004). *1421: The year China discovered America*. NYC: HarperCollins.

Menzies, G. (2008). *1434: The year a magnificent Chinese fleet sailed to Italy and ignited the renaissance*. NYC: HarperCollins.

Miles, J. W. (1948). The coupling of a cylindrical tube to a half-infinite space. *Journal of the Acoustical Society of America, 20*(5), 652–664.

Miller, D. A. (1991). Huygens's wave propagation principle corrected. *Optics Letters, 16*(18), 1370–2.

Montgomery, C., Dicke, R., & Purcell, E. (1948). *Principles of microwave circuits*. New York: McGraw-Hill Inc.

Moralee, D. (1974, 1980, 1995). *The first ten years*. Cambridge University Physics Society, Cambridge, UK.

Morse, P. M. (1948). *Vibration and sound (468 pages)*. New York, NY: McGraw-Hill.

Morse, P. M., & Feshbach, H. (1953). *Methods of theoretical physics* (Vols. I, II). New York: McGraw-Hill Inc.

Nyquist, H. (1924). Certain factors affecting telegraph speed. *Bell System Technical Journal, 3*, 324.

Olson, H. F. (1947). *Elements of acoustical engineering* (2nd ed., 539 pages). D. Van Nostrand Company.

O'Neill, M. E. (2009). The genuine sieve of eratosthenes. *Journal of Functional Programming, 19*(1), 95–106.

Palmerino, C. (1999). Infinite degrees of speed: Marin mersenne and the debate over Galileo's law of free fall. *Early Science and Medicine, 4*(4), 269–328. www.jstor/stable/4130144.

Papasimakis, N., Raybould, T., Fedotov, V. A., Tsai, D. P., Youngs, I., & Zheludev, N. I. (2018). Pulse generation scheme for flying electromagnetic doughnuts. *Physical Review B, 97*, 201409.

Papoulis, A. (1962). *The Fourier integral and its applications*. New York: McGraw-Hill Book Co.

Papoulis, A., & Pillai, S. U. (2002). *Probability, random variables, and stochastic processes*. Tata McGraw-Hill Education.

Pierce, A. D. (1981). *Acoustics: An introduction to its physical principles and applications (678 pages)*. New York: McGraw-Hill.

Pipes, L. A. (1958). *Applied mathematics for engineers and physicists*. New York, NY: McGraw-Hill.

Ramo, S., Whinnery, J. R., & Van Duzer, T. (1965). *Fields and waves in communication electronics*. New York: Wiley.

Rasetshwane, D. M., Neely, S. T., Allen, J. B., & Shera, C. A. (2012). Reflectance of acoustic horns and solution of the inverse problem. *The Journal of the Acoustical Society of America, 131*(3), 1863–1873.

Rayleigh, J. W. (1896). *Theory of sound* (Vol. I 480 pages, Vol. II 504 pages). New York: Dover.

Riemann, B. (1859). On the number of primes less than a given magnitude. *Monatsberichte der Berliner Akademie* (pp. 1–10). Translated by David R. Wilkins-1998.

Rotman, J. J. (1996). A first course in abstract algebra. Prentice-Hall.

Salmon, V. (1946). Generalized plane wave horn theory. *Journal of the Acoustical Society of America, 17*(3), 199–211.

Schelkunoff, S. (1943). *Electromagnetic waves* (6th ed.). Toronto, New York and London: Van Nostrand Co., Inc.

Schwinger, J. S., & Saxon, D. S. (1968). *Discontinuities in waveguides: Notes on lectures by Julian Schwinger*. New York: Gordon and Breach.

Scott, A. (2002). *Neuroscience: A mathematical primer*. Berlin: Springer Science & Business Media.

Shaw, E. (1970). Acoustic horns with spatially varying density or elasticity. *Journal of the Acoustical Society of America, 50*(3), 830–840.

Sommerfeld, A. (1949). *Partial Differential Equations in Physics, Lectures on Theoretical Physics* (Vol. I). New York: Academic Press INC.

Sommerfeld, A. (1952). *Electrodynamics, Lectures on Theoretical Physics* (Vol. III). New York: Academic Press INC.

Sondhi, M. and Gopinath, B. (1971). Determination of vocal-tract shape from impulse responses at the lips. *Journal of the Acoustical Society of America, 49*(6:2), 1867–1873.

Stewart, G. W. (1922). Acoustic wave filters. *Physical Review, 20*(6), 528.

Stewart, G. W., & Lindsay, R. B. (1930). *Acoustics: A text on theory and applications*. New Jersey, NJ: D. Van Nostrand Company Incorporated.

Stewart, J. (2012). *Essential calculus: Early transcendentals*. Boston, MA: Cengage Learning. https://books.google.com/books?id=AcQJAAAAQBAJ.

Stillwell, J. (2002). *Mathematics and its history; Undergraduate texts in mathematics* (2d ed.). New York: Springer.

Stillwell, J. (2010). *Mathematics and its history; Undergraduate texts in mathematics* (3d ed.). New York: Springer.

Strang, G., Strang, G., Strang, G., & Strang, G. (1993). *Introduction to linear algebra* (Vol. 3). Cambridge, MA: Wellesley-Press.

Tisza, L. (1966). *Generalized thermodynamics* (Vol. 1). Cambridge: MIT Press.

Tolman, R. C. (1948). Consideration of the Gibbs theory of surface tension. *The Journal of Chemical Physics, 16*(8), 758–774.

Van Valkenburg, M. (1960). *Introduction to modern network synthesis*. New York: Wiley Inc.

Van Valkenburg, M. (1964a). *Network analysis* (2nd ed.). Englewood Cliffs, N.J.: Prentice-Hall.

Van Valkenburg, M. E. (1964b). *Modern network synthesis*. New York, NY: Wiley Inc.

Vanderkooy, J. (1989). A model of loudspeaker driver impedance incorporating eddy currents in the pole structure. *JAES, 37*(3), 119–128.

Webster, A. G. (1919). Acoustical impedance, and the theory of horns and of the phonograph. *Proceedings National Academe of Science*, *5*, 275–282.

White, M. (1999). *Isaac Newton: The last sorcerer*. Basic Books (AZ).

Wikipedia: Speed of Light (2019). Wikipedia citation. *Wikipedia*. How speed of light was first measured: https://en.wikipedia.org/wiki/Speed_of_light.

Winchester, S. (2009). *The man who loved China: The fantastic story of the eccentric scientist who unlocked the mysteries of the middle kingdom*. Harper Collins.

Youla, D. (1964). Analysis and synthesis of arbitrarily terminated lossless nonuniform lines. *IEEE Transactions on Circuit Theory*, *11*(3), 363–372.

Index

J. Allen, *An Invitation to Mathematical Physics and Its History*,
https://doi.org/10.1007/978-3-030-53759-3

F

G

H

Printed in the United States
by Baker & Taylor Publisher Services